Microbiology and Technology of Fermented Foods

The *IFT Press* series reflects the mission of the Institute of Food Technologists—advancing the science and technology of food through the exchange of knowledge. Developed in partnership with Blackwell Publishing, *IFT Press* books serve as essential textbooks for academic programs and as leading edge handbooks for industrial application and reference. Crafted through rigorous peer review and meticulous research, *IFT Press* publications represent the latest, most significant resources available to food scientists and related agriculture professionals worldwide.

Microbiology and Technology of Fermented Foods

Robert W. Hutkins

©2006 Blackwell Publishing

Blackwell Publishing Professional
2121 State Avenue, Ames, Iowa 50014, USA

Orders: 1-800-862-6657
Office: 1-515-292-0140
Fax: 1-515-292-3348
Web site: www.blackwellprofessional.com

Blackwell Publishing Ltd
9600 Garsington Road, Oxford OX4 2DQ, UK
Tel.: 144 (0)1865 776868

Blackwell Publishing Asia
550 Swanston Street, Carlton, Victoria 3053, Australia
Tel.: 161 (0)3 8359 1011

Authorization to photocopy items for internal or personal use, or the internal or personal use of specific clients, is granted by Blackwell Publishing, provided that the base fee is paid directly to the Copyright Clearance Center, 222 Rosewood Drive, Danvers, MA 01923. For those organizations that have been granted a photocopy license by CCC, a separate system of payments has been arranged. The fee codes for users of the Transactional Reporting Service are ISBN-13: 978-0-8138-0018-9/2006.

First edition, 2006

Library of Congress Cataloging-in-Publication Data

Hutkins, Robert W. (Robert Wayne)
    Microbiology and technology of fermented foods / Robert W. Hutkins.
  —1st ed.
       p. cm.
    Includes bibliographical references and index.
    ISBN-13: 978-0-8138-0018-9 (alk. paper)
    ISBN-10: 0-8138-0018-8 (alk. paper)
    1. Fermented foods—Textbooks. 2. Fermented foods—Microbiology—Textbooks. I. Title.
  TP371.44.H88 2006
  664′.024—dc22   2006002149

The last digit is the print number: 9 8 7 6 5 4 3 2

Titles in the *IFT Press* series

- *Biofilms in the Food Environment* (Hans P. Blaschek, Hua Wang, and Meredith E. Agle)
- *Food Carbohydrate Chemistry* (Ronald E. Wrolstad)
- *Food Irradiation Research and Technology* (Christopher H. Sommers and Xuetong Fan)
- *High Pressure Processing of Foods* (Christopher J. Doona, C. Patrick Dunne, and Florence E. Feeherry)
- *Hydrocolloids in Food Processing* (Thomas R. Laaman)
- *Multivariate and Probabilistic Analyses of Sensory Science Problems* (Jean-Francois Meullenet, Hildegarde Heymann, and Rui Xiong)
- *Nondestructive Testing of Food Quality* (Joseph Irudayaraj and Christoph Reh)
- *Preharvest and Postharvest Food Safety: Contemporary Issues and Future Directions* (Ross C. Beier, Suresh D. Pillai, and Timothy D. Phillips, Editors; Richard L. Ziprin, Associate Editor)
- *Regulation of Functional Foods and Nutraceuticals: A Global Perspective* (Clare M. Hasler)
- *Sensory and Consumer Research in Food Product Development* (Howard R. Moskowitz, Jacqueline H. Beckley, and Anna V.A. Resurreccion)
- *Thermal Processing of Foods: Control and Automation* (K.P. Sandeep)
- *Water Activity in Foods: Fundamentals and Applications* (Gustavo V. Barbosa-Canovas, Anthony J. Fontana Jr., Shelly J. Schmidt, and Theodore P. Labuza)

Contents

Preface

This project started out innocently enough, with the simple goal of providing a resource to students interested in the microbiology of fermented foods. Since 1988, when I first developed a course in fermentation microbiology at the University of Nebraska, there has not been a suitable student text on this subject that I could recommend to my students. *Pederson's Microbiology of Food Fermentations* had last been published in 1979 and *Fermented Foods,* by A.H. Rose, was published in 1982. Brian Wood's two volume *Microbiology of Fermented Foods,* published in 1998 (revised from an earlier 1985 edition), is an excellent resource and is considered to be one of the most thorough texts on fermented foods, but it and other handbooks are generally beyond the scientific scope (and budget) of most students in a one-semester-long course. Finally, there are many excellent resources devoted to specific fermented foods. The recently published (2004) *Cheese Chemistry, Physics and Microbiology* (edited by Fox, McSweeney, Cogan, and Guinee) is an outstanding reference text, as are *Jackson's Wine Science Principles, Practice, and Perceptions* and Steinkraus' *Industrialization of Indigenous Fermented Foods.* However, their coverage is limited to only those particular foods.

I hope this effort achieves the dual purposes for which it is intended, namely to be used as a text book for a college course in fermentation microbiology and as a general reference on fermented food microbiology for researchers in academia, industry, and government.

In organizing this book, I have followed the basic outline of the course I teach, Microbiology of Fermented Foods. Students in this course, and hopefully readers of this text, are expected to have had a basic course in microbiology, at minimum, as well as courses in food microbiology and food science. An overview of microorganisms involved in food fermentations, their physiological and metabolic properties, and how they are used as starter culture provides a foundation for the succeeding chapters. Nine chapters are devoted to the major fermented foods produced around the world, for which I have presented both microbiological and technological features for the manufacture of these products. I confess that some subjects were considered, but then not included, those being the indigenous fermented foods and the natural fermentations that occur during processing of various "non-fermented" foods, such as cocoa beans and coffee beans. These topics are thoroughly covered in the above mentioned texts.

One of my goals was to provide a historical context for how the manufacture of fermented foods evolved, while at the same time emphasizing the most current science. To help accomplish this goal I have included separate entries, called "Boxes," that describe, in some detail, current topics that pertain to the chapter subjects. Some of these boxes are highly technical, whereas others simply provide sidebar information on topics somewhat apart from microbiology or fermentation. Hopefully, the reader will find them interesting and a pleasant distraction from the normal text.

Finally, in an effort to make the text easier to read, I made a conscious decision to write the narrative portion of the book with minimal point-by-point referencing.

Each chapter includes a bibliography from which most source materials were obtained. The box entries, however, are fully referenced.

Acknowledgments

I am grateful to the many colleagues who reviewed chapters and provided me with excellent suggestions and comments. Any questionable or inaccurate statements, however, are due solely to the author (and please let me know). To each of the following reviewers, I thank you again: Andy Benson, Larry Beuchat, Lloyd Bullerman, Rich Chapin, Mark Daeschel, Lisa Durso, Joe Frank, Nancy Irelan, Mark Johnson, Jake Knickerbocker, David Mills, Dennis Romero, Mary Ellen Sanders, Uwe Sauer, Randy Wehling, and Bart Weimer.

For the generous use of electron micrographs, photos, and other written materials used in this text, I thank Kristin Ahrens, Andreia Bianchini, Jeff Broadbent, Lloyd Bullerman, Rich Chapin (Empyrean Ales), Lisa Durso, Sylvain Moineau, Raffaele de Nigris, John Rupnow, Albane de Vaux, Bart Weimer, Jiujiang Yu, and Zhijie Yang.

For their encouragement and support during the course of this project, special thanks are offered to Jim Hruska, Kari Shoaf, Jennifer Huebner, Jun Goh, John Rupnow, and the SMB group. The editorial staff at Blackwell Press, especially Mark Barrett and Dede Pederson, have been incredibly patient, for which I am very appreciative.

I thank my wife, Charla, and my kids, Anna and Jacob, for being such good sports during the course of this project. At least now you know why I was busier than usual these past two years.

Finally, I would not be in a position of writing an acknowledgment section, much less this entire text, were it not for my graduate mentors, Robert Marshall, Larry McKay, Howard Morris, and Eva Kashket. Role models are hard to find, and I was fortunate to have had four. My greatest inspiration for writing this book, however, has been the many students, past and present, that have made teaching courses and conducting research on fermented foods microbiology such a joy and privilege.

Microbiology and Technology of Fermented Foods

1

Introduction

"When our souls are happy, they talk about food."
Charles Simic, poet

Fermented Foods and Human History

Fermented foods were very likely among the first foods consumed by human beings. This was not because early humans had actually planned on or had intended to make a particular fermented food, but rather because fermentation was simply the inevitable outcome that resulted when raw food materials were left in an otherwise unpreserved state. When, for example, several thousands of years ago, milk was collected from a domesticated cow, goat, or camel, it was either consumed within a few hours or else it would sour and curdle, turning into something we might today call buttermilk. A third possibility, that the milk would become spoiled and putrid, must have also occurred on many occasions. Likewise, the juice of grapes and other fruits would remain sweet for only a few days before it too would be transformed into a pleasant, intoxicating wine-like drink. Undoubtedly, these products provided more than mere sustenance; they were also probably well enjoyed for aesthetic or organoleptic reasons. Importantly, it must have been recognized and appreciated early on that however imperfect the soured milk, cheese, wine, and other fermented foods may have been (at least compared to modern versions), they all were less perishable and were usually (but not always) safer to eat and drink

than the raw materials from which they were made. Despite the "discovery" that fermented foods tasted good and were well preserved, it must have taken many years for humans to figure out how to control or influence conditions to consistently produce fermented food products. It is remarkable that the means for producing so many fermented foods evolved independently on every continent and on an entirely empirical basis. Although there must have been countless failures and disappointments, small "industries," skilled in the art of making fermented foods, would eventually develop. As long ago as 3000 to 4000 B.C.E., for example, bread and beer were already being mass produced by Egyptian bakeries and Babylonian breweries. Likewise, it is clear from the historical record that the rise of civilizations around the Mediterranean and throughout the Middle East and Europe coincided with the production and consumption of wine and other fermented food and beverage products (Box 1-1). It is noteworthy that the fermented foods consumed in China, Japan, and the Far East were vastly different from those in the Middle East; yet, it is now apparent that the fermentation also evolved and became established around the same time.

Fermentation became an even more widespread practice during the Roman Empire, as

Although the very first fermentations were certainly inadvertent, it is just as certain that human beings eventually learned how to intentionally produce fermented foods. When, where, and how this discovery occurred have been elusive questions, since written records do not exist. However, other forms of archaeological evidence do indeed exist and have made it possible to not only establish the historical and geographical origins of many of these fermentations, but also to describe some of the techniques likely used to produce these products.

For the most part, investigations into the origins of food fermentations have focused on alcoholic fermentations, namely wine and beer, and have been led primarily by a research group at the University of Pennsylvania Museum of Archaeology and Anthropology's Museum Applied Science Center for Archaeology (http://masca.museum.upenn.edu). These "biomolecular archaeologists" depend not so much on written or other traditional types of physical evidence (which are mostly absent), but rather on the chemical and molecular "records" obtained from artifacts discovered around the world (McGovern et al., 2004).

Specifically, they have extracted residues still present in the ancient clay pottery jars and vessels found in excavated archaeological sites (mainly from the Near East and China). Because these vessels are generally porous, any organic material was adsorbed and trapped within the vessel pores. In a dehydrated state, this material was protected against microbial or chemical decomposition. Carbon dating is used to establish the approximate age of these vessels, and then various analytical procedures (including gas chromatography-mass spectroscopy, Fourier transform infrared spectrometry, and other techniques) are used to identify the chemical constituents.

The analyses have revealed the presence of several marker compounds, in particular, tartaric acid, which is present in high concentrations in grapes (but is generally absent elsewhere), and therefore is ordinarily present in wine, as well (Guash-Jané et al., 2004; McGovern, 2003). Based on these studies (and others on "grape archaeology"), it would appear that wine had been produced in the Near East regions around present-day Turkey, Egypt, and Iran as long ago as the Neolithic Period (8500 to 4000 B.C.E.).

Recent molecular archaeological analyses have revealed additional findings. In 2004, it was reported that another organic marker chemical, syringic acid (which is derived from malvidin, a pigment found in red wines), was present in Egyptian pottery vessels. This was not a real surprise, because the vessels were labeled as wine jars and even indicated the year, source, and vintner. What made this finding especially interesting, however, was that one of the vessels had originally been discovered in the tomb of King Tutankhamun (King Tut, the "boy king"). Thus, not only does it now appear that King Tut preferred red wine, but that when he died (at about age 17), he was, by today's standards, not even of drinking age.

new raw materials and technologies were adopted from conquered lands and spread throughout the empire. Fermented foods also were important for distant armies and navies, due to their increased storage stability. Beer and wine, for example, were often preferred over water (no surprise there), because the latter was often polluted with fecal material or other foreign material. During this era, the means to conduct trade had developed, and cheese and wine, as well as wheat for bread-making, became available around the Mediterranean, Europe, and the British Isles.

Although manufacturing guilds for bread had existed even during the Egyptian empire, by the Middle Ages, the manufacture of many fermented foods, including bread, beer, and cheese, had become the province of craftsmen and organized guilds. The guild structure involved apprenticeships and training; once learned, these skills were often passed on to the next generation. For some products, particularly beer, these craftsmen were actually monks operating out of monasteries and churches, a tradition that lasted for hundreds of years. Hence, many of the technologies and

Box 1–1. Where and When Did Fermentations Get Started? Answers from Biomolecular Archaeologists (*Continued*)

As noted above, the origins of wine making in the Near East can be reliably traced to about 5400 B.C.E. The McGovern Molecular Archaeology Lab group has also ventured to China in an effort to establish when fermented beverages were first produced and consumed (McGovern et al., 2004). As described in Chapter 12, Asian wines are made using cereal-derived starch rather than grapes. Rice is the main cereal used. Other components, particularly honey and herbs, were apparently added in ancient times.

As had been done previously, the investigators analyzed material extracted from Neolithic (ca. 7000 B.C.E.) pottery vessels. In this case, the specific biomarkers would not necessarily be the same as for wine made from grapes, but rather would be expected to reflect the different starting materials. Indeed, the analyses revealed the presence of rice, honey, and herbal constituents, but also grapes (tartaric acid). Although domesticated grape vines were not introduced into China until about 200 B.C.E., wild grapes could have been added to the wine (as a source of yeast). Another explanation is that the tartaric acid had been derived from other native fruits and flowers. Additional analyses of "proto-historic" (ca. 1900 to 700 B.C.E.) vessels indicate that these later wines were cereal-based (using rice and millet). Thus, it now appears clear that fermented beverage technology in China began around the same time as in the Near East, and that the very nature of the fermentation evolved over several millennia.

References

Guasch-Jané, M.R., M. Ibern-Gómez, C. Andrés-Lacueva, O. Jáuregui, and R.M. Lamuela-Raventós. 2004. Liquid chromatography with mass spectrometry in tandem mode applied for the identification of wine markers in residues from ancient Egyptian vessels. Anal. Chem. 76:1672–1677.

McGovern, P.E. 2003. Ancient Wine: The Search for the Origins of Viniculture. Princeton University Press. Princeton, New Jersey.

McGovern, P.E., J. Zhang, J. Tang, Z. Zhang, G.R. Hall, R.A. Moreau, A. Nuñez, E.D. Butrym, M.P. Richards, C.-S. Wang, G. Cheng, Z. Zhao, and C. Wang. 2004. Fermented beverages of pre- and proto-historic China. Proc. Nat. Acad. Sci. 101:17593–17598.

manufacturing practices employed even today were developed by monks. Eventually, production of these products became more privatized, although often under some form of state control (which allowed for taxation).

From the Neolithic Period to the Middle Ages to the current era, fermented foods have been among the most important foods consumed by humans (Figure 1-1). A good argument can be made that the popularity of fermented foods and the subsequent development of technologies for their production directly contributed to the cultural and social evolution of human history. Consider, after all, how integral fermented foods are to the diets and cuisines of nearly all civilizations or how many fermented foods and beverages are consumed as part of religious customs, rites, and rituals (Box 1-2).

Fermented Foods: From Art to Science

It is difficult for the twenty-first century reader to imagine that fermented foods, whose manufacture relies on the intricate and often subtle participation of microorganisms, could have been produced without even the slightest notion that living organisms were actually involved. The early manufacturers of fermented foods and beverages obviously could not have appreciated the actual science involved in their production, since it was only in the last 150 to 200 years that microorganisms and enzymes were "discovered." In fact, up until the middle of the nineteenth century, much of the scientific community still believed in the concept of spontaneous generation. The very act of fermentation was a subject for philosophers

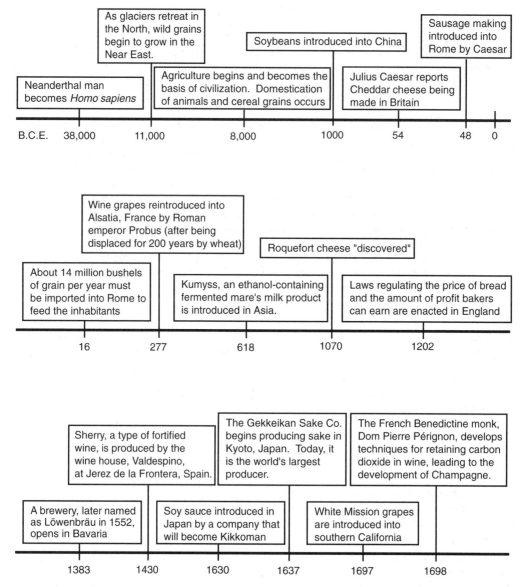

Figure 1–1. Developments in the history of fermented foods. From Trager, J., 1995. *The Food Chronology.* Henry Holt and Co., New York, and other sources.

and alchemists, not biologists. Although the Dutch scientist Antonie van Leeuwenhoek had observed microorganisms in his rather crude microscope in 1675, the connection between Leeuwenhoek's "animalcules" and their biological or fermentative activities was only slowly realized. It was not until later in the next cen-

tury that scientists began to address the question of how fermentation occurs.

Initially it was chemists who began to study the scientific basis for fermentation. In the late 1700s and early 1800s, the chemists Lavoisier and Gay-Lussac independently described the overall equations for the alcoholic fermenta-

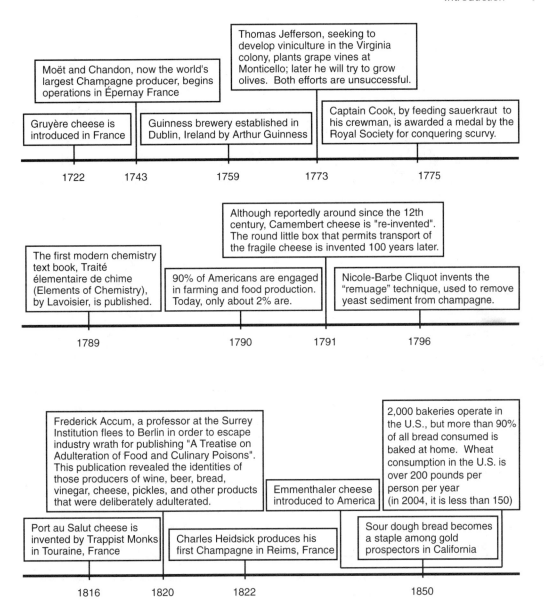

Figure 1–1. *(Continued)*

tion. Improvements in microscopy led Kützing, Schwann, and others to observe the presence of yeast cells in fermenting liquids, including beer and wine. These observations led Schwann to propose in 1837 (as recounted by Barnett, 2003) that "it is very probable that, by means of the development of the fungus, fermentation is started." The suggestion that yeasts were actually responsible for fermentation was not widely accepted, however; and instead it was argued by his contemporaries (namely Berzelius, Liebig, and Wöhler) that fermentation was caused by aerobic chemical reactions and that yeasts were inert and had nothing to do with fermentative

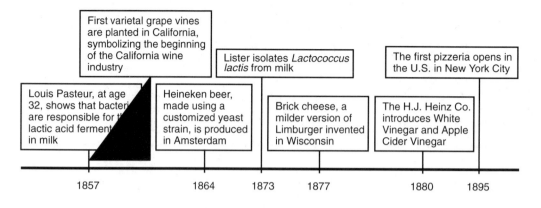

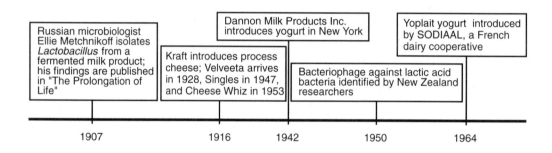

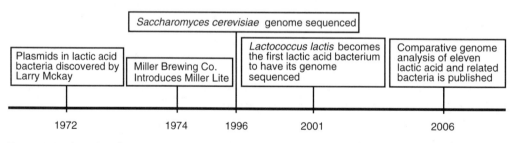

Figure 1–1. *(Continued)*

processes. The debate over the role of microorganisms in fermentation was brought to an unequivocal conclusion by another chemist, Louis Pasteur, who wrote in 1857 that "fermentation, far from being a lifeless phenomenon, is a living process" which "correlates with the development of . . . cells and plants which I have prepared and studied in an isolated and pure state" (Schwartz, 2001). In other words, fermentation could only occur when microorganisms were present. The corollary was also true—that when fermentation was observed, growth of the microorganisms occurred.

In a series of now famous publications, Pasteur described details on lactic and ethanolic fermentations, including those relevant to milk

Box 1–2. Fermented foods and the Bible.

The importance of fermented foods and beverages to the cultural history of human societies is evident from many references in early written records. Of course, the Bible (Old and New Testaments) and other religious tracts are replete with such references (see below). Fermented foods, however, also serve a major role in ancient Eastern and Western mythologies.

The writers , who had no scientific explanation for the unique sensory and often intoxicating properties of fermented foods, described them as "gifts of the gods." In Greek mythology, for example, Dionysus was the god of wine (Bacchus, according to Roman mythology). The Iliad and the Odyssey, classic poems written by the Greek poet Homer in about 1150 B.C.E., also contain numerous references to wine, cheese, and bread. Korean and Japanese mythology also refers to the gods that provided miso and other Asian fermented foods (Chapter 12).

Fermented foods and the Bible

From the Genesis story of Eve and the apple, to the dietary laws described in the books of Exodus and Leviticus, food serves a major metaphoric and thematic role throughout the Old Testament. Fermented foods, in particular, are frequently mentioned in biblical passages, indicating that these foods must have already been well known to those cultures and civilizations that lived during the time at which the bible was written.

In Genesis (9:20), for example, one of the first actions taken by Noah after the flood waters had receded was to plant a vineyard. In the very next line, it is revealed that Noah drank enough wine to become drunk (and naked), leading to the first, but certainly not last, episode in which drunkenness and nakedness occur. Later in Genesis (18:8), Abraham receives three strangers (presumably angels), to whom he offers various refreshments, including "curds."

Perhaps the most relevant reference to fermentation in the Bible is the Passover story. As described in Exodus (12:39), once Moses had secured the freedom of the Hebrew slaves, they were "thrust out of Egypt, and could not tarry." Thus, the dough could not rise or become leavened, and was baked instead in its "unleavened" state. This product, called matzoh, is still eaten today by people of the Jewish faith to symbolically commemorate the Hebrew exodus.

Ritual consumption of other fermented foods is also prescribed in Judaism. Every Sabbath, for example, the egg bread, Challah, is to be eaten, and grapes or wine is to be drunk, preceded by appropriate blessings of praise.

Fermented foods are also featured prominently in the New Testament. At the wedding in Cana (John 2:1-11), Jesus' first miracle is to turn water into wine. Later (John 6:1-14), another miracle is performed when five loaves of bread (and two fish) are able to feed 5,000 men. The Sacrament of Holy Communion (described by Jesus during the Last Supper) is represented by bread and wine.

fermentations, beer, and wine. He also identified the organism that causes the acetic acid (i.e., vinegar) fermentation and that was responsible for wine spoilage. The behavior of yeasts during aerobic and anaerobic growth also led to important discoveries in microbial physiology (e.g., the aptly named Pasteur effect, which accounts for the inhibitory effect of oxygen on glycolytic metabolism). Ultimately, the recognition that fermentation (and spoilage) was caused by microorganisms led Pasteur to begin working on other microbial problems, in particular, infectious diseases. Future studies on fermentations would be left to other scientists who had embraced this new field of microbiology.

Once the scientific basis of fermentation was established, efforts soon began to identify and cultivate microorganisms capable of performing specific fermentations. Breweries such as the Carlsberg Brewery in Copenhagen and the Anheuser-Busch brewery in St. Louis were among the first to begin using pure yeast strains, based on the techniques and recommendations of Pasteur, Lister, and others. By the early 1900s, cultures for butter and other dairy products had also become available. The dairy industry was soon to become the largest user of commercial cultures, and many specialized culture supply "houses" began selling not only cultures, but also enzymes, colors, and other products necessary for the manufacture of cheese and cultured milk products (Chapter 3). Although many cheese factories continued to propagate their own cultures throughout the first half of the century, as factory size and product throughput increased, the use of dairy starter cultures eventually became commonplace. Likewise, cultures for bread, wine, beer, and fermented meats have also become the norm for industries producing those products.

The Modern Fermented Foods Industry

The fermented foods industry, like all other segments of the food processing industry, has changed dramatically in the past fifty years. Certainly, the average size of a typical production facility has increased several-fold, as has the rate at which raw materials are converted to finished product (i.e., throughput). Although small, traditional-style facilities still exist, as is evident by the many microbreweries, small wineries, and artisanal-style bakery and cheese manufacturing operations, the fermented foods industry is dominated by producers with large production capacity.

Not only has the size of the industry changed, but so has the fundamental manner in which fermented foods are produced (Table 1-1). For example, up until the past forty or so years, most cheese manufacturers used raw, manufacturing (or Grade B) milk, whereas pasteurized Grade A milk, meeting higher quality standards, is now more commonly used. Manufacturing tanks or vats are now usually enclosed and are constructed from stainless steel or other materials that facilitate cleaning and even sterilization treatments. In fact, modern facilities are designed from the outset with an emphasis on sanitation requirements, so that exposure to air-borne microorganisms and cross-contamination is minimized.

Many of the unit operations are mechanized and automated, and, other than requiring a few keystrokes from a control panel, the manufacture of fermented foods involves minimal human contact. Fermented food production is now, more than ever before, subject to time and scheduling demands. In the so-called "old days," if the fermentation was slow or sluggish, it simply meant that the workers (who were probably family members) would be late for supper, and little else. In a modern production operation, a slow fermentation may mean that the workers have to stay beyond their shift (requiring that they be paid overtime), and in many cases, it could also affect the entire production schedule, since the production vat could not be turned over and refilled as quickly as needed. Although traditional manufacturing practices may not have always yielded consistent products, lot sizes were small and economic losses due to an occasional misstep were not likely to be too se-

Table 1.1. Fermented foods industry: past and present

Traditional	Modern
Small scale (craft industry)	Large scale (in factories)
Non-sterile medium	Pasteurized or heat-treated medium
Septic	Aseptic
Open	Contained
Manual	Automated
Insensitive to time	Time-sensitive
Significant exposure to contaminants	Minimal exposure to contaminants
Varying quality	Consistent quality
Safety a minor concern	Safety a major concern

rious. Besides, for every inferior cask of wine or wheel of cheese, there may have been an equally superior lot that compensated for the one that turned out badly. Even the absolute worst case scenario—a food poisoning outbreak as a result of an improperly manufactured product—would have been limited in scope due to the small production volume and narrow distribution range.

Such an attitude, today, however, is simply beyond consideration. A day's worth of product may well be worth tens, if not hundreds of thousands of dollars, and there is no way a producer could tolerate such losses, even on a sporadic basis. Food safety, in particular, has become an international priority, and there is generally zero tolerance for pathogens or other hazards in fermented foods. Quality assurance programs now exist throughout the industry, which strive to produce safe and consistent products. In essence, the fermented foods industry has evolved from a mostly art- or craft-based practice to one that relies on modern science and technology. Obviously, the issues discussed above—safety, sanitation, quality, and consistency—apply to all processed foods, and not just fermented foods. However, the fermented foods industry is unique in one major respect—it is the only food processing industry in which product success depends on the growth and activity of microorganisms. The implications of this are highly significant.

Microorganisms used to initiate fermentations are, unlike other "ingredients," not easily standardized, since their biochemical activity and even their concentration (number of cells per unit volume) may fluctuate from lot to lot. Although custom-made starter cultures that are indeed standardized for cell number and activity are readily available, many industrial fermentations still rely, by necessity, on the presence of naturally-occurring microflora, whose composition and biological activities are often subject to considerable variation. In addition, microorganisms are often exposed to a variety of inhibitory chemical and biological agents in the food or environment that can compromise their viability and activity. Finally, the culture

organisms are often the main means by which spoilage and pathogenic microorganisms are controlled in fermented foods. If they fail to perform in an effective and timely manner, the finished product will then be subject to spoilage or worse. Thus, the challenge confronting the fermented foods industry is to manufacture products whose very production is subject to inherent variability yet satisfy the modern era demands of consistency, quality, line-speed, and safety.

Properties of Fermented Foods

As noted in the previous discussion, fermented foods were among the first "processed" foods produced and consumed by humans. Their popularity more than 5,000 years ago was due to many of the same reasons why they continue to be popular today (Table 1-2).

Preservation

The preservation aspect of fermented foods was obviously important thousands of years ago, when few other preservation techniques existed. A raw food material such as milk or meat had to be eaten immediately or it would soon spoil. Although salting or smoking could be used for some products, fermentation must have been an attractive alternative, due to other desirable features. Preservation was undoubtedly one of the main reasons why fermented foods became such an integral part of human diet. However, even today, preservation, or to use modern parlance, shelf-life (or extended shelf-life), is still an important feature of fermented foods. For example, specialized cultures that contain organisms that produce

Table 1.2. Properties of fermented foods

Enhanced preservation
Enhanced nutritional value
Enhanced functionality
Enhanced organoleptic properties
Uniqueness
Increased economic value

specific antimicrobial agents in the food are now available, providing an extra margin of safety and longer shelf-life in those foods.

Nutrition

The nutritional value of fermented foods has long been recognized, even though the scientific bases for many of the nutritional claims have only recently been studied. Strong evidence that fermentation enhances nutritional value now exists for several fermented products, especially yogurt and wine. Fluid milk is not regularly consumed in most of the world because most people are unable to produce the enzyme β-galactosidase, which is necessary for digestion of lactose, the sugar found naturally in milk. Individuals deficient in β-galactosidase production are said to be lactose intolerant, and when they consume milk, mild-to-severe intestinal distress may occur. This condition is most common among Asian and African populations, although many adult Caucasians may also be lactose intolerant.

Many studies have revealed, however, that lactose-intolerant subjects can consume yogurt without any untoward symptoms and can therefore obtain the nutritional benefits (e.g., calcium, high quality protein, and B vitamins) contained in milk. In addition, it has been suggested that there may be health benefits of yogurt consumption that extend beyond these macronutrients. Specifically, the microorganisms that perform the actual yogurt fermentation, or that are added as dietary adjuncts, are now thought to contribute to gastrointestinal health, and perhaps even broader overall well-being (Chapter 4).

Similarly, there is now compelling evidence that wine also contains components that contribute to enhanced health (Chapter 10). Specific chemicals, including several different types of phenolic compounds, have been identified and shown to have anti-oxidant activities that may reduce the risk of heart disease and cancer. That wine (and other fermented foods) are widely consumed in Mediterranean countries where mortality rates are low has led to the suggestion that a "Mediterranean diet" may be good for human health.

Functionality

Most fermented foods are quite different, in terms of their functionality, from the raw, starting materials. Cheese, for example, is obviously functionally different from milk. However, functional enhancement is perhaps nowhere more evident than in bread and beer. When humans first collected wheat flour some 10,000 years ago, there was little they could do with it, other than to make simple flat breads. However, once people learned how to achieve a leavened dough via fermentation, the functionality of wheat flour became limitless. Likewise, barley was another grain that was widespread and had use in breadmaking, but which also had limited functionality prior to the advent of fermentation. Given that barley is the main ingredient (other than water) in beer manufacture, could there be a better example of enhanced functionality due to fermentation?

Organoleptic

Simply stated, fermented foods taste dramatically different than the starting materials. Individuals that do not particularly care for Limburger cheese or fermented fish sauce might argue that those differences are for the worse, but there is little argument that fermented foods have aroma, flavor, and appearance attributes that are quite unlike the raw materials from which they were made. And for those individuals who partake of and appreciate Limburger cheese, the sensory characteristics between the cheese and the milk make all the difference in the world.

Uniqueness

With few exceptions (see below), there is no way to make fermented foods without fermentation. Beer, wine, aged cheese, salami, and sauerkraut simply cannot be produced any other way. For many fermented products, the very procedures used for their manufacture are unique and require strict adherence. For example, Parmesan cheese must be made in a defined region of Italy, according to traditional and established procedures, and then aged under specified conditions. Any deviation results in forfeiture of the name Parmesan. For those "fermented" foods made without fermentation (which includes certain fresh cheeses, sausages, and even soy sauce), their manufacture generally involves direct addition of acids and/or enzymes to simulate the activities normally performed by fermentative microorganisms. These products (which the purist might be inclined to dismiss from further discussion) lack the flavor and overall organoleptic properties of their traditional fermented counterparts.

Economic value

Fermented foods were the original members of the value-added category. Milk is milk, but add some culture and manipulate the mixture just right, age it for a time, and the result may be a fine cheese that fetches a price well above the combined costs of the raw materials, labor, and other expenses. Grapes are grapes, but if grown, harvested, and crushed in a particular environment and at under precise conditions, and the juice is allowed to ferment and mature in an optimized manner, some professor may well pay up to $6 or $7 (or more!) for a bottle of the finished product. Truly, the economic value of fermented foods, especially fermented grapes, can reach extraordinary heights (apart from the professor market). As noted in Chapter 10, some wines have been sold for more than $20,000 per bottle. Even some specialty vinegars (Chapter 11) sell for more than $1,000 per liter. It should be noted that not all fermented foods command such a high dollar value. In truth, the fermented foods market is just as competitive and manufacturers are under the same market pressures as other segments of the food industry. Fermented foods are generally made from inexpensive commodities (e.g., wheat, milk, meat, etc.) and most products have very modest profit margins (some products, such as "current" or un-

aged cheese, are sold on commodity markets, with very tight margins). There is a well-known joke about the wine business that applies to other products as well, and that summarizes the challenge in making fermented foods: "How do you make a million dollars in the wine business? Easy, first you start with two million dollars." Finally, on a industry-wide basis, fermented foods may have a significant economic impact on a region, state or country. In California, for example, the wine industry was reported to contribute more than $40 billion to the economy in 2004 (according to a Wine Institute report; www.wineinstitute.org). A similar analysis of the U.S. beer industry (www.beerinstitute.org) reported an overall annual impact of more than $140 billion to the U.S. economy.

Fermented Foods in the Twenty-first Century

For 10,000 years, humans have consumed fermented foods. As noted above, originally and throughout human history, fermentation provided a means for producing safe and well-preserved foods. Even today, fermented foods are still among the most popular type of food consumed. No wonder that about one-third of all foods consumed are fermented. In the United States, beer is the most widely consumed fermented food product, followed by bread, cheese, wine, and yogurt (Table 1–3). Global statistics are not available, but it can be estimated that alcoholic products head the list of most popular fermented foods in most of the world. In Asia, soy sauce production and consumption ranks at or near the top. Collectively, sales of fermented foods on a global basis exceeds a trillion dollars, with an even greater overall economic impact.

Although fermented foods have been part of the human diet for thousands of years, as the world becomes more multicultural and cuisines and cultures continue to mix, it is likely that fermented foods will assume an even more important dietary and nutritional role. Foods such as kimchi (from Korea), miso (from Japan), and kefir (from Eastern Europe) are fast becoming part

Table 1.3.   U.S. production and consumption of selected fermented foods[a]

Food	Production	Consumption[b]
Wine	2.3×10^9 L	9 L
Beer	23×10^9 L	100 L
Cheese	4×10^9 Kg	14 Kg
Yogurt	1.2×10^9 Kg	3.4 Kg
Fermented meats	na[c]	0.3 Kg
Bread	7×10^9 Kg	25 Kg
Fermented vegetables	0.8×10^9 Kg	2.8 Kg

[a]Sources: 2001–2004 data from USDA, WHO, and industry organizations
[b]Per person, per year
[c]Not available

of the Western cuisine. Certainly, the desirable flavor and sensory attributes of traditional, as well as new-generation fermented foods, will drive much of the interest in these foods.

Consumption of these products also will likely be increased as the potential beneficial effects of fermented foods on human health become better established, scientifically and clinically. As noted above, compelling evidence now exists to indicate that red wine may reduce the risk of heart disease and that live bacteria present in cultured milk products may positively influence gastrointestinal health. Armed now with extensive genetic information on the microorganisms involved in food fermentations that has only become available in the last century, it is now possible for researchers to custom-produce fermented foods with not only specific flavor and other functional characteristics, but that also impart nutritional properties that benefit consumers.

References

Barnett, J.A. 2003. Beginnings of microbiology and biochemistry: the contribution of yeast research. Microbiol. 149:557–567.

Bulloch, W. 1960. The History of Bacteriology. Oxford University Press, London.

Cantrell, P.A. 2000. Beer and ale. *In* K.F. Kiple, K.C. Ornelas (ed). Cambridge World History of Food, p. 619–625. Cambridge University Press, Cambridge, United Kingdom

Steinkraus, K.H. 2002. Fermentations in world food processing. Comp. Rev. Food Sci. Technol. 1: 23–32.

2

Microorganisms and Metabolism

"We can readily see that fermentations occupy a special place in the series of chemical and physical phenomena. What gives to fermentations certain exceptional characters, of which we are only now beginning to suspect the causes, is the mode of life in the minute plants designated under the generic name of ferments, a mode of life which is essentially different from that of other vegetables, and from which result phenomena equally exceptional throughout the whole range of the chemistry of living beings."

From The Physiological Theory of Fermentation *by Louis Pasteur, 1879*

When one considers the wide variety of fermented food products consumed around the world, it is not surprising that their manufacture requires a diverse array of microorganisms. Although lactic acid-producing bacteria and ethanol-producing yeasts are certainly the most frequently used organisms in fermented foods, there are many other bacteria, yeast, and fungi that contribute essential flavor, texture, appearance, and other functional properties to the finished foods. In most cases, more than one organism or group of organisms is involved in the fermentation.

For example, in the manufacture of Swiss-type cheeses, thermophilic lactic acid bacteria from two different genera are required to ferment lactose, produce lactic acid, and acidify the cheese to pH 5.2, a task that takes about eighteen hours. Weeks later, another organism, *Propionibacterium freudenreichii* subsp. *shermani*, begins to grow in the cheese, producing other organic acids, along with the carbon dioxide that eventually forms the holes or eyes that are characteristic of Swiss cheese.

Even for those fermented foods in which only a single organism is responsible for performing the fermentation, other organisms may still play inadvertent but important supporting roles. Thus, tempeh, a fermented food product popular in Indonesia, is made by inoculating soybeans with the fungal organism *Rhizopus oligosporus*. The manufacturing process lends itself, however, to chance contamination with other microorganisms, including bacteria that synthesize Vitamin B_{12}, making tempeh a good source of a nutrient that might otherwise be absent in the diet of individuals who consume this product.

A Primer on Microbial Classification

For many readers, keeping track of the many genus, species, and subspecies names assigned to the organisms used in fermented foods can be a challenging task. However, knowing which organisms are used in specific fermented foods is rather essential (to put it mildly) to understanding the metabolic basis for how microbial fermentations occur. Therefore, prior to describing the different groups of microorganisms involved in food fermentations, it is first necessary to review the very nature of microbial taxonomy (also referred to as systematics) and how microbiologists go about classifying, naming, and identifying microorganisms.

Although this might seem to be a thankless task, it is, after all, part of human nature to sort or order things; hence, the goal of taxonomy is

15

to achieve some sense of order among the microbial world. Specifically, taxonomy provides a logical basis for: (1) classifying or arranging organisms into related groups or taxa; (2) establishing rules of nomenclature so that those organisms can be assigned names on a rational basis; (3) identifying organisms based on the accepted classification scheme and nomenclature rules; and (4) understanding evolutionary relationships of species, one to another.

As will be noted later in this and successive chapters, rules for classification are not permanently fixed, but rather can be amended and re-defined in response to new, more discriminating methods. For the most part, these new classification methods are based on molecular composition and genetic properties, which can also be used to determine phylogenetic or evolutionary relationships between related organisms.

The three domains of life

According to modern taxonomy, life on this planet can be grouped into three branches or domains—the *Eukarya*, the *Bacteria*, and the *Archaea* (Figure 2–1). This organization for classifying all living organisms was proposed and established in the 1980s by Carl Woese and is based on the relatedness of specific 16S rRNA sequences using a technique called oligonucleotide cataloging. This three-branch tree of life displaced the classical taxonomy that had recognized only two groups, eukaryotes and prokaryotes, and that was based primarily on morphology and biochemical attributes. All of the microorganisms relevant to fermented foods (and food microbiology, in general) belong to either the *Eukarya* or *Bacteria*. The *Archaea*, while interesting for a number of reasons, consists of organisms that generally live and grow in

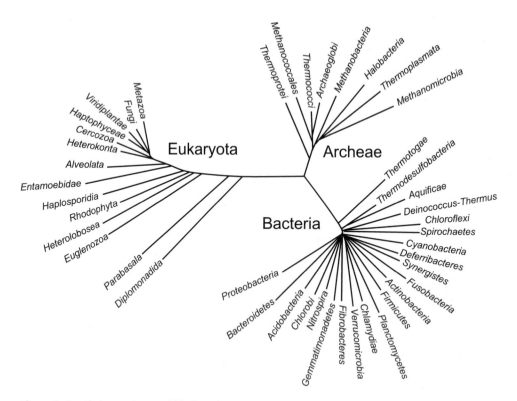

Figure 2–1. Phylogenetic tree of life (based on 16S rRNA sequences). Courtesy of the Joint Genome Institute (U.S. Department of Energy).

extreme environments (e.g., very high temperature, very low pH, very high salt), but rarely are they associated with foods.

Classification of organisms as eukaryotic or prokaryotic is based on a variety of characteristics (*Bergey's Manual* lists more than 50 cytological, chemical, metabolic, molecular, and reproductive properties), but the traditional distinguishing feature is the presence of a nuclear membrane in eukaryotes. Included within the *Eukarya* domain are animals, plants, protists, and fungi. The latter are represented by the Kingdom Fungi, which includes both yeasts and molds.

Fungi

The Kingdom Fungi is very large, containing many species, and although the exact number is unknown, it is thought to possibly number as many as 1.5 million. Only about 10% of these have been observed and described. In the current system of taxonomy (2004), four major groups or phyla are recognized among the true fungi or Eumycota. These are *Chytridiomycota, Zygomycota, Ascomycota,* and *Basidiomycota.*

The *Chytridiomycota* are primitive fungi and may be a link to ancestral fungi. They are considered true fungi based upon metabolic patterns, the presence of chitin in their cell walls, and small subunit rDNA sequence comparisons with other fungi. The chytrids, as they are sometimes called, live in wet soil, fresh water and marine environments, and are not common in foods, nor are they involved in food fermentations. They may be unicellular or may form simple branching chains of cells or primitive hyphae, which are coenocytic, that is, without cross walls.

The *Zygomycota* group, sometimes called zygomycetes, also produces coenocytic hyphae and reproduce sexually by fusion of two cells known as gametangia that form a zygosporangium containing zygospores. This group also produces asexual sporangiospores, stolons, and rhizoids. Sporangiospores are produced in a sac-like structure on an aerial stalk that extends into the air from the point of attachment to the substrate by the rhizoid. The rhizoids are specialized root-like hyphae that anchor the mold to its substrate, secrete enzymes, and absorb nutrients. The stolons are rapidly growing hyphae that run over the surface of the substrate from one rhizoid to another. The *Zygomycota* may contain as many as 900 species, but those important in foods and food fermentations are found primarily in the *Rhizopus* and *Mucor* genera.

The *Ascomycota* is a very large phylum of fungi, containing as many as half of all fungal species. The ascomycetes (members of the *Ascomycota*) have septate mycelia, with cross walls or septa, and produce sexual spores, called ascospores within a structure known as an ascus (plural asci). The ascomycetes also produce asexual reproductive states, where reproduction occurs by production of asexual spores known as conidia, which are spores borne on the end of special aerial fertile hyphae (conidiophores). Conidia are produced free, often in chains and not in any enclosing structure. The asexual state or stage is referred to as the anamorphic state and the fungus is called an anamorph. This state is sometimes called the "imperfect" stage or state of the fungus. The sexual state (ascogenous) is referred to as the teleomorphic state, and the ascospore-producing fungus as a teleomorph. The sexual stage is also referred as the "perfect" stage or state of the fungus.

Another group of fungi that are related to the ascomycetes, but which are not given phylum status, are the deuteromycetes, sometimes referred to as Deuteromycota. This is a group of fungi that are similar to members of the ascomycetes, but which have no observed sexual stage in their life cycle. They are anamorphs and are sometimes (in older literature) referred as the "imperfects" or Fungi Imperfecti.

The zygomycetes, ascomycetes, and deuteromycetes are fungi that are also called molds, or micro fungi, because of their small size and because they include the fungal species important in foods and food fermentations. One group of ascomycetes, also known as hemiascomycetes because they do not produce a separate ascoma, are primarily single celled organisms known as yeasts. Yeasts may produce ascospores within the yeast cell, but they also reproduce asexually by a process known as budding, and are very important in foods and food fermentations.

The *Basidiomycota* includes certain plant pathogens, such as rusts and smuts, and macro fungi such as mushrooms, puffballs, and bracket fungi that grow on trees and fallen logs, as well as ectomycorrhizal fungi that are associated with certain plant roots. Another group of fungi that form the most common type of mycorrhizal associations with plant roots are the Glomales, sometimes referred to as a fifth phylum, the *Glomeromycota*.

Bacteria

Within the *Bacteria* domain, there has been less consensus among taxonomists on how to organize bacteria into higher taxa (e.g., Kingdom, Sub-Kingdom, etc.). The *Ninth Edition of Bergey's Manual of Determinative Bacteriology* (published in 1994) described three major categories of bacteria and one of archaeobacteria, and then further divided the bacteria categories into thirty different descriptive groups. According to the most current taxonomy (Garrity, et al., 2004), the *Bacteria* are now divided into twenty-four different phyla. Nearly all of the bacteria important in food fermentations, including lactic acid bacteria, belong to a single phylum, the firmicutes. Beyond the phyla, bacteria can be further divided into classes (and sub-classes), orders (and sub-orders), families, and genera. Other details on their taxonomic positions will be described below.

Nomenclature

Like all living organisms, microorganisms are named according to Latinized binomial nomenclature, meaning they are assigned two names, a genus and a species. By convention, both the genus and species names are italicized, but only the genus is capitalized. Thus, the name of the common food yeast is written as *Saccharomyces cerevisae*. For some organisms, a trinomial system is applied to indicate a subspecies epithet, as is the case for the dairy organism *Lactococcus lactis* subsp. *lactis*.

Microorganisms are named according to the rules established by the appropriate governing body. For bacteria and fungi, the International Committee on Systematic Bacteriology and the International Association for Plant Taxonomy, respectively, define the nomenclature rules. These rules are then published in the respective "codes," the International Code of Nomenclature of Bacteria and the International Code for Botanical Nomenclature. In addition, there is a "running list," called the List of Bacterial Names with Standing in Nomenclature (www.bacterio.cict.fr), that provides updated bacterial nomenclature. To be included in these lists and considered as "valid," a name must have been published in the scientific literature, along with a detailed description and relevant supporting data.

In some cases, a validly named organism may be referred to by another name, indicated as a synonym. In other instances, the name of an organism may have been replaced by a new name, in which case the original name is indicated as a basonym. In situations where a name was "unofficially" assigned to an organism and does not appear on the list, that name is considered to be illegitimate and its use should be discontinued. However, even if the name assigned to a particular organism is supported by a valid publication, this does not mean that the name is or must be accepted by the scientific community.

Although names are indeed based on official rules (where each taxon has a valid name), the utility of a given classification scheme, on which a given organism is named, is left up to the scientists who use it. That is, a researcher may propose that a given organism be assigned a "new" name, and have the supporting evidence published in a valid journal, but other microbiologists are entitled to disagree with the taxonomy and reject the proposed classification.

It is relevant to raise these issues, because many of the organisms used in food fermentations have either undergone nomenclature revisions or have been reclassified into new taxa. For example, the official name of the dairy organism mentioned above, *L. lactis* subsp. *lactis* was originally *Bacterium lactis* (the first organism isolated in pure culture and named by Joseph Lister in 1873). It was renamed *Streptococcus lactis* in 1909, which is how this organism was known for more than 70 years (and which still shows up occasionally in "current" texts), before the new genus, *Lactococcus*, was adopted. In other cases, a name was proposed, then rescinded (see below). There are also instances of organisms that had been assigned "unofficial" names (see above), and through frequent use, had acquired some level of validity, however undeserved. A good example of the latter situation was for Lactobacillus sporogenes, an organism that is properly classified as *Bacillus subtilis*. Yeast nomenclature, although under the authority of the International Code for Botanical Nomenclature, is also subject to taxonomical challenges and changes in classification (see below).

Microbial taxonomy and methods of analysis

If microbiology began with Pasteur in the middle of the nineteenth century, then for the next 120 years, microbial classification was based primarily on phenotypic characteristics. Although many of these traits remain useful as diagnostic tools, by far, the most powerful means of classifying microorganisms is now via molecular techniques.

Originally, routine tests were based on nucleic acid composition (mol% G + C) and DNA-DNA and DNA-RNA hybridization. The latter has long been considered the gold standard for defining a species, in that organisms sharing high DNA homology (usually greater than 70%) are regarded as members of the same species. However, more recently, the nucleic acid sequence of the 16S rRNA region has become the most common way to distinguish between organisms, to show relatedness, and ultimately to classify an organism to the genus or species level.

These sequences can be easily determined from the corresponding DNA after PCR amplification. By comparing the sequence from a given organism to those obtained from reference organisms in the 16S rRNA database, it is possible to obtain highly significant matches or to infer relationships if the sequence is unique. In addition, DNA probes, based on genus- or species-specific 16S rRNA sequences from a particular reference organism, are also useful for identifying strains in mixed populations.

Other techniques that essentially provide a molecular fingerprint and that can be used to distinguish between strains of the same species include restriction fragment length polymorphism (RFLP), randomly amplified polymorphic DNA (RAPD), pulsed field gel electrophoresis (PFGE), and DNA microarray technology. Finally, it should be noted that nucleic acid-based methods for classification, whether based on 16S rRNA sequences, DNA-DNA hybridization values, or even whole genome analyses, are not necessarily the be-all, end-all of microbial taxonomy. Rather, a more holistic or integrative taxonomy, referred to as polyphasic taxonomy, has been widely adopted and used to classify many different groups of bacteria. Polyphasic taxonomy is based on genotypic as well as phenotypic and phylogenetic information and is considered to represent a "consensus approach" for bacterial taxonomy.

Bacteria Used in the Manufacture of Fermented Foods

Despite the diversity of bacteria involved directly or indirectly in the manufacture of fermented foods, all are currently classified in one of three phyla, the *Proteobacteria*, *Firmicutes*, and the *Actinobacteria*. Within the *Firmicutes* are the lactic acid bacteria, a cluster of Gram-positive bacteria that are the main organisms used in the manufacture of fermented foods.

This phylum also includes the genera *Bacillus* and *Brevibacterium* that contain species used in the manufacture of just a few selected fermented foods.

The *Proteobacteria* contains Gram negative bacteria that are involved in the vinegar fermentation, as well as in spoilage of wine and other alcoholic products. The *Actinobacteria* contains only a few genera relevant to fermented foods manufacture, and only in a rather indirect manner. These include *Bifidobacteriim*, *Kocuria*, *Staphylococcus*, and *Micrococcus*. In fact, *Bifidobacterium* do not actually serve a functional role in fermented foods; rather they are added for nutritional purposes (see below).

While species of *Kocuria* and the *Staphylococcus/Micrococcus* group are used in fermented foods, they are used for only one product, fermented meats, and for only one purpose, to impart the desired flavor and color. It is worth emphasizing that fermented foods may contain many other microorganisms, whose presence occurs as a result of inadvertent contamination. However, the section below describes only those bacteria whose contribution to fermented foods manufacture is known.

The Lactic Acid Bacteria

From the outset, it is important to recognize that the very term "lactic acid bacteria" has no official status in taxonomy and that it is really just a general term of convenience used to describe a group of functionally and genetically related bacteria. Still, the term carries rather significant meaning among microbiologists and others who study food fermentations, and, therefore, will be used freely in this text. Accordingly, the lactic acid bacteria are generally defined as a cluster of lactic acid-producing, low %G+C, non-spore-forming, Gram-positive rods and cocci that share many biochemical, physiological, and genetic properties (Table 2-1). They are distinguished from other Gram positive bacteria that also produce lactic acid (e.g., *Bacillus*, *Listeria*, and *Bifidobacterium*) by virtue of numerous phenotypic and genotypic differences.

Table 2.1.   Common characteristics of lactic acid bacteria

Gram positive
Fermentative
Catalase negative
Facultative anaerobes
Non-sporeforming
Low mol% G + C
Non-motile
Acid-tolerant

In addition to the traits described above, other important properties also characterize the lactic acid bacteria, but only in a general sense, given that exceptions occasionally exist. Most lactic acid bacteria are catalase-negative, acid-tolerant, aerotolerant, facultative anaerobes. In terms of their carbon and energy needs, they are classified as heterotrophic chemoorganotrophs, meaning that they require pre-formed organic carbon both as a source of carbon and energy.

Until recently, it was thought that all lactic acid bacteria lack cytochrome or electron transport proteins and, therefore, could not derive energy via respiratory activity. This view, however true for the majority of lactic acid bacteria, has been revised, based on recent findings that indicate some species may indeed respire, provided the medium contains the necessary nutrients (Box 2-1). Still, substrate level phosphorylation reactions that occur during fermentative pathways (see below) are the primary means by which ATP is obtained.

The lactic acid bacteria as a group are often described as being fastidious with complex nutritional requirements, and indeed, there are species that will grow only in nutrient-rich, well-fortified media under optimized conditions. However, there are also species of lactic acid bacteria that are quite versatile with respect to the growth environment and that grow reasonably well even when the nutrient content is less than ideal. Furthermore, some lactic acid bacteria are actually known for their ability to grow in inhospitable environments, including those that often exist in fermented foods. This is reflected by the diverse habitats

Box 2–1. Lactic acid bacteria learn new tricks

Look up "lactic acid bacteria" in any older (or even relatively recent) text, and in the section on physiological characteristics, it will likely be stated that these bacteria "lack cytochromes and heme-linked electron transport proteins" and are, therefore, "unable to grow via respiration." In fact, this description was considered to be dogma for generations of microbiology students and researchers who studied these bacteria, despite the occasional report that suggested otherwise (Ritchey and Seeley, 1976; Sijpesteijn, 1970). In the past few years, however, it has become apparent that this section on physiology of lactic acid bacteria will have to be re-written, because biochemical and genetic evidence, reported by researchers in France, now supports the existence of an intact and functional respiratory pathway in *Lactococcus lactis* and perhaps other lactic acid bacteria (Duwat et al., 2001; Gaudu et al., 2002).

Respiratory metabolism

Respiration is the major means by which aerobic microorganisms obtain energy. During respiration, electrons generated during carbon metabolism (e.g., citric acid cycle) are transported or carried across the cytoplasmic membrane via a series of electron carrier proteins. These proteins are arranged such that the flow of electrons (usually in the form of $NADH^+$) is toward increasing oxidation-reduction potentials. Along the way, protons are translocated across the membrane, effectively converting an oxidation-reduction potential into a proton electrochemical potential. This proton potential or proton motive force (PMF) can drive transport, operate flagella motors, or perform other energy-requiring reactions within the membrane. It can also be used to make ATP directly via the ATP synthase. This reaction (called oxidative phosphorylation) provides aerobes with the bulk of their ATP and is the coupling step between the respiratory electron transport system and ATP formation.

Respiration in lactic acid bacteria

For respiration to occur in lactococci, several requirements must be met. First, the relevant genes encoding for electron transport proteins must be present, and then the functional proteins must be made. Based on physiological and genetic data, these genes are present and the respiratory pathway is intact in a wild-type strain of *Lactococcus lactis* (Vido et al., 2004). In this and presumably other strains of lactococci, this pathway is comprised of several proteins, including dehydrogenases, menaquinones, and cytochromes. In particular, cytochrome oxidase (encoded by *cydAB*) serves as the terminal oxidase and is essential (Gaudu et al., 2002). Missing, however, is a system for making porphyrin groups. The latter is combined with iron to make heme, which is required for activity of cytochrome proteins, as well as the enzyme catalase. Thus, heme (or a heme precursor) must be added to the medium for respiration to occur. Respiratory growth follows a fermentative period and occurs primarily during the later stages of growth. Interestingly, expression of the heme transport system is subject to negative regulation, mediated by the catabolite control protein CcpA (Gaudu et al., 2003). Recall that the activity of this regulator is high during rapid growth on glucose and that genes for other catabolic pathways are repressed. Thus, it appears that CcpA might be the metabolic conduit between fermentation and respiration.

Importantly, it was observed that when lactococci were grown under respiring conditions (i.e., aerobic, with heme added), not only did cells perform respiratory metabolism, they also grew better (Duwat et al., 2001). That is, they grew for a longer time and reached higher cell densities compared to non-respiring cells. Moreover, during prolonged incubation at 4°C, respiring cells remained viable for a longer time, in part because the pH remained high (near 6.0), since less lactic acid was produced via fermentation. These findings are of considerable industrial importance, given that one of the main goals of the starter culture industry (Chapter 3)

(continued)

Box 2–1.   Lactic acid bacteria learn new tricks *(Continued)*

is to maximize cell biomass while maintaining cell viability. Thus, the respiration story has certainly captured the attention of the starter culture industry (Pedersen et al., 2005). Finally, since lactococci do not have an intact citric acid cycle, the electrons that feed the respiratory chain cannot be supplied from the citric acid cycle-derived NADH pool (Vido et al., 2004). Instead, they must be generated though other pathways.

These recent findings on lactococci have not yet been extended to other lactic acid bacteria. However, based on several sequenced genomes, it appears that *cyd* genes may be present in other lactococci as well as some oenococci and leuconostocs (but are absent in *Streptococcus thermophilus* and *Lactobacillus delbrueckii* subsp. *bulgaricus*). If, and under what circumstances these bacteria actually perform respiration, is not known. As had initially been reported thirty years ago, and re-confirmed by these more recent studies, molar growth and ATP yields are higher during respiratory growth. Thus, it may well be that the respiratory pathway provides these bacteria an alternative and efficient life-style choice, were they to find themselves back out in nature rather than in the confines of food environments (Duwat et al., 2001).

References

Duwat, P., S. Sourice, B. Cesselin, G. Lamberet, K. Vido, P. Gaudu, Y. Le Loir, F. Violet, P. Loubière, and A. Gruss. 2001. Respiration capacity of the fermenting bacterium *Lactococcus lactis* and its positive effects on growth and survival. J. Bacteriol. 183:4509–4516.

Gaudu, P., G. Lamberet, S. Poncet, and A. Gruss. 2003. CcpA regulation of aerobic and respiration growth in *Lactococcus lactis*. Mol. Microbiol. 50:183–192.

Gaudu, P., K. Vido, B. Cesselin, S. Kulakauskas, J. Tremblay, L. Rezaïki, G. Lamberet, S. Sourice, P. Duwat, and A. Gruss. 2002. Respiration capacity and consequences in *Lactococcus lactis*. Antonie van Leeuwenhoek 82:263–269.

Pedersen, M.B., S.L. Iversen, K.I. Sørensen, and E. Johansen. 2005. The long and winding road from the research laboratory to industrial applications of lactic acid bacteria. FEMS Microbiol. Rev. 29:611–624.

Ritchey, T.W., and H.W. Seely. 1976. Distribution of cytochrome-like respiration in streptococci. J. Gen. Microbiol. 93:195–203.

Sijpesteijn, A.K. 1970. Induction of cytochrome formation and stimulation of oxidative dissimilation by hemin in *Streptococcus lactis* and *Leuconostoc mesenteroides*. Antonie Van Leeuwenhoek 36:335–348.

occupied by lactic acid bacteria, which include not only plant material, milk, and meat, but also salt brines, low pH foods, and ethanolic environments.

Perhaps the most relevant properties of lactic acid bacteria are those related to nutrient metabolism. Specifically, the main reason why lactic acid bacteria are used in fermented foods is due to their ability to metabolize sugars and make lactic and other acid end-products. Two fermentative pathways exist. In the homofermentative pathway, more than 90% of the sugar substrate is converted exclusively to lactic acid. In contrast, the heterofermentative pathway results in about 50% lactic acid, with the balance as acetic acid, ethanol, and carbon dioxide. Lactic acid bacteria possess one or the other of these two pathways (i.e., they are obligate homofermentative or obligate heterofermentative), although there are some species that have the metabolic wherewithal to perform both (facultative homofermentative). These pathways will be described in detail later in this chapter.

The genera of lactic acid bacteria

According to current taxonomy, the lactic acid bacteria group consists of twelve genera (Table 2-2). All are in the phylum *Firmicutes*, Order, *Lactobacillales*. Based on 16S rRNA sequencing and other molecular techniques, the lactic acid bacteria can be grouped into a broad phylogenetic cluster, positioned not far from other low G + C Gram positive bacteria (Figure 2–2). Five sub-clusters are evident from this tree, including: (1) a *Streptococcus-Lactococcus* branch (Family *Streptococcaceae*), (2) a *Lactobacillus* branch (Family *Lactobacillaceae*), (3) a separate *Lactobacillus-Pediococcus* branch (Family *Lactobacillaceae*); (4) an *Oenococcus-Leuconostoc-Weisella* branch (Family *Leuconostocaceae*), and (5) a *Carnobacterium-Aerococcus-Enterococcus-Tetragenococcus-Vagococcus* branch (Families *Carnobacteriaceae*, *Aerococcaceae*, and *Enterococcaceae*).

It is worth noting that this phylogeny is not entirely consistent with regard to the morphological and physiological characteristics of these bacteria. For example, *Lactobacillus* and *Pediococcus* are in the same sub-cluster, yet the former are rods and include heterofermentative species, whereas the latter are homofermenting cocci. Likewise, *Carnobacterium* are

obligate heterofermentative rods, and *Enterococcus* and *Vagococcus* are homofermentative cocci.

Seven of the twelve genera of lactic acid bacteria, *Lactobacillus*, *Lactococcus*, *Leuconostoc*, *Oenococcus*, *Pediococcus*, *Streptococcus*, and *Tetragenococcus*, are used directly in food fermentations. Although *Enterococcus* sp. are often found in fermented foods (e.g., cheese, sausage, fermented vegetables), except for a few occasions, they are not added directly. In fact, their presence is often undesirable, in part, because they are sometimes used as indicators of fecal contamination and also because some strains may harbor mobile antibiotic-resistance genes.

Importantly, some strains of *Enterococcus* are capable of causing infections in humans. Likewise, *Carnobacterium* are also undesirable, mainly because they are considered as

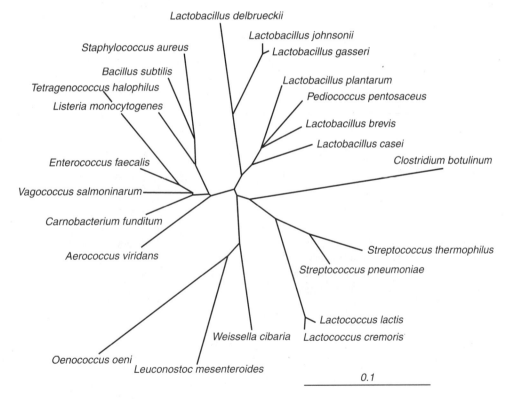

Figure 2–2. Phylogeny of lactic acid and other Gram positive bacteria (based on 16s rRNA). The tree (un-rooted) was generated using the neighbor-joining method.

Table 2.2.   Genera of lactic acid bacteria and their properties[1,2]

Genus	Cell Morphology	Fermentation route[3]	Growth at: 10°C	45°C	Growth in NaClat: 6.5%	18%	Growth at pH: 4.4	9.6	Lactic acid isomer
Lactobacillus	rods	homo/hetero[3]	±[4]	±	±	−	±	−	D,L,DL[5]
Lactococcus	cocci	homo	+	−	−	−	±	−	L
Leuconostoc	cocci	hetero	+	−	±	−	±	−	D
Oenococcus	cocci	hetero	+	+	±	−	±	−	D
Pediococcus	cocci (tetrads)	homo	±	±	±	−	+	−	D,L,DL
Streptococcus	cocci	homo	−	+	−	−	−	−	L
Tetragenococcus	cocci (tetrads)	homo	+	−	+	+	−	+	L
Aerococcus	cocci (tetrads)	homo	+	−	+	−	−	+	L
Carnobacterium	rods	hetero	+	−	−	−	−	−	L
Enterococcus	cocci	homo	+	+	+	−	+	+	L
Vagococcus	cocci	homo	+	−	−	−	±	−	L
Weissella	coccoid	hetero	+	−	±	−	±	−	D,L,DL

[1]Adapted from Axelsson, 2004

[2]Refers to the general properties of the genus; some exceptions may exist

[3]Species of *Lactobacillus* may be homofermentative, heterofermentative, or both

[4]This phenotype is variable, depending on the species

[5]Some species produce D-, L-, or a mixture of D- and L-lactic acid.

spoilage organisms in fermented meat products. Finally, species of *Aerococcus, Vagococcus*, and *Weisella* are not widely found in foods, and their overall significance in food is unclear. In the section to follow, the general properties, habitats, and practical considerations of the genera that are relevant to food fermentations will be described; descriptions of the important species will also be given. Information on the genetics of these bacteria, based on genome sequencing and functional genomics, will be presented later in this chapter.

Lactococcus

The genus *Lactococcus* consists of five phylogenetically-distinct species: *Lactococcus lactis, Lactococcus garviae, Lactococcus piscium, Lactococcus plantarum*, and *Lactococcus raffinolactis* (Figure 2–3). They are all non-motile, obligately homofermentative, facultative anaerobes, with an optimum growth temperature near 30°C. They have a distinctive microscopic morphology, usually appearing as cocci in pairs or short chains.

One species in particular *L. lactis*, is among the most important of all lactic acid bacteria (and perhaps one of the most important organ-

isms involved in food fermentations, period). This is because *L. lactis* is the "work horse" of the dairy products industry—it is used as a starter culture for most of the hard cheeses and many of the cultured dairy products produced around the world. There are actually three *L. lactis* subspecies: *L. lactis* subsp. *lactis, Lactococcus lactis* subsp. *cremoris*, and *Lactococcus lactis* subsp. *hordinae*. Only *L. lactis* subsp. *lactis* and *L. lactis* subsp. *cremoris*, however, are used as starter cultures; *L. lactis* subsp. *hordinae* has no relevance in fermented food manufacture. Another variant of *L. lactis* subsp. *lactis*, formerly named *L. lactis* subsp. *diacetilactis* (or *Lactococcus lactis* subsp. *lactis* biovar. *diacetylactis*), is distinguished based on its ability to metabolize citrate. This species is not included in the current List of Bacterial Names, and instead is encompassed within *L. lactis* subsp. *lactis* (see below). However, *Lactococcus lactis* subsp. *diacetilactis* is listed in *Bergey's Taxonomical Outline of the Prokaryotes* (2004 version).

Plant material has long been considered to be the "original" habitat of both *L. lactis* subsp. *lactis* and *L. lactis* subsp. *cremoris*. The suggestion, however, that milk is now their "new" habitat is supported by several observations.

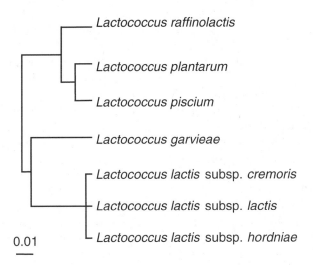

Figure 2–3. Phylogeny of *Lactococcus* based on 16S rRNA sequence analysis.

First, they are readily isolated from raw milk; in fact, it is difficult to find *L. lactis* subsp. *cremoris* anywhere but milk. Second, both species grow rapidly in milk, producing lactic acid and lowering the pH to below 4.5. Thus, they generally out-compete most potential competitors. Finally, the genes required for growth in milk are located on plasmids (extra-chromosomal DNA), indicating they were acquired recently (relatively, speaking).

Specific plasmid-borne genes in lactococci encode for proteins involved in lactose transport and metabolism and casein hydrolysis and utilization. There is considerable selective pressure, therefore, for milk-borne strains to retain these plasmids, since plasmid-cured derivatives grow poorly in milk. Indeed, some lactococcal plasmids are readily exchanged among different strains (via conjugal transfer), and in other cases, plasmid DNA can integrate within the chromosome, resulting in stable, highly adapted derivative strains. Plasmids, however, are common in lactococci, even in strains isolated from non-dairy sources.

Although *L. lactis* subsp. *lactis* and *L. lactis* subsp. *cremoris* differ in just a few seemingly minor physiological respects, at least two of these differences are significant during milk fermentations. For example, the highest temperature at which most strains of *L. lactis* subsp. *lactis* are able to grow is 40°C, whereas most *L. lactis* subsp. *cremoris* strains do not grow above 38°C. In addition, *L. lactis* subsp. *lactis* has greater tolerance to salt (up to 4%) than does *L. lactis* subsp. *cremoris*.

What makes these differences relevant is that both temperature and salt are among the primary means by which the activity of the starter culture is controlled during cheese manufacture. Moreover, *L. lactis* subsp. *cremoris* is generally considered to be, overall, less tolerant to the environmental stresses encountered during both culture production and during dairy fermentations. Thus, if *L. lactis* subsp. *cremoris* is used as a culture for cheese making (and it is generally accepted that this organism makes better quality products), then temperature and salt concentrations must be adjusted to account for this organism's particular growth requirements.

Complicating this discussion on the differences between these two organisms, it has recently been noted that there are strains of *L. lactis* subsp. *lactis* that have a "*lactis*" genotype, but a "*cremoris*-like" phenotype. Likewise, there are strains of *L. lactis* subsp. *cremoris* that have a "*cremoris*" genotype, but a "*lactis*-like" phenotype (Box 2–2). The use of genomics (discussed later in the chapter), will help to determine the genetic basis for these industrially relevant differences.

Streptococcus

The genus *Streptococcus* contains many diverse species with a wide array of habitats. Included in this genus are human and animal pathogens, oral commensals, intestinal commensals, and one (and only one) species, *Streptococcus thermophilus*, that is used in the manufacture of fermented foods. In general, streptococci are non-motile, facultative anaerobes, with an obligate homofermentative metabolism.

Since the mid-1980s, there have been several major taxonomical revisions within this genus. Some of these changes were especially relevant for food microbiologists. For the previous fifty years, the streptococci were divided into four main groups: pyogenic, enterococcus, viridans, and lactic. These groupings, which were based on the so-called Sherman scheme (published in 1937, as described in Chapter 3), were revised in subsequent years, but generally served as the primary means for organizing streptococcal species. Starting in 1984, however, two major revisions were proposed and adopted. First, the enterococcus (or enteric streptococci), which included *Streptococcus faecalis*, *Streptococcus faecium*, and *Streptococcus durans*, were moved to a new genus, *Enterococcus*. Then, in 1985, two species that had been referred to as "lactic streptococci" (*Streptococcus lactis* and *Streptococcus cremoris*) were also assigned to a new genus, *Lactococcus* (see above). Thus, *S. thermophilus* is now the only member of this

Box 2–2. *Lactococcus lactis*—Differences Between the Subspecies

As might be expected, *Lactococcus lactis* subsp. *lactis* and *Lactococcus lactis* subsp. *cremoris* share many phenotypic properties. In fact, these organisms are so similar that in the 1986 *Bergey's Manual of Systematic Bacteriology*, the lactis and cremoris strains were considered as belonging to a single species (Mundt, 1986). However, several phenotypic differences do exist, as described earlier, and these differences have traditionally been used to distinguish between the two subspecies (Table 1). As noted previously, differences also exist at the molecular level. Currently, assignment of a strain to the appropriate subspecies is based on the sequence of a 9 to 10 base pair region of the VI segment in the 16S rRNA gene. Moreover, the genomes of both organisms have now been sequenced, and other differences are likely to become apparent as comparative analyses of these genome are completed (Klaenhammer et al., 2002; Makarova et al., 2006).

Table 1. Distinguishing characteristics of dairy *Lactococcus*[1].

Property	*Lactococcus lactis* subsp. *lactis*	*Lactococcus lactis* subsp. *cremoris*
Growth at 40°C	+	−
Growth in 4.0% NaCl	+	−
Growth at pH 9.2	+	−
Growth in 0.1% methylene blue	+	−
Arginine hydrolysis	+	−
Acid from maltose	+	−
Acid from ribose	+	−

[1]Adapted from Teuber, 1992

Interestingly, it now appears that classification of *L. lactis* by both phenotypic and genotypic criteria is not so clear-cut. In fact, one research group recently proposed that *L. lactis* strains can be organized into one of five different groups (Kelly and Ward, 2002). Group 1 consists of strains having a *L. lactis* subsp. *lactis* genotype and a "lactis" phenotype. These strains may be considered as the common or typical *L. lactis* subsp. *lactis*. Group 2 strains also have a *L. lactis* subsp. *lactis* genotype, but have one particular distinguishing phenotypic property, namely the ability to ferment citrate and to produce diacetyl. (As noted previously, these citrate-citrate-fermenting strains had, for a long time, held subspecies status.) Strains in Group 3 have a cremoris phenotype, but a *L. lactis* subsp. *lactis* genotype. They are not commonly found. In contrast, Group 4 strains, having a *L. lactis* subsp. *cremoris* genotype but a lactis phenotype, are frequently isolated from milk and plant sources. Their lactic-like phenotype makes it difficult to distinguish them from other lactis strains. Finally, Group 5 strains have a *L. lactis* subsp. *cremoris* genotype and phenotype. They are found almost exclusively in milk and dairy environments, and are the most common and preferred strains for manufacture of cheese.

On a practical basis, it is probably easier to establish the genotype of a particular lactococcal strain than it is to determine its phenotype (at least with regard to traits relevant to fermentations). However, it is the phenotypic properties of a given strain that dictate whether that strain will be useful in fermented foods manufacture. In some cases, properties beyond those normally considered as part of a phenotype will be relevant, such as bacteriophage sensitivity or the ability to produce good cheese flavor. A genetic method recently was devised to identify *L. lactis* strains to the subspecies level; it also provided a basis for determining phenotypes (Nomura et al., 2002). The method was based on the presence of glutamate decarboxylase activity in *L. lactis* subsp. *lactis*, which contains a functional *gadB* gene, and its absence in *L. lactis* subsp. *cremoris*, which contains insertions, deletions, and point mutations in *gadB*. Although the sample set was small, the size of specific DNA fragments from the *gadB* gene (amplified by

(continued)

Box 2–2. *Lactococcus lactis*—Differences Between the Subspecies *(Continued)*

PCR) from *L. lactis* subsp. *lactis* and *L. lactis* subsp. *cremoris* strains correlated with both their genotypes and phenotypes. Ultimately, this and other similar procedures could provide a relatively rapid and efficient means to screen lactococcal strains with the desired phenotype.

References

Bolotin, A., P. Wincker, S. Mauger, O. Jaillon, K. Malarme, J. Weissenbach, S.D. Ehrlich, and A. Sorokin. 2001. The complete genome sequence of the lactic acid bacterium *Lactococcus lactis* ssp. *lactis* IL1403. Genome Res. 11:731-753.

Goffeau, A., B. G. Barrell, H. Bussey, R. W. Davis, B. Dujon, H. Feldmann, F. Galibert, J. D. Hoheisel, C. Jacq, M. Johnston, E. J. Louis, H. W. Mewes, Y. Murakami, P. Philippsen, H. Tettelin, and S. G. Oliver. 1996. Life with 6000 genes. Science 274:546-567.

Kelly, W., and L. Ward. 2002. Genotypic vs. phenotypic biodiversity in *Lactococcus lactis*. Microbiol. 148:3332-3333.

Klaenhammer et al. (36 other authors). 2002. Discovering lactic acid bacteria by genomics. Antonie van Leeuwenhoek 82:29-58.

Makarova, K., Y. Wolf, et al., (and 43 other authors) 2006. Comparative genomics of the lactic acid bacteria. 2006. Proc. Nat. Acad. Sci. In Press.

Mundt, J.O. 1986. Lactic acid streptococci. In Sneath, P.H.A., N.S. Mair, M.E. Sharpe, J.G. Holt. *Bergey's Manual of Systematic Bacteriology, Volume 2*. Williams and Wilkins, Baltimore, Maryland. 1065-1071.

Nomura, M., M. Kobayashi, and T. Okamoto. 2002. Rapid PCR-based method which can determine both phenotype and genotype of *Lactococcus lactis* subspecies. Appl. Environ. Microbiol. 68:2209-2213.

Teuber, M. 1992. The genus Lactococcus. *In* B.J.B. Wood and W.H. Holzapfel, ed. *The Lactic Acid Bacteria, Volume 2, The genera of lactic acid bacteria*. Blackie Academic and Professional pp 173—234.

genus used in food fermentations (mainly yogurt and cheese).

That is not to say that this organism has been exempt from taxonomical considerations. Originally, this species was part of the Sherman viridans group (which included oral streptococci), and in the 1986 edition of Bergey's manual, it was listed as an "Other Streptococci." Its physiological, structural, and genetic similarities to *Streptococcus salivarius* led to the recommendation in 1984 that it be reclassified as a subspecies of *S. salivarius*. Although this name change was indeed adopted (and *"Streptococcus salivarius* subsp. *thermophilus"* is sometimes still seen in the literature), subsequent DNA-DNA homology studies eventually led to the restoration of its original name. Thus, labels on yogurt products can justifiably claim the presence of *Streptococcus thermophilus* rather than the less appetizing *Streptococcus salivarius*.

In several respects, *S. thermophilus* is not that different from the mesophilic dairy lactococci (*L. lactis* subsp. *lactis* and *L. lactis* subsp. cremoris), as is evident by their close phylogenetic position (Figure 2-2). Like the lactococci, *S. thermophilus* is highly adapted to a milk environment in that it ferments lactose rapidly and produces lactic acid in homolactic fashion. The route by which lactose is metabolized by *S. thermophilus*, however, is quite different from how *L. lactis* ferments this sugar (discussed later). Also, *S. thermophilus* has a higher temperature optima (40°C to 42°C), a higher maximum growth temperature (52°C), and a higher thermal tolerance (above 60°C). Its nutritional requirements are somewhat more demanding than lactococci, in that *S. thermophilus* is weakly proteolytic and, therefore, needs pre-formed amino acids. Salt tolerance, bile sensitivity, and a limited metabolic diversity are also characteristic of *S. thermophilus*. In fact, the statement made by Sherman in 1937 that *S. thermophilus* "is marked more by the things which it cannot do than by its positive actions" describes this organism quite well. Finally, in contrast to lactococci, *S. thermophilus* strains contain few plasmids. When plasmids are found, they are generally small and cryptic (i.e., having no known function).

Leuconostoc

The *Leuconostoc* belong to the *Leuconostocaceae* Family, which also contains the closely related genera *Weissella* and *Oenococcus* (see below). Leuconostocs are mesophilic, with optimum growth temperatures ranging from 18°C to 25°C. Some species are capable of growth at temperatures below 10°C. Microscopically, they appear coccoid or even somewhat rod-like, depending on the composition and form of the growth medium (liquid versus solid). The leuconostocs, in contrast to the obligate homofermenting lactococci and streptococci, are obligately heterofermentative. Accordingly, they are missing an intact glycolytic pathway, and instead rely on the phosphoketolase pathway for metabolism of sugars (see below). Although leuconostocs can grow in an ambient atmosphere, a reduced or anaerobic environment generally enhances growth. Plasmids are common in *Leuconostoc* species and, when present, may encode for important functions including lactose and citrate metabolism and bacteriocin production.

The genus *Leuconostoc* has also been subject to taxonomical revision, because several species have been moved to other existing or newly-formed genera. For example, *Oenococcus oeni* was formerly classified as *Leuconostoc oenos*, and other *Leuconostoc* species have been reclassified as *Weisella*. The genus currently consists of thirteen species (Figure 2-4).

Speciation is based on both genetic and phenotypic characteristics. The latter include carbohydrate fermentation profiles, the ability to produce the polysaccharide dextran, resistance to the antibiotic vancomycin, and various physiological properties (Table 2-3). Most species are associated with particular habitats, including plant and vegetable material, milk and dairy environments, and meat products. In addition, some species are involved in food spoilage (e.g., *Leuconostoc gasicomitatum*), whereas others are used in food fermentations. The latter include *Leuconostoc mesenteroides* subsp. *cremoris* and *Leuconostoc lactis*, which are used in dairy fermentations, and *Leuconostoc mesenteroides* subsp. *mesenteroides*, *Leuconostoc kimchii*, and *Leuconostoc fallax*, that are used in vegetable fermentations. In general, the habitats occupied by these strains are reflected by the particular carbohydrates they ferment. Plant-associated strains, such as *L. mesenteroides* subsp. *mesenteroides*, ferment plant sugars (i.e., fructose, sucrose, arabinose, trehalose), whereas dairy strains (*L. mesenteroides* subsp. *cremoris* and *L. lactis*) are more likely to ferment lactose, galactose, and glucose. It is also possible for otherwise useful species to cause spoilage, particularly in the case of polysaccharide-producing strains that form slime.

Heterofermentative metabolism of sugars by leuconostocs results in formation of a mixture of end-products, including lactic acid, acetic acid, ethanol, and carbon dioxide. The presence of CO_2 is readily detected and can be used diagnostically to separate these bacteria from homofermentative streptococci, lactococci, and pediococci. Also, while leuconostocs generally decrease the pH in the growth medium to between 4.5 and 5.0, acid production by some species may be relatively modest, especially when compared to homofermentative lactobacilli and other lactic acid bacteria. Thus, acidification is not necessarily the major function of these bacteria during fermentations.

In sauerkraut and other vegetable fermentations, for example, they lower the pH during the very early manufacturing stages and produce enough CO_2 to reduce the redox potential in the food environment. These metabolic events then create an atmosphere conducive for growth of other microorganisms that are responsible for more significant acid development. In addition, the heterofermentative end-products result in a more diverse flavor and aroma profile in fermented foods. The *Leuconostoc* species used in dairy fermentations are particularly important for flavor, in part, for the same reasons as the vegetable-associated strains. However, dairy leucononstocs also produce the four-carbon volatile, diacetyl, which imparts a desirable buttery aroma to cultured dairy products (discussed below).

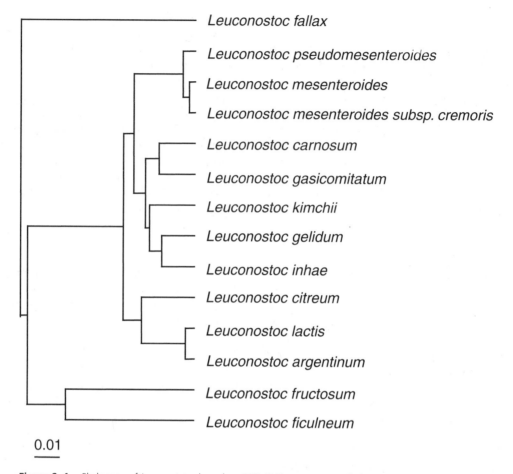

Leuconostoc fallax

Leuconostoc pseudomesenteroides

Leuconostoc mesenteroides

Leuconostoc mesenteroides subsp. cremoris

Leuconostoc carnosum

Leuconostoc gasicomitatum

Leuconostoc kimchii

Leuconostoc gelidum

Leuconostoc inhae

Leuconostoc citreum

Leuconostoc lactis

Leuconostoc argentinum

Leuconostoc fructosum

Leuconostoc ficulneum

0.01

Figure 2–4. Phylogeny of *Leuconostoc* based on 16S rRNA sequence analysis.

Oenococcus

This genus was established as recently as 1995, and contains only one species, *O. oeni*. Not surprisingly, *O. oeni* is located, phylogenetically, within the *Leuconostoc*aceae branch; however, it is somewhat distant from *Leuconostoc* and *Weissella* (Figure 2-2). Although *O. oeni* shares many phenotypic properties with *Leuconostoc* sp. (e.g., heterofermentative metabolism, mesophilic growth range), several important physiological differences exist. In particular, *O. oeni* is much more acid-tolerant than *Leuconostoc* as it is able to commence growth in media with a pH below 5.0. In addition, *O. oeni* is one of the most ethanol-tolerant

of all lactic acid bacteria and can grow in the presence of 10% ethanol. In general, however, most strains of *O. oeni* are slow-growing and ferment a limited number of sugars.

As implied by the etymology of its name, the use of *O. oeni* in fermented foods is restricted to only one application, namely wine making (oenos is the Greek word for wine). Despite its limited use, however, the importance of *O. oeni* during the wine fermentation cannot be overstated. This is because *O. oeni* has the ability to de-acidify wine via the malolactic fermentation, whereby malic acid is decarboxylated to lactic acid (described in detail in Chapter 10). Moreover, given the ability of *O. oeni* to fer-

Table 2.3. Characteristics of *Leuconostoc* and *Oenococcus*[1]

Property	*Leuconostoc mesenteroides* subsp.			*Leuconostoc lactis*	*Oenococcus oeni*
	cremoris	*mesenteroides*	*dextranicum*		
% G+C	38–44	28–32	28–32	37–42	37–42
Growth at 37°C	−	±[2]	+	+	±
Growth at pH 4.8	−	−	−	−	+
Dextran from sucrose	−	+	+	−	−
Growth in 10% ethanol	−	−	−	+	+
Acid from:					
arabinose	−	+	−	−	±
fructose	−	+	+	+	+
maltose	±	+	+	+	−
melibiose	±	+	+	±	±
salicin	−	±	±	±	±
sucrose	−	+	+	+	−
trehalose	−	+	+	−	+

[1]Adapted from *Bergey's Manual of Determinative Bacteriology*
[2]Variable reaction, depending on strain

ment glucose and fructose, but few other sugars, and its high tolerance to low pH and high ethanol, wine or juice would appear to be the natural habitat for this organism.

Pediococcus

The pediococci are similar, in many respects, to other coccoid-shaped, obligate homofermentative lactic acid bacteria, with one main exception. When these bacteria divide, they do so in two "planes" (and in right angles). Thus, tetrads are formed, which can be observed visually. Cells may appear as pairs (and always spherical in shape), but chains are not formed, as they are for lactococci, streptococci, and leuconostocs. Pediococci, like other lactic acid bacteria, are facultative anaerobes, with complex nutritional requirements. They have optimum growth temperatures ranging from 25°C to 40°C, but some species can grow at temperatures as high as 50°C. Several of the pediococci are also distinguished from other lactic acid bacteria by their ability to tolerate high acid (growth at pH 4.2) and high salt (growth at 6.5% NaCl) environments (Table 2-4).

The pediococci can be found in diverse habitats, including plant material, milk, brines, animal urine, and beer. There are six recognized species of *Pediococcus* (Figure 2–5); several are important in food fermentations. Two species, *Pediococcus acidilactici* and *Pediococcus pentosaceus,* are naturally present in raw vegetables, where, under suitable conditions, they play a key role in the manufacture of sauerkraut and other fermented vegetables. These same species may also be added to meat to produce fermented sausages. Despite their inability to ferment lactose, *P. acidilactici* and *P. pentosaceus* are frequently found in cheese, where they may participate in the ripening process. Pediococci are also important as spoilage organisms in fermented foods, in particular, beer, wine, and cider. One species, *Pediococcus damnosus,* is especially a problem in beer, where it produces diacetyl, which in beer is a serious defect.

Plasmids are frequently present in pediococci. The genes located on these plasmids may encode for functions involved in sugar metabolism (e.g., raffinose and sucrose) and production of bacteriocins. The latter (described in Chapter 6) are defined as anti-microbial proteins that inhibit closely related bacteria, including the meat-associated pathogens *Listeria monocytogenes, Staphylococcus aureus,* and *Clostridium botulinum.* Bacteriocin

Table 2.4.   Characteristics of *Pediococcus*[1]

Property	Pediococcus acidilactici	Pediococcus pentosaceus	Pediococcus damnosus
% G+C	38–44	37	37–42
Growth at:			
35°C	+	+	−
40°C	+	+	−
50°C	+	−	−
Optimum growth temperature (°C)	40	28–32	28–32
Growth at:			
pH 4.2	+	+	+
pH 7.0	+	+	−
Growth in:			
4.0% NaCl	+	+	−
6.5% NaCl	+	+	−
Arginine hydrolysis	+	+	−

[1]Adapted from *Bergey's Manual of Determinative Bacteriology*

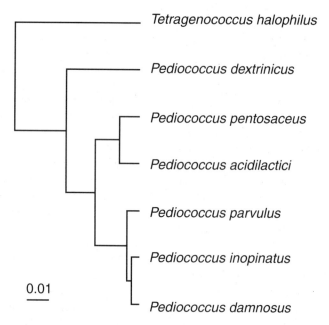

Figure 2–5. Phylogeny of *Pediococcus* based on 16S rRNA sequence analysis.

production is a particularly important trait in pediococci, since these bacteria are used as starter cultures for the sausage fermentation. Thus, bacteriocin-producing pediococci could provide an additional level of preservation when used in sausage manufacture.

Tetragenococcus

Like the pediococci, *Tetragenococcus* are homofermentative, tetrad-forming, facultative anaerobes. They are mesophilic and neutraphilic, with temperature optima generally between 25°C and 30°C and pH optima between 6.5 and 8.0. The genus contains only three species, *Tetragenococcus halophilus, Tetragenococcus muriaticus,* and *Tetragenococcus solitarius.* Based on 16S rRNA sequences, these bacteria are phylogenetically more closely related to *Lactobacillus* and *Enterococcus,* than to *Pediococcus.*

The genus was first recognized in 1990, when *Pediococcus halophilus* was re-classified as *T. halophilus.* A second species, *T. muriaticus,* isolated from salty fish sauce, was proposed in 1997, and in 2005, *Enterococcus solitarius* was reclassified as *T. solitarius.* All of these species are characterized by their remarkable tolerance to salt. Not only is growth possible in media containing up to 25% salt, but maximum growth rates require the presence of 3% to 10% NaCl, and no growth occurs when salt is absent. They also grow well when the water activity is low, presumably due to their ability to maintain osmotic balance by accumulating betaine, carnitine, and other compatible solutes.

Lactobacillus

The genus *Lactobacillus* consists of more than eighty species (Figure 2–6). In the last decade, microbial taxonomists have been very busy, proposing and validating new taxa. In one single month (January 2005), seven new species and subspecies were described in the published literature. About all that is common to the species within this genus is that they are all non-sporing rods, but even this description is not wholly satisfactory. There are species that appear rather short (<1.5 μm), whereas others are more than 5 μm in length (some are reportedly up to 10 μm). They may also have a slender, curved, or bent appearance. When viewed microscopically, it is difficult to be sure that a given strain is even a rod. Colony morphology is also variable on agar plates, with some strains producing large round colonies and others producing small or irregular colonies. Not only does this genus contain far more species, with extensive morphological heterogeneity, than any other genera of lactic acid bacteria, but this group is also the most ecologically, physiologically, biochemically, and genetically diverse.

Ecologically, lactobacilli occupy a wide range of habitats. Except for very extreme environments, there are few locales where lactobacilli are not found, and they are often described as being ubiquitous in nature. Some species are normal inhabitants of plant and vegetable material, and they are frequently found in dairy and meat environments, in juice and fermented beverages, and in grains and cereal products. Their presence in the animal and human gastrointestinal tract (as well as in the stomach, mouth and vagina) has led to the suggestion (which has gained substantial scientific support) that these bacteria have broad "probiotic activity," meaning they promote intestinal as well as extra-intestinal health. In foods, they are involved not only in many important fermentations, but are also frequently implicated in spoilage of fermented and non-fermented foods.

The ability of lactobacilli to grow and persist in so many diverse environments and conditions reflects their diverse physiological properties. Although most species are mesophilic, the genus also contains species that are psychrotrophic, thermoduric, or thermophilic. Temperature optima varies widely, from 30°C to 45°C. Some species show high tolerance to salt, osmotic pressure, and low water activity. Acid-tolerance is a common trait of lactobacilli (many strains actually prefer an acidic environment), and some also are ethanol-tolerant or bile-tolerant. Most species are aerotolerant, whereas others require more strict anaerobic conditions.

Like all lactic acid bacteria, lactobacilli are fermentative, but, again, they are more metabolically diverse than other lactics. In fact, one of the major ways in which the genus is subdivided into groups is based on the pathways they use to ferment sugars. As noted previously, lactic acid bacteria, in general, ferment sugars in either homofermentative or heterofermentative fashion. However, some species of *Lactobacillus* have the genetic and physiological wherewithal to ferment sugars by either pathway. These species, therefore, are referred to as facultative heterofermentative.

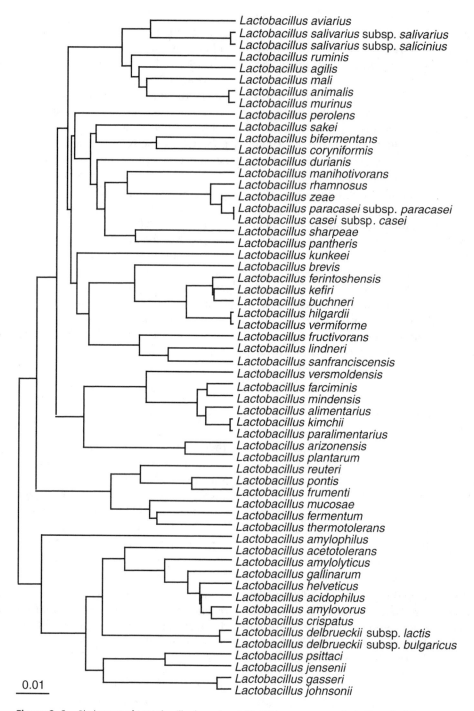

Figure 2–6.   Phylogeny of *Lactobacillus* based on 16S rRNA sequence analysis. Not all of the species are listed.

Accordingly, Group I lactobacilli consist of obligate homofermenting species, Group II contains facultative heterofermenting species, and Group III contains obligate heterofermenting species (Table 2–5). These groups can be further divided on the basis of other biochemical properties.

Despite their metabolic diversity, lactobacilli generally are quite fastidious, and many species require nutrient-rich environments. They are not especially proteolytic or lipolytic, and amino acids, peptides, and fatty acids are usually required for rapid growth. Some strains are particularly demanding, requiring an array of vitamins, nucelotides, and other nutrients. Lactobacilli, of course also need fermentable carbohydrates, and, depending on their source or habitat, they ferment a wide array of sugars. Thus, not only are most common sugars (e.g., glucose, fructose, lactose) fermented, but many plant-derived carbohydrates, such as cellobiose, amygdalin, and trehalose, are also used. Several species even ferment starch (Table 2–5).

Numerous species of *Lactobacillus* are relevant in fermented foods. Some are added directly in the form of starter cultures (Chapter 3), but there are even more endogenous lactobacilli present in the raw material or equipment surfaces that indirectly contribute to the manufacture or to the finished properties of fermented foods.

Starter culture lactobacilli are used primarily in dairy and sausage applications. There are two main species used as dairy starter cultures (mainly for cheese and yogurt), *Lactobacillus helveticus* and *Lactobacillus delbrueckii* subsp. *bulgaricus*. Other common dairy-related species include *Lactobacillus casei* and *Lactobacillus acidophilus* (both used frequently as probiotics). Starter cultures for sausage fermentations contain *Lactobacillus plantarum* or *Lactobacillus sakei* subsp. *sakei*.

Sourdough breads are also made using heterofermentative *Lactobacillus sanfranciscensis*, *Lactobacillus brevis*, and other lactobacilli. Pure starter cultures for sourdough breads are available, but wild cultures, propagated "in house," are still commonly used (Chapter 8).

For the production of sauerkraut, kimchi, and other fermented vegetables, the natural microflora is all that is necessary to initiate and perform the fermentation, although pure starter cultures containing *L. plantarum* and *L. brevis* and related species have become more common for so-called "controlled pickle" fermentations.

Finally, it is very important to emphasize, despite the general association of a particular species with a particular fermented food, that a given species will often be used in or appear in quite dissimilar situations. In many cases, the species name may even add to the confusion. Thus, *L. sakei* was originally isolated from sake, Japanese rice wine, yet this same organism can be isolated from fermented sausage and is even used in sausage starter cultures. Likewise, *L. plantarum* is not just found in plant material or in pickle brines, as its name might indicate, but also appears to be a normal inhabitant of the human intestinal tract.

As one would expect, the lactobacilli are genetically very diverse. Although the lactic acid bacteria, as a group, generally have a low %G+C (35 to 40), some lactobacilli have %G+C as low as 32 and others are as high as 55. The lactobacilli also are positioned phylogenetically in one of two distinct branches. For example, *L. plantarum* and *L. brevis* belong to the "*Lactobacillus-Pediococcus*" branch, whereas *L. delbrueckii*, *L. helveticus* and *L. acidophilus* are located in a separate "*Lactobacillus delbrueckii*" branch (Figures 2–2 and 2–6). Plasmids are common in the lactobacilli, although for many of these plasmids, the functions of the encoded genes are not established. Lactose metabolism is encoded on at least one plasmid, and several bacteriocin production and antibiotic-resistant systems are also plasmid-encoded.

Other Bacteria Important in Food Fermentations

In addition to the lactic acid bacteria, several other genera are involved in fermented foods. In most cases, these bacteria are used for a singular purpose, that is, they are involved in just

Table 2.5. Fermentation characteristics of *Lactobacillus*[1]

Representative strains	% G+C	Growth at 15°C	Cel[2]	Fru	Gal	Lac	Mal	Starch	Suc	Arginine Hydrolysis
Group I **Obligate homofermentative**										
Lactobacillus acidophilus	34–37	–	+	+	+	+	+	nd[3]	+	–
Lactobacillus delbrueckii	49–51	–	±[4]	+	+	+	±	nd	±	±
Lactobacillus helveticus	38–40	–	+	±	+	+	±	nd	–	–
Lactobacillus amylophilus	44–46	+	–	+	+	–	+	+	–	nd
Lactobacillus amylovorus	40–41	–	+	+	+	–	+	+	+	nd
Lactobacillus crispatus	35–38	–	+	+	+	+	+	nd	+	–
Lactobacillus gasseri	33–35	–	+	+	+	±	±	nd	+	–
Lactobacillus jensenii	35–37	–	+	+	+	–	±	nd	+	+
Group II **Facultative heterofermentative**										
Lactobacillus paracasei	46	nd	+	+	+	±	+	nd	+	–
Lactobacillus curvatus	43	nd	+	+	+	±	+	nd	–	–
Lactobacillus plantarum	45	+	+	+	+	+	+	nd	+	–
Lactobacillus sakei	43	nd	+	+	+	+	+	nd	+	–
Lactobacillus bavaricus	43	nd	+	+	+	+	+	nd	+	–
Lactobacillus homobiochii	36	+	±	+	–	–	+	nd	–	–
Lactobacillus coryniformis	36	+	–	+	+	±	+	nd	+	–
Lactobacillus alimentarius	36	nd	+	+	+	–	+	nd	+	–
Group III **Obligate heterofermentative**										
Lactobacillus fermentum	53	+	±	+	+	+	+	nd	+	+
Lactobacillus sanfranciscensis	37	nd	–	–	–	+	+	nd	–	–
Lactobacillus reuteri	41	nd	–	+	+	+	+	nd	+	+
Lactobacillus buchneri	45	nd	–	±	±	±	+	nd	±	+
Lactobacillus brevis	45	nd	–	±	±	+	+	nd	±	+
Lactobacillus kimchii	35	+	+	+	w[5]	–	+	–	+	–
Lactobacillus kefiri	41	nd	–	–	–	+	+	nd	–	+
Lactobacillus divergens	34	+	+	+	±	+	+	nd	+	+

[1] Adapted from *Bergey's Manual of Systematic Bacteriology, Volume 2* (1986)
[2] Cel = cellulose; Fru = fructose; Gal = galactose; Lac = lactose; Mal = maltose; Suc = sucrose
[3] Not determined
[4] Variable reaction, depending on strain or subspecies
[5] Weak positive reaction

one application and perform only one major function. These non-lactic acid bacteria represent several different genera and include both Gram positive and Gram negative bacteria. Their taxonomical position and general properties are given below. How they are actually used and the particular functions they perform in fermented foods are described in detail in the relevant chapters.

Acetobacter, Gluconobacter *and* Gluconoacetobacter

The only Gram negative bacteria used in the manufacture of fermented foods are the acetic acid-producing rods belonging to the genera *Acetobacter*, *Gluconobacter*, and *Gluconoacetobacter* (in the *Proteobacteria* phylum, Family *Acetobacteraceae*). These bacteria are obligate aerobes, with a respiratory-only metabolism. They make acetic acid via oxidation of ethanol; some species may also have the capacity to further oxidize acetic acid completely to CO_2 and water. The acetic acid bacteria are mesophilic, with optimum growth temperatures about 25°C to 30°C. Although acetobacteria are acid-tolerant, their preferred growth pH is generally between 5.3 and 6.3. Some species of *Acetobacteraceae* produce surface film and pigments. Recent changes in the taxonomy of these bacteria has resulted in the transfer of several important species from *Acetobacter* to *Gluconoacetobacter*.

It is important to note that other bacteria, including species of *Acetobacterium* and *Clostridium*, also make acetic acid as a primary metabolic end-product, but these bacteria are obligate anaerobes and rely on reductive pathways (i.e., reduction of CO_2). In wine, beer, cider, and other ethanol-containing products, *Acetobacter*, *Gluconobacter*, and *Gluconoacetobacter* can act as spoilage organisms. However, they are also the bacteria used in the manufacture of vinegar.

Industrially, vinegar can be produced using either pure culture or natural fermentations. In general, *Acetobacter aceti* is considered to be the "vinegar bacterium," because it is the most commonly used species in pure culture processes and is also found in natural vinegar fermentations. Other species, however, are also frequently isolated from, or are used in vinegar fermentations, including *Acetobacter orleanensis*, *Acetobacter pasteurianus* subsp. *pasteurianus*, *Gluconoacetobacter europaeus*, *and Gluconoacetobacter xylinus*.

Bacillus

Species of the genus *Bacillus* are ubiquitous in foods, serving either as benign contaminants, as spoilage organisms, or occasionally as the cause of food poisoning syndromes. Strains of *Bacillus subtilis*, however, are also used for the production of an Asian fermented soybean product called natto (Chapter 12). These natto strains are closely related to the widely used wild-type laboratory strain, *B. subtilis* Marburg 168, except that the former contain several biochemical and physiological properties that are required for the natto fermentation. Specifically, the natto strains produce a capsular polysaccharide material that is responsible for the viscous texture and sweet flavor of this product.

Bifidobacterium

It is arguable whether species of *Bifidobacterium* should be considered "involved" in food fermentations. Although they have a fermentative metabolism, these bacteria are not used in the manufacture of any fermented food, nor are they even found in most raw food materials. Rather, bifidobacteria are added to certain foods, mostly milk and fermented dairy products, strictly for their probiotic functions. The intestinal tract is their primary habitat, and their elevated presence in the human gastrointestinal tract is correlated with a reduced incidence of enteric infections and overall intestinal health. Bifidobacteria are now so frequently used as probiotic adjuncts in foods that they have become a commercially important product line for starter culture companies as ingredients in yogurt and other dairy culture formulations (Chapter 3).

There are more than twenty-five recognized species of *Bifidobacterium* (Figure 2–7), although only about ten are ordinarily used commercially as probiotics. These include *Bifidobacterium bifidum*, *Bifidobacterium adolescentis*, *Bifidobacterium breve*, *Bifidobac-* *terium infantis*, *Bifidobacterium lactis*, and *Bifidobacterium longum*. For many years (until the 1970s), bifidobacteria were classified in the genus *Lactobacillus*. It is now clear that they are phylogenetically distinct from the lactic acid bacteria (Figure 2.2), and are in an en-

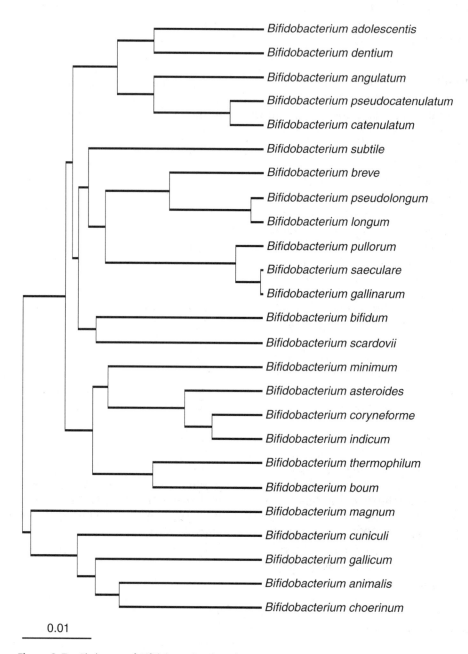

Figure 2–7. Phylogeny of *Bifidobacterium* based on 16S rRNA sequence analysis.

tirely different phylum (*Actinobacteria*). Bifidobacteria are Gram positive, non-motile, non-sporing rods with a high G+C content (55% to 67 mol%). Cells often occur in pairs with a V- or Y-like appearance. They are strictly anaerobic and catalase negative, with a temperature optima between 37°C and 41°C and a pH optima (for growth initiation) between 6.5 and 7.0. Bifidobacteria are nutritionally fastidious and require vitamins and other nutrients for growth. Their ability to use a wide array of carbohydrates, including non-digestible oligosaccharides that reach the colon, may provide selective advantages in the colonic environment (discussed in Chapter 4). Sugar metabolism occurs primarily via the well-studied "bifidum" fermentation pathway that yields acetic acid and lactic acid. Bifidobacteria (except for *B. longum*) rarely contain plasmids.

Brevibacterium

The genus *Brevibacterium* (phylum, *Actinobacteria*) are described as non-motile, non-sporing, non-acid-fast, irregular-shaped organisms that belong to the "coryneform" group. Like other coryneform bacteria, *Brevibacterium* sp. are Gram positive rods, but both their staining pattern and shape can vary, depending on the age and condition of the cells. They also are high G+C organisms (60 to 67 mol%). *Brevibacterium* are strictly aerobic, catalase-positive mesophiles, with an optimum growth temperatures between 20°C and 35°C. Most species are salt-tolerant (>10%) and able to grow over a wide pH range. There are currently eighteen recognized species; the phylogeny of twelve representative species is shown in Figure 2–8.

Several are of medical importance and have been considered as opportunistic pathogens. One species, *Brevibacterium linens*, is important in fermented foods, mainly because it is involved in the manufacture of bacterial, surface-ripened cheeses, such as Limburger and Muenster. In these products, *B. linens* produces a yellow-orange-red pigment on the cheese surface that gives these cheeses their characteristic appearance. Their ability to hydrolyze proteins and metabolize amino acids, especially the sulfur-containing amino acids, also contributes to cheese ripening and flavor development in a variety of cheeses.

Kocuria, Micrococcus, *and* Staphylococcus

These closely related genera (Phylum Firmicutes) consist of Gram positive, catalase positive, non-motile, non-sporeforming, aerobic cocci. Only a few species are relevant to fermented foods, and their involvement is limited to the manufacture of fermented sausages (although some may show up as inadvertent contaminants in cheese and other products). The main species include *Kocuria varians* (formerly *Micrococcus varians*), *Micrococcus luteus*, *Staphylococcus xylosus*, and *Staphylococcus carnosus*. In fermented meats, these bacteria are not used for acid formation (because they are non-fermentative), but rather to enhance flavor and color development. In particular, they produce aroma compounds, hydrolyze lipids and proteins, and reduce nitrate to nitrite. The latter property is important (or essential) in European- or traditional-style dry-cured sausages in which the curing agent contains nitrate salts, rather than the active, nitrite form used in conventional sausage manufacture. This function is discussed in more detail in Chapter 6.

Propionibacterium

The propionibacteria are in the phylum *Actinobacteria*. They are high G+C (53 to 68 mol%), non-sporing, Gram positive, non-motile rods, often with a characteristic club-shaped appearance. They may also appear coccoidal or branched. Propionibacteria are catalase positive (or variable), anaerobic to aerotolerant mesophiles. They are neutraphilic and grow slowly at pH 5.0 to 5.2 (the pH that exists in Swiss-type cheese).

In general, the propionibacteria are associated with two quite different habitats, the human skin and dairy products. The former

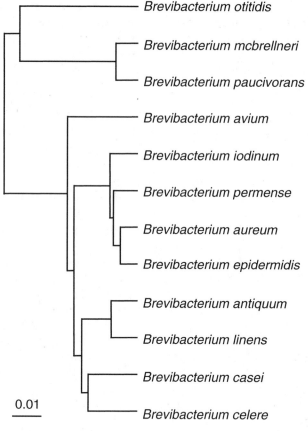

Figure 2–8. Phylogeny of *Brevibacterium* based on 16S rRNA sequence analysis.

group contains the cutaneous species *Propionibacterium acnes* and *Propionibacterium avidum*, the organisms that cause acne. The dairy group consists of several species that are important in food fermentations, due to their use in the manufacture of Swiss-type cheeses. The most frequently used dairy species include *Propionibacterium freudenreichii* subsp. *shermanii*, *Propionibacterium freudenreichii* subsp. *freudenreichii*, *Propionibacterium acidopropionici*, and *Propionibacterium jensenii*. They have a mol% G+C content of 66% to 68% and are distinguished from other propionibacteria based mainly on morphological and biochemical

characteristics, in that they are usually shorter rods and ferment fewer sugars.

The ability to ferment lactose varies among strains, but in the absence of lactose, dairy strains are quite capable of using lactate as a carbon and energy source. Metabolic end-products include propionic acid, acetic acid, and carbon dioxide. Small amounts of Vitamin B_{12} may also be produced. During growth in a peptide-rich medium, such as cheese, propionibacteria release the amino acid proline via peptide hydrolysis. They also are lipolytic, releasing free fatty acids from triglycerides. Despite their role as "eye-formers" in Swiss-type cheese, they are not particularly prolific CO_2-producers.

Yeasts and Molds Used in the Manufacture of Fermented Foods

From 80,000 to 1.5 million species of fungi are believed to exist, but, fortunately for this discussion, relatively few are involved in food fermentations. As described previously, yeasts and molds are in the Eukarya domain and belong to the Kingdom Fungi. Of the phyla within this kingdom, most of the fungi relevant to fermented foods are classified as *Zygomycota, Ascomycota*, or as deuteromycetes. These groups contain several important genera of yeasts, including *Saccharomyces, Kluyveromyces*, and *Zygosaccharomyces*, as well as the mold genera *Aspergillus, Penicillium*, and *Rhizopus*.

Fungi, as eukaryotes, differ from prokaryotes in several major respects. In particular, fungi have a much more complex physical structure and they reproduce in an entirely different manner than bacteria. Although molds and yeasts are both fungi, molds are multicellular and filamentous, whereas yeast are unicellular and non-filamentous. In addition, molds can produce asexual and sexual spores, while yeast produce only sexual spores or are non-sporulating.

The most common asexual spores produced by molds are the sporaniga or conidiospores type and can be further divided as conidia, arthrospores, and sporangiospores.

The most common molds used in fermented foods, *Penicillium* and *Aspergillus*, produce conidia, and *Rhizopus* and *Mucor* produce sporangia. Yeasts, in addition to producing ascospores, also reproduce asexually by budding, a process in which part of the cytoplasm bulges from the cell and eventually separates as a new cell. Budding can be either multilateral or polar, and is characteristic of the species.

Saccharomyces

It can reasonably be argued that the yeasts belonging to the genus *Saccharomyces* are among the most important of all organisms used in fermented foods, perhaps more so than even the lactic acid bacteria. These yeasts are required, after all, for the production of beer, wine, and spirits (not to mention bread), products that have a combined, world-wide economic impact in the trillions of dollars. In addition, the main species, *Saccharomyces cerevisiae*, is widely used as a model organism in biology and genetics, and its physiological and biochemical properties, as well as its genome sequence, have been well studied. At the gene level, *Saccharomyces* and other yeasts are much more complicated than are the bacteria. They have sixteen chromosomes, they can be diploidal or polyploidal (i.e., more than one set of chromosomes), and the size of their genomes is several times larger (more than 6,000 protein-encoding genes compared to about 2,000 in most lactic acid bacteria).

The taxonomy and nomenclature of *Saccharomyces* has been subject to rather regular and frequent revisions for more than a century. Although *S. cerevisiae* has long been one of the major species used in wine, brewing, and bread-making applications, the specific strains involved in the manufacture of these products are clearly different from each other and from laboratory strains of *S. cerevisiae*. The frequent changes in nomenclature have made it even more difficult to keep track of the particular species associated with a given fermentation.

Several examples illustrate this point. Prior to the 1970s, *Saccharomyces ellipsoideus* was recognized as one of the primary wine yeasts and was accorded species status. It was then reclassified as *S. cerevisiae* (and demoted to "synonym" status), although the earlier name continues to be used. Similarly, another group of wine strains classified as *Saccharomyces uvarum* were reclassified as *Saccharomyces bayanus* (or sometimes listed as *Saccharomyces bayanus* subsp. *uvarum*). Finally, for many years, the species used in the lager fermentation (where lager is one of the two styles of beer) was classified as *Saccharomyces carlsbergensis*, which was also referred to as its synonym, *Saccharomyces uvarum*. Both species then merged as *S. uvarum*, only to be reclassified yet again, first as *Saccharomyces bayanus* and then as *Saccharomyces pastorianus*, which is now the valid species name for the lager yeast. In the most recent volume (1998) of *The Yeast, A Taxonomic Guide*, there were fourteen accepted species of *Saccharomyces*. However, in a 2003 report (FEMS Yeast Research 4:223-245), only seven species were recognized: *S. bayanus*, *Saccharomyces cariocanus*, *S. cerevisiae*, *Saccharomyces kudriavzevii*, *Saccharomyces mikatae*, *Saccharomyces paradoxus*, and *S. pastorianus*.

Distinguishing between species of *Saccharomyces* is based primarily on morphological, physiological, and biochemical properties. These yeasts usually have a round spherical or ovoid appearance, but they may be elongated with a pseudohyphae. The sugar fermentation patterns and the assimilation of carbon sources are key factors for speciation (Table 2-6; also see Chapter 9 for beer yeast speciation). Other specific diagnostic tests include hyphae formation, ascospore formation, resistance to cycloheximide, and growth temperatures. Several physiological traits vary among the *Saccharomyces* and are useful not only for classification, but are important for strain selection. Some strains, for example, are very osmophilic and halotolerant, and can grow in foods containing high concentrations of carbohydrates (e.g., high sugar grapes) or salt (soy sauce).

Nowhere, however, does the species of yeasts influence the fermentation as much as in brewing. As noted above, there are two general styles of beer, ales and lagers. Each requires a specific yeast, *S. cerevisiae* for ales and *S. pastorianus* for lagers. These yeasts vary in several respects (reviewed in Chapter 9), but mainly on the basis of where they settle or flocculate in the beer. Hence, ale yeasts are referred to as top-fermenting yeasts because they tend to flocculate at the top of the fermentation vessel, and lager yeasts are referred to as bottom-fermenting because they settle at the bottom of the fermentor. It is important to note that however useful these phenotypic characteristics are for classification, they are not always consistent with species assignments based on the sequences of 18S rRNA and other regions.

Ecologically, *Saccharomyces* and other yeasts are ubiquitous in foods. In some fermented

Table 2.6. Distinguishing characteristics of *Saccharomyces* species[1]

Species	Fermentation of: Gal[2]	Mal	Mel	Me-glu	Assimilation of: Gal	Mal	Tre	Mel	Inu	Rib	Gtl	Me-glu	Growth on 10% NaCl/5% Glu
S. bayanus	−	−	−	−	−	+	+	−	−	+	−	+	−
S. cariocanus	+	−	−	−	+	−	−	−	L[3]	−	−	+	L
S. cerevisiae	−	−	−	−	−	+	+	−	−	+	−	+	−
S. kudriavzevii	−	s[4]	−	+	+	−	−	−	+	L	+	+	−
S. mikatae	+	−	+	+	+	+	L	L	−	+	+	+	−
S. paradoxus	−	−	−	−	+	+	+	−	−	+	−	+	−
S. pastorianus	−	−	−	−	+	+	+	−	−	+	−	+	+

[1] Adapted from Naumov et al., 2000. Int. J. Syst. Evol. Microbio. 50:1931-1942
[2] Abbreviations: Gal, galactose; Mal, maltose; Mel, melibiose; Me-glu, methyl glucoside; Tre, trehalose; Inu, inulin; Rib, ribitol; Gtl, galactitol; Glu, glucose
[3] Latent (delayed positive)
[4] Slow

foods, such as wine, their natural presence on grapes and equipment is sufficient to initiate a fermentation. Several different yeasts may be involved in "natural" or "spontaneous" wine fermentations, usually in a successive manner, where one species is dominant for a time, then gives way to others (Chapter 10). However, in most modern wine fermentations, as well as beer and bread fermentations, starter culture yeasts are used, selected on the basis of their physiological and biochemical properties. Yeasts are also important in spoilage of fermented (and non-fermented) foods. For example, one of the serious defects in sauerkraut is caused by the pink pigment-producing yeast *Rhodotorula*.

Penicillium *and* Aspergillus

As previously mentioned, these molds are among the most common and widespread in foods. In the older literature, they were often referred to as "Fungi Imperfecti," due to the absence of a sexual stage in their life cycle. *Penicillium* and *Aspergillus* are mainly of concern due to their role in food spoilage and as potential producers of mycotoxins; however, some species of *Penicillium* and *Aspergillus* are also used to produce fermented foods. In fact, one of the most famous of all organisms used in fermented foods is *Penicillium roqueforti*. This mold gives Roquefort and other blue-veined cheeses their characteristic color. It is also largely responsible for the flavor and aroma properties of these cheeses. A related white mold species, *Penicillium camemberti*, is equally important (and no less famous), due to its involvement in the manufacture of Camembert and Brie cheeses.

Although the role of *Aspergillus* in fermented foods manufacture is somewhat less prominent (at least in foods popular in Western cultures), species of *Aspergillus* are involved in some of the world's most widely-consumed fermented foods. Specifically, two species, *Aspergillus oryzae* and *Aspergillus sojae*, are used in the manufacture of soy sauces and soy pastes and sake and other rice wines, so-called Oriental or

Asian fermented foods that are consumed literally by billions of people (Chapter 12).

There are many different species of *Penicillium* and *Aspergillus*. As with the yeasts, speciation is based on morphology, structure, and spore type. Biochemical and physiological properties of the fungi, however, are generally less diagnostic. These fungi are heterotrophic and saprophytic, and are usually very good at breaking down complex macromolecules (e.g., protein, polysaccharides, lipids) via secretion of proteinases, amylases, and lipases. The hydrolysis products can then be used as substrates for growth.

Despite their readily observed structural differences, *Penicillium* and *Aspergillus* do share some common features. Both are filamentous and produce conidiospores, and are probably related. Typical spore-bearing structures of *Penicillium* and *Aspergillus* are shown in Figure 2–9. Finally, *Rhizopus oligosporus* species are used to produce a fermented soybean product known as tempeh that is popular in Indonesia and other Southeast Asian countries (Chapter 12). Typical morphological structures of *Rhizopus* are shown in Figure 2–9.

Fermentation and Metabolism Basics

If one looks up fermentation in a biochemistry textbook, the definition that appears is usually something like this: "energy-yielding reactions in which an organic molecule is the electron acceptor . . . ." Thus, in the lactic acid fermentation, pyruvic acid that is generated by the glycolytic pathway serves as the electron acceptor, forming lactic acid. Likewise, in the ethanolic pathway, acetaldehyde, formed by decarboxylation of pyruvate, is the electron recipient (forming ethanol). So although this definition certainly is true for many of the fermentations that occur in foods, it is rather narrow.

For several fermented foods, important end-products are produced via non-fermentative pathways (as classically defined). For example, the malolactic fermentation that is very important in wine making is really just a decarboxylation reaction that does not conform, strictly

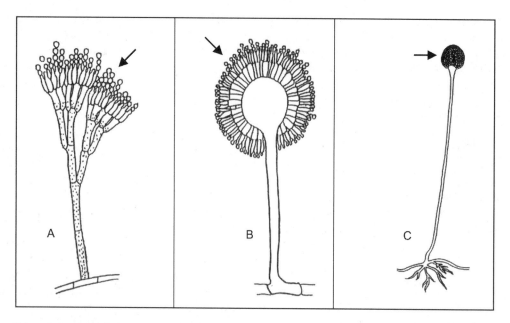

Figure 2–9. Principle fungi used in food fermentations. Shown schematically are *Penicillium* (A), *Aspergillus* (B), and *Rhizopus* (C). The spore structures are indicated by arrows (conidia in *Penicillium* and *Aspergillus* and sporangiospores in *Rhizopus*). Adapted from Bullerman, 1993.

speaking, to the definition stated above. Neither does this definition apply in the manufacture of tempeh and other fungal-fermented foods, in which soy proteins and polysaccharides are degraded and metabolized by fungi, but without production of glycolytic end-products. Therefore, in this text, the term fermentation will be used in a broader sense, accounting for the many metabolic processes that occur during the course of a food fermentation.

From the point of view of the microorganisms (which care even less about definitions), fermentation is merely the means by which they obtain energy. Microorganisms, after all, need energy to perform work (e.g., nutrient transport and biosynthesis), to maintain chemical and physical homeostasis (e.g., ionic and osmotic), and to grow and replicate. For the most part, the energy generated by fermentation is in the form of ATP, and usually it is produced via metabolism of sugars (although, exceptions for both of these claims exist, as suggested above). Whether sugar metabolism by microorganisms also results in the conversion of milk into yo-

gurt or juice into wine is of no concern to the organism. Of course, the same metabolic pathway may also lead to formation of end-products that cause milk or juice to become spoiled, which also matters little to the offending organism. However, these outcomes do matter to yogurt and wine manufacturers.

Knowing how to control and manipulate microorganisms and their metabolic activities can mean the difference between a pleasant-tasting container of yogurt or soured milk that is tossed down the drain, between success and failure, and between profit and loss. Therefore, understanding the biochemical basis for metabolism of sugars, as well as other substrates, is essential for consistent production of fermented foods with the expected biological, physical, chemical, nutritional, and sensory characteristics.

Sugar Metabolism

The biochemical means by which microorganisms metabolize sugars was first studied by Pasteur, Buchner, Schwann, and other early microbiologists and biochemists. That food-related sugar fermentations, in particular, had attracted the attention of these scientists was actually quite reasonable. After all, cultured milk products, beer, and wine were industrially-important products whose manufacture depended on fermentation of sugars. Thus, identifying the microorganisms and pathways was essential to understanding how the manufacture of these products could be improved and controlled and spoilage prevented. Metabolism and transport of sugars by lactic acid bacteria and ethanol-producing yeasts will be reviewed in the sections to follow.

Homofermentation

Lactic acid bacteria are obligate fermentors, and cannot obtain energy by oxidative or respiratory processes (with the exception noted previously; Box 2-1). Technically, the precursor-product exchange systems, described below, provide an alternate way for these organisms to earn ATP "credits" by conserving the energy that would ordinarily be used to perform metabolic work. However, the substrate level phosphorylation reactions that occur during fermentation are by far the major means by which these cells make ATP. For homofermentative lactic acid bacteria, hexoses are metabolized via the enzymes of the glycolytic Embden-Meyerhoff pathway.

One of the key enzymes of this pathway is aldolase, which commits the sugar to the pathway by splitting fructose-1,6-diphosphate into the two triose phosphates that eventually serve as substrates for ATP-generating reactions. The Embden-Meyerhoff pathway yields two moles of pyruvate and two moles of ATP per mole of hexose (Figure 2-10). The pyruvate is then reduced to L- or D-lactate by the enzyme, lactate dehydrogenase. More than 90% of the substrate is converted to lactic acid during homofermentative metabolism.

Importantly, the NADH formed during the glyceraldehyde-3-phosphate dehydrogenase reaction must be re-oxidized by lactate dehydrogenase, so that the $[NADH]/[NAD^+]$ balance is maintained. Homofermentative lactic acid bacteria include *Lactococcus lactis*, *Streptococcus thermophilus*, *Lactobacillus helveticus*, and *L. delbrueckii* subsp. *bulgaricus* (used as dairy starter organisms); *Pediococcus* sp. (used in sausage cultures); and *Tetragenococcus* (used in soy sauces).

Heterofermentation

Heterofermentative lactic acid bacteria metabolize hexoses via the phosphoketolase pathway (Figure 2-11). In obligate heterofermentative bacteria, aldolase is absent, and instead the enzyme phosphoketolase is present. Approximately equimolar amounts of lactate, acetate, ethanol and CO_2 are produced, along with only one mole of ATP per hexose. Oxidation of NADH and maintenance of the $[NADH]/[NAD^+]$ balance occurs via the two reductive reactions catalyzed by acetaldehyde dehydrogenase and alcohol dehydrogenase. Many of the lactic acid bacteria used in food fermentations are heterofermentative. Included are *L. mesenteroides* subsp. *cremoris* and *Leuconostoc lactis* (used in dairy fermentations), *L. mesenteroides* subsp. *mesenteroides* and *Leuconostoc kimchii* (used in fermented vegetables), *O. oeni* (used in wine fermentations), and *Lactobacillus sanfranciscensis* (used in sourdough bread).

In general, the product yields for both pathways may vary during actual fermentation processes, and depend on the type and concentration of substrate, the growth temperature and atmospheric conditions, and the growth phase of the cells. Certainly some of the carbohydrate carbon is incorporated into cell mass. However, it has also been shown that under certain conditions, such as sugar-limitation or aerobiosis, homofermentative lactococci can divert some of the pyruvate away from lactate and toward other alternative end-products, such as acetate and CO_2. This so-called heterolactic fermentation may provide cells with additional ATP or serve as a way to deal with excess pyruvate (Box 2-3).

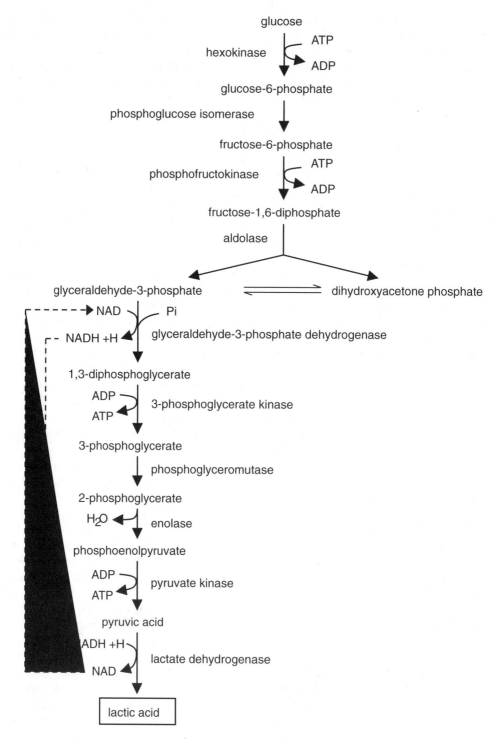

Figure 2–10. The Embden-Meyerhoff pathway used by homofermentative lactic acid bacteria. The dashed line indicated the NAD/NADH oxidation-reduction part of the pathway.

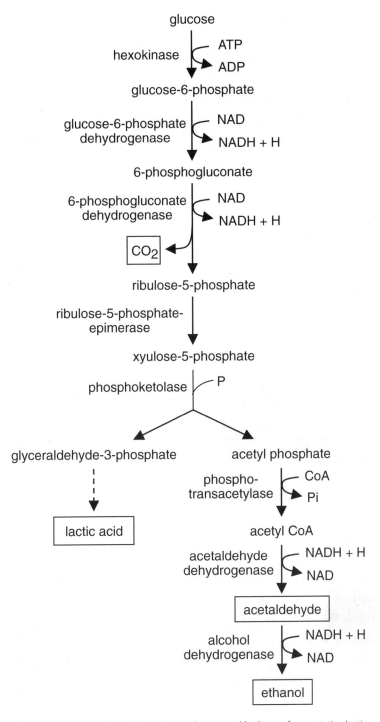

Figure 2–11. The phosphoketolase pathway used by heterofermentative lactic acid bacteria.

Box 2–3. The Heterolactic Fermentation: Dealing with Pyruvate

Lactic acid bacteria, as previously noted, are either homofermentative, heterofermentative, or facultative heterofermentative (where both pathways are present). However, even obligate homofermentative strains have the potential to produce acetic acid, ethanol, acetoin, CO_2 and end-products other than lactic acid. These alternative fermentation end-products are formed, however, only under conditions in which pyruvate concentrations are elevated.

Such a scenario occurs when the rate of intracellular pyruvate formation exceeds the rate at which pyruvate is reduced to lactate (via lactate dehydrogenase). The pyruvate can arise from sugars (see below), but also from amino acids. In either case, the excess pyruvate must be removed because it could otherwise become toxic. Moreover, when excess pyruvate is generated as a result of sugar metabolism, the cells must also have a means for re-oxidizing the NADH formed from glycolysis (upstream in the pathway). Several different pathways in lactic acid bacteria appear to serve this function (Figure 1; Cocaign-Bousquet et al., 1996; Garrigues et al., 1997). In addition, at least one of these alternative pathways include a substrate level phosphorylation reaction and, therefore, provides the cells with additional ATP.

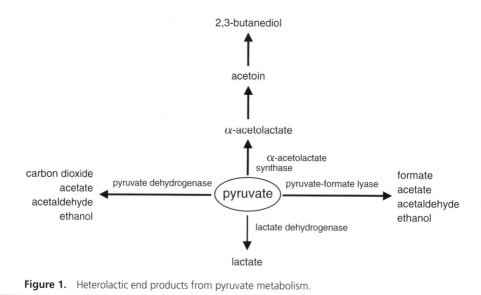

Figure 1. Heterolactic end products from pyruvate metabolism.

Sugar Transport by Lactic Acid Bacteria

Although the catabolic pathways used by lactic acid bacteria to ferment sugars constitute a major part of the overall metabolic process, the first step (and perhaps an even more important one) involves the transport of the substrate across the cytoplasmic membrane. Transport is important for several reasons, not the least of which is that the cell membrane is impermeable to polar solutes (e.g., sugars, amino acids, peptides). In the absence of transport systems, these solutes would be unable to transverse the membrane and gain entry into the cell. Second, the cell devotes a considerable amount of its total energy resources to support active transport. Third, some sugars are phosphorylated during the transport event, which then dictates the catabolic pathway used by that organism. Finally, transport systems may serve a regulatory role, influencing expression and activity of other transport systems.

Box 2–3. The Heterolactic Fermentation: Dealing with Pyruvate *(Continued)*

What are the conditions or environments that result in pyruvate accumulation and induction of alternate pathways? There are several possible situations where these reactions occur. First, glycolysis is subject to several levels of regulation, such that when fermentation substrates are limiting, the glycolytic flux tends to be diminished (Axelsson, 2004). Specifically, when the concentration of fructose-1,6-diphosphate (a glycolytic regulator that forms early during glycolysis), is low, the activity of lactate dehydrogenase (an allosteric enzyme that occurs at the end of glycolysis) is reduced. Thus, pyruvate accumulates. The glycolytic flux also may be decreased during growth on less preferred carbon sources, such as galactose, again resulting in excess pyruvate. At the same time that the lactate dehydrogenase activity is decreased, the enzyme, pyruvate-formate lyase, is activated. This enzyme splits pyruvate to form formate and acetyl CoA. The latter is subsequently reduced to ethanol or phosphorylated to acetyl phosphate (both reactions releasing CoA). Importantly, the acetyl phosphate can be used as part of a substrate level phosphorylation reaction (via acetate kinase), resulting in the formation of an ATP.

Under aerobic environments, however, pyruvate-formate lyase is inactive, and other pathways become active. In the pyruvate dehydrogenase pathway, for example, pyruvate is decarboxylated by pyruvate dehydrogenase, and acetate and CO_2 are formed. NADH that would normally reduce pyruvate also is oxidized directly by molecular oxygen when the environment is aerobic, rendering it unavailable for the lactate dehydrogenase reaction. Finally, excess pyruvate can be diverted to end-products that may have a more functional role in certain fermented dairy products. Specifically, pyruvate can serve as a substrate for α-acetolactate synthase to form α-acetolactate. The α-acetolactate may then be further oxidized to form diacetyl, which has desirable aroma properties.

References

Axelsson, L. 2004. Lactic acid bacteria: classification and physiology, pp. 1—66. *In* Salminen, S., A. von Write, and A. Ouwehand. *Lactic Acid Bacteria Microbiological and Functional Aspects Third Edition.* Marcel Dekker, Inc. New York, New York.

Cocaign-Bousquet, M., C. Garrigues, P. Loubière, and N. Lindley. 1996. Physiology of pyruvate metabolism in *Lactococcus lactis.* Antonie Van Leeuwenhoek 70:253–267.

Garrigues, C., P. Loubière, N. Lindley, and M. Cocaign-Bousquet. 1997. Control of shift from homolactic acid to mixed-acid fermentation in *Lactooccus lactis*: predominant role of the NADH/NAD$^+$ ratio. J. Bacteriol. 179:5282–5287.

The phosphoenolpyruvate-dependent phosphotransferase system

In general, several different systems are used by lactic acid bacteria to transport carbohydrates, depending on the species and the specific sugar. The phosphoenolpyruvate (PEP)-dependent phosphotransferase system (PTS) is used by most mesophilic, homofermentative lactic acid bacteria, including lactococci and pediococci for transport of lactose and glucose. In contrast, other lactic acid bacteria transport sugars via symport-type or ATP-dependent systems (described below). Another group of transporters are the precursor-product exchange systems that provide an altogether different (and more efficient) means for transporting sugars (Box 2-4).

The PTS consists of both cytoplasmic and membrane-associated proteins that sequentially transfer a high-energy phosphate group from PEP, the initial donor, to the incoming sugar (Figure 2-12). The phosphorylation step coincides with the vectorial movement of the sugar substrate across the membrane, resulting in the intracellular accumulation of a sugar-phosphate. Two of the cytoplasmic proteins of the PTS, Enzyme I and HPr (for histidine-containing protein), are non-specific and serve all of the sugar

Box 2–4.   Something For Nothing, or How Lactic Acid Bacteria Conserve Energy Via "Precursor-product" Exchange Systems

Transport of nutrients from the environmental medium across the cytoplasmic membrane and into the cytoplasm is one of the most important of all microbial processes. For many microorganisms, including lactic acid bacteria, nearly 20% of the genome is devoted to transport functions, and cell membranes are literally crammed with a hundred or more transport permeases, translocators, and accessory proteins (Lorca et al., 2005). The function of transporters, of course, is to provide a means for the cell to selectively permit solutes to move, back and forth as the case may be, across a generally impermeable membrane.

For most nutrients (and other transport substrates), however, transport is not cheap, in that energy (e.g., ATP) is usually required to perform "vectorial" work (i.e., to move molecules from one side of a membrane to another). This is because nutrient transport generally occurs against a concentration gradient, meaning that the concentration of the substrate is usually much lower in the outside medium than it is on the inside (i.e., within the cytoplasm). Although passive or facilitated diffusion, where the concentration gradient (higher outside than inside) drives transport occurs occasionally, this is not the normal situation, and instead some sort of active, energy-requiring process is required.

The metabolic cost of transport is especially high for lactic acid bacteria, given the limited and generally inefficient means by which these bacteria make energy. After all, glycolysis generates only two molecules of ATP per molecule of glucose or hexose fermented. If ATP or its equivalent (e.g., phosphoenolpyruvate or the proton motive force) is required to "move" mono- and disaccharides across the cytoplasmic membrane, the net energy gain by the cell may be reduced by as much as 50%. On the other hand, if the driving force for solute transport does not depend on an energy source, then the cell can conserve that un-spent energy or use it for other reactions.

By what means can cells transport solutes without having to spend energy? In many lactic acid bacteria, metabolism of substrate (or precursor) molecules results in a large amount of metabolic products that must be excreted from the cell. In some cases, these products are not further metabolized; in other cases, they may even be toxic if allowed to accumulate intracellularly. In any event, a rather large concentration gradient is formed, where the inside concentration of "product" is much higher than the outside concentration. Efflux of "product" molecules, therefore, is driven by the concentration gradient in a downhill manner. The carrier for efflux in these situations, however, actually has affinity not just for the product, but also for the substrate. Moreover, it operates in opposing directions, such that product excretion (via the downhill concentration gradient) can drive uphill uptake of the precursor substrate (against the concentration gradient). These so-called precursor-product exchange systems, therefore, represent a novel means by which cells can accumulate nutrients and metabolic substrates without having to consume much-needed sources of energy.

Examples of Precursor-product Exchange Systems in Lactic Acid Bacteria

Precursor-product exchange systems are widely distributed in lactic acid bacteria, and are used to transport fermentation substrates, amino acids, and organic acids (Konings, 2002). These systems can be electroneutral, without a net change in the electric charge across the cell membrane, or electrogenic, where a charge is generated across the membrane. For example, a neutral sugar precursor (lactose) exchanged for a neutral sugar product (galactose) is electroneutral, as is the exchange of the amino acid arginine for ornithine (Figure 1, panels A and C). However, a di-anionic precursor (citrate) exchanged for a mono-anionic product (lactate) is electrogenic, since it results in an increase of the transmembrane electric charge (Figure 1, panel B).

Box 2–4. Something For Nothing, or How Lactic Acid Bacteria Conserve Energy Via "Precursor-product" Exchange Systems *(Continued)*

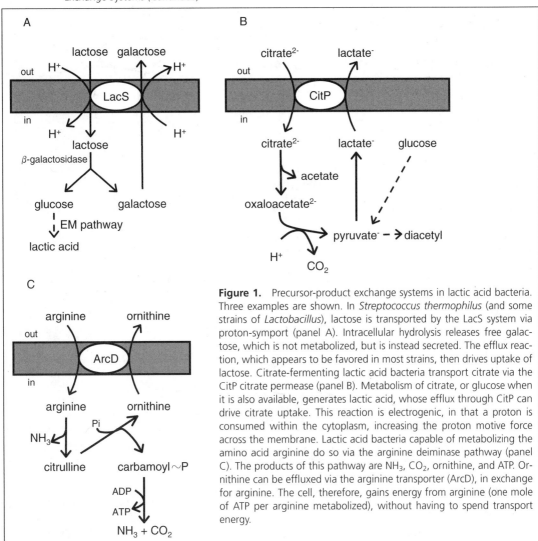

Figure 1. Precursor-product exchange systems in lactic acid bacteria. Three examples are shown. In *Streptococcus thermophilus* (and some strains of *Lactobacillus*), lactose is transported by the LacS system via proton-symport (panel A). Intracellular hydrolysis releases free galactose, which is not metabolized, but is instead secreted. The efflux reaction, which appears to be favored in most strains, then drives uptake of lactose. Citrate-fermenting lactic acid bacteria transport citrate via the CitP citrate permease (panel B). Metabolism of citrate, or glucose when it is also available, generates lactic acid, whose efflux through CitP can drive citrate uptake. This reaction is electrogenic, in that a proton is consumed within the cytoplasm, increasing the proton motive force across the membrane. Lactic acid bacteria capable of metabolizing the amino acid arginine do so via the arginine deiminase pathway (panel C). The products of this pathway are NH_3, CO_2, ornithine, and ATP. Ornithine can be effluxed via the arginine transporter (ArcD), in exchange for arginine. The cell, therefore, gains energy from arginine (one mole of ATP per arginine metabolized), without having to spend transport energy.

References

Konings, W.N. 2002. The cell membrane and the struggle for life of lactic acid bacteria. Antonie van Leeuwenhoek 82:3–27.

Lorca, G.L., R.D. Barabote, V. Zlotopolski, C. Tran, B. Winnen, R.N. Hvorup, A.J. Stonestrom, E. Nguyen, L.-W. Huang, D. Kimpton, and M.H. Saier. 2006. Transport capabilities of eleven Gram-positive bacteria: comparative genomic analyses. In press.

A

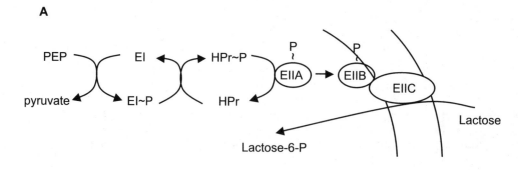

B

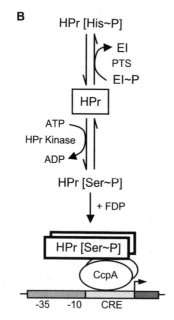

Figure 2–12. The phosphotransferase system (PTS) in Gram positive bacteria. As shown in panel A, the PTS cascade is initiated by the cytoplasmic proteins Enzyme I (EI) and HPr. Phosphorylated HPr (HPr~P) then transfers the phosphoryl group (obtained originally from PEP) to the substrate-specific Enzyme II complex. The latter consists of several proteins or domains, shown here as EIIA, EIIB, and EIIC. However, depending on the organism and the substrate, EII complexes may be organized differently, for example, as EIIA and EIIBC or as EIIABC. Regulation of the PTS is mediated, in part, via the phosphorylation state of HPr (panel B). Phosphorylation by HPr kinase results in formation of HPr[Ser~P], which, along with CcpA and fructose diphosphate (FDP), form a dimeric complex that recognizes and binds to CRE sites and prevents transcription of catabolic genes.

phosphotransferase systems present in the cell. The other PTS components, however, are substrate-specific. The latter proteins form an Enzyme II complex, which most often consists of three protein domains: EIIA, EIIB, and EIIC. These EII domains may be present as individual proteins or are combined or fused as one or two proteins. For example, the lactose PTS Enzyme II complex in *Lactococcus lactis* contains three lactose-specific protein domains that exist in the form of two proteins, i.e., EIIA^{lac} and EIIBC^{lac} (where "lac" denotes lactose-specificity). As for all of the EII proteins, the EIIB and EIIC domains are membrane-associated.

In the case of hexoses, the product of the PTS is hexose-6-phosphate, which can then feed directly into the glycolytic pathway. The PTS has the advantage, therefore, of sparing the cell of the ATP that ordinarily would be required to phosphorylate the free sugar. When lactose is the substrate, the product is lactose-phosphate (or more specifically, glucose-β-1,4-galactosyl-6-phosphate). Lactose-phosphate is not hydrolyzed by the enzyme β-galactosidase, which is widespread in the microbial world, but rather by phospho-β-galactosidase. The products of this reaction are glucose and galactose-6-phosphate. The glucose is subsequently phosphorylated by hexokinase, and the glucose-6-phosphate that forms feeds into the glycolytic pathway, as described earlier. The galactose-6-phosphate, in contrast, is metabolized by the tagatose pathway (Figure 2–13), eventually leading to the formation of the same triose phosphates, glyceraldehyde-3-phosphate and dihydroxyacetone phosphate, that form during glycolysis. It is perhaps not unexpected that the structural (and regulatory) genes coding for lactose transport and hydrolysis are located on the same operon as the galactose/tagatose genes (Figure 2–14).

Symport and ABC Transport Systems in Lactic Acid Bacteria

Although the PTS is widely distributed among lactic acid bacteria, several species rely on other active transport systems to transport sugars. The latter include symport systems, driven

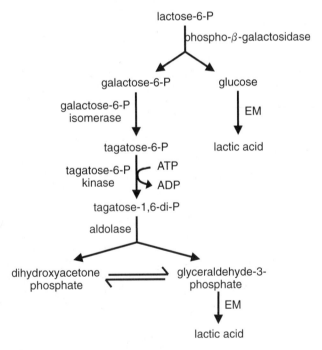

Figure 2–13. Tagatose pathway in lactococci. Galactose-6-phosphate is formed from hydrolysis of lactose-phosphate, the product of the lactose PTS. Isomerization and phosphorylation form tagatose-1,6-diphosphate, which is split by an aldolase, yielding the triose phosphates that feed into the EM pathway.

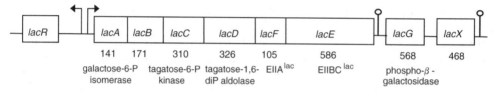

Figure 2–14.   The *lac* operon in lactococci. The operon consists of four structural genes (*lacABCD*) coding for enzymes of the tagatose pathway, two genes, *lacFE*, coding for lactose-specific PTS proteins, and the *lacG* gene coding for phospho-β-galactosidase. A divergently transcribed *lacR* gene codes for a repressor protein. The function of *lacX* is not known. Promoter sites and directions are shown by the arrows, and potential transcriptional terminators are shown as hairpin loops. The number of amino acid residues for each protein is given. Adapted from de Vos et al., 1990.

by ion gradients, and ATP-binding cassette (ABC) systems, fueled by ATP (Figure 2-15). Moreover, it is often the case that an organism uses a PTS for one sugar and a symport or ABC system for another. According to the Transport Classification system (www.tcdb.org), symport system transporters belong to a class of secondary carriers within the Major Facilitator Superfamily.

These symport systems consist of a membrane permease that has binding sites for both the substrate and a coupling ion, usually protons. Transport of the substrate is driven by the ion gradient across the membrane. The most common ion gradient in bacteria is the proton gradient (called the proton motive force or PMF), although there are also symport systems driven by sodium ion gradients (such as the melibiose system that exists in *Lactobacillus casei*).

The PMF is comprised of the sum of two components: (1) the charge difference ($\Delta\psi$) across the membrane, where the outside is positive and the inside is negative; and (2) the chemical difference (ΔpH) across the membrane, where the proton concentration is high on the outside and low inside. The positively charged protons then flow "down" this gradient (i.e., toward the inside) in symport with the "uphill" intracellular accumulation of the solute. In lactic acid bacteria, symport systems exist for several sugars (based on biochemical and genetic evi-

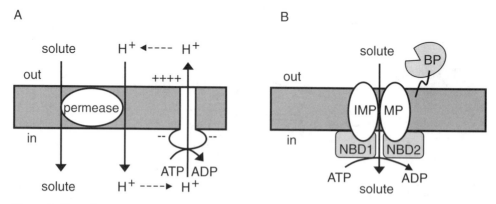

Figure 2–15.   Solute transport in lactic acid bacteria. In Panel A, a proton symport system, fueled by the proton gradient, drives solute uptake. The proton gradient is formed via the proton translocating ATPase. In panel B, ATP hydrolysis drives transport directly via nucleotide binding domains (NBD). The solute is first captured by a binding protein (BP) which delivers it to the integral membrane proteins (IMP).

dence), including lactose (in *Lactobacillus brevis, L. delbrueckii, L. acidophilus*), galactose (*L. lactis*), raffinose (*P. pentosaceus*), melibiose (*L. lactis*), and xylose (*Lactobacillus pentosus*). Although generation of the PMF is mediated via the proton-translocating ATPase (also called the F_0F_1-ATPase), which requires ATP, lactic acid bacteria may also make a PMF by alternative routes that spare the cell of ATP (Box 2–4).

For those organisms that transport lactose by a symport system, the intracellular product is lactose (chemically unaltered, in contrast to the lactose PTS, where the product is lactose-phosphate). Intracellular hydrolysis of the free lactose then occurs via β-galactosidase, yielding glucose and galactose. The former is phosphorylated by hexokinase to form glucose-6-phosphate, which feeds directly into the Embden-Meyerhoff pathway.

To convert galactose into glucose-6-phosphate requires the presence of the Leloir pathway (Figure 2–16), a three enzyme pathway whose expression is subject to strong negative regulation in some lactic acid bacteria. In particular, *S. thermophilus* and *Lb. delbruecki* subsp. *bulgaricus* transport and hydrolyze lactose by a PMF and a β-galactosidase, respectively, but ferment only the glucose and not the galactose moiety. Although these bacteria have the genes encoding for the Leloir pathway enzymes, the genes are poorly transcribed or expressed, due to mutations within the promoter region of the operon. As a result, the intracellular galactose is largely unfermented and is excreted via the lactose transport system (LacS in *S. thermophilus*), a seemingly wasteful process. What makes this process especially interesting, however, is that LacS can catalyze a lactose:galactose exchange reaction, such that galactose efflux drives lactose uptake (Box 2–4). Thus, what might initially appear to be an inefficient and wasteful act (secreting a perfectly good energy source into the medium), is, instead, an efficient, energy-saving means of exchanging a readily fermentable sugar for one that is slowly metabolized, if at all.

The other major group of transport systems used by lactic acid bacteria for accumulating sugars are the ABC transport systems. These systems consist of several proteins or protein domains (Figure 2–15). Two of these proteins or domains are membrane-spanning (i.e., integrated within the membrane) and serve as porters. Two intracellular proteins/domains bind and hydrolyze ATP. An additional protein is extracellular, but is tethered to the cell

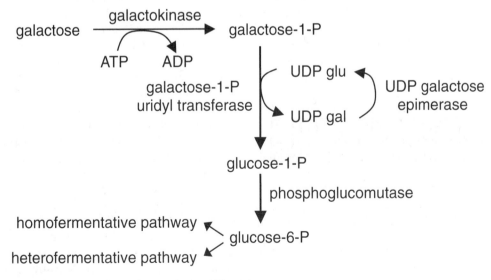

Figure 2–16. The Leloir pathway in lactic acid bacteria.

surface and has substrate-binding activity. The energy released from ATP hydrolysis drives transport. In general, ABC systems are used to transport amino acids, peptides, and osmoprotectants. However, several ABC systems are involved in sugar transport, including the maltose ABC systems in *L. plantarum* and the oligosaccharide-multiple sugar transport system in *L. acidophilus*.

Regulation of Transport Systems

As noted above, regulation of sugar metabolism in lactic acid bacteria is intimately connected to the transport machinery. In *L. lactis*, for example, lactose fermentation is regulated biochemically and genetically at the level of transcription through a signal transduction cascade mediated via the lactose transport system. Biochemical regulation occurs in two major ways. First, several glycolytic enzymes are allosteric, and their activities are subject to the intracellular concentration of specific glycolytic metabolites. When lactose is plentiful, the high intracellular concentration of fructose-1,6-diphosphate (and low level of inorganic phosphate) activates pyruvate kinase, the enzyme that catalyzes substrate level phosphorylation and ATP synthesis, and the NADH-dependent lactate dehydrogenase, which oxidizes NADH$^+$. In contrast, when lactose is limiting, pyruvate kinase activity decreases, causing a metabolic bottleneck in glycolysis. This cessation of glycolysis results in the accumulation of PEP, which primes the cell for PTS-mediated transport when sugars ultimately become available.

A second biochemical mechanism for regulating lactose metabolism involves the PTS transport proteins, and the cytoplasmic component HPr, in particular (Figure 2-12). As outlined above, the PTS cascade starts when HPr is phosphorylated by Enzyme I. Ordinarily, phosphorylation occurs at the histidine-15 [His-15] residue of HPr. However, an ATP-dependent HPr kinase can also phosphorylate HPr, but at a serine residue [Ser-46]. When HPr is in the latter phosphorylation state (i.e., HPr [Ser-46-P]),

phosphorylation at the His-15 site does not occur and the PTS cannot function. At the same time, HPr [Ser-46-P] can bind directly to the interior side of other sugar permeases and prevent transport of those sugars, a process known as inducer exclusion.

It is interesting to note that HPr kinase is activated by fructose-1,6-diphosphate, which reaches high concentrations only when the substrate is plentiful. That is, HPr is evidently converted to HPr [Ser-46-P] at just the time when one would expect Hpr to be in a non-phosphorylated state (i.e., in the active transport mode). Thus, it seems counterintuitive that the cells would slow down transport when substrates are so readily available.

Why, one might ask, would the cell down-regulate the activity of the transport system in times of plenty? One possible answer is that the cell is simply modulating or controlling its appetite by not transporting more substrate than it can reasonably consume. This hypothesis is supported by the observation that HPr not only exerts biochemical control on transport, but that HPr [Ser-46-P] also regulates transcription of sugar transport genes (Figure 2-12, Panel B). Specifically, HPr [Ser-46-P] interacts with a DNA-binding, trans-acting protein called CcpA (Catabolite control protein A). The HPr [Ser-46-P]-CcpA complex (along with fructose-1,6-diphosphate), binds to 14 base pair DNA regions called Catabolite Responsive Elements (CRE). These CRE sites are located just upstream of the transcription start sites of certain catabolic genes. When these CRE regions are occupied by the HPr [Ser-46-P]-CcpA complex, transcription by RNA polymerase is effectively blocked or reduced and mRNA is not made. This process, whereby transcription of catabolic genes, including other PTS genes, is blocked in the presence of a preferred sugar, is called catabolite repression.

Finally, sugar metabolism in lactic acid bacteria can also be genetically regulated directly via repressor proteins. For example, in *L. lactis*, the lactose PTS operon is negatively regulated by LacR, a repressor protein encoded by the *lacR* gene (Figure 2-14). When lactose is present, *lacR* expression is itself repressed, and transcription of the *lac* operon is induced. However, when lactose is unavailable or when cells are grown on glucose, LacR is expressed and transcription of the *lac* genes is repressed. Although a CRE site is also located near the transcriptional start site of the *lac* operon, suggesting that the *lac* genes are subject to CcpA-mediated repression, it seems that LacR may have primary responsibility for regulating sugar metabolism.

Sugar Metabolism by *Saccharomyces cerevisiae*

Metabolism of carbohydrates by *S. cerevisiae* and related yeasts is, in general, not dramatically different from the lactic acid bacteria. These yeasts are facultative anaerobes that are able to use a wide range of carbohydrates. In *S. cerevisiae*, as in the homofermentative lactic acid bacteria, fermentable sugars are metabolized via the glycolytic pathway to pyruvate, yielding two moles of ATP per hexose. However, *S. cerevisiae*, rather than making lactic acid, instead decarboxylates the pyruvate to form acetaldehyde, which is then reduced to ethanol. Thus, the end-products of sugar fermentation by *S. cerevisiae* are ethanol and CO_2. Furthermore, whereas most lactic acid bacteria are obligate fermentors and have limited means for respiratory metabolism, *S. cerevisiae* has an intact citric acid cycle and a functional electron transport system and can readily grow and respire under aerobic conditions.

Glucose transport is the rate-limiting step in glycolysis by *S. cerevisiae*. Remarkably, *S. cerevisiae* has as many as seventeen functional glucose transporters (referred to as hexose transporters or Hxt proteins). Equally surprising, perhaps, is the fact that glucose transport by these hexose transport systems occurs by facilitated diffusion. These are energy-independent systems, driven entirely by the concentration gradient. Some of these transporters are constitutive or inducible, some have low affinity or high affinity for their substrates, and others

function not as transporters, per se, but rather as molecular sensors (see below). Thus, the seemingly excessive redundancy inferred by such a large number of transporters instead provides these organisms with the versatility necessary to grow in environments with a wide range of glucose concentrations.

This metabolic flexibility has practical significance. During beer or wine fermentations, for example, the initial fermentable carbohydrate concentration is high, but then decreases substantially as the fermentation proceeds. Having transport systems that can function at high, low, and in-between substrate concentrations is essential for *Saccharomyces* to ferment all of the substrate. Complete fermentation of sugars ensures that these products can reach full attenuation or dryness, qualities important in beer and wine manufacture.

Regulation of sugar metabolism in *S. cerevisiae* is similar, in principle, to that in lactic acid bacteria, but dramatically different with regard to the actual mechanisms and the extent to which other catabolic pathways are affected. In general, catabolite repression occurs when two or more fermentable sugars are present in the medium and the cell must decide the order in which these sugars are to be fermented. Since glucose is usually the preferred sugar, genes that encode for metabolism of other sugars are subject to catabolite repression. In lactic acid bacteria, catabolite repression is mediated via the phosphorylation state of HPr, in concert with CcpA (see above). Repression of catabolic genes for other sugars occurs at *cre* sites located near promoter regions. In *S. cerevisiae*, catabolite repression (more frequently referred to as glucose repression) assumes a much broader function. Glucose not only represses transcription of genes for sugar use, but also turns off genes coding for pathways involved in the citric acid cycle, the glyoxylate cycle, mitochondrial oxidation, and glycerol utilization.

How does glucose mediate this global response in *S. cerevisiae*? Although the answer is not fully established, it appears that glucose generates a specific signal that is transmitted via protein cascades and that ultimately results in protein-promoter interactions. Several such cascades actually exist, accounting for situations in which glucose is present at high or low concentrations. The main pathway involves the repressor protein, Mig1 (Figure 2–17). This is a DNA-binding protein, produced when glucose is present. In the presence of glucose, Mig1 is localized in the nucleus, where it (in conjunction with accessory proteins) binds to promoter regions, blocking transcription of the downstream genes. When glucose is absent, however, Mig1 becomes phosphorylated by Snf1, a protein kinase, and is translocated out of the nucleus, resulting in de-repression of the regulated genes (i.e., those genes are then expressed). A second protein cascade pathway is also involved in glucose induction. In this system, the membrane proteins Snf3 and Rgt2 sense the presence and level of glucose and then generate a signal that either activates or inactivates specific glucose transporters (whose activities depend on the glucose concentration).

Protein Metabolism

Like many bacteria, lactic acid bacteria are unable to assimilate inorganic nitrogen and instead must rely on preformed amino acids to satisfy their amino acid requirements. Since most foods contain only a small pool of free amino acids, this means that lactic acid bacteria must first be able to degrade proteins and large peptides and then also be able to transport the free amino acids and small peptides released during proteolysis. The proteolytic system in lactic acid bacteria, especially those species involved in dairy fermentations, has been well studied. For these bacteria, the milk protein casein serves as the primary substrate. Casein metabolism by lactic acid bacteria not only is important nutritionally for the organisms, but its degradation during cheese manufacture has major implications for flavor and texture development (Chapter 5). However, in vegetable, bread, and other fermentations, protein metabolism is somewhat less important.

The casein utilization system in lactic acid bacteria involves three main steps. First, casein is hydrolyzed by proteinases to form peptides. Next, the peptides are transported into cells via peptide transport systems. Finally, the peptides are hydrolyzed by intracellular peptidases to form free amino acids (Figure 2–18). Each of these steps are described below.

1. The proteinase system

Hydrolysis of casein by lactic acid bacteria occurs via a cell envelope-associated serine proteinase called PrtP. This enzyme is actually synthesized as a large inactive pre-pro-proteinase (>200 kDa). The "pre" portion contains a leader sequence, whose function is to direct the protein across the cytoplasmic membrane. The "pro" sequence presumably stabilizes the protein during its synthesis. After both of these regions are removed, the now mature proteinase remains anchored to the cell envelope. With casein as a substrate, more than 100 products are formed by PrtP. The majority are large oligopeptides (up to thirty amino acid residues), but most are between four and ten residues.

2. Peptide transport systems

Despite their rather large size and bulkiness, lactococci and other lactic acid bacteria transport peptides directly into the cell without further extracellular hydrolysis. In addition, although there are a myriad of peptides formed by PrtP, relatively few transporters can deliver these peptides into the cell, and even fewer are actually essential for growth in milk. For example, both DtpT and DtpP transport di- and tripeptides, respectively, but given that these peptides are generally not released during casein proteolysis, the role of these transporters in nitrogen nutrition is probably not very critical. Indeed, lactococcal strains with mutations in the genes encoding for DtpT and DtpP (i.e., *dtpT* and *dtpP*) grow fine in milk. Lactococci and other lactic acid bacteria instead rely on an oligopeptide transport system (Opp) to satisfy

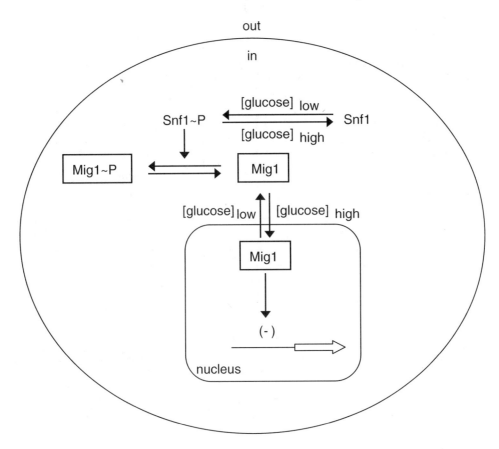

Figure 2–17.   Catabolite repression in *Saccharomyces*. The key regulatory protein is Mig1, whose presence in the nucleus or the cytoplasm depends on the glucose concentration. When glucose is low, Snf1~P phosphorylates Mig1, trapping it in the cytoplasm, where it is inactive as a repressor, allowing expression of catabolic genes. In the presence of glucose, Mig1 is not phosphorylated and remains in the nucleus, where it is able to repress catabolic gene expression.

their amino acid requirements. The Opp system transports about ten to fourteen different peptides of varying size (between four and eleven amino acid residues). In contrast to *dtpT* and *dtpP* mutants, strains unable to express genes coding for the Opp system do not grow in milk.

3. Peptidases

In the final step of protein metabolism, the peptides accumulated in the cytoplasm by the Opp system are hydrolyzed by intracellular peptidases. There are more than twenty different peptidases produced by lactococci and lactobacilli that ultimately generate the pool of amino acids necessary for biosynthesis and cell growth. Included are endopeptidases (that cleave internal peptide bonds) and exopeptidases (that cleave at terminal peptide bonds). The latter group consist entirely of aminopeptidases, as carboxypeptidases in lactic acid bacteria have not been reported (although carboxypeptidases may be present). For lactic acid bacteria to fully use peptides accumulated by the Opp system requires the combined action of endopeptidases, aminopeptidases, dipeptidases, and tripeptidases.

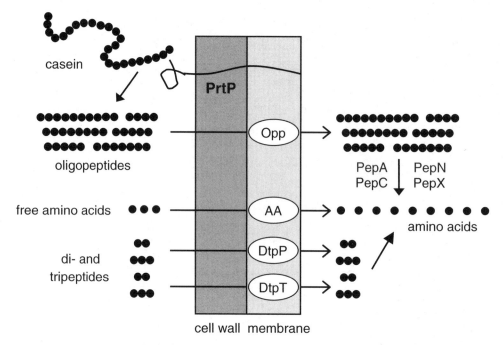

Figure 2–18. Proteolytic system in lactococci. Milk casein is hydrolyzed by a cell envelope-associated proteinase (PrtP) to form oligopeptides. These oligopeptides are then transported across the membrane by the oligopeptide transport system (Opp). The intracellular oligopeptides are then hydrolyzed by cytoplasmic peptidases (e.g., PepA, PepC, PepN, and PepX) to form amino acids. Extracellular di- and tripeptides and free amino acids in milk are transported by di- and tripeptide transporters (DtpT, DtpP) and amino acid (AA) transporters. The intracellular di- and tripeptides are then hydrolyzed to amino acids. Adapted from Hutkins, 2001.

Other Metabolic Systems

Although sugar metabolism, and the lactic and ethanolic fermentations in particular, form the basis for the manufacture of most fermented foods and beverages, other metabolic activities are also important. As noted above, protein metabolism serves the needs of the cells and provides flavor attributes in cheese. In other dairy products, such as buttermilk and sour cream, lactic acid bacteria perform fermentations that yield diacetyl, a compound that imparts the characteristic flavor of these products. In Swiss-type cheeses, the propionic acid fermentation results in formation of several metabolic end-products that likewise give these cheese their unique features. Finally, while fungi do not ferment substrates in the classical sense (see above), they nonetheless convert proteins and fats into an array of products with strong aromatic and flavor properties.

The citrate fermentation

Several species of lactic acid bacteria can ferment citrate and produce the flavor compound diacetyl. Initially, citrate is transported by a pH-dependent, PMF-mediated citrate permease (CitP). The intracellular citrate is split by citrate lyase to form acetate and oxaloacetate (Figure 2–19). The acetate is released directly into the medium, whereas the oxaloacetate is decarboxylated to form pyruvate and CO_2. The pyruvate concentration, however, may increase beyond the reducing capacity of lactate dehydrogenase, causing a glycolytic bottleneck due a shortage of NADH. Therefore, the excess pyruvate is removed by a decarboxylation reaction catalyzed by thiamine pyrophosphate (TPP)-dependent pyruvate decarboxylase. The product, acetaldehyde-TPP, then condenses with another molecule of pyruvate, via α-acetolactate synthase, forming α-acetolactate. The latter is unstable in the presence of oxygen and is non-enzymatically decarboxylated to form diacetyl.

Because the citrate-to-diacetyl pathway does not generate ATP, it would appear to hold no metabolic advantage for the organism. In fact, it costs the cell energy (supplied by the PMF) to transport citrate. Despite these observations, it is now clear that the cell does derive energy during the citrate fermentation (Box 2–4). First, during the intracellular oxaloacetate decarboxylation reaction, a cytoplasmic proton is consumed, causing the cytoplasmic pH and the ΔpH component of the PMF to increase. Second, when both citrate and a fermentable sugar are present, the efflux of lactate can drive uptake of citrate. Moreover, since lactate is monovalent and citrate (at physiological pH) is divalent, the citrate permease acts as an electrogenic precursor-product exchanger (i.e., making the inside more negative and the outside more positive), resulting in a net increase in the $\Delta\psi$ or electrical component of the PMF. Collectively, therefore, citrate fermentation results in an increase in the metabolic energy available to the cell.

The propionic acid fermentation

The propionic acid fermentation is performed by *P. freudenreichii* subsp. *shermanii* and related species. In fermented foods, the relevance of this pathway is limited to the manufacture of

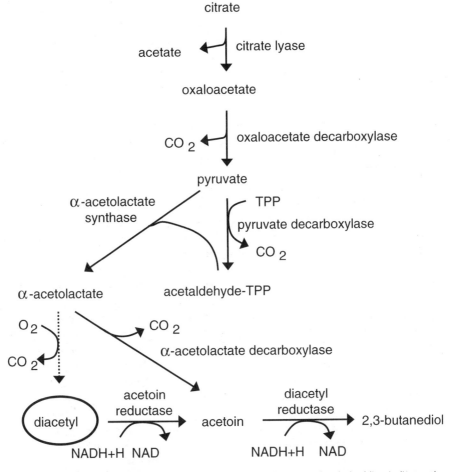

Figure 2–19. Citrate fermentation pathway in lactic acid bacteria. The dashed line indicates the non-enzymatic, oxidative decarboxylation reaction.

Swiss-type cheeses. Although some strains of propionibacteria can ferment lactose, none is available during the time at which these bacteria are given the opportunity to grow (i.e., several weeks after the primary lactose fermentation is complete). Instead, lactate is the only energy source available. Lactate fermentation occurs via the propionate pathway, which yields two moles of propionate, and one each of acetate and CO_2 per three moles of lactate fermented. The cell nets one mole of ATP per lactate. In cheese, the actual amount of end-products varies as a result of condensation reactions, co-metabolism with amino acids, or strain variation. The propionic acid pathway is quite complex and requires several metal-containing enzymes and vitamin co-factors (Chapter 5).

Metabolism of molds

As noted above, the metabolic activities of *Penicillium*, *Aspergillus*, and other fungi are quite unlike those of bacteria and yeasts. The latter have a mostly fermentative metabolism, growing on simple sugars and producing just a few different end-products. In contrast, fungal metabolism is characterized by secretion of numerous proteolytic, amylolytic, and lipolytic enzymes. These enzymatic end-products then serve as substrates for further metabolism.

Often, the metabolic pathways used by fungi result in an array of unique products. For example, when *P. roqueforti*, the blue mold organism, grows in cheese, substantial proteolysis occurs through elaboration of several extracellular proteinases, endopeptidases, and exopeptidases. The resulting amino acids are subsequently metabolized via deaminases and decarboxylases releasing amines, ammonia, and other possible flavor compounds.

More end-products that are characteristic of blue cheese flavors are generated, however, from lipid metabolism. About 20% of triglycerides in milk are initially hydrolyzed by *P. roqueforti*-produced lipases, releasing free fatty acids, including short chain, volatile fatty acids, such as butyric and caproic acids. Metabolism of free fatty acids via β-oxidation pathways then yields a variety of methylketones, compounds that are responsible for the characteristic flavor and aroma of blue cheese. Similarly, growth of the Brie cheese mold, *P. camemberti*, results in a similar sequence of metabolic events. Proteinases and lipases diffuse through the cheese (since the mold grows only at the surface), generating amino and free fatty acids. Subsequent metabolism of the amino acids results in formation of ammonia, methanethiol, and other sulfur compounds, presumably derived from sulfur-containing amino acids. Lipolysis of triglycerides and fatty acid metabolism by *P. camemberti* are just as important in Brie-type cheeses as in blue-veined cheese, and methylketones are abundant. In the cheese environment, both *P. roqueforti* and *P. camemberti* can use lactic acid as a carbon source, which causes the pH to rise, often to near-neutral levels.

Metabolic Engineering

As described earlier in this chapter as well as throughout this text, the key to a successful fermentation is control. The difference between fermentation and spoilage really boils down to controlled growth and metabolism versus uncontrolled growth and metabolism. Therefore, one way to improve a given fermentation process would be to impose either greater control on that process or, alternatively, simply engineer the preferred metabolic route directly into the organism of interest. The latter strategy, referred to as metabolic engineering, provides for a more precise and consistent metabolic result. For example, if increased diacetyl production by a lactic acid bacterium is the goal, rather than manipulate substrate or oxygen levels, the pathway can be genetically manipulated, such that carbon flow is directed to diacetyl rather than lactate (e.g., by inactivating lactate dehydrogenase). The availability of sequenced genomes, along with computational tools, now makes it possible to screen those genomes, not just for a particular enzyme, but for entire pathways or clusters of

Box 2–5. Fermentation Microbiology: From Pasteur's Ferment to Functional Genomics

In 1999, the American Society of Microbiology published a chronology of "Microbiology's fifty most significant events during the past 125 years" (ASM News, 1999). The list started with Pasteur's discovery in 1861 that yeast produced more ethanol during anaerobic fermentative growth compared to growth under aerobic, respiring conditions (the aptly called "Pasteur effect"). The list ended with the report in 1995 that described, for the first time, the complete genome sequence of a bacterium (Fleischmann et al., 1995).

The latter event marked, perhaps, the end of one era and the start of another. From 1995 to the present (June, 2005), more than 230 microbial genomes have been sequenced (and published), and another 370 are nearly finished (according to the National Center for Biotechnology Information, www.ncbi.nlm.nih.gov). Sequencing genomes has evolved from an expensive, multi-year, labor-intensive endeavor to one that is affordable, fast, and highly automated. At the same time, the development of bioinformatics and various computational tools used to analyze and compare genome information has also advanced at an equally rapid rate.

Genomes of Food Fermentation Organisms

Among organisms involved in food fermentations, the genome of the common yeast, *Saccharomyces cerevisiae,* was published in 1996 (Goffeau et al., 1996). Although the sequence strain was a lab strain, and was quite different from the typical bakers' or brewers' yeasts used in industry, the sequence nonetheless provided valuable information on yeast biology and genetics. Several years later, the genome sequence of the lactic acid bacterium, *Lactococcus lactis* subsp. *lactis,* was reported (Bolotin et al., 2001). In the past several years, groups in the United States, Europe, and New Zealand have completed sequencing projects for more than 30 other lactic acid and related bacteria (Table 1). A major collaborative effort was also begun in 2002, when the Lactic Acid Bacteria Genome Consortium (in collaboration with the U.S. Department of Energy's Joint Genome Institute) was organized (Klaenhammer et al., 2002). The genomes of eleven commercially important bacteria were sequenced as part of this project (Figure 1; Makarova et al., 2006).

Genome Sequences are Only the Beginning

The actual output of a genome sequencing project is really just the linear order of nucleotides arranged as a continuous circle. It is not until the nucleotide sequences (i.e., strings of Gs, Cs, As, and Ts) are analyzed computationally, that open reading frames (orf) and putative genes can be inferred or predicted. Comparison of those orfs (or the translated proteins) to genome data bases provides a basis for assigning probable functions to those genes. The latter process, called genome annotation, usually results in about 70% of the orfs having assigned functions, although as databases grow, so does the likelihood a given orf will have a homolog (i.e., find a match to a previously assigned protein).

The genome may also contain regions encoding for insertion sequence elements, prophages, and tRNA and rRNA. Of course, predicted function is just that—it is based on the statistical probability that a gene codes for a particular protein. To confirm an actual biological function for a gene requires some form of biochemical demonstration (e.g., creating loss-of-function mutations having the expected phenotype). Therefore, genome sequencing and annotation are really only the first steps toward understanding how pathways are constructed, how regulatory networks function, and how these bacteria ultimately behave during food fermentations. Functional and comparative genomics not only addresses these questions, but also provides insight into their phylogeny and evolution.

(Continued)

Box 2–5. Fermentation Microbiology: From Pasteur's Ferment to Functional Genomics *(Continued)*

Table 1. Sequenced genomes of lactic acid bacteria[1].

Organism	Strain(s)
Lactobacillus acidophilus	NCFM
Lactobacillus brevis	ATCC 367
Lactobacillus delbrueckii subsp. *bulgaricus*	ATCC 11842, DN-100107, ATCC BAA-365
Lactobacillus casei	ATCC 334, BL23
Lactobacillus gasseri	ATCC 33323
Lactobacillus helveticus	CNRZ 32, DPC 4571
Lactobacillus johnsonii	NCC 533
Lactobacillus plantarum	WCRS1
Lactobacillus reuteri	100-23, DSM 20016T
Lactobacillus rhamnosus	HN001
Lactobacillus sakei	23K
Lactococcus lactis subsp. *cremoris*	MG 1363, SK11
Lactococcus lactis subsp. *lactis*	IL 1403
Leuconostoc mesenteroides	ATCC 8293
Leuconostoc citreum	KM20
Oenococcus oeni	PSU-1
Pediococcus pentosaceus	ATCC 25745
Streptococcus thermophilus	LMG 18311, CNRZ 1066, LMD-9

[1]Adapted from Klaenhammer et al., 2002

What have the genomes revealed?

The genomes of the lactic acid bacteria are generally small, with several less than 2 Mb. Accordingly, the genomes reflect the rather specialized metabolic capabilities of these bacteria and the specific environmental niches in which they live (Makarova et al., 2006). For example, most species contain genes encoding for only a few of the carbon-utilization pathways (i.e., homo- and heterofermentation; pyruvate dissimulation; and pentose, citrate, and malate fermentation). These bacteria are auxotrophic for several amino acids, and, therefore, contain many genes encoding for protein and peptide catabolism. In fact, given their overall dependence on obtaining nutrients from the environment, it is not surprising that the genomes are replete (between 13% and 18%) with transport system genes (Lorca et al., 2006).

Genome evolution analyses have revealed that the loss of biosynthetic genes over time, and the corresponding acquisition of catabolic genes, are consistent with the view that the lactic acid bacteria have recently adapted from nutritionally-poor, plant-type habitats to nutritionally-complex food environments. This suggestion is supported by several observations. First, some of the lactic acid bacteria contain a large number of plasmids and transposons that accelerate horizontal gene transfer (genetic exchange between related organisms) and that promote adaptation to new environments. Second, there are a large number of non-functional pseudogenes (containing mutations or truncation) within the genomes that likely represent "gene decay" or loss of genes that are no longer needed (Makarova et al., 2006). For example, 10% of the genome of the nutritionally-fastidious *Streptococcus thermophilus* are classified as pseudogenes, with many encoding specifically for energy metabolism and transport functions (Bolotin et al., 2004). As lactic acid bacteria have evolved, therefore, significant genome reduction, with modest genome expansion, has apparently become the norm (Makarova et al., 2006).

References

ASM News. Microbiology's fifty most significant events during the past 125 years. Poster supplement. 65.

Box 2–5. Fermentation Microbiology: From Pasteur's Ferment to Functional Genomics *(Continued)*

Bolotin, A., P. Wincker, S. Mauger, O. Jaillon, K. Malarme, J. Weissenbach, S.D. Ehrlich, and A. Sorokin. 2001. The complete genome sequence of the lactic acid bacterium *Lactococcus lactis* ssp. *lactis* IL1403. Genome Res. 11:731-753.

Fleischmann, R.D., M.D. Adams, et al., (38 other authors). 1995. Whole-genome random sequencing and assembly of *Haemophilus influenza* Rd. Science 269:496-512.

Goffeau, A., B. G. Barrell, H. Bussey, R. W. Davis, B. Dujon, H. Feldmann, F. Galibert, J. D. Hoheisel, C. Jacq, M. Johnston, E. J. Louis, H. W. Mewes, Y. Murakami, P. Philippsen, H. Tettelin, and S. G. Oliver. 1996. Life with 6000 genes. Science 274:546-567.

Klaenhammer, T., E. Altermann, et al., (35 other authors). 2002. Discovering lactic acid bacteria by genomics. Antonie van Leeuwenhoek 82:29-58.

Lorca, G.L., R.D. Barabote, V. Zlotopolski, C. Tran, B. Winnen, R.H. Hvorup, A.J. Stonestrom, E. Nguyen, L.-W. Huang, D. Kempton, and M.H. Saier. 2006. Transport capabilities of eleven Gram-positive bacteria: comparative genomic analyses. In press.

Makarova, K., Y. Wolf, et al., (and 43 other authors) 2006. Comparative genomics of the lactic acid bacteria. 2006. Proc. Nat. Acad. Sci. In Press.

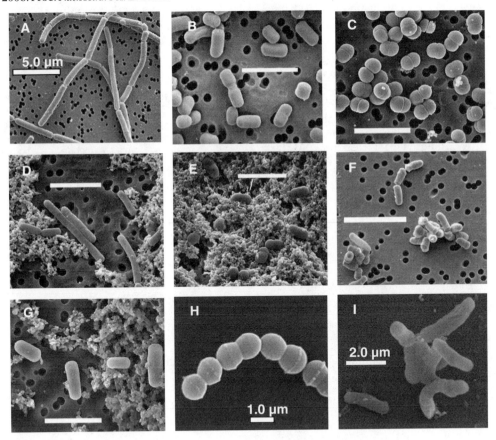

Figure 1. Electron micrographs of lactic acid and related bacteria: A, *Lactobacillus delbrueckkii* subsp. *bulgaricus*; B, *Lactobacillus brevis*; C, *Pediococcus pentosaceus*; D, *Lactobacillus casei*; E, *Lactococcus lactis*; F, *Brevibacterium linens*; G, *Lactobacillus helveticus*; H, *Streptococcus thermophilus*; and I, *Bifidobacterium longum*. Scale bars are 3.0 μm, unless otherwise indicated. Not shown: *Lactobacillus gasseri* and *Oenococcus oeni*. Photos courtesy of J. Broadbent and B. McManus, Utah State University, and D. O'Sullivan, University of Minnesota, and with permission from the American Society for Microbiology (ASM News, March, 2005, p. 121–129)

pathways (Box 2-5). Moreover, a variety of molecular tools now exist that can be used to inactivate some pathways and activate others.

Bibliography

Axelsson, L. 2004. Lactic acid bacteria: classification and physiology, pp. 1-66. *In* Salminen, S., A. von Write, and A. Ouwehand. Lactic Acid Bacteria Microbiological and Functional Aspects Third Edition Marcel Dekker, Inc. New York, New York.

Bullerman, L.B. 1993. Biology and health aspects of molds in foods and the environment. J. Korean Soc. Food Nutr. 22:359-366.

Dellaglio, F., L.M.T. Dicks, and S. Torriani. 1992. The genus *Leuconostoc*. *In* B.J.B. Wood and W.H. Holzapfel, ed. The Lactic Acid Bacteria, Volume 2, The genera of lactic acid bacteria. Blackie Academic and Professional pp 235-268.

de Vos, W.M., I. Boerrigter, R.J. van Rooyen, B. Reiche, and W. Hengstenberg. 1990. Characterization of the lactose-specific enzymes of the phosphotransferase system in Lactococcus lactis. J. Biol. Chem. 265:22554-22560.

Gancedo, J.M. 1998. Yeast carbon catabolite repression. Microbiol. Mol. Biol. Rev. 62:334-361.

Garrity, G.M., J. Bell, and T.G. Lilburn. 2004. Taxonomic Outline of the Prokaryotes. *Bergey's Manual of Systematic Bacteriology, Second Edition*. Release 5.0, Springer-Verlag. DOI: 10.1007/bergeysonline.

Hemme, D., C. Foucaud-Scheunemann. 2004. *Leuconostoc*, characteristics, use in dairy technology and prospects in functional foods. Int. Dairy J. 14:467-494.

Hutkins, R.W. 2001. Metabolism of starter cultures, p. 207. In E.H. Marth and J.L. Steele (ed.), *Applied Dairy Microbiology*. Marcel Dekker, Inc., New York, NY.

Makarova, K., Y. Wolf, et al., (and 43 other authors) 2006. Comparative genomics of the lactic acid bacteria. 2006. Proc. Nat. Acad. Sci. In Press.

Moore, D. and L. Novak Frazer. 2002. *Essential Fungal Genetics*. Springer-Verlag. New York, NY.

Naumov, G.I., S.A. James, E.S. Naumova, E.J. Louis, and I.N. Roberts. 2000. Three new species in the *Saccharomyces sensu stricto* complex: *Saccharomyces cariocanus, Saccharomyces kudriavzevii* and *Saccharomyces mikatae*. Int. J. Syst. Evol. Microbiol. 50:1931-1942.

Pitt, J.I. and A.D. Hocking. 1999. *Fungi and Food Spoilage*. Aspen Publishers, Inc. Gaithersburg, MD.

Samson, R.A., E.S. Hoekstra, J.C. Frisvad and O. Filtenborg. 2000. *Introduction to Food and Airborne Fungi. Sixth Ed*. Centraalbureau voor Schimmelcultures. Utrecht, The Netherlands.

3

Starter Cultures

"The connection between wine fermentation and the development of the sugar fungus is not to be underestimated; it is very probable that, by means of the development of the sugar fungus, fermentation is started."
From A Preliminary Communication Concerning Experiments on Fermentation of Wine and Putrefaction *by Theodor Schwann, 1837 (as recounted by Barnett, 2003.)*

Introduction

The successful manufacture of all fermented products relies on the presence, growth, and metabolism of specific microorganisms. In reality, however, it is possible to produce non-fermented counterparts of some fermented products. For example, sour cream, cottage cheese, and summer sausage can be produced in the absence of microorganisms simply by adding food-grade acidulants to the raw material. Similarly, carbon dioxide can be produced in dough by adding chemical leavening agents. Even products as complicated biochemically and microbiologically as soy sauce can be made by chemical means. However, despite the technical feasability of producing these products without fermentation, these non-fermented versions generally lack the desired organoleptic qualities that are present, and that consumers expect, in the fermented products. This is because microorganisms are responsible for producing an array of metabolic end-products and textural modifications, and replicating those effects by other means is simply not possible.

If microorganisms are indeed necessary for converting raw materials into fermented foods with the desired characteristics and properties, then what is the best way to ensure that the relevant organisms are present in the starting material? In other words, how are fermentations started? There are essentially three ways to induce or initiate a food fermentation. The oldest method simply relies on the indigenous microorganisms present in the raw material. Raw milk and meat, for example, usually harbor the very bacteria necessary to convert these materials into cheese and sausage. Grapes and grape crushing equipment, likewise, contain the yeasts responsible for fermenting sugars into ethanol and for transformation of juice into wine.

For these natural fermentations to be successful, however, requires not only that the "correct" microorganisms be present, but also that suitable conditions for their growth are established. Even if these requirements are satisfied, however, there is no guarantee that the product will meet the quality expectations, be safe to consume, or even be successfully produced. Still, many foods are produced by natural fermentation, including some sausages (Chapter 6), wines (Chapter 10), and pickles and other fermented vegetables (Chapter 7).

Once a successful fermentation has been achieved, a portion of that product could be transferred to fresh raw material to initiate a fermentation. This method, called backslopping, is probably nearly as ancient as the natural fermentation practice. It works for almost any fermented food, and is still commonly used for beer, some cheeses and cultured dairy products, sour dough bread, and vinegar. In addition, these methods are still practiced today for small-scale production facilities, as well as in less developed countries and in home-made type products. The principle, regardless of product, is the same. Any successfully fermented product should contain the relevant number and type of microorganisms, and, given a fresh opportunity, they will perform much the same as they had the previous time. Despite the detractors of the backslopping technique (see below), it can be argued that this practice actually selects for hardy and well-acclimated organisms with many of the desired traits necessary for successful production.

The demonstration by Pasteur that fermentation (as well as spoilage) was caused by microorganisms led Lister, Orla-Jensen, and other early microbiologists, more than a century ago, to isolate and identify the responsible organisms. Koch's postulates regarding the germ-disease connection could then be applied to fermentation science. Thus, an organism isolated from soured or fermented milk could be purified, re-introduced into fresh milk causing the expected fermentation, and then re-isolated from the newly fermented product. The implication of this discovery—that pure cultures could be obtained and used to start fermentations, did not go unnoticed. Indeed, these observations resulted in a third way to produce fermented foods, namely via the use of a starter culture containing the relevant microorganisms for that particular product.

Role of Starter Cultures

It is often argued by advocates of traditional manufacturing methods that natural fermentations, whether initiated by the endogenous flora or by backslopping, yield products that have unique or singular quality attributes. Naturally-fermented wines, for example, are often claimed to be superior to wines made using a starter culture. Even if a slow or "stuck" fermentation occurs occasionally, it would certainly be worth it (the argument goes) to end up with a truly exceptional product. This attitude may be perfectly fine on a small scale basis, given the inherent flexibility in terms of time and quality expectations. In contrast, however, modern large-scale industrial production of fermented foods and beverages demands consistent product quality and predictable production schedules, as well as stringent quality control to ensure food safety. These differences between traditional and modern fermentations, in terms of both how fermented products are manufactured and their quality expectations, are summarized in Chapter 1, Table 1-1.

Simply defined, starter cultures consist of microorganisms that are inoculated directly into food materials to overwhelm the existing flora and bring about desired changes in the finished product. These changes may include novel functionality, enhanced preservation, reduced food safety risks, improved nutritional or health value, enhanced sensory qualities, and increased economic value. Although some fermented foods can be made without a starter culture, as noted above, the addition of concentrated microorganisms, in the form of a starter culture, ensures (usually) that products are manufactured on a timely and repeatable schedule, with consistent and predictable product qualities. In the case of large volume fermentations, specifically, cheese fermentations, there is also a volume factor that must be considered. In other words, there is no easy way to produce the amount of culture necessary for large scale cheese production without the use of concentrated starter cultures (see below). For all practical purposes (but with a few notable exceptions), starter cultures are now considered an essential component of nearly all commercially-produced fermented foods.

History

As noted above, microbiology was not established as a scientific discipline until the 1860s and '70s. And although Pasteur, Lister, Koch, Ehrlich, and other early microbiologists were concerned about the role of microorganisms as a cause of infectious disease, many of the issues addressed by these scientists dealt with foodstuffs, including fermented foods and beverages. In fact, many of the members of the microbiology community at the turn of the twentieth century were essentially food microbiologists, working on very applied sorts of problems. The discovery that bacteria and yeast were responsible for initiating (as well as spoiling) food fermentations led to the realization that it was possible to control and improve fermentation processes. Although the brewing, baking, and fermented dairy industries were quick to apply this new knowledge and to adopt starter culture technologies, other fermented food industries did not adopt starter cultures until relatively recently. And some still rely on natural fermentations.

One of the first such efforts to purify a starter culture was initiated in 1883 at the Carlsberg Brewery in Copenhagen, Denmark. There, Emil Christian Hansen used a dilution method to isolate pure cultures of brewing yeast, derived from a mixed culture that occasionally produced poor quality beer. Subsequently, he was able to identify which strain produced the best (or worst) beers. Hansen also was the first to isolate the two types of brewing yeast, the top (or ale) yeast and the bottom (or lager yeast). Eventually, all of the beer produced by the Carlsberg Brewery was made using the Hansen strain, *Saccharomyces carlsbergensis* (later reclassified as Saccharomycer pastonanus), which also became the lager type strain.

Although Pasteur had also proven that wine fermentations, like those of beer, were performed by yeast, there was little interest at the time in using pure yeast cultures for winemaking. That satisfactory wine could be made using selected strains was demonstrated by the German scientist Hermann Müller-Thurgau in 1890; however, it wasn't until the 1960s that wine yeasts became available (and acceptable) as starter cultures for wine fermentations.

At around the same time and place, another Hansen, Christian Ditlev Ammentorp Hansen, was working on the extraction of enzymes from bovine stomach tissue. This work led to the isolation of the enzyme chymosin, an essential ingredient in cheese manufacture. Prior to this time, chymosin had been prepared and used as a crude paste, essentially ground-up calf stomachs. Using the Hansen process, chymosin could be partially purified in a stable, liquid form, and the activity standardized. Factories dedicated to chymosin production were built in Copenhagen in 1874 and in New York in 1878. Later, Hansen developed procedures to produce natural coloring agents for cheese and other dairy products. By the end of the century, the Chr. Hansen's company began producing dairy starter cultures, thus establishing a full-service business that continues even today to be a world-wide supplier of starters cultures and other products for the dairy, meat, brewing, baking, and wine industries.

In the United States, a culture industry devoted to bakers' yeast arose in the 1860s and '70s. Two immigrant brothers from Europe, Charles and Max Fleischmann, began a yeast factory in Cincinnati, Ohio, producing a compressed yeast cake for use by commercial as well as home baking markets. Bread produced using this yeast culture was far superior to breads made using brewing yeasts, which was the common form at the time. More than forty years later, in 1923, another bakers' yeast production facility was built in Montreal, Canada, by Fred Lallemand, another European immigrant.

With the exception of brewers' and bakers' yeasts, all of the initial cultures developed and marketed prior to the 1900s consisted of bacteria that were used by the dairy industry. There was particular interest in flavor-producing cultures that could be used for butter manufacture. Most of the butter produced in the United States and Europe was cultured butter, made from soured cream (in contrast to sweet cream butter which now dominates the U.S. market). The cream was soured either via a natural or spontaneous fermentation or by addition of previously soured cream or buttermilk. Both methods, however, often resulted in inconsistent product quality. By the 1880s, researchers in

Europe (Storch in Denmark and Weigman in Germany) and the United States (Conn) showed that strains of lactic acid bacteria could be grown in pure culture, and then used to ripen cream.

The first dairy starter cultures were liquid cultures, prepared by growing pure strains in heat-sterilized milk. The main problem with these cultures was that they would become over-acidified and lose viability (discussed in more detail later in this chapter). To maintain a more neutral pH, calcium carbonate was often added as a buffer. Still, liquid cultures had a relatively short shelf life, which eventually led to the development of air-dried cultures. The latter were rather crude preparations and were produced simply by passing liquid cultures through cheese cloth, followed by dehydration at 15°C to 18°C. These cultures were more stable, yet required several transfers in milk to revive the culture and return it to an active state. Freeze dried dairy cultures had also become available by the 1920s, but cell viability remained a problem, and even these products required growth in intermediate or mother cultures to activate the cells. Frozen cultures, which are now among the most popular form for dairy cultures, were introduced in the 1960s. In the last twenty years, significant improvements in both freezing and freeze-drying technologies have made these types of cultures the dominant forms in the starter culture market.

The technologies described above were largely developed by starter culture companies and by researchers at universities and research institutes. Most of the culture houses started out as family-owned enterprises and then grew to rather large companies with significant research and production capabilities. By the 1980s, large pharmaceutical-based corporations acquired many of these companies, although a few smaller ones remain, selling mostly specialized products.

As noted above, cultures are not the only product the industry sells. There is a rather large amount of culture media that is used to propagate starter cultures and this represents a substantial market, as do the enzymes and coloring agents used for cheese production. Finally, due to the demanding nature of the cheese and cultured dairy products industries—successful product manufacture requires consistent culture performance—starter culture companies often provide nearly round-the-clock technical service and support to their customers. If a fermentation is delayed, slow, or is otherwise defective, the culture (and the culture supplier) will often be blamed. Cultures, after all, are about the only food ingredient supplied in a biologically active form. Enzymes also fall into the category, but are generally more easily standardized and stabilized.

A discussion of the history of starter culture science and technology would be incomplete without noting what has been perhaps the main driving force for much of the basic and applied research on lactic starter cultures. The observation that bacteriophages, viruses whose hosts are bacteria, could infect and then kill starter culture bacteria was first made in the 1930s. Since that time, but especially in the past thirty years (coinciding with the advent of molecular biology), several research groups in the United States and from around the world have devoted significant attention to understanding and controlling bacteriophages (Box 3–1). This research has led to countless other important and fundamental discoveries about lactic acid bacteria.

Starter Culture Microorganisms

Chapter 2 described many of the microorganisms that are important in fermented foods. Not all of these organisms, however, are produced or marketed as starter cultures. Some of the products do not lend themselves to starter culture applications, and for others, the market is simply too small or too specialized. For example, sauerkraut and pickle fermentations are mediated by the natural microflora; therefore, even commercial manufacturers can produce these products without a starter culture. Although cultures and manufacturing procedures now exist and can be used for controlled starter culture-mediated pickle fermentations,

Box 3-1. Bacteriophages of Lactic Acid Bacteria

The physical structure, biology, taxonomy, and life cycle of bacteriophages are quite unlike those of the bacteria, yeasts, and molds. In fact, bacteriophages, defined simply as viruses that attack bacteria, look and behave like nothing else in the biological world. In essence, bacteriophages are just packages of DNA (or RNA) contained within a protective "head" which is attached to structures that enable them to adhere to and ultimately parasitize host cells.

Despite their simplicity, their importance in food microbiology cannot be overstated. Bacteriophages that infect *Escherichia coli* O157:H7, *Vibrio cholera,* and other pathogens, for example, may carry genes encoding for toxins and other virulence factors, which are then transferred to and expressed by the host cell. In fermented foods made using lactic starter cultures, bacteriophages are often responsible for causing manufacturing delays, quality defects, and other detrimental effects. The economic consequences of bacteriophage infection can be enormous, especially in large scale operations. Thus, understanding the ecology, mode of replication, and transmission of bacteriophages are essential for minimizing the incidence of bacteriophage problems and for developing strategies for their control.

Bacteriophage classification

Because bacteriophages are parasites and cannot grow outside of a host, they have no real physiological activities that can be used as a basis for systematic classification. That is, bacteriophages are neither aerobic or anaerobic, they have no temperature or moisture optima, per se (aside from attachment kinetics or host-dependent functions), and biochemical or metabolic pathways do not exist. Thus, classification is based on other criteria, including morphology, structural composition, serology, host range, and DNA and genome analyses (McGrath et al., 2004). Morphological characteristics of bacteriophages that infect lactic acid bacteria are summarized in Table 1.

Table 1. Morphotypes of lactic acid bacteriophage.

Morphotypes	Group	Description	Representative types
Myoviridae	A	Contractile tails	3-B1, 7E1
Siphoviridae	B1	Long non-contractile tails; small isometric head	r1t, P335, sk1, blL170
	B2	Long non-contractile tails; prolate head	c2, blL67
	B3	Long non-contractile tails; elongated head	JCL 1032
Podoviridae	C	Short non-contractile tails	PO34, KSY1

From Brusow, 2001; Desiere et al., 2002; McGrath et al., 2004; and Stanley et al., 2003

Based on morphological distinctions, there are three main types of bacteriophages: morphotypes A, B, and C, with the B or Siphoviridae group being the most common type that infect lactic acid bacteria. The group B phages have long tails with heads that have either a small isometric (B1 sub-group), prolate (B2), or elongated (B3) shape (Figure 1). The lactococcal phages generally belong to the B1 or B2 group, whereas the phages that infect *Streptococcus thermophilus* are all in the B1 group. Furthermore, most of the lactococcal phages that have been characterized share structural similarity to three specific lactococcal phages: c2 (a B2 morphotype), 936, and P335 (morphotype B1). Therefore, it is common to refer to other phages as being represented by one of these three types. For example, phages of the P335 type are considered to be the most common type infecting industrial fermentations that rely on lactococci (Durmaz and Klaenhammer, 2000).

Bacteriophage genomics has now become one of the most powerful means for characterizing and classifying lactococcal and related phages. Sequence analyses of bacteriophage

(Continued)

Box 3-1.   Bacteriophages of Lactic Acid Bacteria *(Continued)*

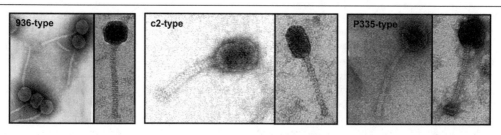

Figure 1.   Electron photomicrographs of representative lactococcal bacteriophages. All were all isolated from cheese factories. Photos provided courtesy of Sylvain Moineau.

genomes (more than thirty have been sequenced) have revealed that gene clusters are arranged in modular fashion (Figure 2). That is, the genome consists of modules that code for particular functions, such as replication, structure, or assembly. In addition, these modules are transcribed either during the early or late stages of the phage replication cycle. In some phages, a middle gene cluster also exists. Phage classification, then, can be based on similarities in the genetic organization of different phages. Moreover, it is now recognized that the ability of lytic bacteriophages to undergo frequent recombination is likely due to the modular nature of the phage genome. Importantly, recombination by phage may result in the acquisition of new genetic information, including genes that enable the phage to counter host defense systems (Rakonjac et al., 2005).

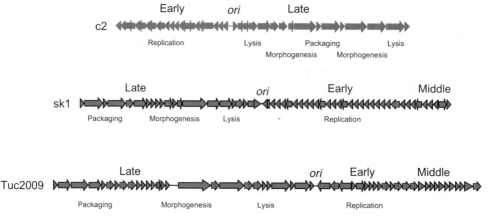

Figure 2.   Alignment of the three main classes of lactococcal bacteriophages: the prolate headed c2, the small isometric headed sk1 (936 class), and the small isometric headed Tuc2009 (P335 class). Early, middle, and late expressed regions and putative functions are indicated. Adapted from Desiere et al., 2002 and Stanley et al., 2003.

Bacteriophage biology

Not all bacteriophages infect and lyse their host cells. Some bacteriophages have two routes of infection or "lifestyle" choices: lytic or lysogenic. Lytic phages infect a host, replicate inside that host cell, and eventually lyse the cell, releasing new infectious phage particles. In contrast, a

Box 3-1. Bacteriophages of Lactic Acid Bacteria *(Continued)*

lysogenic infection occurs when so-called temperate bacteriophages infect the host cell, but rather than initiating a lytic cycle, they can instead integrate their genome within the host chromosome. The phage then exists as a dormant prophage, and is replicated along with the host genome during cell growth.

Importantly, lysogenic cells may be immune to subsequent infection. Although the mechanisms dictating whether a temperate phage enters a lytic or lysogenic phase are not completely known (for lactic phages, that is), it appears that lytic and lysogenic genes are subject to transcriptional regulation. Specifically, repressors of these two pathways (i.e., lysis or lysogeny) apparently exist that prevent transcription of the relevant genes. It has also been demonstrated experimentally that prophages can be induced, such that the phage is excised from the chromosome and is converted from a lysogenic state to a lytic phase. The observation that prophage induction may occur during cheese manufacture suggests that prophages may be a source of new phages within fermentation environments (e.g., cheese plants). In addition, the ability to manipulate prophage induction by temperature, osmolarity, and other environmental cues, makes it possible to accelerate lysis of starter culture cells and release of enzymes necessary for cheese ripening (Feirtag and McKay, 1987; Lunde et al., 2005).

References

Brüssow, H. 2001. Phages of dairy bacteria. Annu. Rev. Microbiol. 55:283–303.

Desiere, F., S. Lucchini, C. Canchaya, M. Ventura, and H. Brüssow. 2002. Comparative genomics of phages and prophages in lactic acid bacteria, p. 73—91. *In* Siezen, R.J., J. Kok, T. Abee, and G. Schaafsma (ed.). *Lactic Acid Bacteria: Genetics, Metabolism and Applications.* Antonie van Leeuwenhoek 82:73–91. Kluwer Academic Publishers. The Netherlands.

Feirtag, J.M., and L.L. McKay. 1987. Thermoinducible lysis of thermosensitive *Streptococcus cremoris* strains. J. Dairy Sci. 70:1779–1784.

Lunde, M., A.H. Aastveit, J.M. Blatny, and I.F. Nes. 2005. Effects of diverse environmental conditions on φLC3 prophage stability in *Lactococcus lactis.* Appl. Environ. Microbiol. 71:721–727.

McGrath, S., G.F. Fitzgerald, and D. van Sinderen. 2004. Starter cultures: bacteriophage, p. 163—189. *In* P.F. Fox, P.L.H. McSweeney, T.M. Cogan, and T.P. Guinee (ed.). *Cheese: Chemistry, Physics, and Microbiology, Third ed., Volume 1: General Aspects.* Elsevier Ltd., London.

Rakonjac, J., P.W. O'Toole, and M. Lubbers. 2005. Isolation of lactococcal prolate phage-phage recombinants by an enrichment strategy reveals two novel host range determinants. J. Bacteriol. 187:3110–3121.

Stanley, E., S. McGrath, G.F. Fitzgerald, and D. van Sinderen. 2003. Comparative genomics of bacteriophage infecting lactic acid bacteria p. 45—94. *In* B.J.B. Wood, and P.J. Warner (ed.). *Genetics of Lactic Acid Bacteria.* Kluwer Academic/Plenum Publishers. New York, New York.

culture systems for sauerkraut have not been as readily adopted (Chapter 7). In other cases, the manufacturer may prefer to maintain and propagate their own proprietary cultures. Many breweries, for example, have the necessary laboratory facilities and personnel to support internal culture programs. The production of fungal-fermented foods (described in Chapter 12) also frequently relies on house strains.

When starter culture technology first developed (and for many years thereafter), the actual organisms contained within most starter cultures were generally not well established in terms of strain or even species identity. Rather, a culture was used simply because it worked, meaning that it produced a good product with consistent properties. Such undefined cultures, however, are now less frequently used. Instead, the organisms present in modern commercial starter culture preparations are usually very well defined, often to a strain level. They are also carefully selected based on the precise phenotypic criteria relevant for the particular product (Table 3-1).

Bacterial starter cultures

The most important group of bacteria used as starter organisms are the lactic acid bacteria (LAB); (Table 3-2). In fact, only a few non-lactic acid bacteria are commercially available as starter cultures (see below). As described in Chapter 2, the LAB consist of a cluster of Gram positive cocci and rods that share several physiological and biochemical traits. In general, they

are Gram positive, catalase negative heterotrophs that metabolize sugars via either homofermentative or heterofermentative metabolism. The LAB grow over a wide temperature range, although most LAB are either mesophiles or moderate thermophiles (with temperature optima of about 30°C and 42°C, respectively). Likewise, they vary with respect to salt tolerance, osmotolerance, aerobiosis, and other environmental conditions, accounting, in part, for the diversity of habitats with which they are associated.

Although twelve genera of LAB are now recognized (Chapter 2), starter culture LAB belong to one of four genera, with the dairy LAB representing the largest group. The latter include species of *Lactococcus*, *Streptococcus*, *Leuconostoc*, and *Lactobacillus*. The dairy starters are generally grouped as mesophiles or thermophiles, depending on the product application, but even this designation has become somewhat blurry. In fact, thermophilic strains of *Streptococcus thermophilus* are now sometimes included in mesophilic starter cultures (e.g., for Cheddar cheese), and mesophilic *Lactococcus lactis* susbsp. *lactis* are occasionally incorporated in thermophilic cultures (e.g., for Mozzarella cheese).

In general, dairy cultures perform three main functions: (1) to ferment the milk sugar, lactose, and acidify the milk or cheese; (2) to generate flavor or flavor precursors; and (3) to modify the texture properties of the product. Lactic acid bacteria used as starter cultures for other fermented food products have similar functions, although acidification is by far the most important. In sourdough bread, for example, starter culture strains of *Lactobacillus sanfranciscansis* ferment maltose and lower the dough pH via production of lactic acid, but the culture also produces acetic acid and other flavor and aroma compounds.

In contrast, one of the lactic acid bacteria included in sausage starter cultures, *Pediococcus acidilactici*, essentially performs one job—to make lactic acid and reduce the meat pH to a level inhibitory to undesirable competitors (including pathogens). For some applications, the

Table 3.1. Desirable properties of starter cultures.

Culture	Property
Dairy cultures	Controlled lactic acid production rates
	Short lag phase
	Phage resistance
	Ease of manufacture
	Stability and consistency
	Produce desired flavor and texture
	Preservation tolerance
	Lack of off-flavors
Meat cultures	Fast acidification
	Produce desired flavor
	Antimicrobial activity
Beer cultures	Rapid fermentation
	Produce desired flavor
	Preservation tolerance and stability
	Flocculation
	Lack of off-flavors
	Proper attenuation
	Growth at wide temperature range
	Tolerant to osmotic, temperature, and handling stresses
Wine cultures	Osmotolerant
	Ethanol tolerant
	Flocculation, sedimentation
	Growth at low temperature
	Produce consistent flavor
	Malolactic fermentation
Bread cultures	Freeze tolerant
	Produce desired flavor
	Produce adequate leavening

Table 3.2. Lactic acid bacteria used as starter cultures.

Organism	Application
Lactococcus lactis subsp. *lactis*	Cheese, cultured dairy products
Lactococcus lactis subsp. *cremoris*	Cheese, cultured dairy products
Lactococcus lactis subsp. *lactis* biovar. *diacetylous*	Cheese, cultured dairy products
Lactobacillus helveticus	Cheese, cultured dairy products
Lactobacillus delbrueckii subsp. *bulgaricus*	Cheese, yogurt
Lactobacillus sanfranciscensis	Sourdough bread
Lactobacillus casei	Cheese, cultured dairy products
Lactobacillus sakei	Sausage
Lactobacillus plantarum	Sausage, fermented vegetables
Lactobacillus curvatus	Sausage
Streptococcus thermophilus	Cheese, yogurt
Pediococcus acidilactici	Sausage, fermented vegetables
Pediococcus pentosaceus	Sausage
Tetragenococcus halophila	Soy sauce
Oenococcus oeni	Wine
Leuconostoc lactis	Cheese, cultured dairy products
Leuconostoc mesenteroides subsp. *cremoris*	Cheese, cultured dairy products, fermented vegetables

lactic starter culture may serve a purpose quite removed from those described above. The function of *Oenococcus oeni*, an organism used during the wine fermentation, is to convert malic acid to lactic acid. This reaction results in a slight but critical increase in pH, and is necessary to soften or de-acidify overly acidic wines (Chapter 10).

Although LAB are the most important and most widely-used group of starter culture bacteria, other non-lactic bacteria also are included in various starter cultures (Table 3-3). In the cheese industry, *Propionibacterium freudenreichii* subsp. *shermanii* is used to manufacture Emmenthaler and other Swiss-type cheeses and *Brevibacterium linens* is used in the manufacture of so-called surface-ripened cheeses, such as Limburger, Muenster, and brick cheeses (Chapter 5). In Europe (and occasionally in the United States), manufacturers of dry fermented sausages add *Micrococcus* spp. to the meat mixtures. This organism produces the enzyme nitrate reductase which reduces the curing salt, sodium (or potassium) nitrate, to the nitrite form, and in so doing, enhances development of flavor and color attributes (Chapter 6). Finally, the manufacture of

vinegar involves oxidation of ethanol by *Acetobacter aceti* and related species (Chapter 11).

Yeast starter cultures

Bread manufacturers are by far the largest users of yeast starter cultures. Several different forms of yeast starter cultures are available, ranging from the moist yeast cakes, used exclusively by the baking industry, to the active dry yeast packages sold at retail to consumers. Yeasts used for the bread fermentation (i.e., bakers' yeast) are classified as *Saccharomyces cerevisae*, and are selected, in large part, on their ability to produce large amounts of carbon dioxide in rapid fashion. Wine, beer, distilled spirits, and nearly all other alcoholic beverages are also made using yeast cultures (Table 3-3).

Many of these ethanol-producing yeasts are also classified as *S. cerevisae*, although they differ markedly from those strains used for bread manufacture. As noted above, many of the wines produced in Europe are made via natural fermentations, relying on the naturally-occurring yeasts present on the grape surface and winery equipment. The trend for most wine manufacturers, however, has been to use

Table 3.3. Other organisms used as starter cultures.

Organism	Application
Bacteria	
Brevibacterium linens	Cheese: pigment, surface
Propionibacterium freudenreichii subsp. *shermanii*	Cheese: eyes in Swiss
Staphylococcus carnosus subsp. *carnosus*	Meat: acid, flavor, color
Mold	
Penicillium camemberti	Cheese: surface ripens white
Penicillium roqueforti	Cheese: blue veins, protease, lipase
Penicillium chrysogenum	Sausage
Aspergillus oryzae	Soy sauce, miso
Rhizopus microsporus subsp. *oligosporus*	Tempeh
Yeast	
Saccharomyces cerevisiae	Bread: carbon dioxide production
Saccharomyces cerevisiae	Ale beers
Saccharomyces pastorianus	Lager beers
Saccharomyces cerevisiae	Wine

starter culture yeasts. As with other starter cultures, strain selection is based primarily on the desired flavor and sensory attributes of the final product. However, other production traits, including the ability to flocculate, to grow at high sugar concentrations, and to produce adequate ethanol levels, are also relevant and are taken into consideration. The brewing industry is another potential user of yeast starter cultures, although large breweries generally maintain their own proprietary cultures. As for wine, strain selection is based on characteristics relevant to the needs of the particular brewer, and include factors such as flocculation, flavor development, and ethanol production rates.

Mold starter cultures

Despite the relatively few products made using fungi, manufacturers of these products often prefer to use mold starter cultures rather than to propagate their own cultures. Thus, fungal starter cultures are available for several types of cheeses, including both the blue-veined types (e.g., Roquefort and Gorgonzola), as well as the white surface mold-ripened cheeses (e.g., Brie and Camembert). The blue mold cultures consist of spore suspensions of *Penicillium roqueforti* and white mold cheese cultures contain *Penicillium camemberti*. Fungal cultures are also used in the production of many of the Asian or soy-derived fermented foods (Chapter 12). The main examples include tempeh, made using *Rhizopus microsporus subsp. oligosporus*, and miso and soy sauce, made using *Aspergillus sojae* and *Aspergillus oryzae*. There is also a small but important market for fungal starter cultures used to produce European-style sausages and hams.

Strain identification

Although it may seem obvious that the precise identities of the organisms present in a starter culture should be known, strain identification is not a trivial matter. In fact, microbial systematics is an evolving discipline that has been significantly influenced by advances in genetics, genomics, and various molecular techniques (Box 3-2). Thus, the movement or reassignment of a given organism from one species to another is not uncommon. In recent years, for example, strains of lactic acid bacteria, brewers' yeasts, and acetic acid-producing bacteria used in vinegar production have all been reclassified into different species or assigned new names altogether.

Box 3–2. Who's Who in Microbiology: Identifying the Organisms in Starter Cultures

In the early days of microbiology, Pasteur, Lister, and their contemporaries could view an organism microscopically, note its appearance, perhaps even describe a few physiological properties, and then assign a name to it (Krieg, 1988). Thus, classification was initially based on a limited number of phenotypic properties or traits. However, this situation changed in the early 1900s as more biochemical, physiological, and genetic characteristics became known and as objective methods for assigning particular organisms to specific groups were adopted. Thus was borne the science of microbial taxonomy and the development of classification schemes for bacteria.

Although classification and nomenclature are generally important for all biologists, how an organism is named and the taxonomical group into which it belongs are especially relevant for starter culture microbiologists. For many years, up until the 1970s, identification of lactic acid bacteria relied mainly on ecological, biochemical, and physiological traits (see below). The first comprehensive description of these bacteria was provided in 1919 by Orla-Jensen, and 20 years later the Sherman scheme was published (Orla-Jensen, 1919 and Sherman, 1937). As an aside, both of these historical reviews make for excellent reading; the Sherman paper was the first review article in the very first issue of the prestigious Bacteriological Reviews journal series. The Sherman scheme, which is still often used today, systematically placed streptococci into four distinct groups: pyogenic, viridan, lactic, and enterococci (Table 1).

Table 1. Sherman scheme for classification of *streptococci*[1].

Division	Representative species[2]	Growth at: 10°C	45°C	6.5% NaCl	Growth in presence of: methylene blue (0.1%)	pH 9.6	Survival at 60°C, 30 minutes	NH₃ from peptone
Pyogenic	*Streptococcus pyogenes*	−	−	−	−	−	−	+
Viridans	*Streptococcus thermophilus*	+	−	−	−	−	+	−
Lactic	*Lactococcus lactis* subsp. *lactis*	−	+	−	−	−	+	+
Enterococcus	*Enterococcus faecium*	+	+	+	+	+	+	+

[1]Adapted from Sherman, 1937
[2]Names reflect current nomenclature

These early papers also reflect the changing nature of bacterial taxonomy and nomenclature. In the Orla-Jensen treatise, for example, a description is given of the dairy lactic acid bacterium that was originally called *Bacterium lactis* (by Lister in 1878), then changed, successively to *Streptococcus acidi lactici, Bacterium lactis acidi, Bacterium Güntheri, Streptococcus lacticus*, and then *Streptococcus lactis*. In 1985, *Streptococcus lactis* became *Lactococcus lactis* subsp. *lactis* (Schleifer et al., 1985). Interestingly, Orla-Jensen noted that "it would be tempting to employ the name *Lactococcus*" for these bacteria (which was the genus name first used by the Dutch turn-of-the-century microbiologist, Beijerinck).

Although the initial classification schemes for lactic acid bacteria were based on phenotypic properties, serological reactions have also been used to differentiate streptococci. The Lancefield reactions (named after Rebecca Lancefield) were based on the antigenic properties of cell wall-associated components, resulting in more than thirteen distinct groups. The dairy streptococci (now *L. lactis* subsp. *lactis* and *L. lactis* subsp. *cremoris*) were found to possess the group

(Continued)

N antigen and became referred to as the Group N streptococci. Although the other major dairy streptococci, *Streptococcus thermophilus*, was not antigenic and could not be serologically grouped, the Lancefield groupings were and still are useful for classifying pathogenic streptococci, including *Streptococcus pyogenes* and *Streptococcus pneumonia* (Group A) and *Streptococcus agalactiae* (Group B), as well as enteric streptococci (Group D; now referred to as *Enterococcus*). However, one taxonomist suggested that "serology is best forgotten when working with dairy streptococci" (Garvie, 1984).

As noted above, phenotypic traits are still used successfully to classify lactic acid bacteria. The most informative distinguishing characteristics include: (1) temperature and pH ranges of growth; (2) tolerance to sodium chloride, methylene blue, and bile salts; (3) production of ammonia from arginine; and (4) carbohydrate fermentation patterns. Some of these are physiologically relevant in fermented foods (e.g., salt tolerance and growth at low or high temperature), whereas others are simply diagnostic (e.g., inhibition by methylene blue).

Based on these criteria (as well as microscopic morphology), it is, experimentally, rather easy to perform the relevant tests and to obtain a presumptive identification of a given lactic acid bacterium. Kits based on sugar fermentation profiles are also available that can be used to identify species of lactic acid bacteria. Specific classification schemes for lactic acid bacteria, based primarily on phenotypes, are described in *Bergey's Manual* (Holt et al., 1994). Other identification schemes that rely on membrane fatty acid composition, enzyme structure similarities, and other properties also exist, although they are now less often used.

Despite the value of traditional identification schemes, there is no doubt that the methods described above lack the power and precision of genome-based techniques. In fact, advances in nucleic acid-based bacterial fingerprinting methods have led, not only to new identification tools, but also to renewed interest in bacterial taxonomy. The ability to distinguish between strains of the same species is important not only for identification purposes, but also because it provides a basis for understanding the phylogenetic and evolutionary relationships between starter culture organisms. Although morphological, biochemical, and other phenotypic characteristics remain useful for genus and species identification, molecular approaches that rely on nucleotide sequences have proven to be more reliable, more reproducible, and more robust. Several techniques, in particular, are widely used, including pulsed field gel electrophoresis

Despite the numerous changes that have occurred in microbial classification and nomenclature (especially since the early 1980s), it is still essential that the microbial contents of a starter culture be accurately described and that species identification be based on the best available taxonomical information. First, culture suppliers need to include the correct species names on their products, since Generally Regarded As Safe (GRAS) status is affirmed only for specific organisms. A cheese culture claiming to contain species of *Streptococcus thermophilus*, but actually containing closely related strains of *Enterococcus*, would be mis-labeled. Second, accurate identification is necessary for the simple reason that

culture propagation and production processes require knowledge of the organism's nutritional and maintenance needs, which are species-dependent. The growth requirements of *Lactococcus lactis* subsp. *lactis* are different, albeit only slightly, from the closely related *Lactococcus lactis* subsp. *cremoris*. Finally, should the product manufacturer prefer to include species information on the label of a fermented product, the species name should be correct. (This would be voluntary, since it is not required.) For example, yogurt and other cultured dairy products that contain probiotic bacteria often include species information, since some consumers may actually look for and recognize the names of particular species.

Box 3–2. Who's Who in Microbiology: Identifying the Organisms in Starter Cultures *(Continued)*

(PFGE), restriction fragment length polymorphism (RFLP), ribotyping, 16s ribosomal RNA sequence analysis, and various PCR-based protocols. These precise methods of strain identification have also become important for commercial reasons because proprietary organisms, with unique characteristics and production capabilities, are developed and used in fermentation processes.

Finally, the reader should be aware, as authors of *Bergey's Manual* have warned, that bacteria do not really care into what group they are classified or what they are called—rather, classification is done for the sole benefit of microbiologists (Staley and Krieg, 1986). Therefore, there is a certain degree of arbitrary decision-making involved in the classification process, despite efforts by taxonomists to be totally objective. In fact, as these authors noted, in bold type, "There is no 'official' classification of bacteria." Names of bacteria can certainly be 'valid' (as per the List of Bacterial Names with Standing in Nomenclature, available online at www.bacterio.citc.fr), but the value of a given classification system depends entirely on its acceptance by the microbiology community.

References

Garvie, E.I. 1984. Taxonomy and identification of bacteria important in cheese and fermented dairy products, p. 35–66. *In* F.L. Davies and B.D. Law (ed.). *Advances in the Microbiology and Biochemistry of Cheese and Fermented Milk.* Elsevier Science Publishers. London.

Holt, J.G., N.R. Krieg, P.H.A. Sneath, J.T. Staley, and S.T. Williams. *1994. Bergey's Manual of Determinative Bacteriology. Ninth Edition.* Williams and Wilkins. Baltimore, Maryland.

Krieg, N.R., 1988. Bacterial classification: an overview. Can. J. Microbiol. 34:536–540.

Orla-Jensen, S. 1919. The lactic acid bacteria. Copenhagen.

Schleifer, K.H., J. Kraus, C. Dvorak, R. Kilpper-Bälz, M.D. Collins, and W. Fischer. 1985. Transfer of *Streptococcus lactis* and related streptococci to the genus *Lactococcus* gen. nov. System. Appl. Microbiol. 6:183–195.

Sherman, J.M. 1937. The streptococci. Bacteriol. Rev. 1:3–97.

Staley, J.T., and N.R. Krieg. 1986. Classification of procaryotic organisms: an overview, p. 965–968. *In* P.H.A. Sneath, N.S. Mair, M.E. Sharpe, and J.G. Holt (ed.). *Bergey's Manual of Systematic Bacteriology, Volume 2.* Williams and Wilkins. Baltimore, Maryland.

For such products, the name of the organism (and perhaps even the strain) is especially important because probiotic activity depends on the actual species or strain. Unfortunately, species declarations on consumer products are often incorrect or are outdated.

Starter Culture Math

Using starter cultures not only ensures a consistent and predictable fermentation, it also addresses a more fundamental problem, namely how to produce enough cells to accommodate the inocula demands of large-scale fermentations. For most fermented foods, the first requirement of a starter culture is that it initiate a fermentation promptly and rapidly. Although exceptions exists (e.g., in the case of the secondary, carbon dioxide-evolving fermentation that occurs late in the Swiss cheese process), it is usually necessary that the fermentation commence shortly after the culture is added. And while a short lag phase may certainly be acceptable or even expected, a long lag phase is generally a sign that the culture has suffered a loss of cell viability. Thus, for a starter culture to function effectively, it must contain a large number of viable microorganisms.

As the mass or volume of the food (or liquid) starting material increases, either larger starter culture volumes or greater starter culture cell concentrations are required. For example, if a 1% starter culture inoculum is ordinarily used for a given product, then 1 kg (or 1 L) of starter culture would be necessary to inoculate 100 kg (or 100 L) of the food substrate material. This modest-sized inoculum could easily be produced from a colony or stock culture, simply by one or two successive transfers through an intermediate culture (e.g., 0.1 ml or one colony into 10 ml, followed by 10 ml into 1 L).

When pure cultures, rather than backslopping techniques, were first introduced a century ago in the dairy industry, this approach of making intermediate and mother cultures was generally how cultures were prepared and used. As the size of the industry increased, however, such that larger and larger starter culture volumes were required, it was no longer feasible for cheese manufacturers to prepare cultures in this manner. In other words, a cheese plant receiving 1 million L of milk per day would need 10,000 L of culture, plus all of the intermediate cultures (and incubations) necessary to reach this volume.

To address this situation, two general types of cultures are manufactured and sold to the fermented food and beverage industries. The first type, often referred to in the dairy industry as bulk cultures, are used to inoculate a bulk tank. The bulk starter culture essentially is the equivalent of several intermediate cultures that traditionally have been required to build up the culture. After a suitable incubation period in the appropriate culture medium, the fully-grown bulk culture (which is akin to a mother culture) is then used to inoculate the raw material. The starter culture organisms comprising the bulk culture will remain viable for many hours, provided they are protected against acid damage, oxygen, hydrogen peroxide, or other inhibitory end-products. In the cheese and fermented dairy products industry, bulk cultures are routinely used to inoculate production vats throughout a manufacturing day. Maintaining culture viability is still an important issue, however, as will be discussed below.

A bulk culture is not warranted for many fermented foods simply because raw material volumes are more modest. That is, the amount of culture necessary to inoculate the fermentation substrate can easily be met using the culture as supplied directly from the manufacturer. For example, bakers' yeasts may be supplied as yeast cakes, which are added directly to the dough ingredients just prior to mixing. Similarly, meat starter cultures, whether in frozen can form or lyophilized packets, are added directly to the meat mixture. Even dairy cultures that are designed to be inoculated directly into the food material are now available, thus eliminating the need to prepare bulk cultures.

In the dairy industry, these cultures are referred to as direct-to-vat set cultures. They have several advantages that have made them very popular. They eliminate the labor, hardware, and capital costs associated with the construction, preparation, and maintenance of bulk starter culture systems. They also eliminate leftover or wasted bulk culture. Although they were initially produced as frozen concentrates, packaged in cans, they are now available as pourable pellets or lyophilized powders, making it easy to dispense the exact amount necessary for inoculation.

As described below, eliminating the bulk culture fermentation also means that bacteriophages have one less opportunity to infect the culture and cause trouble. The use of these cultures can also reduce mixed strain compositional variability. However, when direct-to-vat set cultures are used to inoculate large volumes, the culture must be highly concentrated to deliver a sufficient inoculum into the raw material. Freezing, centrifugation, filtration, lyophilization, and other concentration steps may indirectly reduce culture viability, ultimately leading to slow-starting fermentations. Improvements in concentration technologies have minimized some of these problems. Still, direct-to-vat cultures are generally more expensive than bulk cultures, and, despite their convenience, may not be economical for all operations.

Culture Composition

Mixed or undefined cultures

Mixed or undefined cultures were once the main type of culture used throughout the fermented foods industry. They typically contain blends of different organisms, representing several genera, species, or even strains. The actual identities of the organisms in a mixed culture are often not known, and the individual species may or may not have been characterized microbiologically or biochemically. Even the proportion of different organisms in a mixed culture is not necessarily constant from one product lot to another.

Nevertheless, undefined cultures are still used as starter cultures for many applications because they have a proven history of successful use. In Italy, the famous Parmigiano Reggiano cheese is made using a mixed culture obtained by overnight incubation of the whey collected from that day's cheese production. Moreover, depending on the product, mixed cultures may have a particular advantage. For example, mixed starter cultures are commonly used in the Netherlands for the manufacture of Gouda, Edam, and related cheese varieties. The fermentation part of the cheese-making process can be quite long, and it is not unusual for a given strain to be infected and subsequently killed by bacteriophages that inhabit the cheese plant environment (or that are present within the culture itself). However, given the diversity of lactic acid bacteria present in these mixed cultures, there are likely strains that are resistant to that particular bacteriophage and that can then complete the fermentation. In fact, it is well established that frequent exposure to different bacteriophages provides a natural and effective mechanism for ensuring that phage-resistant strains will be present in repeatedly propagated mixed cultures.

Despite their proven track record, undefined mixed cultures are not without problems. In particular, product quality may be inconsistent and fermentation rates may vary widely, affecting production schedules. For small-scale operations, where time is somewhat flexible and where quality variations may be more tolerable, these issues are not so serious. For large production facilities, however, where precise schedules are critical and consistent product quality is expected, mixed cultures have become less common. Starter cultures comprised of defined strains, with more precise biological and biochemical properties, are now prevalent. These cultures, referred to as defined cultures, simply refer to cultures that contain microbiologically characterized strains.

Defined cultures

Defined starter cultures can be comprised of a single individual strain or as a blend of two or more strains. The origin of the defined strains that are used commercially varies. Some were simply present in traditional mixed cultures and others have been isolated from natural sources. For example, in the case of dairy starters, milk production habitats and cheese serve as good sources; in the case of wine starters, grapes and wine-making equipment are good locations to find suitable yeasts. Defined strains must be identified and characterized for relevant metabolic and physiological properties, phage-resistance (in the case of dairy strains), and other desirable traits. For fungal starters, safety is an important issue and the inability of a putative culture strain to produce mycotoxins is an essential criteria. In addition, when multiple strains are assembled into a culture blend, all strains must be compatible. That is, one strain must not produce inhibitory agents (e.g., bacteriocins, hydrogen peroxide, or killer factors) that would affect growth of other organisms or that would cause one strain to dominate over others.

Ever since the defined strain concept for dairy starters was introduced in the 1970s, there has been debate about whether single, paired, or multiple strain blends are preferred (Box 3-3). Multiple strains are used for some applications, such as yogurt cultures, which

Box 3-3. Defined Strain Cultures: How Many Strains Are Enough?

Most large cheese factories in the United States rely on defined multiple strain starter cultures. These cultures, which typically contain up to five different strains, perform in a timely and predictable manner and generally yield products with consistent quality attributes. Perhaps most importantly, they provide the manufacturer with a reasonable level of assurance that phages will not be a problem, because if one of the strains was suddenly attacked and lysed, other strains in the culture would act as back-ups and would see the fermentation through to completion.

However popular the multiple strain approach has become, it is limited for several reasons. First, in the case of Cheddar cheese manufacture, there are not that many phage-unrelated strains of *Lactococcus lactis* subsp. *lactis* that have the appropriate cheese-making properties. There are even fewer distinct strains of *Lactococcus lactis* subsp. *cremoris*, which is generally considered to be the "better" cheese-making strain. Although it is relatively easy to isolate spontaneous phage-insensitive mutants (derived from good cheese-making strains), these derivative strains often revert to a phage-sensitive phenotype or else lose plasmids or acquire other mutations that make them unsuitable for cheese-making.

Second, it is not so easy to produce and then maintain multiple strain blends that contain the correct strains and in the correct ratio. In theory, strains present in the multiple strain culture are selected based on good cheese-making properties. However, in reality, there may still be variation in product quality using multiple culture blends, at least more so than would occur if only one or two strains were used. Also, whenever multiple strain starters are used, phage diversity within the production environment would likely increase, whereas when a single or paired strain starter is used, the types of phages that one would expect to see in the cheese plant would be limited.

Given the advantages of single or paired strain starters, why didn't such cultures quickly become the norm in the cheese industry? Indeed, the very idea of defined strain cultures (first conceived in the 1930s by researchers in New South Wales, and later employed by New Zealand researchers) was based on the use of phage-resistant, single-strain cultures (recounted by O'Toole, 2004). A simple test was also devised to provide a reliable basis for predicting whether a strain will remain resistant to bacteriophages during repeated cheese making trials (Figure 1; Heap and Lawrence, 1976).

Obviously, reliance on a single strain carries a perceived risk (despite the advantages noted above), because the appearance of an infective phage could quickly decimate the culture and ruin the fermentation. Even if the single strain was known to be highly phage resistant, it would be hard to imagine that the cheese plant manager would get much sleep. Indeed, in the United States and Europe, cheese manufacturers were initially quite reluctant to adopt single strain starters (Sandine in 1977 described this as a possibly "frightening" practice). However, research in the 1980s from several groups supported the earlier results from New Zealand and demonstrated that defined single or paired strains could be used successfully without phage infections (Hynd, 1976; Richardson et al., 1980; Thunell et al., 1984; Timmons et al., 1988).

Another way to assuage the worries of the sleepless cheese manager would be to use a strain genetically configured to resist bacteriophage infection (reviewed in Walker and Klaenhammer, 2003). Specifically, genes encoding for phage resistance could be introduced into the genome of the selected organism, conferring a bacteriophage-resistant phenotype. As described later in this chapter (and in Chapter 5), there are multiple mechanisms by which lactic acid bacteria can become phage-resistant. These strains can be constructed such that a single strain is stacked with several different phage-resistant systems and is therefore insensitive to a variety of phage. Alternatively, isogenic derivatives, each carrying a different phage resistance system, can be constructed and used in a rotation program.

Box 3-3. Defined Strain Cultures: How Many Strains Are Enough? *(Continued)*

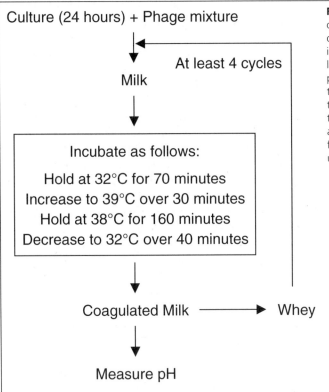

Culture (24 hours) + Phage mixture

At least 4 cycles

Milk

Incubate as follows:

Hold at 32°C for 70 minutes
Increase to 39°C over 30 minutes
Hold at 38°C for 160 minutes
Decrease to 32°C over 40 minutes

Coagulated Milk ⟶ Whey

Measure pH

Figure 1. Heap-Lawrence Test. The principle of this simple but highly useful test is based on the premise that strains capable of growing and producing acid after repeated challenges against bacteriophages have innate phage resistance. Furthermore, by performing the phage challenge in a laboratory situation that mimics cheese making, it is predicted that these strains will behave similarly in an actual cheese production environment. Therefore, strains that "pass" the test could be used in industrial cheese manufacture.

References

Heap, H.A., and R.C. Lawrence. 1976. The selection of starter strains for cheesemaking. N.Z.J. Dairy Sci. Tech. 11:16–20.

Hynd, J. 1976. The use of concentrated single strain cheese starters in Scotland. J. Soc. Dairy Technol. 29:39–45.

O'Toole, D.K. 2004. The origin of single strain starter culture usage for commercial Cheddar cheesemaking. Int. J. Dairy Technol. 57:53–55.

Richardson, G.H., G.L. Hong, and C.A. Ernstrom. 1980. Defined single strains of lactic streptococci in bulk culture for Cheddar and Monterey cheese manufacture. J. Dairy Sci. 63:1981–1986.

Sandine, W.E. 1977. New techniques in handling lactic cultures to enhance their performance. J. Dairy Sci. 60:822–828.

Thunell, R.K., F.W. Bodyfelt, and W.E. Sandine. 1984. Economic comparisons of Cheddar cheese manufactured with defined-strain and commercial mixed-strain cultures. J. Dairy Sci. 67:1061–1068.

Timmons, P., M. Hurley, F. Drinan, C. Daly, and T.M. Cogan. 1988. Development and use of a defined strain starter system for Cheddar cheese. J. Soc. Dairy Technol. 41:49–53.

Walker, S.A., and T.R. Klaenhammer. 2003. The genetics of phage resistance in *Lactococcus lactis. In* B.J.B. Wood and P.J. Warner (ed.), p. 291—315. *The Lactic Acid Bacteria, Volume 1, Genetics of Lactic Acid Bacteria.* Kluwer Academic/Plenum Publishers. New York, New York.

require the presence of two different organisms, *Streptococcus thermophilus* and *Lactobacillus delbrueckii* subsp. *bulgaricus*. Sour cream cultures, likewise, contain acid-producing lactococci and flavor-producing *Leuconostoc* sp. Many sausage cultures similarly contain species of *Lactobacillus* and *Pediococcus*. However, even for products requiring only a single organism (e.g., *Lactococcus lactis* used for cheddar cheese manufacture), paired or multiple strains are often desired.

The rationale for this preference is based mainly on bacteriophage concerns. Multiple strain starters are blended such that they contain a broad spectrum of strains with dissimilar phage sensitivity patterns. That is, each strain is resistant to different phage types, so that if one strain is infected, the other strains can complete the fermentation. This approach is reminiscent of the natural mixed strain cultures described above, except that the defined strain blends behave in a more predictable and consistent manner. Identifying which strains have become sensitive to the indigenous phages in that particular plant requires regular monitoring of phage levels in the cheese whey, and simple in-plant tests have been developed for this purpose (along with more sophisticated titer assays conducted by the culture supplier laboratories). When phage titers reach some critical level or a particular strain begins to show sensitivity, that strain is removed from the blend and replaced with a new phage-resistant strain or derivative. Multiple strain cultures can also be rotated in a particular order to achieve an even greater level of security (see below).

Manufacture of Starter Cultures

Most starter cultures, including those produced for the dairy, meat, baking, and other fermented foods industries, are mass-produced in modern, large-volume fermenters. The manufacturing process actually begins with a single colony isolate or stock culture, which is then grown in a small volume prior to inoculation into the production fermentor (Figure 3–1).

These fermentors operate under aseptic conditions, not unlike those used in the pharmaceutical industry for production of biomedical products. They almost always operate in a batch mode, because continuous cell production systems have not yet been adopted by the industry. The size of the fermentors may vary, from as little as 10 liters to several thousand liters. However, the basic operational parameters and control features are essentially the same for both small and large batches.

The choice of media also varies, depending on the organisms being grown and the nutrient requirements necessary for optimum cell production. For example, dairy starter cultures are frequently produced using milk- or whey-based media, whereas molasses or corn syrup can be used as the basal medium for other lactic cultures. Specialized nutrients are also often added to the medium. Water-soluble vitamins, such as the B vitamins, are required for optimum growth of lactococci and lactobacilli, and some species of *Streptococcus*, *Leuconostoc*, and *Lactobacillus* also require specific amino acids. Other compounds, such as the surfactant Tween 80, are added to the growth medium to promote membrane stability of lactobacilli and other LAB during subsequent frozen storage and lyophilization.

To achieve high cell density and maximize biomass production and cell viability, it is important that inhibitory metabolic end-products be removed or neutralized. In particular, lactic acid can decrease the medium pH to a level low enough to significantly inhibit cell growth. Although the optimum pH for cell growth depends on the specific organism (e.g., lactobacilli prefer a slightly more acidic medium pH than lactococci), lactic acid bacteria, in general, grow best within a pH range of 5 to 6. Therefore, neutralization of the accumulated acid via addition of alkali, usually gaseous NH_3, NH_4OH, Na_2CO_3, or KOH, is essential. Similarly, hydrogen peroxide can also accumulate to inhibitory levels during culture production as a result of peroxide-forming oxidation-reduction reactions. Hydrogen peroxide formation can be reduced or removed by minimizing incor-

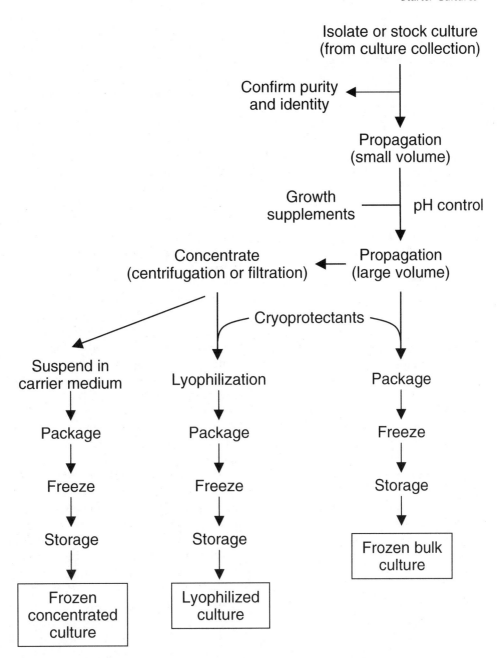

Figure 3–1. Production of industrial starter cultures. Adapted from Buckenhüskes, 1993.

poration of oxygen and by adding catalase directly to the fermentor during cell production.

Provided that optimum growth conditions are provided, as described above, maximum cell densities of 10^9 to 10^{10} cells per ml are typically obtained. In general, lactic acid bacteria are harvested either at late log phase or early stationary phase. However, the optimum harvest time depends on the specific organism. At this point, the cells can be directly packaged as frozen liquid cultures, concentrated and then frozen, or lyophilized (freeze-dried).

Frozen liquid cultures are dispensed into metal cans or plastic containers with volumes ranging from 100 ml to 500 ml, then rapidly frozen in liquid nitrogen ($<-196°C$). This is the most common form for lactic acid bacterial cultures, and it is how bulk dairy starters and meat cultures are usually packaged. Concentrated frozen cultures, in contrast, are used for direct-to-vat set cheese applications. For both types of frozen cultures, it is critically important that the cultures remain frozen ($<-45°C$) and that temperature fluctuations be avoided during transport and during storage at the manufacturing site. This is because freeze-thaw cycles result in the formation of ice crystals, which can then puncture and kill cells, reducing cell number and viability when the cells are thawed and used.

To produce concentrated frozen cultures, cells are first grown to high cell densities (as described above) and then concentrated either by continuous centrifugation or cross-flow membrane filtration. The cells may then be washed to remove the spent medium and re-suspended in a suspension solution containing stabilizing agents (see below). The concentrated cells are then packaged into cans (usually 300 g to 500 g) at cell densities as high as 10^{11} to 10^{12} cells per gram and are rapidly frozen. Thus, the concentrated culture contains about the equivalent amount of cells that would normally be present in a fully grown bulk culture. Typically, a single 500 g can is sufficient to inoculate 5,000 kg of milk. Another way to freeze concentrate cells is via a pellet technology, whereby concentrated cells are dropped directly into liquid nitrogen. The cells freeze instantaneously in the form of pellets, which can be collected and dispensed into paperboard containers or foil or plastic pouches. Provided they are maintained in a frozen state, the pellets are pourable and are easy to weigh out and dispense.

Finally, concentrated cells can also be obtained by one of several different dehydration processes. Spray drying, for example, can be very effective for some biological materials; however, it is inapplicable for lactic acid bacteria, due to substantial loss in cell number and viability. In contrast, lyophilization, or freeze drying, has long been known to be a more gentle process for dehydrating and preserving starter culture lactic acid bacteria. Although some cell inactivation or injury may still occur, lyophilized cells are generally more stable than cells concentrated by other means.

The lyophilization process involves removal of water from a frozen material by applying sufficient vacuum (<1.0 Torr) such that sublimation of ice occurs. For lyophilization of cultures, cells are first grown to high cell density, harvested, and concentrated, as described above, and then subjected to the freeze-drying process (usually about 5 Pascal to 20 Pascal for eighteen to thirty hours). Freeze-dried cultures can contain from 10^9 to 10^{12} cells/gram. The cells are usually packaged in foil pouches or other oxygen-impermeable material. Although lyophilized cells are best maintained at $-20°C$, they show good stability even at refrigeration temperature. Lyophilized cultures are now as popular as frozen direct-to-vat set starter cultures, especially for yogurt and other cultured milk products.

On occasion, growth of both frozen and lyophilized cells may be sluggish, with a prolonged lag phase, after inoculation into the food substrate. To maintain cell viability, various cryoprotectant agents are usually added to protect the cells against freeze and stress damage. Depending on the cell species, cryoprotectant agents include glycerol, lactose, sucrose, trehalose, ascorbate, and glutamate. Moreover, some microorganisms simply do not lyophilize well, and suffer a more serious loss in cell number and viability. For example, there are several reports indicating that dairy lactobacilli, including *L. delbrueckii* susbp. *bulgaricus, L. helveticus,* and *L. acidophilus*, may be particularly sensitive to lyophilization, and that special efforts must be made to ensure that these organisms remain viable (Box 3-4). This is important because one or more of these organisms are included in commercial

Box 3-4. Improving Cell Viability During Freezing and Lyophilization Processes

Although lyophilization is generally considered as one of the more gentle means for long-term preservation of starter cultures, the initial freezing and subsequent sublimation steps do take a toll on microorganisms (Carvalho et al., 2004). Freezing, in particular, is detrimental to cells for several reasons. First, freezing causes ice crystals that can damage membranes and puncture cells. As water turns to ice, water activity decreases and the intracellular solutes become concentrated, both of which cause osmotic stress. Within the lactic acid bacteria, lactobacilli are especially sensitive to the freezing and lyophilization process (Monnet et al., 2003). In part, this may be due to geometry—large rods have more surface area than small spherical cells, and are more prone to crystal-induced damage. The drying step also influences survival and viability, because cells are prone to collapse as vacuum is applied and as the temperature increases in the freeze drying chamber (Fonseca et al., 2004).

Growth conditions

Several factors must be considered to achieve high survival rates and to enhance cell viability during freezing and lyophilization. The first is how the cells are grown (Table 1; Carvalho et al., 2004). It is known that many bacteria, including lactic acid bacteria, when exposed to a decrease in temperature, alter their membrane lipid composition by synthesizing and incorporating specific fatty acids into the membrane (Béal et al., 2001; Murga et al., 2003; Wang et al., 2005). This change in membrane composition provides fluidity at low temperature and helps to minimize membrane damage. Therefore, lower, sub-optimal growth temperatures (e.g., 25°C) may enhance resistance to subsequent freezing conditions (Murga et al., 2000). The addition of Tween 80 (an oleate ester of sorbitan polyethoxylate) to the growth medium also may promote this shift in fatty acid synthesis and similarly improve stability during freezing (Goldberg and Eschar, 1977).

Table 1. Factors affecting cell survival and viability during lyophilization.

Growth conditions
Osmoprotectants
Cryoprotectants
Freezing
Drying medium
Storage
Rehydration

Another means by which cells protect themselves against freezing and dehydration is by accumulating cryoprotectant and osmoprotectant molecules. These compounds, also referred to as compatible solutes, can reach high intracellular concentrations (above 1 M) and function, in part, by preventing denaturation of macromolecules within the cytoplasm and cell membrane. They also lower intracellular water activity and prevent water loss and eventual plasmolysis.

The most common protectant molecules used by lactic acid bacteria are betaine, carnitine, and proline. They are usually directly transported into the cell, but in some organisms, they can be synthesized from precursor molecules. Including these substances in the growth medium allows the bacteria to stock up so that when freezing occurs, they will be loaded with protectant molecules. Importantly, compatible solutes accumulate in response to shock conditions other than freezing and dehydration, so a quick heat shock may also activate their uptake. Although glycerol is also known to enhance cryotolerance, it is not clear that this compound is actively accumulated in lactic acid bacteria (Fonseca et al., 2001).

(Continued)

Sub-lethal stress

In addition to activating transport systems for compatible solutes, cold shock, heat shock, and osmotic shock may induce expression of genes involved in the general stress response system in lactic acid bacteria. The genetic response to these (and other) shock conditions is mediated by specific sigma factors that regulate transcription of genes whose protein products are necessary for defending the cell against stress damage or death.

The function of these so-called stress proteins varies, but, in general, involves repair or degradation of misfolded proteins and stabilization of macromolecules and membranes. Thus, greater cell survival and stability rates might be achieved during lyophilization by prior induction of the stress-response system.

This approach might be counter-intuitive in at least one respect. It is well established, for example, that rapid freezing is better than slow freezing, since the latter generates large ice crystal that are very damaging to cells. However, cells of *L. acidophilus*, adapted to low temperature either by slow cooling or by exposure to a pre-freezing stress, were reported to have greater survival rates compared to rapidly frozen cells, presumably because the stress-response system had been induced in the adapted cells (Bâati et al., 2000). Induction of a cold shock response and identification of cold shock genes were also reported for other lactic acid bacteria, although cryotolerance was strain-dependent (Broadbent and Lin, 1999; Kim and Dunn, 1997, Wouters et al., 1999).

Freeze-drying medium

After cells are grown and harvested, they are usually resuspended in a suitable medium prior to freezing and lyophilization. For dairy applications, the basal medium is typically milk-based. However, because the suspension medium will remain a component of the final cell mixture, it should not only be optimized to maintain viability during freeze-drying, the medium should also contain components that enhance survival and preservation while the cells are in the dehydrated state. Among the compounds that appear to be most effective are various sugars and sugar alcohols, including lactose, mannose, glucose, trehalose, and sorbitol, as well as amino acids and phenolic antioxidants (Carvalho et al., 2003). The latter may be especially important, because exposure to oxygen or air is known to reduce viability of dehydrated cells. Membrane components, in particular, are very sensitive to oxidation reactions (Castro et al., 1996).

Storage conditions

Although lyophilized cells are generally considered to be quite stable, they are sensitive to environmental abuses that may occur during storage. High storage temperatures (above 20°C) and temperature fluctuations significantly reduce survival rates and cell viability. Aerobiosis, as noted above, may cause detrimental oxidation reactions. Likewise, high relative humidity conditions increase the water activity of the cell material and promote undesirable lipid oxidation reactions (Castro et al., 1995).

Rehydration

The conditions by which lyophilized cells are rehydrated can have a significant influence on the viability of the culture. In general, the best suspension medium approximates the medium in which the cells were originally dehydrated. Suspension media that are hypotonic or that are too warm or cold should be avoided (Carvalho et al., 2004).

Cryotolerant mutants

As described above, there is both a genetic as well as a physiological basis for cryotolerance (the ability to remain viable during freezing). Genes that encode for compatible solute uptake systems and for shock-response systems have been identified in several lactic acid bacteria

Box 3-4. Improving Cell Viability During Freezing and Lyophilization Processes *(Continued)*

(Kim and Dunn, 1997; Wouters et al., 1999, 2000b). It has also been reported that differences in cryotolerance exists even between different strains of the same species. These observations suggest that it may be possible to identify mutants that have enhanced crytolerance (Monnet et al., 2003). Indeed, when genes encoding for known cold-shock proteins were overexpressed in a strain of *Lactococcus lactis*, five- to ten-fold increases in survival rates were observed after cells were exposed to multiple freeze-thaw cycles (Wouters et al., 2000a).

In another study, spontaneous mutants were obtained by passing *Lactobacillus delbrueckii* subsp. *bulgaricus* cells through thirty growth and freeze-thaw cycles (Monnet et al., 2003). In theory, these treatments would have selected for cryotolerant mutants, although there were subpopulations with varying cryotolerant phenotypes. Although it is not clear what the nature of the mutation(s) may have been that conferred enhanced crytolerance, this approach appears to be a useful means to generate natural cryotolerant strains.

References

Bâati, L., C. Fabre-Gea, D. Auriol, and P.J. Blanc. 2000. Study of the crytolerance of *Lactobacillus acidophilus*: effect of culture and freezing conditions on the viability and cellular protein levels. Int J. Food Microbiol. 59:241–247.

Béal, C., F. Fonseca, and G. Corrieu. 2001. Resistance to freezing and frozen storage of *Streptococcus thermophilus* is related to membrane fatty acid composition. J. Dairy Sci. 84:2347–2356.

Broadbent, J.R., and C. Lin. 1999. Effect of heat shock or cold shock treatment on the resistance of *Lactococcus lactis* to freezing and lyophilization. Cryobiology 39:88–102.

Carvalho, A.S., J. Silva, P. Ho, P. Teixeira, F.X. Malcata, and P. Gibbs. 2003. Protective effect of sorbitol and monosodium glutamate during storage of freeze-dried lactic acid bacteria. Le lait. 83:203–210.

Carvalho, A.S., J. Silva, P. Ho, P. Teixeira, F.X. Malcata, and P. Gibbs. 2004. Relevant factors for the preparation of freeze-dried lactic acid bacteria. Int. Dairy J. 14:835–846.

Castro, H.P., P.M. Teixeira, and R. Kirby. 1995. Storage of lyophilized cultures of *Lactobacillus bulgaricus* under different relative humidities and atmospheres. Appl. Microbiol. Biotechnol. 44:172–176.

Castro, H.P., P.M. Teixeira, and R. Kirby. 1996. Changes in the cell membrane of *Lactobacillus bulgaricus* during storage following freeze-drying. Biotechnol. Lett. 18:99–104.

Fonseca, F., C. Béal, and G. Corrieu. 2001. Operating conditions that affect the resistance of lactic acid bacteria to freezing and frozen storage. Cryobiology 43:189–198.

Fonseca, F., S. Passot, O. Cunin, and M. Marin. 2004. Collapse temperature of freeze-dried *Lactobacillus bulgaricus* suspensions and protective media. Biotechnol. Prog. 20:229–238.

Kim, W.S., and N.W. Dunn. 1997. Identification of a cold shock gene in lactic acid bacteria and the effect of cold shock on cryotolerance. Curr. Microbiol. 35:59–63.

Monnet, C., C. Béal, and G. Corrieu. 2003. Improvement of the resistance of *Lactobacillus delbrueckii* ssp. *bulgaricus* to freezing by natural selection. J. Dairy Sci. 86:3048–3053.

Murga, M.L.F., G.M. Cabrera, G.F. de Valdez, A. Disalvo, and A.M. Seldes. 2000. Influence of growth temperature on cryotolerance and lipid composition of *Lactobacillus acidophilus*. J. Appl. Microbiol. 88:342–348.

Wang, Y., G. Corrieu, and C. Béal. 2005. Fermentation pH and temperature influence the cryotolerance of *Lactobacillus acidophilus* RD758. J. Dairy Sci. 88:21–29.

Wouters, J.A., B. Jeynov, F.M. Rombouts, W.M. de Vos, O.P. Kuipers, and T. Abee. 1999. Analysis of the role of 7 kDa cold-shock proteins of *Lactococcus lactis* MG1363 in cryoprotection. Microbiol. 145:3185–3194.

Wouters, J.A., F.M. Rombouts, O.P. Kuipers, W.M. de Vos, and T. Abee. 2000a. The role of cold-shock proteins in low-temperature adaptation of food-related bacteria. System. Appl. Microbiol. 23:165–173.

Wouters, J.A., M. Mailhes, O.P. Kuipers, W.M. de Vos, and T. Abee. 2000b. Physiological and regulatory effects of controlled overproduction of five cold shock proteins of *Lactococcus lactis* MG1363. Appl. Environ. Microbiol. 66:3756–3763.

yogurt and other dairy cultures, which are increasingly produced in lyophilized forms.

Evaluating Culture Performance

Simply stated, the job of the starter culture is to carry out the desired fermentation, promptly and consistently, and to generate the expected flavor and texture properties relevant to the specific food product. The particular requirements for a given strain, however, depend entirely on the application for that culture (Table 3–1). As customers have come to expect particular and oftentimes exacting product specifications, the specific traits and properties expressed by culture organisms have also become quite demanding. Thus, strains of lactic acid bacteria used as dairy starter cultures are now selected based not just on lactic acid production rates, but also on flavor- and texture-producing properties, salt sensitivity, compatibility with other strains, phage resistance, and durability during production and storage.

For example, strains of *S. thermophilus* and *L. delbrueckii* subsp. *bulgaricus* used as yogurt starter cultures should have, at minimum, rapid acid development rates and good phage resistance, but other factors are just as important. That is, how much acetaldehyde the culture produces, whether or not it synthesizes exopolysaccharides, and to what extent it contributes to post-fermentation acidity are all factors that influence strain selection. Moreover, strains of these bacteria that make them suitable as thermophilic starter cultures for yogurt manufacture does not necessarily mean they will be useful as starters for Mozzarella or Swiss cheese manufacture, since the requirements for the latter products are different from those of yogurt. Likewise, traits desirable for brewing strains of *Saccharomyces cerevisiae* (e.g., flocculation and ethanol tolerance) are quite different from those important for bakers' yeasts strains of *S. cerevisiae*.

Several tests are routinely performed to assess functional properties of starter cultures. For lactic acid bacteria, acid-production rates are relevant, whereas for yeast cultures, rates of CO$_2$ evolution (or gassing rates) may be appropriate. Fermentation performance for dairy cultures, for example, can be easily determined by inoculating heat-treated milk with a standardized inoculum (e.g., 0.1% to 1.0%), incubating the material at an appropriate temperature, and then either measuring the decrease in pH as a function of time or simply by determining the terminal pH or titratable acidity after a specific incubation period (e.g., 6 or 18 hours).

For some cultures, especially frozen or lyophilized products, lag times may be important and can also be estimated from growth curve data. Similar tests are performed for meat starter cultures, whose main function is also to produce lactic acid.

Another critical test used specifically for dairy cultures involves the determination of bacteriophage resistance. In one specific procedure, the Heap-Lawrence challenge test, conditions that simulate cheese making are used (Box 3–3). Cells are inoculated into culture tubes of milk, a representative bank or factory phage mixture is added, and the tubes are incubated according to a typical cheese production time-temperature profile. If the pH at the end of the incubation cycle has reached the expected target value (as easily judged by milk coagulation), then it is assumed that the culture was not affected by the added phage. The entire process is then repeated up to six times (adding a portion of the whey from the previous cycle). The absence of a sufficient pH decrease for two successive cycles indicates that the culture has become sensitive to bacteriophages and should be removed from use.

Similar to lactic cultures, yeast cultures used for wine, beer, and bread should also ferment rapidly, have good sensory properties, and be stable during storage. Depending on the product, these yeasts must also possess other specific properties. For example, a critical physiological characteristic of wine and beer yeasts is their ability to flocculate or sediment at the end of the fermentation. Cells that flocculate will sediment much faster relative to free cells. Sedimentation of cells enhances their separation from the wine or beer and reduces autolysis and subsequent release of intracellular constituents. Importantly, flocculation is an inheritable trait, meaning that the flocculation phenotype is subject to genetic control and can be lost or acquired. Identification of several flocculation genes that encode for proteins involved in cell aggregation has made it possible to improve brewing yeast strains (Chapter 9).

Other performance characteristics of yeast starters also depend on the specific product. Lager beer yeasts, for example, should be able to grow at low temperature and produce flavorful end-products. The ability of brewing yeasts to ferment all available sugars (at least those that are fermentable), a property known as attenuation, is a critical property. In wine manufacture, wine yeasts are also often selected based on growth at low temperatures (10°C to 14°C), as well as resistance to sulfiting agents. For some applications, wine yeast should also have high ethanol tolerance and osmotolerance. Finally, the main requirements for bakers' yeasts are to have a short lag phase and to produce carbon dioxide rapidly in the dough (gassing rate). For some applications, such as frozen doughs, the yeasts should also have good cryotolerance, the ability to withstand freezing conditions, so that the frozen dough will rise after thawing.

Compatibility issues

One other absolute requirement for all multiple-strain starter cultures is that each organism must be compatible with all the other organisms within a particular culture. For example, lactic acid bacteria are well-known producers of bacteriocins, peroxides, and other antimicrobial substances that can inhibit other organisms (see below). Even individual strains of a single species may inhibit other strains of that same species. If a multiple strain *L. lactis* culture was assembled without regard to compatibility issues, it would be entirely possible that one strain may produce a bacteriocin that was inhibitory to all other *L. lactis* strains in that culture. In much the same manner, some wine and beer yeasts produce inhibitory substances called

"killer" factors that inhibit other yeast, therefore influencing how mixed strain cultures are prepared. Compatibility also extends to the growth characteristics of the organisms in a given culture. Organisms that produce acids and lower the pH quickly may prevent growth of their culture partners, a situation encountered in yogurt products containing probiotic adjuncts.

How Starter Cultures Are Used

There are two general ways that starter cultures are used. As described above, frozen or lyophilized starter cultures can simply be inoculated directly into the food substrate (i.e., direct-to-vat). This is the normal means by which active bakers' yeast cultures, fermented meat starter cultures, and many dairy cultures are added to the starting food material. In the case of frozen cultures, the cans are first thawed in cold (and usually chlorinated) water immediately prior to use. Frozen pellets can be added directly. Lyophilized cultures can also be added directly, but may require additional stirring or mixing in the vat to facilitate hydration.

Bulk cultures

In contrast to the direct-to-vat type cultures, some fermented foods are inoculated with bulk cultures (Figure 3-2). Bulk cultures are produced to increase the size of the starter culture (in terms of total cell number and volume) by several orders of magnitude (2,000 L to 4,000 L is about an average size tank).

The dairy industry is by far the main user of bulk cultures (although the liquid cream yeasts used by large bread manufacturers may be considered bulk cultures). When bulk dairy cultures are used, several additional steps are necessary to ensure that the organisms have reached high cell densities and that they are active and viable. First, specialized culture media, formulated to provide optimum growth conditions, are almost always used for bulk starter preparation. Bulk culture media may also contain phosphate buffers and other agents that protect the cells from acid damage

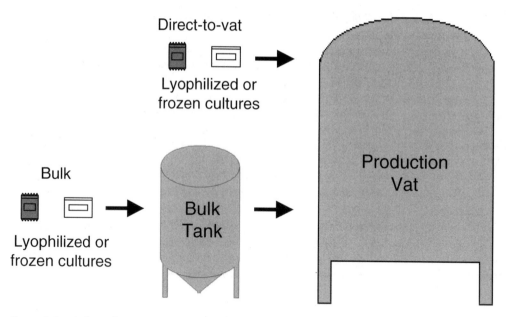

Figure 3–2.   Bulk vs. direct-to-vat culture preparation.

and from infection by lytic bacteriophages (see below).

In general, bulk culture media are not that different from the fermentation media used by culture manufacturers for mass production of starter culture cells. They contain a basal medium consisting of a fermentable carbohydrate (usually lactose, but glucose or sucrose can also be used) and a nitrogen source (usually proteins derived from milk or whey). Culture media are also supplemented with additional sources of vitamins, minerals, and other nutrients. Yeast extract and corn steep liquor are excellent sources of many of these materials.

Buffering activity is usually supplied in the form of carbonate, citrate, or phosphate salts. The latter also provides a phage-inhibitory function via calcium chelation (see below). In some cases, lactose concentrations may be lowered to limit acid production and reduce the risk of acid injury to the cells (see below). All ingredients used in the bulk culture medium must be food grade.

Growth and activity of the starter culture are also influenced by the incubation conditions. In general, the incubation temperature for a given culture is established based on optimum culture performance, and not necessarily on optimum growth rates. This is because most cultures usually contain more than one strain, and often more than one species. Thus, the priority is maintaining the desired balance between multiple strains.

For example, Mozzarella cheese cultures contain a mixture of two thermophilic lactic acid bacteria, *S. thermophilus* and *Lactobacillus helveticus*. The final ratio of these organisms in the bulk culture tank depends primarily on the incubation temperature. If the culture is incubated at 35°C to 41°C, growth of *S. thermophilus* is favored, whereas incubation at 42°C to 46°C favors *L. helveticus*.

These organisms also differ physiologically, especially with respect to sugar and protein metabolism, such that the strain ratio in the culture will have a major impact on the functional properties of the finished cheese. Higher

S. thermophilus levels generally yield a Mozzarella cheese having high residual lactose and galactose concentrations and lower levels of protein hydrolysis. By adjusting the incubation temperature the manufacturer can control these compositional characteristics, which ultimately determine how the cheese will perform when used in pizza manufacture (Chapter 5).

Multiple strain mesophilic cultures are also used for the manufacture of sour cream and cultured buttermilk. These cultures contain two general types of organisms, flavor-producers and acid-producers. The flavor-producing organisms consist of one or more strains of citrate-fermenting, diacetyl-forming, heterofermentative *Leuconostoc mesenteroides* subsp. *cremoris*. Their temperature optimum is about 18°C to 25°C. The acid-producers include homofermentative *L. lactis* subsp. *lactis*, *L. lactis* subsp. *cremoris*, and *L. lactis* subsp. *diacetilactis* (that has both acid-forming and flavor-forming ability). The latter group have an optimum growth temperature of 28°C to 32°C. If a greater ratio of flavor-to-acid producers is desired in the finished culture, low incubation temperatures (<25°C) should be used, whereas a high incubation temperature (>25°C) favor the acid producers.

Controlling pH

Growth of lactic acid bacteria in milk or whey is accompanied by a marked decrease in pH. Although lactic acid bacteria are copious producers of lactic acid, most species are, in fact, true neutraphiles and grow best at pH levels above 6.0. Cells held at pH 5.0 or below for even a modest period of time will lose viability and behave sluggishly when inoculated into the starting material. Therefore, preventing acid injury or damage during the production and propagation of starter cultures is essential.

In general, there are two ways to minimize or prevent acid damage to lactic starter cultures. Both involve controlling the pH as the culture grows in the bulk culture tank. One approach relies on the addition of alkaline solutions into the bulk culture tank to neutralize the acid produced during fermentative growth. Neutralizing agents may include sodium or ammonium hydroxide or ammonia gas. In actual practice, these so-called external pH control systems consist of a tank fitted with a pH electrode and pH monitoring device (some systems may control multiple bulk tanks). When the pH has decreased below a critical pre-set threshold, a signal is sent to a pump that then adds neutralizing agent to the tank until an upper pH limit is reached. As cells grow and produce acid, the pH never falls below the set threshold. The optimum pH range depends on the specific organisms; most external control systems for mesophilic dairy starter bacteria maintain a pH between 5.8 and 6.2. These systems yield bulk starter cultures with cell densities as much as ten times greater than uncontrolled cultures, and with enhanced viability. Although lactate salts may accumulate as a by-product, resulting in higher media osmolality and ionic strength, they do not have significant effects on cell viability.

It is also possible to neutralize the bulk culture tank via a manual one- or two-time addition of neutralizing agent (sodium or potassium hydroxide) to the bulk culture tank. This approach reduces some of the hardware expenses (pH electrodes, pumps, pH monitoring devices) associated with automated external pH control systems. Bulk cultures, produced under pH control conditions, may subsequently be cooled (or not) and used for up to several days.

An alternative approach to the external pH control systems is to incorporate buffer salts directly into the bulk culture medium. These so-called internal pH control systems have one major advantage—they do not require the purchase of the expensive external pH control hardware. The development of internal pH control media (in the late 1970s), however, was not quite so simple.

Although phosphate or citrate salts are common ingredients in culture medium, it is not possible to add enough of these salts to provide the buffering power necessary to maintain a near-neutral pH during growth by LAB. This is because high phosphate concentrations can inhibit LAB used as starter cultures. Thus, the buffer agents must be incorporated into the medium in either an insoluble or encapsulated form. Internal pH control buffers now include sodium carbonate and trimagnesium phosphate in forms that are solubilized and released into the medium as a function of low pH, such that the pH of the culture medium is maintained above 5.0. It is necessary to provide some agitation to prevent settling of the buffer agents during bulk culture growth. Compared to the external pH control systems, internal pH control media provides similar advantages. Since higher cell densities are achieved, and cell viability is enhanced, less culture is needed. In addition, because the cells are maintained in a viable state, they can be used for a longer period of time.

Bacteriophages and Their Control

As noted earlier, problems caused by bacteriophages have been the driving force for much of the research on starter cultures. Bacteriophages capable of infecting the culture bacteria will likely be present in open environments, where lactic fermentations generally occur. Their ability to multiply and spread rapidly (a single replication cycle can be as short as 30 minutes) in a cheese or yogurt factory may lead to decimation of the starter culture.

During the production of bulk cultures, bacteriophage problems have become relatively rare for several reasons. First, the culture media is ordinarily prepared by heating to pasteurization temperatures high enough to inactivate bacteriophages. Also, most modern manufacturing facilities employ aseptic conditions within the culture preparation areas, including air handling systems to prevent entry of airborne phages. Finally, closed, sterilizable tanks are usually used for culture preparation.

In contrast, cheese manufacturing conditions are not always so restrictive and opportunities for phage infection commonly exist. For

example, milk used for cheese manufacture is often unpasteurized or may receive only a relatively modest heat treatment. Moreover, the cheesemaking process, in general, is often conducted in an open manner. Thus, the milk and curds are readily accessible to air- or whey-borne phages. Furthermore, in high-throughput cheese production factories, where vats are filled and re-filled several times per day, there is ample time for phages that may initially be present at low levels to propagate and reach populations high enough to cause slow or arrested fermentations and poor quality product.

Ultimately, the lactic starter culture and the phages that infect them are in a sort of race— can the culture do its job (i.e., bring about a successful fermentation) before the phage causes culture inhibition? In general, for Cheddar-type cheeses, it only takes about 3 to 3.5 hours for the fermentation to be completed (the so-called set-to-salt time). However, bearing in mind that a single phage can infect a host cell and replicate within 30 minutes, releasing fifty or more new phage particles into the environment, it may not take long for the phage population to exceed that of the culture (Table 3–4). Although this exact scenario in which the fermentation is actually halted prematurely by infectious bacteriophages may be uncommon, there are more frequent occasions when phage infections either cause production delays or downgrading of cheese quality, both of which are costly to the manufacturer.

For many years, lactic starter cultures comprised of mesophilic lactococci were the "workhorse" cultures. They were used in the manufacture of the most popular cheeses and cultured dairy products, including Cheddar and Cheddar-like cheeses, mold-ripened cheeses, Dutch-type cheeses, cottage cheese, cultured buttermilk, and sour cream (all described, in detail, in Chapters 4 and 5). Thus, bacteriophages that attacked lactococci, especially *L. lactis* susbsp. *lactis*, were really the only concern. Reports of phage problems occurring for products made using other lactic acid bacterial cultures were relatively rare.

Table 3.4. Dynamics of a phage infection[1].

Time (h:m)	Phage/ml	Cells/ml
0	1	1,000,000
0:40	50	$2,000,000 - 50 = 1,999,950$
1:20	2,500	$3,999,900 - 2,500 = 3,997,400$
2:00	1.25×10^5	$7,994,800 - 1.25 \times 10^5 = 7,869,800$
2:40	6.25×10^6	$15,739,600 - 6.25 \times 10^6 = 9,489,600$
3:20	3.13×10^8	$18,979,200 - 3.13 \times 10^8 = \text{less than } 1$

[1]Given the following assumptions:
Initial phage level = 1 phage/ml of milk
Initial cell level = 1,000,000 cells/ml
Phage latent period = 40 minutes
Cell generation time = 40 minutes
Average burst size = 50

In the past twenty years, however, the incidence of phage infections against thermophilic cultures (consisting of *S. thermophilus* and either *L. bulgaricus* or *Lactobacillus helveticus*) has increased significantly. It is no coincidence that there has been a huge increase in the production of Mozzarella cheese, yogurt, and other products that rely on these thermophilic starter cultures. The emergence of thermophilic bacteriophage underscores the problem faced by the starter culture industry and its customers: wherever and whenever lactic acid bacteria are used on a large scale, phages that infect those organisms will undoubtedly appear as potential adversaries.

In response to the bacteriophage problem, the dairy starter culture industry, as well as the cheese and cultured dairy products industries, have adopted several strategies (Table 3-5; also reviewed in Chapter 5). First, and perhaps most importantly, high standards of hygiene must be applied. In fact, sanitation and asepsis, appropriate plant design, and phage exclusion programs are the first lines of defense against phages. Sanitizing agents that are ordinarily used in dairy plants (e.g., chlorine and hypochlorite solutions) are also effective against bacteriophages.

All areas where starter cultures are handled and grown should be isolated from the rest of the processing facility. Because phages are frequently transmitted via air or airborne droplets, the contained areas should be maintained under positive pressure, and filtered air should be used within the starter culture room, as well as in the starter tanks. It is also important to locate the actual production facilities down stream from the starter preparation area to ensure that waste flow (i.e., whey) does not contaminate the culture area. The latter point is critical, because whey is considered to be the major reservoir of phages within a dairy plant, and the primary vehicle by which phage dissemination occurs. In fact, in other lactic acid fermentations where opportunities for phage transmission are limited, infection of the culture by phages does not occur. For example, in the

Table 3.5. Phage control strategies.

Method	Purpose or Function
Sanitation	Kill and remove phages in plant environment
Plant design	Keep phages out of production area, prevent cross contamination
Phage inhibitory media	Prevent phages from attaching to and infectingculture cells
Phage resistant cultures	Design starter cultures cells that will grow and perform well even in presence of phages
Culture rotation	Prevent proliferation of phages by limiting access to suitable host

manufacture of fermented meats (Chapter 6), fermentation takes place within the individually encased sausages, and spread of phages from one link to another is not possible.

Another way to control phage is to use phage inhibitory media during the culture propagation step. As part of the infection process, calcium ions are required for phages to attach to and subsequently invade their bacterial hosts. The development of phage inhibitory media nearly fifty years ago was based on the principle that by reducing ionic calcium or by incorporating calcium chelating agents into the culture media, calcium ions would then be unavailable for phage attachment. The main chelating agents are phosphate and citrate salts, which have the added advantage of providing buffering capacity (see above). On the flip side, however, some phages exist that can effectively attach to and infect host cells even in the absence of calcium.

A third approach to reduce infection by bacteriophages takes advantage of the innate ability of some starter culture bacteria to defend themselves against phage attack (discussed below and in Chapter 5). Several types of natural phage-resistance mechanisms exist, including inhibition of adsorption, restriction of phage DNA, and abortive infection. Thus, it is possible to isolate strains that possess one or more of these systems and to use them in industrial fermentations. This approach, as described previously, forms the basis of most dairy starter culture systems. In particular, the multiple defined strain starter cultures used in cheese manufacture often contain as many as five different phage unrelated strains. These cultures can be used either on a continuous basis (the same culture every day) or rotated such that strains with the same phage sensitivity pattern are not used for consecutive fermentations. Rotations programs can also be performed even with single or paired strain starter. In either case, it is important to monitor the phage levels in the whey or milk on a regular basis.

When phage titers reach a particular threshold, signifying a strain has become phage-sensitive, that strain is removed from the mixture and replaced by a resistant strain. Since phage proliferation requires susceptible host strains, when those strains are removed from the production environment, the background level and accumulation of phages will be reduced such that normal fermentation rates can be achieved. This practice, however, is constrained by the limited availability of phage-unrelated strains (discussed previously). Also, frequent exchange of one strain with another within a given culture may result in undesirable variations in product quality. In fact, many of the defined, phage-resistant cultures now contain only two or three strains to maintain greater product consistency. In theory, it should be possible to isolate new strains from nature that are both phage-resistant and that have good cheese making properties, a strategy that is now standard industry practice (Box 3-5).

Of course, if some strains are naturally resistant to phage infections, then there must be a genetic basis for the phage-resistant phenotypes described above. Indeed, genes responsible for these phage-resistant phenotypes have been identified and characterized. Importantly, these genes can be introduced into phage-sensitive strains, making them phage-resistant. In addition, it is also possible to obtain spontaneous phage-resistant mutants by simply exposing sensitive, wild-type strains to lytic bacteriophages. The phage-resistant derivatives that are then selected must be evaluated for cheese-making properties before they are reintroduced into a starter culture, because pleiotropic mutations frequently occur that render them defective as cheese cultures (discussed previously).

Engineered Phage Resistance

Genes in *Lactococcus lactis* subsp. *lactis* that conferred a phage-resistance phenotype were identified in the early 1980s by researchers at North Carolina State University. These genes were located on a 46 kb, self-transmissible plasmid, named pTR2030. Initially, it was thought that pTR2030 contained a single phage-

Box 3-5. Looking Far and Wide for Lactococci

Over the thousands of years that lactic acid bacteria have been used in the manufacture of fermented foods, and the 100 years that lactococci, in particular, have been used as dairy starter cultures, these organisms have undoubtedly been exposed to considerable selective pressure. To grow and compete well in a milk environment, for example, an organism must transport and metabolize lactose rapidly and also be able to hydrolyze milk proteins and use the resulting peptides efficiently. Other traits, specific for cheese fermentations and for which strains are routinely screened, include the ability to produce good cheese flavor and texture, to grow within the relevant temperature range, to resist bacteriophage infection, and to be amenable to large-scale propagation, handling, and storage.

Because of these strict phenotypic requirements, it has been suggested that relatively few distinct strains of *Lactococcus lactis* subsp. *lactis* and *Lactococcus lactis* subsp. *cremoris* actually exist and that the overall genetic diversity of dairy lactococci is limited (Salama et al., 1995). Moreover, many of the strains that are marketed commercially, even those from disparate geographical locations, are derived from common stock cultures or dairy products (Ward et al., 2004). It is reasonable to assume that in many cases, these cultures, even if marketed by different culture companies, may not be very different from one another. Therefore, the discovery of new strains that satisfy culture requirements would be quite valuable, especially since these strains might provide a rich source of new genetic information, including genes encoding for novel phage resistance mechanisms.

In an effort to obtain new lactococcal strains that could potentially be used as starter cultures, researchers at Oregon State University obtained environmental and dairy samples from diverse geographical locations (Salama et al., 1993, 1995). Samples (more than 200) were obtained from local plant sources, including vegetables, wildflowers, weeds, and grasses, as well as milk and dairy products from China, Morocco, Ukraine, and the former Yugoslavia. Their screening method was modeled after procedures used for large, rare clone library screening, and involved an enrichment step followed by colony hybridization using lactococci- or *L. lactis* subsp. *cremoris*-specific 16rRNA probes. The selected isolates were then characterized based on various phenotypic characteristics (growth at 10°C, 40°C, and 45°C or in 10% salt or at pH 9.2, and arginine hydrolysis). In general, *L. lactis* subsp. *lactis* were isolated from both plant and dairy sources, whereas *L. lactis* subsp. *cremoris* could only be found in dairy samples.

A second level of screening was then performed, based on rapid acid production and flavor profiles (i.e., tasting milk fermented with each strain). These results indicated that 61 out of 120 strains grew well in milk and produced acceptable flavor characteristics. Additional investigations indicated that ten new strains were phage resistant and suitable for cheesemaking (Urbach et al., 1997).

Collectively, these studies suggest that new strains, with the necessary phenotypes for dairy fermentations, certainly do exist in nature. However, having a rational isolation and screening plan is also required if such "bio-prospecting" efforts are to be successful.

References

Salama, M.S., W.E. Sandine, and S.G. Giovannoni. 1993. Isolation of *Lactococcus lactis* subsp. *lactis* from nature by colony hybridization with rRNA probes. Appl. Environ. Microbiol. 59:3941–3945.

Salama, M.S., T. Musafija-Jeknic, W.E. Sandine, and S.G. Giovannoni. 1995. An ecological study of lactic acid bacteria: isolation of *Lactococcus lactis* subsp. *cremoris*. J. Dairy Sci. 78:1004–1017.

Urbach, E., B. Daniels, M.S. Salama, W.E. Sandine, and S.G. Giovannoni. 1997. The *ldb* phylogeny for environmental isolates of *Lactococcus lactis* is consistent with rRNA genotypes but not phenotypes. Appl. Environ. Microbiol. 63:694–702.

Ward, L.J.H., H.A. Heap, and W.J. Kelly. 2004. Characterization of closely related lactococcal starter strains which show differing patterns of bacteriophage sensitivity. J. Appl. Microbiol. 96:144–148.

resistance determinant (responsible for a heat-sensitive abortive infection phenotype), but it was later discovered that other genes, encoding for a restriction and modification system, were also present. When the plasmid was transferred via a simple conjugal mating procedure into a phage-sensitive, industrial cheese-making strain, transconjugants with resistance to a broad range of lytic industrial phages were obtained. This was a noteworthy accomplishment, in part, because it represented the first application of biotechnology to improve dairy starter cultures, but also because the actual technique of gene transfer did not involve recombinant DNA technology. Thus, these modified strains could be used commercially.

Indeed, this approach was successful in actual cheese manufacturing environments. However, with prolonged use, bacteriophages eventually appeared in cheese plants that were able to circumvent the resistance of these strains. Nonetheless, these early efforts marked the beginning of a new era in starter culture technology and led to the development of other molecular strategies aimed at controlling bacteriophages. One approach, for example, involved introducing different phage-resistance genes into a single strain, on an individual basis, thereby generating several isogenic phage-resistant derivatives. When used in a rotation scheme, the properties of the parental strain remain the same, while the resistance pattern against different phage types is expanded. Other examples of engineered phage resistance are described in Chapter 5.

Starter Culture Technology in the Twenty-first Century

Phage is not the only concern of the dairy starter culture industry. How fast a given strain grows, what sugars it ferments, and what end-products are formed are also important issues that influence culture performance and that have been addressed by various research groups. In particular, much of the research on lactic acid bacteria in the past thirty-five years was based on the pioneering work of McKay

and co-workers at the University of Minnesota (Box 3-6). It was the McKay lab that showed that most of the phenotypic traits necessary for growth and activity of lactic starter cultures, including lactose fermentation, casein hydrolysis, and diacetyl formation, were encoded by plasmid DNA. In addition, genes encoding for nisin production and immunity, for phage resistance, and for conjugal transfer factors were also identified by the McKay group.

These discoveries, and the development of gene transfer and gene exchange techniques, have made it possible to manipulate the physiological, biochemical, and genetic properties of lactic starter cultures in ways hardly imaginable a generation ago. Cultures are now customized to satisfy individual customer demands. Not only are improved strains of lactic acid bacteria now used in dairy, meat, vegetable, soy, and wine fermentations to satisfy modern manufacturing requirements and enhance product quality, they also are being used for novel applications. Lactic acid bacteria, by virtue of their ability to survive digestive processes and reach the intestinal tract, have been found to serve as excellent delivery agents for vaccines and antigens. The use of lactic acid bacteria as probiotics has advanced to the point where the medical community is now actively involved in evaluating these bacteria, in well-designed clinical trials, for their ability to treat important chronic diseases (Chapter 4). Certainly, the information mined from the "omic" fields of genomics, proteomics, and metabolomics is bound to lead to many other new applications (Chapter 2).

Of course, research on starter culture microorganisms is not limited to lactic acid bacteria. The genome of *Saccharomyces cerevisiae* was sequenced in 1996, and efforts aimed at manipulating the physiological and biochemical properties of bread, beer, and wine yeasts have been ongoing for many years (Chapters 8, 9, and 10). Despite the apparent simplicity of these fermentations, there are many opportunities for improvement. Bakers' yeasts, for example, can be made more cryotolerant so that leavening will still occur when

Box 3-6. Plasmid Biology, Gene Transfer Technology, Microbial Starter Cultures, and Softball: Life in the Larry McKay Laboratory

It can be fairly argued that research interest in lactic acid bacteria is at an all-time high. Nearly twenty genomes have been sequenced, sophisticated culture improvement programs are underway throughout the world, and pharmaceutical companies have developed strains of lactic acid bacteria that are now being used to deliver human and animal vaccines.

Of course, lactic acid bacteria were among the very first groups of bacteria studied by the very first microbiologists more than a century ago. Pasteur, Lister, Metchnikoff, and Koch all focused (literally) their microscopes and attention on these bacteria. As the applied significance of these bacteria became evident, researchers devoted their entire careers to understanding their physiological and biochemical properties and their particular performance characteristics in fermented dairy foods. Indeed, great discoveries were made by Orla-Jensen, Sherman, and Whitehead, and research groups led by Paul Elliker, Bob Sellars, Marvin Speck, Bill Sandine, Robert Lawrence, and many others.

Research on lactic acid starter cultures moved into an entirely new, uncharted, and ultimately revolutionary direction in the early 1970s. It was at that time when Larry McKay, a new assistant professor at the University of Minnesota, embarked on a research career that essentially created the field of starter culture genetics. The McKay lab, managed by his extraordinary assistant, Kathy Baldwin, also became the main teaching laboratory for the next thirty years, responsible for training a cadre of students and future scientists that has continued to make advances in starter culture research.

McKay had been a graduate student in the Sandine lab at Oregon State University when he first became interested in lactic acid bacteria and when he developed a keen sense of observation. It was his ability to understand what those observations meant that led him to discover the biochemical pathway by which lactic acid bacteria metabolized lactose, a key finding that was of fundamental as well as applied significance.

When McKay arrived at Minnesota in 1971, he began to address another question that had long puzzled dairy microbiologists—specifically, why is it that some lactococci appear to lose their ability to ferment lactose (referred to as a lac⁻ phenotype, in contrast to a lactose-fermenting or lac+ phenotype). Similarly, there were also strains that had spontaneously lost the ability to degrade milk proteins (a *prt⁻* phenotype). This loss of function phenotype had been observed by Orla-Jensen as long ago as 1919, by Sherman in 1937, and Hirsch in 1951, but as McKay noted, "The question remains, however, as to how the loss of lactose metabolism is induced in cultures of lactic streptococci" (McKay et al., 1972).

The answer, while obvious today, was not so obvious in 1971, when McKay rightly hypothesized that these bacteria "could be carrying a genetic element which is responsible for the cells' ability to ferment lactose" and that "the loss of this element would cause the cell to become lac⁻" (McKay et al., 1972). Shortly thereafter, the McKay lab provided the first experimental proof that these genetic elements were indeed plasmids (Cords et al., 1974; the reader is reminded that plasmids had only been "discovered" in 1969). The McKay group then published an entire series of seminal papers demonstrating that other essential traits in lactic acid bacteria, including protein metabolism, citrate fermentation, nisin production and resistance, conjugation factors, and phage resistance, were also encoded by plasmid DNA. These (and subsequent) reports all had the McKay signature—they were written in a simple and concise style; the experimental approach was straightforward, precise, and well controlled; and the results were presented clearly and definitively.

As it was becoming clear that important plasmid-encoded functions were widespread in lactic acid bacteria and lactococci in particular, McKay quickly realized the implications. Already, in 1974, McKay and Baldwin had shown that strains which had lost the ability to ferment lactose or produce proteinase could be transduced by tranducing bacteriophages, generating "*Lac⁺*"

(Continued)

Box 3-6. Plasmid Biology, Gene Transfer Technology, Microbial Starter Cultures, and Softball: Life in the Larry McKay Laboratory *(Continued)*

and "*Lac*$^+$, *Prt*$^+$" transductants (McKay, Cords, and Baldwin, 1973; McKay and Baldwin, 1974). The development of a conjugation or cell-to-cell mating system (Kempler and McKay, 1979; McKay et al., 1980; and Walsh and McKay, 1981) provided a basis for transferring genes from one organism to another.

Although transformation, the direct uptake of DNA into the cells, proved to be more elusive, Kondo and McKay reported the first successful use of a transformation technique for lactococci in 1982 and later described optimized procedures in 1984. Over the next several years, studies on gene cloning, construction of food-grade cloning vectors, and mechanisms of gene integration and conjugation were reported (Harlander et al., 1984; Froseth et al., 1988; Mills et al., 1996; Petzel and McKay, 1992). Strain improvement strategies were also devised, including efforts aimed at isolating strains that could accelerate cheese ripening, over-produce bacteriocins, and resist bacteriophage infection (Feirtag and McKay, 1987; McKay et al., 1989; Scherwitz-Harmon and McKay, 1987).

During his career, McKay enjoyed nothing more than perusing the lab, examining plates, looking at gels, or asking questions of the lab personnel. He was just as passionate about teaching and mentoring. His course on microbial starter cultures, which was taught every other year, was so valuable that students would often take the class a second time, just to stay caught up in the field. Upon leaving the lab, McKay's students and postdoctorates were always well prepared, scientifically and professionally, and assumed prestigious positions in academia and industry. Of course, as productive as the McKay lab was for more than three decades, the lab did have other important interests. There are stories, perhaps apocryphal, that upon interviewing a potential new graduate student, McKay's final question would have something to do with the applicant's skills on the softball field and whether they could contribute to the department's chances for the next season. While there was no substitute for having a solid background in biochemistry, genetics, and microbiology, having a good arm at shortstop would certainly have helped your chances at joining one of the best research laboratories in the world.

References

Cords, B.R., L.L. McKay, and P. Guerry. 1974. Extrachromosomal elements in group N streptococci. J. Bacteriol. 117:1149–1152.

Harlander, S.K., L.L. McKay, and C.F. Schachtele. 1984. Molecular cloning of the lactose-metabolizing genes from *Streptococcus lactis*. Appl. Environ. Microbiol. 48:347–351.

Feirtag, J.M., and L.L. McKay. 1987. Isolation of *Streptococcus lactis* C2 mutants selected for temperature sensitivity and potential in cheese manufacture. J. Dairy Sci. 70:1773–1778.

Froseth, B.R., R.E. Herman, and L.L. McKay. 1988. Cloning of nisin resistance determinant and replication origin on 7.6-kilobase EcoRI fragment of pNP40 from *Streptococcus lactis* subsp. *diacetylactis* DRC3. Appl. Environ. Microbiol. 54:2136–2139.

Kempler, G.M., and L.L. McKay. 1979. Genetic evidence for plasmid-linked lactose metabolism in *Streptococcus lactis* subsp. *diacetylactis*. Appl. Environ. Microbiol. 37:1041–1043.

Kondo, J.K., and L.L. McKay. 1982. Transformation of *Streptococcus lactis* protoplasts by plasmid DNA. Appl. Environ. Microbiol. 43:1213–1215.

Kondo, J.K., and L.L. McKay. 1984. Plasmid transformation of *Streptococcus lactis* protoplasts: optimization and use in molecular cloning. Appl. Environ. Microbiol. 48:252–259.

McKay, L.L., and K.A. Baldwin. 1974. Simultaneous loss of proteinase- and lactose-utilizing enzyme activities in and reversal of loss by transduction. Appl. Microbiol. 28:342–346.

McKay, L.L., K.A. Baldwin, and P.M. Walsh. 1980. Conjugal transfer of genetic information in Group N streptococci. Appl. Environ. Microbiol. 40:84–91.

Box 3-6. Plasmid Biology, Gene Transfer Technology, Microbial Starter Cultures, and Softball: Life in the Larry McKay Laboratory *(Continued)*

McKay, L.L., M.J. Bohanon, K.M. Polzin, P.L. Rule, and K.A. Baldwin. 1989. Localization of separate genetic loci for reduced sensitivity towards small isometric-headed bacteriophage sk1 and prolate-headed bacteriophage c2 on pGBK17 from *Lactococcus lactis* subsp. *lactis* KR2. Appl. Environ. Microbiol. 55: 2702–2709.

McKay, L.L., K.A. Baldwin, and E.A. Zottola. 1972. Loss of lactose metabolism in lactic streptococci. Appl. Microbiol. 23:1090–1096.

McKay, L.L., B.R. Cords, and K.A. Baldwin. 1973. Transduction of lactose metabolism in *Streptococcus lactis* C2. J. Bacteriol. 115:810–815.

Mills, D.A., L.L. McKay, and G.M. Dunny. 1996. Splicing of a group II intron involved in the conjugative transfer of pRSO1 in lactococci. J. Bacteriol. 178:3531–3538.

Petzel, J.P., and L.L. McKay. 1992. Molecular characterization of the integration of the lactose plasmid from *Lactococcus lactis* subsp. *cremoris* SK11 into the chromosome of *Lactococcus lactis* subsp. *lactis*. Appl. Environ. Microbiol. 58:125–131.

Scherwitz-Harmon, K.M., and L.L. McKay. 1987. Restriction enzyme analysis of lactose and bacteriocin plasmids from *Streptococcus lactis* subsp. *diacetylactis* WM4 and cloning of *Bcl*I fragments coding for bacteriocin production. Appl. Environ. Microbiol. 53:1171–1174.

Walsh, P.M., and L.L. McKay. 1981. Recombinant plasmid associated with cell aggregation and high-frequency conjugation in *Streptococcus lactis* ML3. J. Bacteriol. 146:937–944.

frozen dough is thawed. Likewise, increasing the metabolic capacity of these yeasts to ferment maltose and other sugars, whose use is ordinarily repressed by glucose, would accelerate the fermentation rate and shorten the leavening step. Modern brewers' yeast strains can be modified such that they produce beers with specific compositional characteristics (e.g., low carbohydrate, low ethanol), that have specific performance traits (e.g., attenuation, flocculating properties), or that simply produce better-tasting beer (e.g., no diacetyl or hydrogen sulfite). Several desirable traits in wine yeasts have also been targeted for improvement. Development of osmophilic wine yeast strains would be desirable due to their ability to grow in high sugar-containing musts. The malolactic fermentation, ordinarily performed by malolactic acid bacteria, could be carried out by *S. cerevisiae* harboring genes encoding for the malic acid transporter and malolactic enzyme.

Despite the opportunities that now exist to modify and improve starter culture organisms, strain improvement programs have been limited by regulatory and public perception considerations. While tools for introducing DNA into food grade starter microorganisms or for altering the existing genetic makeup of these organisms are now widely available, commercialization of genetically modified organisms (GMOs) on a world-wide basis has not occurred. Although GMOs are common in the United States, they are generally not marketed in Europe or the Far East. The debate over GMOs is not likely to abate anytime soon, even as researchers continue to demonstrate the potential benefits GMO technologies may offer (Box 3-7).

Encapsulated and Immobilized Cells

In the fermentation industry, there has been considerable interest in the development and application of encapsulated and immobilized cell technologies. In general, encapsulation refers to a process whereby cells are embedded or enrobed within a gel-containing shell. Encapsulated cells are metabolically active and fully capable of performing fermentations. In addition, encapsulated cells may be either freely suspended in the medium or immobilized to an inert support material. Encapsulated and immobilized cell systems offer several advantages. First, they provide a means for conducting fermentations on a continuous rather

Box 3-7. Genetically Modified Organisms: Current Status and Future Prospects

As noted elsewhere in this chapter, as well as throughout this text, one of the most important developments in the starter culture industry has been the application of molecular biology to improve starter culture microorganisms. It is now rather easy to modify or manipulate the phenotype of starter bacteria or yeasts, either by the introduction of heterologous (i.e., foreign) DNA into selected organisms or by increasing expression or inactivating particular genes. These genetically modified (GM) or genetically engineered organisms may then have traits, that when used to make fermented foods, result in products having better flavor, texture, nutritional value, or shelf-life. In addition, GM organisms may possess other processing advantages, such as immunity to bacteriophages, resistance to biological or chemical agents, and tolerance to low temperature and osmotic and other inhibitory conditions.

Despite the relative ease with which genetically modified organisms (GMOs) or microorganisms can be constructed in the laboratory, there are few actual examples of genetically modified starter culture organisms that are currently used in the fermented foods industry. This is in contrast to the GMO crops that are now fully integrated within the U.S. food supply. For example, most of the soybeans, corn, and canola produced in the United States come from genetically modified seeds. Moreover, even those commercially-available starter cultures that have been "modified" are not really genetically modified(i.e., made via recombinant DNA technology), but rather are derived via classical genetic techniques. They do not contain foreign DNA, and only natural gene exchange systems (i.e., conjugation or hybridization) are used. The notion of "natural" gene transfer is worth emphasizing, since the European Union (EU) defines GMOs as being altered genetically via processes that do not occur naturally (Kondo and Johansen, 2002).

If GM starter culture organisms, possessing highly desirable properties, can be developed in the laboratory, then why aren't these cultures available in the marketplace? This simple question has many answers. First, GMO products must satisfy the regulatory requirements of the country or countries in which the products will be marketed. Because GMO regulations differ throughout the world (and even what constitutes a GMO varies from country to country), gaining approval from one jurisdiction in no way guarantees acceptance elsewhere. Moreover, the ease with which approval for GMOs is obtained varies considerably. For example, in the United States, GM lactic acid bacteria can qualify as GRAS (generally recognized as safe), provided the "bioengineered" organism is not different, in any significant manner, from the original organism (i.e., it satisfies the substantial equivalence concept).

In the EU, however, the GMO application process is much more elaborate, requiring risk assessment and monitoring and detection plans. Labeling also is required in the European Union,

---

than batch mode. This increases throughput and generally reduces overall production costs. In addition, the cells could be recovered after a fermentation and then used repeatedly (within limits) in subsequent fermentations. Because encapsulated cells are surrounded by a protective, inert material, they may be more stable during the fermentation, as well as during storage. In addition, cells that are encapsulated within alginate beads or other matrices may be less sensitive to phage infections due to the limited access phage would have to the cells' surface.

Despite these advantages and extensive research, the number of actual applications of these technologies are relatively few in number. For products such as cheese or yogurt, the non-fluid nature of the food medium places obvious restrictions on the ability to pass substrate through immobilized cell bioreactors or to recover encapsulated cells from fermented products. Still, food fermentation processes are among the applications receiving much of the attention.

One potential application of immobilized cell technology is production of culture bio-

Box 3-7. Genetically Modified Organisms: Current Status and Future Prospects *(Continued)*

whereas no such requirement exists in the United States for most GMO foods. The labeling issue is not a minor point—in fact, it may be a deal-breaker, because many consumers are either opposed to GMOs or confused enough by the GMO declaration on the label that it influences their purchasing decisions.

Although it would appear that the U.S. market (and perhaps others) would be more receptive to GM cultures than the EU market, it must be recognized that most foods or food ingredients, including starter cultures, are marketed globally. Thus, it seems likely that a culture supplier would commit resources to development of GM microorganisms only if approval on a near worldwide basis is expected. Nonetheless, there are some cultures, derived using molecular techniques, that are sold in the United States but are not marketed in Europe.

For example, a high diacetyl-producing derivative strain of *Lactococcus lactis* (MC010), obtained via a recombinant plasmid-mediated mutagenesis procedure, is now widely used in the United States while its approval in the EU is still being considered (Curic et al., 1999; Pedersen et al., 2005). Importantly, this strain contains no foreign DNA and is identical to the parent strain except for a mutation in a single gene. Moreover, there are now a variety of molecular approaches that can be adopted to construct food-grade GM starter culture organisms. Plasmid vectors, for example, that are comprised of only lactococcal DNA with natural selection markers (and devoid of antibiotic-resistance genes), are now available. It remains to be seen, however, whether these so-called "self-cloning" strategies will result in greater acceptance by both regulatory agencies and consumers.

References

Curic, M., B. Stuer-Lauridsen, P. Renault, and D. Nilsson. 1999. A general method for selection of α-acetolactate decarboxylase-deficient *Lactococcus lactis* mutants to improve diacetyl formation. Appl. Environ. Microbiol. 65:1202–1206.

Kondo, J.K., and E. Johansen. 2002. Product development strategies for foods in the era of molecular biotechnology. Antonie van Leeuwenhoek 82:291–302.

Pedersen, M.B., S.L. Iversen, K.I. Sørensen, and E. Johansen. 2005. The long and winding road from the research laboratory to industrial applications of lactic acid bacteria. FEMS Microbiol. Rev. 29:611–624.

mass. Starter culture bacteria are mass produced in a batch mode. Even when pH, atmosphere and oxygen, nutrient feed, and other environmental factors are properly controlled, high cell densities may be difficult to achieve. Continuous processes also do not fare well, mainly because plasmid-borne functions may be lost and because such configurations cannot be used for co-culture or mixed strain situations due to the difficulty in maintaining strain balance. It is possible, however, to produce continuous cell biomass from encapsulated cells. This approach takes advantage of the fact that cell-containing gel beads tend to release free cells at a regular rate. Those free cells can then be collected on a continuous basis while the encapsulated cells are maintained and contained either in an immobilized form or within in a fixed compartment.

Another application for encapsulated or immobilized cells is in liquid fermentations. For example, immobilized lactic acid bacteria can be used in yogurt fermentations to "pre-ferment" the milk, on a continuous basis, prior to the batch incubation. In this approach, the milk pH is reduced to 5.7 during the pre-fermentation step, allowing an overall 50% reduction in the total fermentation time. Likewise, immobilized lactic acid bacteria can be used to produce industrial lactic acid and other organic end-products from whey, whey permeates, or corn-based feedstocks. Other value-added fermentation products that could be produced on a continuous basis by immobilized cells include exopolysaccharides (Chapter 4), bacteriocins, and diacetyl.

Perhaps the encapsulated cell applications with the greatest potential are those involving alcoholic fermentations, especially for wine production. Indeed, several such products, including both encapsulated yeasts and bacteria, are available commercially for use in specific product applications. Encapsulated strains of *S. cerevisiae*, for example, can be used in the manufacture of sweet wines (whose manufacture is described in Chapter 10). These wines, in contrast to dry wines, contain residual or unfermented sugar, due to cessation of the fermentation via a decrease in temperature to near freezing, addition of sulfites, or removal of yeast by filtration. In contrast, bead-encapsulated yeasts can be placed in nylon bags which are then added to the must, such that when the desired sugar concentration is reached, the bags (and yeasts) can simply be removed. Encapsulated yeast cells can also be used to re-start stuck or sluggish wine fermentations and to metabolize malic acid to ethanol. Similarly, encapsulation and immobilization of the malolactic bacterium *Oenococcus oeni* has been proposed to be an effective way to reduce malic acid and deacidify wine on a continuous basis.

Finally, it is relevant to mention that encapsulated starter cultures, in an entirely natural form, have long been used in the manufacture of kefir, a fermented dairy product widely consumed in Russia and Eastern Europe (Chapter 4). The key feature of this product is that it is traditionally made by inoculating milk with a culture contained within inert particles called kefir grains. These irregularly-shaped particles, which can be as large as 15 mm in diameter and consist of polysaccharide and milk protein, harbor a complex microflora. Included are homo- and heterofermentative lactic acid bacteria, as well as several yeasts. The grains can be separated from the fermented kefir by filtration or sieving, and then washed and re-used for subsequent fermentations. Although kefir cultures, which contain pure strains of *Lacto-*

bacillus spp. and *Lactococcus* spp., are now available commercially and are often used in the United States for kefir manufacture, traditional kefir grains still remain popular in many parts of the world.

Probiotics and Cultures Adjuncts

In addition to producing starter cultures for specific food fermentations, the starter culture industry also manufactures microbial products for many other applications. For example, there is now a large market for probiotic microorganisms in both fermented and non-fermented foods. As described in detail in Chapter 4, probiotics are defined as "live microorganisms which when administered in adequate amount confer a health benefit on the host." The main probiotic organisms include species of *Bifidobacterium* and *Lactobacillus,* although *Saccharomyces* and *Bacillus* sp., and even *E. coli*, are also commercially available as probiotics. In general, these organisms are produced industrially in the same manner as starter cultures (i.e., under conditions that maximize cell density). However, because they are not used to carry out a subsequent fermentation, these culture products are of the direct-to-vat type, rather than in bulk culture form.

Another group of culture products used by the cheese industry consists of lactic acid and related bacteria whose function is to accelerate and enhance cheese ripening and maturation. Specifically, these organisms, usually strains of *Lactobacillus helveticus* and *Lactobacillus casei*, produce peptidases and other protein-hydrolyzing enzymes that are necessary for proper flavor and texture development. Citrate-fermenting LAB that produce the flavor compound diacetyl as well as heterofermentative LAB that produce carbon dioxide may also be added as adjunct cultures for specific cheeses.

Finally, adjunct cultures and culture preparations are now available that serve a food safety and preservation function. That is, the microorganisms contained in these products do not necessarily make fermentation acids or modify texture or flavor, but rather they are included in the culture because they inhibit pathogenic or spoilage organisms. The inhibitory activity of these organisms may be due to one of several substances, including hydrogen peroxide, organic acids, diacetyl, and a class of inhibitory compounds called bacteriocins. The latter are proteinaceous, heat-stable materials produced by a given organism that inhibits other closely related organisms (Chapter 6).

Lactic acid bacteria are prolific producers of bacteriocins, with all of the food-related genera capable of producing bacteriocins. Therefore, they can be used either as part of a lactic acid-producing starter culture or as an adjunct in dairy, meat, and other foods to inhibit pathogens and spoilage organisms and to enhance shelf-life. Furthermore, some of these organisms used as adjuncts are capable of producing bacteriocin in the food, but with only minor production of acids or other fermentation end-products. Thus, the sensory characteristics of the product are not affected by the producer organism, a property that would be especially important in non-fermented foods such as ready-to-eat meats.

Another means by which bacteriocins can be introduced into a food without adding live organisms has also been developed. In this process, the producer organism is grown in a dairy- or non-dairy-based medium, and then the spent fermentation medium is harvested, pasteurized, and concentrated. This material would contain the bacteriocin (as well as organic acids), which could then be added to foods as a natural preservative. Although these so-called bioprotective products are mainly effective against Gram positive bacteria, some commercially-available products also inhibit yeasts, mold, and Gram negative spoilage bacteria, including psychrotrophs that spoil refrigerated foods.

The Starter Culture Industry

Although starter cultures are certainly the main product line, the starter culture industry, as noted previously, produces and markets a variety of other products. Dairy culture suppliers, in particular, offer culture media, coagulants, colors, and other ancillary products used in cheese manufacture. Starter culture companies also provide technical services and support, perhaps more so than any other ingredient suppliers. This is because the manufacture of cheese, sausage, wine, and all fermented products depends on the activity of an inherently unstable biological material, i.e., the starter culture. Although there may certainly be quality issues that arise during the manufacture of corn flakes, canned peas, granola bars, or other non-fermented foods, the failure of a biologically-dependent process is not one of them. A cheese plant operations manager who is responsible for converting 2 million Kg of milk into 200,000 kg of cheese (worth at least $250,000) each day will not wait long before calling the culture supplier the minute he or she suspects a culture problem. Thus, a well-trained support staff is a necessary component of the modern culture industry.

Culture suppliers are also often asked to provide customized cultures for specific applications. This means that the phenotype (or in some cases, even the genotype) of each strain in an industrial culture collection must be cataloged. Fermentation rates, product formation and enzyme profiles, sugar fermentation patterns, osmotic- or halo-tolerance behavior, and flocculation properties (for yeasts) should be included in the strain bank data. In the case of lactic acid bacteria and beer and wine yeasts, sensitivities to phages and killer yeasts, respectively, must be determined for each strain. When multiple strains are used in culture blends, production of or sensitivity to antagonism factors must be known.

Finally, the starter culture industry has always been a major contributor to the research on lactic acid bacteria, yeast, and fungi. Several of the large culture companies have maintained active research and development programs that conduct basic and applied research that has led directly to major findings on the genetics and physiology of starter culture organisms. In other cases, their contribution has been indirect, namely by providing research funding, strains, and other materials to investigators at universities and research institutes. Industry sponsorship of conferences and scientific societies have also served an important function in the advancement of starter culture research.

Bibliography

Buckenhüskes, H.J. 1993. Selection criteria for lactic acid bacteria to be used as starter cultures for various food commodities. FEMS Microbiol. Rev. 12:253-272.

Caplice, E., and G.F. Fitzgerald. 1999. Food fermentations: role of microorganisms in food production and preservation. Int. J. Food Microbiol. 50: 131-149.

Cogan, T.M. 1996. History and taxonomy of starter cultures, p. 1—23. *In* T.M. Cogan and J.-P. Accolas (ed.). *Dairy Starter Cultures*. VCH Publishers, Inc., New York, New York.

De Vuyst, L. 2000. Technology aspects related to the application of functional starter cultures. Food Technol. Biotechnol. 38:105-112.

Hammes W.P., and C. Hertel. 1998. New developments in meat starter cultures. Meat Sci. 49 (supplement):S125-S138.

Mäyrä-Mäkinen A, and M. Bigret. 1999. Industrial use and production of lactic acid bacteria, p. 175—198. *In* S. Salminen, A. von Wright, and A. Ouwehand (ed.) *Lactic Acid Bacteria. Microbiological and functional aspects. 3rd Ed.*, Marcel Dekker Inc., New York, New York.

Pedersen, M.B., S.L. Iversen, K.I. Sørensen, and E. Johansen. 2005. The long and winding road from the research laboratory to industrial applications of lactic acid bacteria. FEMS Microbiol. Rev. 29:611-624.

Ross, R.P., C. Stanton, C. Hill, G.F. Fitzgerald, and A. Coffey. 2000. Novel cultures for cheese improvement. Trends Food Sci. Technol. 11:96-104.

Sandine, W.E. 1996. Commercial production of dairy starter cultures, p. 191—206. *In* T.M. Cogan and J.-P. Accolas (ed.) *Dairy Starter Cultures*. VCH Publishers, Inc., New York, New York.

4

Cultured Dairy Products

". . . during recent years, attention has been directed to soured milk to such an extent that it has become necessary for all who are interested in the handling of milk and milk products to have a knowledge of the subject, as it seems clearly demonstrated that, under proper direction, there is every possibility of its forming an important element in the prolongation of life."
From The Bacillus of Long Life *by Loudon Douglas, 1911*

Introduction

Milk fermentations must undoubtedly be among the oldest of all fermented foods. Milk obtained from a domesticated cow or camel or goat, some thousands of years ago, would have been fermented within hours by endogenous lactic acid bacteria, creating a yogurt-like product. In fact, the ability to maintain milk in a fresh state before souring and curdling had occurred would have been quite some trick, especially in warm environments. Of course, fermentation and acid formation would have been a good thing since, in the absence of viable lactic acid bacteria, other bacteria, including pathogens, could have grown and caused unpleasant side effects.

Milk is particularly suitable as a fermentation substrate owing to its carbohydrate-rich, nutrient-dense composition. Fresh bovine milk contains 5% lactose and 3.3% protein and has a water activity near 1.0 and a pH of 6.6 to 6.7, perfect conditions for most microorganisms. Lactic acid bacteria are saccharolytic and fermentative, and, therefore, are ideally suited for growth in milk. In general, they will outcompete other microorganisms for lactose, and by virtue of acidification, will produce an inhospitable environment for would-be competitors. Therefore, when properly made, cultured dairy products have long shelf-lives and,

although growth of acid-tolerant yeast and molds is possible, growth of pathogens rarely occurs.

Given the early recognition of the importance of milk in human nutrition and its widespread consumption around the world, it is not surprising that cultured dairy products have evolved on every continent. Their manufacture was already well established thousands of years ago, based on their mention in the Old Testament as well as other ancient religious texts and writings. Yogurt is also mentioned in Hindu sacred texts and mythology. And although the manufacturing procedures, the sources of milk, and the names of these products may vary considerably, they share many common characteristics. Thus, dahi (India), laban (Egypt, Lebanon), and jugart (Turkey) are all yogurt-like products whose manufacture involves similar milk handling procedures and depends on the same thermophilic culture bacteria. Other products, in particular, kefir and koumiss, evolved from Asia, and are made using various lactose-fermenting yeasts in addition to lactic acid bacteria.

Consumption of Cultured Dairy Products

Yogurt, sour cream (and sour cream-based dips), and cultured buttermilk are the most popular cultured dairy products in the United

States. Yogurt accounts for more than half of all cultured dairy products consumed in the United States (Figure 4-1), with nonfat and low-fat versions being the most popular (about 90% of total yogurt sales). Per capita consumption of buttermilk in the United States has decreased nearly 50% in the past twenty years, but consumption of sour cream (including sour cream-based dips) has doubled and yogurt has tripled during that same time.

Nonetheless, total per capita consumption for all cultured dairy products in 2003 in the United States was less than 6.5 Kg (about 14 pounds), whereas in some European countries, yogurt consumption alone is more than 20 Kg per person per year—the equivalent of ninety 8-ounce cups (Table 4-1). Furthermore, in the Netherlands, France, and other European countries, the dairy sections of food markets and grocery stores contain numerous other traditional as well as new forms of cultured milk and cream products. Many of these new product trends are beginning to catch on in the United States, and now kefir, fluid yogurts,

crème fraîche (a 50% fat sour cream), and other new cultured dairy products are readily available in the market place.

Cultured dairy products and probiotic bacteria

Among the factors contributing to the increased consumption of yogurt and related products are the positive nutritional benefits these products are believed to provide. Throughout the world, yogurt, and specifically, the live cultures present in yogurt, have been promoted on the basis of the many health benefits these bacteria are believed to confer. Although the bacteria that comprise the normal yogurt starter culture, *Streptococcus thermophilus* and *Lactobacillus delbrueckii* subsp. *bulgaricus*, may indeed enhance gastrointestinal health (to be discussed later), yogurt has become widely used as a vehicle for delivery of other microorganisms not ordinarily found in yogurt that may also improve human health. Such health-promoting bacteria are referred to

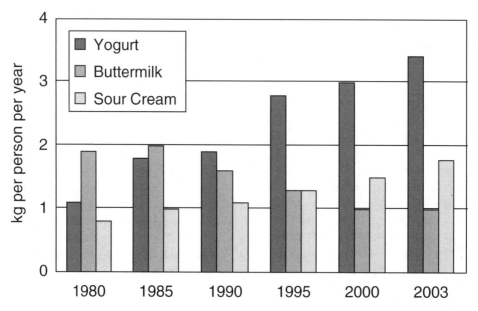

Figure 4–1. Consumption of cultured dairy products in the United States, 1980 to 2003. (Data for sour cream included sour cream dips.) From USDA, Economic Research Service data.

Table 4.1. Per capita consumption of yogurt[1].

Country	Kg per person per year
Australia (2003)	5.9
Austria (1999)	8.2
Bulgaria (2002)	24.7
Canada (2003)	5.8
Denmark (1999)	14.8
Finland (2001)	17.0
France (2001)	14.5
Germany (2004)	16.8
Italy (1999)	6.5
Netherlands (2003)	21.2
Spain (1999)	7.1
Sweden (2002)	28.5
Switzerland (1999)	13.0
Tunisia (2002)	8.0
Turkey (2004)	36.0
Russia (2001)	1.5
United Kingdom (1999)	4.8
United States (2003)	3.4

[1]Data obtained from multiple sources for the given year

as probiotics, organisms that confer a health benefit to the host (Box 4-1).

For example, *Lactobacillus acidophilus* and various species of *Bifidobacterium* are commonly added to yogurt as probiotics, even though these organisms are not part of the yogurt starter culture, nor are they involved in any way in the actual yogurt fermentation. Their only contribution is as a culture adjunct intended to promote good health. Still, according to Dairy Management Inc., about 80% of the yogurt products produced in the United States contain *L. acidophilus*. The mechanisms by which these suggested health benefits actually occur and the evidence to support these claims will be discussed later in this chapter.

In addition to probiotic bacteria, a group of compounds called prebiotics are also now being added to cultured milks—mainly yogurt and kefir in the United States, but many more products in Europe, Japan, and Korea. Prebiotics are oligosaccharide-containing materials that are neither degraded nor absorbed during transit through the stomach or small intestine (Box 4-1). They end up in the colon, where they are preferentially fermented by intestinal strains of lactobacilli

and bifidobacteria. Thus, they enrich the population of those bacteria thought to contribute to gastrointestinal health.

Probiotics can now be found not only in yogurt, but also in a variety of other cultured and non-cultured dairy products (Table 4-2). In some of these products, as noted above, the probiotic bacteria are not involved in the fermentation. However, some new products have been developed to take advantage of their fermentative ability as well as their probiotic activity. Examples include Yakult, a Japanese product made using a special strain of *Lactobacillus casei* (strain Shirota) that was originally isolated from the human intestinal tract, and Cultura, a European "bioyogurt" made with *L. casei* F19 (also a human isolate).

At the same time, there are also dairy products containing probiotic lactic acid bacteria that are not fermented but serve strictly as a carrier. The product known as Sweet Acidophilus Milk, for example, is simply fluid milk supplemented with *L. acidophilus* (added after pasteurization). Similar unfermented products may also contain other probiotic bacteria. Maintaining the product at low refrigeration temperatures (the same as for normal fluid milk products) prevents the bacteria from fermenting and souring the milk.

Fermentation Principles

The principles involved in the manufacture of specific cultured dairy products, which lactic acid bacteria are used for these products, and the attributes and properties desired by these bacteria are the focus of this chapter. Fermented dairy products from around the world, including India, Africa, Iceland, and other countries, will be described, but the emphasis will be on those products most widely consumed in the United States. In theory (and practice), some of these products can be made in the absence of a starter culture simply by adding food-grade acidulants to the milk mixture. Indeed, there are manufacturing advantages for following this practice; however, these acid-set products are not fermented—in fact, they must

Box 4–1. Probiotics and Prebiotics

History

Humans began consuming cultured dairy products thousands of years ago. These yogurt-like products were easy to produce, had good shelf-lives (so to speak), were free of harmful substances, and had a pleasant sensory appeal. At some time, it is theorized, they were also regarded as having therapeutic value, even though there was no scientific basis for this notion (Shortt, 1999). After all, the existence of bacteria and their role in fermentation weren't even recognized until Pasteur's research in the 1860s. Then, at the beginning of the twentieth century, the Russian scientist Elie Metchnikoff (who was working at the Pasteur Institute in Paris) suggested that the health benefits of fermented milk were due to the bacteria involved in the fermentation.

Specifically, Metchnikoff argued that these bacteria (which may have been the yogurt bacteria, *Streptococcus thermophilus* and *Lactobacillus delbrueckii* subsp. *bulgaricus* or other lactobacilli) inhibited putrefactive bacteria in the intestinal tract, thereby influencing the intestinal microflora such that overall health and longevity could be enhanced. At around the same time and shortly thereafter, other scientists (as recounted by Shortt, 1999) reported that consumption of other lactobacilli (including *Lactobacillus acidophilus*) and bifidobacteria also had positive health effects, including reducing the rate of infant diarrhea.

Interestingly, the observation a full century ago that bifidobacteria were present in the fecal contents of breast-fed infants suggested that these bacteria were associated with good intestinal health and possibly foreshadowed the concept of prebiotics (see below). These early reports, by highly regarded scientists and research institutes, attracted the attention of the medical community, and by the 1920s, studies using bacteria therapy (with milk as the carrier vehicle) had begun. Unfortunately, many of these subsequent studies suffered from the absence of established measurement criteria, the use of mis-identified strains, and other design flaws (Shortt, 1999). As described later in this chapter, in more recent years, these experimental limitations have been recognized and addressed and now rigorous and appropriate methodologies are being used (Reid et al., 2003).

Definitions

The term "probiotics," as originally used in 1965, referred to the "growth-promoting factors" produced by one microorganism that stimulated growth of another (i.e., the opposite of antibiotics). This definition went through several permutations and by 1989, probiotics were defined as a "live microbial feed supplement which beneficially affects the host animal by improving its intestinal balance" (Fuller, 1989). The current definition, adopted by the World Health Organization (part of the United Nations' Food and Agriculture Organization), defines probiotics as "live microorganisms which when administered in adequate amounts confer a health benefit on the host." This latest derivation is important because it recognizes both the relevance of a live and sufficient dose as well as the many reports that indicate probiotics may have health benefits that extend beyond the gastrointestinal tract. Whether all of the benefits ascribed to probiotics (Table 1) can be validated is another matter, however, as discussed later in this chapter.

Prebiotics are defined as "non-digestible food ingredient(s) that beneficially affects the host by selectively stimulating the growth and/or activity of one or a limited number of bacteria in the colon, and thus improves host health" (Gibson and Roberfroid, 1995). To re-phrase this definition, but with more detail, prebiotics are carbohydrate substances that escape digestion and adsorption in the stomach and small intestine and instead reach the colon. There, they are selectively metabolized by specific members of the colonic microflora. Prebiotics, therefore, enrich the population of those generally desirable commensal organisms such as lactobacilli and bifidobacteria at the expense (theoretically) of their less desirable competition.

Box 4–1.   Probiotics and Prebiotics *(Continued)*

Table 1.   Suggested health benefits of probiotic bacteria.

Reduce blood cholesterol
Maintain intestinal health
Alleviate intestinal bowel diseases
Modulate immune system
Reduce incidence of gastrointestinal infections
Reduce incidence of urinary and vaginal infections
Alleviate lactose intolerance
Anti-carcinogenic and anti-tumorogenic
Reduce incidence and severity of diarrheal diseases

Most of the prebiotics that are used commercially and that have received the most research attention are either polysaccharides or oligosaccharides. They are either derived from plant materials or are synthesized from natural disaccharide precursors (Figure 1). For example, inulin is

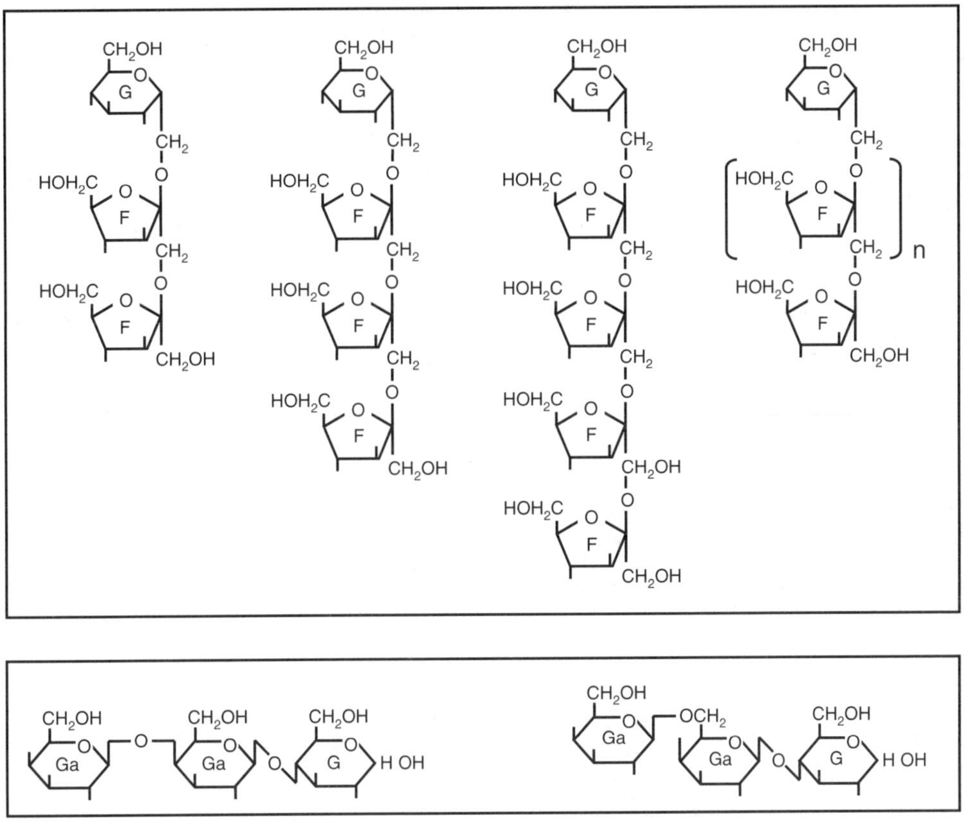

Figure 1.   Structure of prebiotics. Shown (upper panel, left-to-right) are short chain fructooligosaccharides (FOS), containing two, three, or four fructose units linked β-1,2 and with a terminal glucose. On the far right, longer chain FOS are shown, where n can equal up to twenty or more fructose units. The lower panel shows two forms of galactooligosaccharides (GOS), with galactose units linked β-1,4 (left) or β-1,6 (right), both linked to terminal glucose units.

(Continued)

Box 4–1. Probiotics and Prebiotics *(Continued)*

a naturally-occurring plant polysaccharide (consisting of fructose units, linked β-1,2 with a terminal glucose residue) that can be used in its intact form or as a mixture of partially hydrolyzed fructooligosaccharide (FOS) molecules. The latter can also be synthesized from sucrose via a transfructosylating enzyme that adds one, two, or three fructose units to the sucrose backbone.

Another type of prebiotic oligosaccharide that has attracted considerable attention are the galactooligosaccharides (GOS). These oligosaccharides are built from lactose via addition of galactose residues by β-galactosidases with high galactosyltransferase activity. Galactooligosaccharides are arguably the most relevant prebiotics being used in foods, since the GOS molecules closely resemble the oligosaccharides found in human milk. These human milk oligosaccharides (which also exist in milk from other species, but usually at lower levels) are now widely thought to be responsible for the bifidogenic properties associated with human milk. In fact, it had long been suggested that there was something in human milk that promoted growth of bifidobacteria (the so-called "bifidus" factor) and that this factor accounted for the dominance of these bacteria in the colon of nursed infants. That infants fed mother's milk suffered fewer intestinal infections and were generally healthier than formula-fed infants provided circumstantial evidence that having a greater proportion of bifidobacteria (and perhaps lactobacilli) in the colon would be desirable, not just for infants, but for the general population as well.

References

Fuller, R. 1989. Probiotics in man and animals. J. Appl. Bacteriol. 66:365–378.

Gibson, G.R., and M.B. Roberfroid. 1995. Dietary modulation of the human colonic microbiota: introducing the concept of prebiotics. J. Nutr. 125:1401–1412.

Lilly, D.M., and R.H. Stillwell. 1965. Probiotics: growth-promoting factors produced by microorganisms. Science 147:747–748.

Reid, G., M.E. Sanders, H.R. Gaskins, G.R. Gibson, A. Mercenier, R. Rastall, M. Roberfroid, I. Rowland, C. Cherbut, and T.R. Klaenhammer. 2003. New scientific paradigms for probiotics and prebiotics. J. Clin. Gastroenterol. 37:105–118.

Shortt, C. 1999. The probiotic century: historical and current perspectives. Trends Food Sci. Technol. 10:411–417.

Table 4.2. Commercial probiotic organisms used in dairy products[1].

Organism	Supplier or source
Lactobacillus acidophilus NCFM	Danisco, Madison, WI, USA
Lactobacillus acidophilus SBT-2062	Snow Brand Milk Products, Tokyo, Japan
Lactobacillus casei strain Shirota	Yakult, Tokyo, Japan
Lactobacillus casei F19	Arla Foods, Skanderborgvej, Denmark
Lactobacillus fermentum RC-14	Urex Biotech, London, Canada
Lactobacillus gasseri ADH	Danisco, Madison, WI, USA
Lactobacillus johnsonii KA1 (NCC 533)	Nestle, Lausanne, Switzerland
Lactobacillus plantarum 299v	Probi, Lund, Sweden
Lactobacillus reuteri SD2112 (ATCC 55730)	Biogaia, Stockholm, Sweden
Lactobacillus rhamnosus GR-1	Urex Biotech, London, Canada
Lactobacillus rhamnosus GG (ATCC 53103)	Valio Ltd., Helsinki, Finland
Lactobacillus salivarius UCC118	University College, Cork, Ireland
Bifidobacterium longum SBT-2928	Snow Brand Milk Products, Tokyo, Japan
Bifidobacterium longum BB536	Morinaga Milk Industry, Zama City, Japan
Bifidobacterium breve strain Yakult	Yakult, Tokyo, Japan

[1]Adapted from Reid, 2001 and Sanders, 1999.

be labeled as "directly set" to denote this manner of acidification—and often lack the flavor of the fermented versions. They certainly lack the nutritional benefits that may be contributed by live active cultures.

The function of lactic acid bacteria in the manufacture of cultured dairy products is quite simple—they should ferment lactose to lactic acid such that the pH decreases and the isoelectric point of casein, the major milk protein, is reached. By definition, the isoelectric point of any protein is that pH at which the net electrical charge is zero and the protein is at its minimum solubility. In other words, as the pH is reduced, acidic amino acids (e.g., aspartic acid and glutamic acid), basic amino acids (e.g., lysine and arginine), and partial charges on other amino acids become protonated and more positive such that at some point (i.e., the isoelectric point), the total number of positive and negative charges on these amino acids (as well as on other amino acids) are in equilibrium.

For casein, which ordinarily has a negative charge, the isoelectric point is 4.6. Thus, when sufficient acid had been produced to overcome the natural buffering capacity of milk and to cause the pH to reach 4.6, casein precipitates and a coagulum is formed. Along the way, the culture may also produce other small organic molecules, including acetaldehyde, diacetyl, acetic acid, and ethanol. Although these latter compounds are produced in relatively low concentrations, they may still make important contributions to the overall flavor profile of the finished product. The culture may also produce other compounds that contribute to the viscosity, body, and mouth feel of the product (see below). The choice of culture, therefore, is dictated by the product being produced, since different cultures generate flavor and aroma compounds specific to that product (Table 4-3).

The actual manufacture of cultured dairy products requires only milk (skim, lowfat, or whole, or cream, depending on the product) and a suitable lactic starter culture. However, the process is not quite so simple, because de-

Table 4.3. Organisms used as starter cultures in the manufacture of fermented dairy products.

Product	Organisms
Yogurt	*Streptococcus thermophilus* *Lactobacillus delbreckii* subsp. *bulgaricus*
Buttermilk	*Lactobacillus lactis* subsp. *lactis* *Lactobacillus lactis* subsp. *cremoris* *Leuconostoc lactis* *Leuconostoc mesenteroides* subsp. *dextranicum*
Sour Cream	*Lactobacillus lactis* subsp. *lactis* *Lactobacillus lactis* subsp. *cremoris* *Leuconostoc lactis* *Leuconostoc mesenteroides* subsp. *dextranicum*
Kefir	*Lactobacillus kefiri* *Lactobacillus kefiranofaciens* *Saccharomyces kefiri*

fects associated with flavor, texture, and appearance are not uncommon. As will be described later, the most frequent and serious problem in the manufacture of many of these products, especially yogurt, is syneresis.

Syneresis is defined as the separation (or squeezing out) of water from the coagulated milk. To many consumers, the appearance of these pools of slightly yellow-green water (which is actually just whey) from the top of the product is considered unnatural and objectionable. Thus, to minimize syneresis problems, and to improve the body and texture of the finished product, manufacturers perform several steps whose purpose is to enhance the water binding capacity of the milk mixture. First, the milk solids are increased, either by adding dry milk powder or by concentrating milk. Second, the milk mixture is heated well above ordinary pasteurization conditions to denature the whey proteins, exposing more amino acid residues to the aqueous environment. Finally, most manufacturers have incorporated stabilizers, thickening agents, and other ingredients into the formulation to further reduce syneresis. However, a few countries, France, in particular, prohibit many of these additional ingredients and instead rely on other means to reduce syneresis (discussed below).

Yogurt Manufacture

Milk treatment

Yogurt can be made from skim (non-fat), reduced fat, or whole milk. As is true for all dairy products, but especially so for yogurt and other cultured milks, it is important to use good quality milk, free of antibiotics and other inhibitory substances. The first step (Figure 4–2) involves adding nonfat dry milk to the milk to increase the total solids to 12% to 13%, sometimes to as high as 15%. Alternatively, the total milk solids can be increased by concentrating the milk via evaporation. Other permitted ingredients (see below) may be added, and the mix (only if it contains fat) is then usually homogenized, although there are some specialty manufacturers that produce unhomogenized, cream-on-the-top whole milk versions.

Yogurt is considered a fluid milk product by the U.S. Food and Drug Administration (FDA) and must be made using pasteurized milk. However, most yogurt mixes receive a heat treatment well above that required for pasteurization. Thus, instead of pasteurizing milk for 71.7°C for 15 seconds (the minimum required), mixes are heated to between 85°C and 88°C for up to 30 minutes. Other time-temperature conditions can also be used, but kinetically they are usually equivalent. Heating can be done in batch mode (i.e., in vats), but continuous heating in plate or tube type heat exchangers is far more common.

The high temperature treatment not only satisfies all of the normal reasons for pasteurization (i.e., killing pathogens and spoilage organisms and inactivating enzymes), but these severe heating conditions also perform two additional functions. First, even heat-resistant bacteria and their spores are killed, making the mixture essentially free of competing microorganisms. Second, the major whey proteins, α-lactalbumin and β-lactoglobulin, are nearly 100% denatured at the high pasteurization temperatures. These proteins exist in globular form in their native state but once denatured, amino acid residues are exposed and their ability to bind water, via hydrogen bonding, is significantly enhanced. Denatured whey proteins also reduce the Eh and stabilize the milk gel.

Yogurt styles

The pasteurized milk is then cooled via a heat exchanger to the desired incubation temperature, usually between 40°C and 43°C. Alternatively, the milk can be cooled as for conventional processing to 2°C to 4°C, and then warmed to the higher temperature later. The incubation temperature is critical, since it will influence the activity of the culture and ultimately the properties of the finished yogurt (see below).

The route the mix takes next depends on the type or style of yogurt being made. There are two general types of yogurt. Yogurt that is mixed with flavors, fruit, or other bulky ingredients is called stirred or Swiss-style yogurt. For this type, the mix is pumped into vats and the culture is added. The mixture is then incubated such that the entire fermentation occurs in the vat. At the end of the fermentation, the mixture is gently agitated and cooled, and the flavor ingredients are introduced. The mixture is then pumped into containers.

In contrast, mix can be inoculated with culture, pumped immediately into the container, and then fermented directly in the container. If this so-called fermented-in-the-cup style yogurt is to contain fruit or other bulky flavoring (i.e., fruit-on-the-bottom or sundae-style), the fruit or flavoring material is first dispensed into the cup and the yogurt mix added on top, followed by incubation and fermentation. The consumer must do the stirring and mixing to incorporate the flavoring throughout the product.

Yogurt cultures

For either style, the lactic culture represents a critical ingredient and must be carefully selected. Yogurt cultures, in general, and the ac-

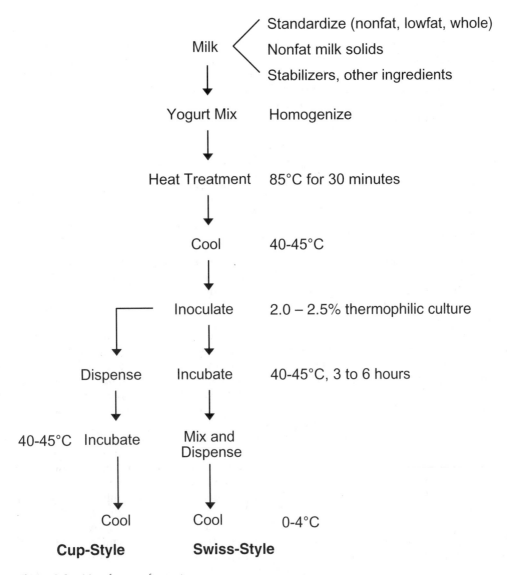

Figure 4–2. Manufacture of yogurt.

tual strains, in particular, have a profound influence on the flavor, texture, appearance, and overall quality attributes of the finished product (Box 4-2). As noted above, starter cultures for yogurt consist of two organisms, *L. delbrueckii* subsp. *bulgaricus* and *S. thermophilus*. Many commercial products contain more than one strain of each organism, but the traditional ratio of *L. delbrueckii* subsp. *bulgaricus* and *S. thermophilus* (or rod to coccus)

has usually been about 1:1. Although this ratio can be adjusted, depending on the desired properties of the final product, it is important that both organisms be present.

Yogurt made with only one of these two organisms will not turn out well, for several reasons. In fact, it has long been known that these bacteria grow faster and perform better when grown as a pair compared to when they are grown separately. The basis for this

Box 4–2. Selecting Yogurt Cultures

Criteria used for selecting yogurt starter culture strains are based on several general performance characteristics, as well as other traits that provide specific functions (Table 1). General requirements include vigorous and rapid growth, such that the culture is able to lower the pH to the target level within four to six hours. The culture should produce good flavor without bitterness or excess acidity. Resistance to bacteriophages is also expected. The starter culture must also be tolerant of storage conditions, whether frozen or lyophilized, so that viability is retained and long lag phases are avoided.

Table 1. Desired properties of yogurt cultures.

Stable during frozen or lyophilized storage	Able to produce the "right" consistency or body, i.e., ropiness
Viable and active after thawing or rehydration	Produces good yogurt flavor, without excess acid or acetaldehyde
Prompt growth and fermentation; short lag phase	No syneresis
Resistant to bacteriophage	No acidity produced during storage (over-acidification)

There are also several specific characteristics expected from the culture, depending on the wants and needs of the manufacturer. For example, cultures may be selected on the basis of whether they produce viscous yogurts with strong body or little viscosity and a relatively mild body. High-body yogurts are generally made using exopolysaccharide-producing cultures, as described elsewhere.

In addition to the use of customized cultures for body characteristics, there is also a preference among yogurt manufacturers in the United States to use cultures that produce so-called mild flavored yogurts (Ott et al., 2000). These cultures make enough acid to cause milk coagulation, but not so much that the pH drops below 4.4 to 4.5. In contrast, traditional acidic yogurts often reach a pH of 4.0 or even lower. The less acidic yogurts are not only preferred by many consumers, but if probiotic adjuncts are added to the yogurt, their rate of survival during storage is generally enhanced at the higher pH.

Of the two organisms used for yogurt fermentations, *Streptococcus thermophilus* and *Lactobacillus delbrueckii* subsp. *bulgaricus*, it is the latter that is usually (but not always) responsible for excessive acid production. This organism grows well at 42°C to 45°C, so one effective way to limit acid production is simply to incubate the yogurt at lower incubation temperatures (around 37°C). Alternatively, *L. delbrueckii* subsp. *bulgaricus* can be omitted altogether from the culture blend and replaced with a different strain (e.g., *Lactobacillus casei*) that may produce less acid. In addition, it is possible to select for natural acid-sensitive derivative strains (that necessarily stop making acid at some particular low pH). Presumably, some of the commercial yogurt cultures promoted as being low acid or mild-flavor cultures contain these strains. Similarly, strains that are sensitive to low temperature could also be used, since they would cease fermentation when the product was shifted to refrigeration temperature. One example of this particular approach was the isolation, via chemical mutagenesis, of a *L. delbrueckii* subsp. *bulgaricus* variant that synthesized a "cold-sensitive" β-galactosidase (Adams et al., 1994).

References

Adams, R.M., S. Yoast, S.E. Mainzer, K. Moon, A.L. Polombella, D.A. Estell, S.D. Power, and R.F. Schmidt. 1994. Characterization of two cold-sensitive mutants of the β-galactosidase from *Lactobacillus delbrueckii* subsp. *bulgaricus*. J. Biol. Chem. 269:5666–5672.

Ott, A., A. Hugi, M. Baumgartner, and A. Chaintreau. 2000. Sensory investigation of yogurt flavor perception: mutual influences of volatiles and acidity. J. Agri. Food Chem. 48:441–450.

synergistic growth has been the subject of considerable study since the 1960s. That *S. thermophilus* always appears to grow first in co-culture experiments suggests the initial milk environment is somehow less conducive for *L. delbrueckii* subsp. *bulgaricus*. It is now thought that *S. thermophilus* lowers the pH and Eh to levels preferred by *L. delbrueckii* subsp. *bulgaricus*. In contrast, *S. thermophilus* is weakly proteolytic and lacks the ability to hydrolyze casein, but can still make do, at least initially, by relying on the small pool of free amino acids present in the milk. Later, when those amino acids are depleted, *S. thermophilus* benefits by its association with *L. delbrueckii* subsp. *bulgaricus*, since the latter produces a proteinase that makes peptides and amino acids available for all organisms that happen to be present.

Eventually, however, during the co-culture environment, *L. delbrueckii* subsp. *bulgaricus* will produce more acid than can be tolerated by *S. thermophilus*. In yogurt, therefore, the *S. thermophilus* population may begin to decrease. If these organisms were continually propagated as a single mixed culture, the *S. thermophilus* would likely be displaced after several transfers. Therefore, in the manufacture of commercial yogurt cultures, the streptococci and lactobacilli are ordinarily grown separately in rich media under species-specific optimum conditions, harvested, and then mixed in the desired ratio.

Depending on the form of the culture and the manufacturer's instructions, the cultures are added to the yogurt mix to give an initial cell concentration of about 10^7 cells per gram. When bulk cultures are used, this translates into an inoculum of about 1.5% to 2% (on a weight basis). Highly concentrated cultures are now available, in either frozen or lyophilized forms, that require a much lower amount of culture to achieve the same starting cell density. The inoculated mixes are then incubated (either in the cup or vat) at 40°C to 45°C for four to six hours or until a titratable acidity (as lactic acid) of 0.8% to 0.9% is reached and the pH is about 4.4 to 4.6. Depending on the culture activity and desired final pH, the fermentation may be even shorter (i.e., three hours or less). The recent trend in the United States toward low acid yogurts means that the fermentation may be judged, by some manufacturers, as complete when the pH reaches 4.8 to 4.9, or just as the mixture begins to coagulate. In contrast, many traditional-style yogurts are fermented until the pH is near 4.0.

As noted above, the incubation temperature can have a profound effect on the fermentation and the overall product characteristics. In general, *L. delbrueckii* subsp. *bulgaricus* has a higher temperature optima than *S. thermophilus*. Growth of *S. thermophilus* is favored at temperatures below 42°C, whereas *L. delbrueckii* subsp. *bulgaricus* is favored above 42°C. Thus, by shifting the incubation temperature by just a few degrees, it is possible to influence the growth rates of the two organisms, as well as the metabolic products they produce. Since, for example, *L. delbrueckii* subsp. *bulgaricus* is capable of producing greater amounts of lactic acid and acetaldehyde compared to *S. thermophilus*, high incubation temperatures may result in a more acidic and flavorful yogurt.

Culture metabolism

S. thermophilus and *L. delbrueckii* subsp. *bulgaricus* both make lactic acid during the yogurt fermentation. They are homofermentative, meaning lactic acid is the primary end-product from sugar metabolism, and both ferment lactose in a similar manner. Moreover, the specific means by which lactose metabolism occurs in these bacteria not only dictate product formation, but also have an important impact on the health-promoting activity these bacteria provide (discussed later).

The first step involves transport of lactose across the cell membrane. As reviewed in Chapter 2, there are two general routes by which this step can occur in lactic acid bacteria. Mesophilic lactococci (i.e., *Lactococcus lactis* subsp. *lactis* and *Lactococcus lactis* subsp. *cremoris*) use the phosphoenolpyruvate (PEP)-dependent phosphotransferase system (PTS), in which lactose is phosphory-lated during its transport across the cytoplasmic membrane. The high energy phosphate group of PEP, following its transfer via a protein cascade, serves as the phosphoryl group donor for the phosphorylation reaction. It also provides the driving force for lactose transport. The product that accumulates in the cytoplasm, therefore, is lactose-phosphate, with the phosphate attached to carbon 6 on the galactose moiety. In contrast, *S. thermophilus* and *L. delbrueckii* subsp. *bulgaricus,* the thermophilic culture bacteria, use a secondary transport system (called LacS) for lactose uptake. Transport of lactose occurs via a symport mechanism, with the proton gradient serving as the driving force (although this is not the entire story, as discussed below). Lactose is not modified during transport and instead accumulates inside the cell as free lactose.

For both the PTS and LacS systems, the next step is hydrolysis, but since the substrates are different, the hydrolyzing enzymes must be different, too. In lactococci, hydrolysis of lactose-phosphate occurs by phospho-β-galactosidase, forming glucose and galactose-6-phosphate. Glucose is subsequently phosphorylated (using ATP as the phosphoryl group donor) by glucokinase (or hexokinase) to form glucose-6-phosphate. The latter then feeds directly into the Embden Meyerhof glycolytic pathway, ultimately leading to lactic acid. The other product of the phospho-β-galactosidase reaction, galactose-6-phosphate, is metabolized simultaneously by a parallel pathway called the tagatose pathway, which also results in formation of lactic acid.

In *S. thermophilus* and *L. delbrueckii* subsp. *bulgaricus,* the intracellular lactose that accumulates is hydrolyzed by a β-galactosidase, releasing glucose and galactose. The glucose is metabolized to lactic acid, the same as in the lactococci. However, most strains of *S. thermophilus* and *L. delbrueckii* subsp. *bulgaricus* lack the ability to metabolize galactose and instead secrete galactose back into the milk. This is despite the observation that a pathway for galactose metabolism indeed exists in these bacteria.

The reason why this pathway does not function, especially in *S. thermophilus*, and why this apparent metabolic defect occurs, has been the subject of considerable study. It now appears that the genes coding for the enzymes necessary to metabolize galactose, the so-called Leloir pathway, are not transcribed or expressed. Rather, these genes are strongly repressed, and the cell responds by excreting the unfermented galactose that it has accumulated. The extracellular galactose is also not metabolized even later, although genes coding for a galactose transport system apparently exist.

Although it would appear that *S. thermophilus* is being wasteful by not making efficient use of both monosaccharide constituents of lactose, this is not the case. When *S. thermophilus* grows in milk, where the lactose concentration is 5% (more than 140 mM), or in yogurt mix, which has an even higher lactose concentration due to added milk solids, sugar limitation is not an issue. Rather, growth cessation in yogurt occurs due to low pH or low temperature, not sugar availability. Interestingly, when *S. thermophilus* excretes galactose it does so via the LacS permease, the same system that is responsible for lactose uptake. Thus, LacS acts as an exchange system, trading an out-bound galactose for an in-coming lactose (Chapter 2). This precursor-product exchange reaction spares the cell of energy that it would normally spend to transport lactose. In fact, *S. thermophilus* may be so well-adapted to growth in milk that the galactose efflux reaction is kinetically favored, even in variant or mutant strains where the galactose genes are de-repressed and galactose pathway enzymes are expressed. In yogurt, galactose accumulation, even at concentrations as high as 0.7% to 1.0%, is ordinarily of no major consequence (the exception may be for individuals prone to cataracts, since dietary galactose may exacerbate that condition). However in cheese manufacture, galactose can be the cause of serious technological problems, and galactose-fermenting strains of *S. thermophilus* may be of considerable value (Chapter 5).

In some yogurt cultures, *Lactobacillus helveticus* is used instead of *L. delbrueckii* subsp. *bulgaricus*. This organism deals with lactose and galactose in a somewhat similar manner. Transport of lactose, as noted above, occurs via a LacS that has nearly the same activity as in *S. thermophilus*. Galactose efflux occurs after hydrolysis by β-galactosidase, but not nearly to the same extent. Instead, some of the galactose is metabolized via the Leloir pathway at the same time as glucose (Chapter 2). In addition, extracellular galactose can be transported and metabolized. However, little of the secreted galactose is fermented because the fermentation period is so short. Even most of the lactose is left unfermented. Since most yogurt is made with high solids milk, the lactose concentration in yogurt is often higher than that of milk, with levels of 6% to 7%.

Post-fermentation

The fermentation is considered complete when the target acidity is reached and the yogurt is cooled quickly to below 4°C. In fact, for all practical purposes, cooling is really the only way to arrest the fermentation and stop further acid production. Cup-set yogurt must be very carefully moved to coolers (0°C to 4°C) to avoid agitation which may disturb the gel, resulting in syneresis. For Swiss style yogurt, where the fermentation occurs in a vat, the yogurt is typically stirred and cooled in the vat, then mixed with fruit or other flavoring, and filled into cups or containers. It is important to recognize that during the cooling period the pH may continue to drop by an additional 0.2 to 0.3 pH units, so initiating the cooling step even when the pH is 4.8 to 4.9 may be warranted. In addition, some culture strains may continue to produce acid during refrigerated storage, albeit slowly. Over-acidification and post-fermentation acidification are major problems, since U.S. consumers generally prefer less acidic yogurts (see below).

Yogurt Flavor and Texture

The most dominant flavor of yogurt is sourness, due to lactic acid produced by the starter culture. Most yogurts contain between 0.8% and 1.0% lactic acid and have a pH below 4.6.

In the absence of sweetener or added flavors, most consumers can detect sourness when the pH is below 5.0. Other organic acids, including formic and acetic, may also be produced by the culture, but at much lower concentrations, and they generally make only modest contributions to yogurt flavor.

There are, however, other metabolic products produced by the culture that accumulate in yogurt and contribute significantly to flavor development. The most important of these is acetaldehyde, a two carbon aldehyde. Although normally present at less than 25 ppm concentration, this is still sufficient to give yogurt its characteristic tart or green apple flavor. Both *S. thermophilus* and *L. delbrueckii* subsp. *bulgaricus* can produce acetaldehyde; however, the rate and amount produced depends on the strain and on the growth conditions.

There appear to be at least two metabolic routes by which acetaldehyde is produced (Figure 4-3). In one pathway, an aldolase enzyme hydrolyzes the amino acid, threonine, and acetaldehyde and glycine are formed directly. It is also possible to form acetaldehyde from pyruvate (generated from glucose metabolism). The pyruvate can either be decarboxylated, forming acetaldehyde directly, or converted first to acetyl CoA, and then oxidized to acetaldehyde. By understanding how these pathways function, it may soon be possible to apply molecular metabolic engineering approaches to enhance or control acetaldehyde levels in yogurt. Finally, diacetyl and acetoin may also be produced by the yogurt culture; however, they are usually present at concentrations below a typical taste threshold.

In the United States, most yogurt is flavored with fruit and other flavorings. Fruits are usually added in the form of a thick puree, with or without real pieces of fruit. These ingredients obviously dilute or mask the lactic acid and acetaldehyde flavors. In fact, the trend in the United States is to produce mild-flavored yogurts with less characteristic yogurt flavor. Thus, strains that produce little acetaldehyde are often used in yogurt cultures.

The texture and rheological properties of yogurt are, perhaps, just as important to

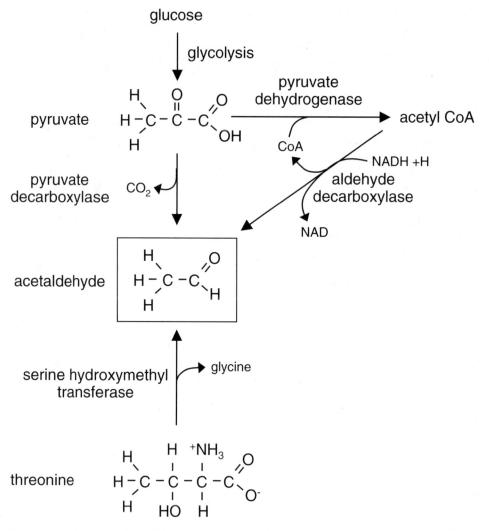

Figure 4–3. Formation of acetaldehyde by yogurt bacteria from pyruvate and threonine. Adapted from Chaves et al. 2002.

consumers as flavor. Coagulated milk is essentially a protein gel that imparts viscosity, mouth feel, and body to the finished product. Formation and maintenance of the gel structure, therefore, is important for yogurt quality. The gel properties are affected by the ingredients in the yogurt mix, how the yogurt mix is processed and produced, culture activity, and post-fermentation factors. For example, proper heat treatment of the milk has a profound effect on gel strength and, specifically, the water holding capacity. If the gel is poorly formed or disrupted syneresis will occur, leading to the defect known as "wheying off" (see below). To control syneresis and maintain a suitable gel structure, yogurt manufacturers in the United States commonly add stabilizers to the mix. Typically, stabilizers are hydrophilic polysaccharides whose purpose is to bind water. The most popular stabilizers are naturally derived gums and starches, including carrageenan, locust bean, and guar gums; corn starch; tapioca;

and pectin. Gelatin is also frequently added as a stabilizer.

In some European countries (e.g., France), stabilizers are not allowed in yogurt. Therefore, the ability of some strains of *S. thermophilus* and *L. delbrueckii* subsp. *bulgaricus* to produce and secrete natural polysaccharide material directly in the yogurt during the fermentation is an especially important trait. This is because many of these extracellular polysaccharides (exopolysaccharides or EPS) have excellent stabilizing and rheological properties. The producer strains are now widely used and included in yogurt starter cultures to provide the desired body characteristics. A variety of EPS produced by these bacteria have been identified (Box 4–3). Some EPS contain only a single sugar (homopolysaccharides). In contrast, other EPS consist of a heterogenous mixture of different sugar monomers (heteropolysaccharides) in varying ratios. The latter contain isomers of galactose and glucose, along with rhamnose and other sugars.

Defects

The stand-up comedian, George Carlin, used to ask how one could tell when yogurt has gone bad. Yogurt is, after all, sour milk, so how could it spoil? In fact, yogurt can suffer from flavor, texture, and appearance defects, and can even become spoiled. In some cases, these defects are caused by microorganisms (including the starter culture), but often they are caused by manufacturing issues. This is especially true for the wheying-off or syneresis defect that is probably the single most serious issue for yogurt consumers.

Although relatively rare, some chemically-derived flavor defects in yogurt can be caused by using poor-quality milk. For example, rancid or oxidized flavors are almost always due to mis-handling of the milk (or nonfat dry milk). The most common flavor defects, however, are caused by microorganisms. In some cases, microbial spoilage is due to the presence of thermophilic sporeforming bacteria, including various species of *Bacillus,* that survived pasteurization and grew in the milk, producing bitter flavors. Yeasts and molds that are present in the fruit flavoring material can also produce off-flavors, although it is usually just the appearance of yeast or mold colonies that cause consumers to reject such products. Of course, bacteria, such as psychrotrophs that grew in the raw milk prior to pasteurization, can produce bitter and rancid flavors.

Despite the acidity and acetaldehyde flavor normally expected in yogurt, these attributes can also be considered as defects if present at high enough concentrations or if they exceed consumer tastes. Yogurts within a range of pH 4.2 to 4.5 (0.8% to 1% lactic acid) are generally acceptable, but the pH can drop to as low as 4.0 (and about 2% lactic acid) under certain circumstances. Excessive acid production, for example, can occur if the yogurt is incubated at too high a temperature or if cooling rates are too slow (both promote growth of *L. delbrueckii* subsp. *bulgaricus*). Even properly produced yogurt can eventually become too acidic during low temperature storage due to the ability of the culture to produce, albeit slowly, lactic acid. This post-fermentation acidification is one of the main reasons why consumers may consider an otherwise perfectly fine yogurt to be spoiled (despite George Carlin's protestations).

One way to prevent over-acidification is to control culture activity. Strains have been identified (either naturally or via molecular interventions) that are sensitive to low pH or low temperatures, such that growth and acid formation are halted when the pH reaches a critical value or when the temperature is sufficiently low (Box 4-2). Of course, one other way to control acidity and to increase shelf-life would be to heat pasteurize the yogurt after the fermentation. This approach has been largely rejected by the industry because it would negate most of the health benefits conferred by the culture and would harm the image yogurt enjoys as a healthy food product. Moreover, the National Yogurt Association, a trade organization representing yogurt manufacturers, has suggested that refrigerated

Box 4–3. Exopolysaccharide production by lactic acid bacteria

The ability of lactic acid bacteria (LAB) to synthesize exopolysaccharides (EPS) is now regarded as particularly important because they enhance texture development in fermented dairy foods. Specifically, they can serve as natural thickeners, stabilizers, bodying agents, emulsifiers, gelling agents, and fat replacers (De Vuyst and Vaningelgem, 2003; Ruas-Madiedo and de los Reyes-Gavilán, 2005).

They are especially applicable in yogurt and other cultured dairy products since they can reduce syneresis and gel fracture, and can increase viscosity and gel strength (Duboc and Mollet, 2001). Thus, incorporating EPS-producing strains in dairy starter cultures could eliminate the necessity for adding other stabilizing agents. This is important in some European countries where the addition of stabilizers is restricted.

Lactic acid bacteria that produce EPS do so in one of two general ways. Some strains synthesize EPS as released or dis-attached material (referred to as "ropy" strains), whereas others produce capsular polysaccharide that remains attached to the cell surface (Broadbent et al., 2003). However, it appears that the EPS from both ropy and capsular strains have useful functional properties in cultured milk products, regardless of how synthesis and attachment occurs. EPS may serve as a defense mechanism against bacteriophages by preventing their adsorption (Forde and Fitzgerald, 1999; Lamothe et al., 2002). In addition, some LAB may produce EPS to adhere or stick to various surfaces, thereby promoting colonization of desirable habitats (Lamothe et al., 2002).

Structure of Exopolysaccharides

The molecular masses of EPS from LAB range from 10 kDa to greater than 1,000 kDa (Welman and Maddox, 2003). The number of monomers present within these EPS range from about 50 to more than 5,000. The most frequently found monosaccharides are glucose and galactose, but rhamnose, mannose, fructose, arabinose and xylose, and the sugar derivatives, *N*-acetylgalactosamine and *N*-acetylglucosamine, are also common (Table 1).

In general, the EPS can be subdivided into two major groups: homopolysaccharides and heteropolysaccharides (Jolly et al., 2002; Ruas-Madiedo et al., 2002). Homopolysaccharides are composed of only one type of monosaccharide, whereas heteropolysaccharides are composed of a repeating unit that contains two or more different monosaccharides. Examples of the homopolysaccharides include: (1) dextrans and mutans, α-1,6 and α-1,3-linked glucose polymers produced by strains of *Leuconostoc mesenteroides* and *Streptococcus mutans*, respectively; and (2) glucans, β-1,3-glucose polymers, produced by strains of *Pediococcus* and *Lactobacillus*; (3) fructans, β-2,6-linked D-fructose polymers produced by strains of *Streptococcus salivarius*, *Lactobacillus sanfranciscensis*, and *Lactobacillus reuteri*; and (4) polygalactans, galactose produced by strains of *Lactococcus lactis*.

The heteropolysaccharides produced by lactic acid bacteria differ considerably between strains in their composition, structure, and physicochemical properties (De Vuyst and Degeest, 1999). Heteropolysaccharides most often contain a combination of glucose, galactose, and rhamnose, as well as lesser amounts of fructose, acetylated amino sugars, ribose, glucuronic acid, and noncarbohydrate constituents such as glycerol, phosphate, pyruvyl, and acetyl groups (Jolly et al., 2002). These EPS are the ones most commonly associated with yogurt, due to their production by *Lactobacillus delbrueckii* subsp. *bulgaricus*, *Lactobacillus helveticus*, and *Streptococcus thermophilus*. The majority of the EPS produced by *S. thermophilus* contain repeating tetrasaccharide units; however, several strains produce EPS consisting of hexasaccharides or heptasaccharides. Strains of *L. delbrueckii* subsp. *bulgaricus* typically produce heptasaccharides composed of glucose, rhamnose, and a high ratio of galactose, whereas *L. helveticus* produces heteropolysaccharides made up almost exclusively of glucose and galactose. Some mesophilic LAB, including *Lactococcus lactis* subsp. *lactis*, *L. lactis* subsp. *cremoris*, *L. sakei*, and *L. rhamnosus*, are also heteropolysaccharide EPS producers.

Box 4–3. Exopolysaccharide production by lactic acid bacteria *(Continued)*

Table 1. Exopolysaccharide composition of various lactic acid bacteria.

Lactic acid bacteria (Strain)	Exoploysaccharide Composition	Ratio of Sugars
Lactobacillus delbrueckii subsp. *bulgaricus*		
LY03	Heptasaccharide of glucose, galactose, rhamnose	Gal:Glu:Rha 5:1:1
rr	Heptasaccharide of glucose, galactose, rhamnose	Gal:Glu:Rha 5:1:1
Lfi5	Heptasaccharide of glucose, galactose, rhamnose	Gal:Glu:Rha 5:1:1
291	Pentasaccharide of glucose, galactose	Glu:Gal: 3:2
Lactobacillus lactis subsp. *cremoris*		
B39	Heptasaccharide of glucose, galactose, rhamnose	Gal:Glu:Rha 3:2:2
H415	Pentasaccharide of galactose	—
SBT 0495	Pentasaccharide glucose, galactose, rhamnose, phosphate	Gal:Glu:Rha:Pho 2.1:1.8:1.3:1
Lactobacillus helveticus TY1-2	Heptasaccharide of glucopyranosyl, galactopyranosyl, 2-acetamido-2-deoxy-D-glucopyranosyl	GluP:GalP:AGluP: 3.0:2.8:0.9
Lactobacillus rhamnosus RW-9595M	Heptasaccharide of glucose, galactose, rhamnose, pyruvate	Glu:Gal:Rha:Pyr 2:1:4:1
Lactobacillus sakei 0-1	Pentasaccharide of glucose, rhamnose, *sn*-glycerol 3-phosphate	Glu:Rha:G3P 3:2:1
Lactobacillus sanfranciscensis LTH2590	Trisaccharide of glucose and fructose	Glu:Fru 1:2
Streptococcus thermophilus		
S3	Hexasaccharide of galactose and rhamnose	Gal:Rha 2:1
Sy89	Tetrasaccharide of glucose and galactose	Gal:Glu 1:1
Sy102	Tetrasaccharide of glucose and galactose	Gal:Glu 1:1
Sfi6	Tetrasaccharide of glucose, and galactose, *N*-acetylgalactosamine	Gal:Glu:GalNAc 2:1:1
Sfi12	Hexasaccharide of galactose, rhamnose, glucose	Gal:Rha:Glu 3:2:1
Sfi39	Tetrasaccharide of glucose and galactose	Gal:Glu 1:1
IMDO1	Tetrasaccharide of glucose and galactose[1]	Gal:Glu 3:1
IMDO2	Tetrasaccharide of glucose and galactose[1]	Gal:Glu 3:1
IMDO3	Tetrasaccharide of glucose and galactose[1]	Gal:Glu 3:1
NCFB 859	Tetrasaccharide of glucose and galactose[1]	Gal:Glu 3:1
CNCMl	Tetrasaccharide of glucose and galactose[1]	Gal:Glu 3:1
MR-lC	Octasaccharide of galactose, rhamnose, fucose	Gal:Rha:Fuc 5:2:1
EU20	Heptasaccharide of glucose, galactose, rhamnose	Glu:Gal:Rha 2:3:2
OR 901	Hepatsaccharide of galactopyranose, rhamnopyranose	GalP:RhaP 5:2

[1]One galactose residue occurs as *N*-acetygalactosamine

(Continued)

Box 4–3. Exopolysaccharide production by lactic acid bacteria *(Continued)*

Production of Exopolysaccharides

Growth-associated EPS biosynthesis has been observed for several strains of LAB. In *Streptococcus thermophilus* LY03 and related strains, for example, there is a direct relationship between EPS yield and optimal growth conditions (de Vuyst et al., 1998). In contrast, EPS production in mesophilic LAB appears to be enhanced by sub-optimal temperatures, with maximum EPS production occurring at 20°C (Degeest and de Vuyst, 1999). The increase in EPS production when cells are growing at lower temperatures may be due to an increase in the number of isoprenoid lipid carrier molecules available for EPS formation, as occurs for Gram negative bacteria (Sutherland, 1972). Another possibility is that EPS synthesis may occur as a response to stress when cells are exposed to harmful or sub-optimal environments (Ophir and Gutnick, 1994). In either case, EPS production is likely correlated to EPS gene expression.

Gene organization for Exopolysaccharides

Genes encoding for EPS production exist as clusters, with specific structural, regulatory, and transport functions (Figure 1). In general, these gene clusters are often located on plasmids in mesophilic LAB and on the chromosome in thermophilic LAB (de Vuyst et al., 2001). Genes or gene clusters involved in LAB exopolysaccharide biosynthesis have been sequenced for several yogurt strains, as well as other mesophilic LAB (Broadbent et al., 2003; van Kranenburg et al., 1999b, 1999c). These clusters reveal a common operon structure, with the genes positioned in a particular order (Figure 1).

At the beginning of the gene cluster is *epsA*, which is suggested to be the region potentially involved in EPS regulation. The central region of the gene cluster, consisting of *epsE*, *epsF*, *epsG*, *epsH*, and *epsI*, encodes for enzymes predicted to be involved in the biosynthesis of the EPS repeating unit. The products of *epsC*, *epsD*, *epsJ*, and *epsK* are responsible for polymerization and export of the EPS. Interestingly, *epsA*, *epsB*, and *epsC* are also present in non-EPS producing

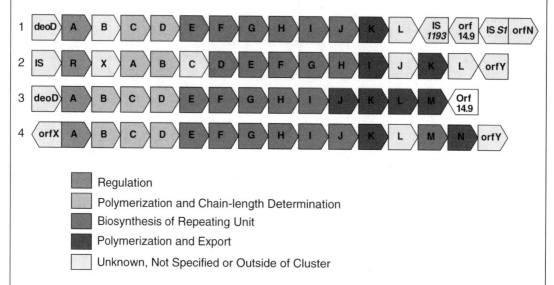

Regulation

Polymerization and Chain-length Determination

Biosynthesis of Repeating Unit

Polymerization and Export

Unknown, Not Specified or Outside of Cluster

Figure 1. Functional comparison of EPS genes from lactic acid bacteria. 1, *Streptococcus thermophilus* NCFB 2393; 2, *Lactobacillus lactis* NIZO B40; 3, *Streptococcus thermophilus* Sfi6; 4, *Lactobacillus delbrueckii* subsp. *bulgaricus* Lif5. Gene maps were adapted from Almirón-Roig et al. 2000; van Kranenburg et al. 1999a; Stingele et al. 1996; and Lamothe et al. 2002.

Box 4–3. Exopolysaccharide production by lactic acid bacteria *(Continued)*

strains of *S. thermophilus,* indicating the possibility that a spontaneous genomic deletion of an ancestral EPS gene cluster may have occurred (Bourgoin et al., 1999). This theory is supported by the observation of genetic instability, including deletions, in the *S. thermophilus* genome.

Regulation of Exopolysaccharide Genes

Production of EPS by LAB is often unstable and yields can be highly variable (Ricciardi and Clementi, 2000). Compared to other Gram positive bacteria, LAB generally produce lower amounts of EPS. Even when grown under optimized conditions, most LAB produce $<$ 3 g per L of EPS, whereas other organisms make as much as 10 g to 15 g per L. Therefore, several approaches have been considered to improve EPS yield and stability (Boels et al., 2003a, 2003b; Levander et al., 2002; Stingele et al., 1999; Kranenburg et al., 1999a).

One strategy involves pathway engineering, such that carbon flow is diverted away from catabolism and toward EPS biosynthesis. For example, synthesis of EPS ordinarily starts when glucose-6-phosphate is isomerized to glucose-1-phosphate by phosphoglucomutase (the product of the *pgm* gene). The glucose-1-phosphate is then converted, via UDP-glucose pyrophosphorylase to UDP-glucose, which serves as a precursor for EPS biosynthesis. Over-expression of *pgm* or *galU* (encoding for UDP-glucose pyrophosphorylase) led to an increase in the EPS yield by *S. thermophilus* LY03 (Hugenholtz and Kleerebezem, 1999).

A second approach involves over-expression of EPS genes. For example, over-expression of the *epsD* gene (encoding the priming glycosyl transferase) in *L. lactis* resulted in a small increase in EPS production (van Kranbenburg et al., 1999c). Cloning the entire EPS gene cluster on a single plasmid with a high copy number could also have an effect of increasing EPS production, as shown for *L. lactis* NIZO B40, which produced four times more EPS than the parent strain (Boels et al., 2003).

References

Almirón-Roig, E., Mulholland, F., Gasson, M.J., and A.M. Griffin. 2000. The complete *cps* gene cluster from *Streptococcus thermophilus* NCFB 2393 involved in the biosynthesis of a new exopolysaccharide. Microbiol. 146:2793–2802.

Boels, I.C., M. Kleerebezem, and W.M. de Vos. 2003a. Engineering of carbon distribution between glycolysis and sugar nucleotide biosynthesis in *Lactococcus lactis*. Appl. Environ. Microbiol. 69:1129–1135.

Boels, I.C., R. van Kranenburg, M.W. Kanning, B.F. Chong, W.M. de Vos, and M. Kleerebezem. 2003b. Increased exopolysaccharide production in *Lactococcus lactis* due to increased levels of expression of the NIZO B40 *eps* Gene Cluster. Appl. Environ. Microbiol. 69:5029–5031.

Bourgoin, F., A. Pluvinet, B. Gintz, B. Decaris, and G. Guedon. 1999. Are horizontal transfers involved in the evolution of the *Streptococcus thermophilus* exopolysaccharide synthesis loci? Gene 233:151–161.

Broadbent, J.R., D.J. McMahon, D.L. Welker, C.J. Oberg, and S. Moineau. 2003. Biochemistry, genetics, and applications of exopolysaccharide production in *Streptococcus thermophilus*: a review. J. Dairy Sci. 86:407–423.

Degeest, B. and L. De Vuyst. 1999. Indication that the nitrogen source influences both amount and size of exopolysaccharides produced by *Streptococcus thermophilus* LY03 and modelling of the bacterial growth and exopolysaccharide production in a complex medium. Appl. Environ. Microbiol. 65:2863–2870.

De Vuyst, L., and B. Degeest. 1999. Heteropolysaccharides from lactic acid bacteria. FEMS Microbiol. Rev. 23:157–177.

De Vuyst, L., and F. Vaningelgem. 2003. Developing new polysaccharides, p. 257—320. *In* B.M. McKenna (ed.), *Texture in food, vol. 2. Semi-solid foods*. Woodhead Publishing Ltd., Cambridge, United Kingdom.

De Vuyst, L., F. De Vin, F. Vaningelgem, and B. Degeest. 2001. Recent developments in the biosynthesis and applications of heteropolysaccharides from lactic acid bacteria. Int. Dairy J. 11:687–707.

(Continued)

Box 4–3. Exopolysaccharide production by lactic acid bacteria *(Continued)*

De Vuyst, L., F. Vanderveken, S. Van de Ven, and B. Degeest. 1998. Production by and isolation of exopolysaccharides from *Streptococcus thermophilus* grown in milk medium and evidence for their growth-associated biosynthesis. J. Appl. Microbiol. 84:1059–1068.

Duboc, P., and B. Mollet. 2001. Applications of exopolysaccharides in the dairy industry. Int. Dairy J. 11: 759–768.

Forde, A., and G.F. Fitzgerald. 1999. Analysis of exopolysaccharide (EPS) production mediated by the bacteriophage adsorption blocking plasmid, pCI658, isolated from *Lactococcus lactis* subsp. *cremoris* HO2. Int. Dairy J. 9:465–472.

Hugenholtz, J. and Kleerebezem, M. 1999. Metabolic engineering of lactic acid bacteria: overview of the approaches and results of pathway rerouting involved in food fermentations. Curr. Opin. Biotechnol. 10:492–497. Jolly, L., S.J.F. Vincent, P. Duboc, and J.-R. Nesser. 2002. Exploiting exopolysaccharides from lactic acid bacteria. Antonie van Leeuwenhoek 82:367–374.

Lamothe, G.T., L. Jolly, B. Mollet, and F. Stingele. 2002. Genetic and biochemical characterization of exopolysaccharide biosynthesis by *Lactobacillus delbrueckii* subsp. *bulgaricus*. Arch. Microbiol. 178:218–228.

Levander, F., M. Svensson, and P. Rådström. 2002. Enhanced exopolysaccharide production of metabolic engineering of *Streptococcus thermophilus*. Appl. Environ. Microbiol. 68:784–790.

Ophir, T. and Gutnick, D.L. 1994. A role for exopolysaccharides in the protection of microorganisms from desiccation. Appl. Environ. Microbiol. 60:740–745.

Ricciardi, A. and F. Clementi. 2000. Exopolysaccharides from lactic acid bacteria: structure, production and technological applications. Ital. J. Food Sci. 12:23–45.

Ruas-Madiedo, P. and C.G. de los Reyes-Gavilán. 2005. Methods for the screening, isolation, and characterization of exopolysaccharides produced by lactic acid bacteria. J. Dairy Sci. 88:843–856.

Ruas-Madiedo, P., J. Hugenholtz, and P. Zoon. 2002. An overview of the functionality of exopolysaccharides produced by lactic acid bacteria. Int. Dairy J. 12:163–171.

Stingele, F., J.-R. Neeser, and B. Mollet. 1996. Identification and characterization of the *eps* (exopolysaccharide) gene cluster from *Streptococcus thermophilus* Sfi6. J. Bacteriol. 178:1680–1690.

Stingele, F., S. Vincent, E.J. Faber, J.W. Newell, J.P. Kamerling, and J.R. Nesser. 1999. Introduction of the exopolysaccharide gene cluster from *Streptococcus thermophilus* Sfi6 into *Lactococcus lactis* MG1363: production and characterization of an altered polysaccharide. Mol. Microbiol. 32:1287–1295.

Sutherland, I.W. 1972. Bacterial exopolysaccharides. Adv. Microbial Physiol. 8:143–213.

van Kranenburg, R., I.C. Boels, M. Kleerebezem, and W.M. de Vos. 1999a. Genetics and engineering of microbial exopolysaccharides for food: approaches for the production of existing and novel polysaccharides. Curr. Opin. Biotechnol. 10:498–504.

van Kranenburg, R., I.I. van Swam, J.D. Marugg, M. Kleerebezem, and W.M. de Vos. 1999c. Exopolysaccharide biosynthesis in *Lactococcus lactis* NIZO B40: functional analysis of the glycosyltransferase genes involved in synthesis of the polysaccharide backbone. J. Bacteriol. 181:338–340.

van Kranenburg, R., H.R. Vos, I.I. van Swam, M. Kleerebezem, and W.M. de Vos. 1999b. Functional analysis of glycosyltransferase genes from *Lactococcus lactis* and other gram-positive cocci: complementation, expression, and diversity. J. Bacteriol. 181:6347–6353.

Welman, A.D., and I.S. Maddox. 2003. Exopolysaccharides from lactic acid bacteria: perspectives and challenges. Trends Biotechnol. 21:269–274.

yogurt contain at least 10^8 viable organisms per gram at the time of manufacture.

Texture, body, and appearance defects are common in yogurt. Two of the more serious defects, weak bodied yogurt and free whey formation (i.e., wheying off), have several causes. They may occur when the solids content is too low or when the mix was pasteurized at too low of a temperature, both of which ultimately influence how much denatured whey proteins are present. The gel structure will shrink and syneresis will occur if the culture produces too much acid too fast during the fermentation, causing the pH to become too low. The same result will also occur if the temperature was lowered too rapidly during cooling. When yogurt is mixed with fruit or other flavorings, in the case of Swiss or stirred style yogurts, or when stirred by consumers prior to consumption, the gel is further disrupted and thinning and syneresis occurs.

Nutritional Benefits of Yogurt

One of great all-time television commercials aired in the late 1970s and featured an interview with a Russian centagenerian who claimed that his longevity was due to his daily consumption of yogurt. When asked who got him started on his yogurt regimen, he proudly stated it was his mother, who's smiling face then moves into the television frame. Indeed, the popularity of yogurt, as implied by this advertisement and as mentioned previously in this chapter (Box 4-1), has long been due, in large part, to the purported health benefits ascribed to yogurt consumption. In fact, this notion of yogurt as an elixir that fends off aging and promotes human health originated at least a century ago, when the Russian immunologist and Nobel laureate Elie (Elia or Ilya) Metchnikoff published *The Prolongation of Life* in 1906. As Metchnikoff and many other microbiologists have since reported, it is the specific bacteria that either conduct the fermentation (i.e., *S. thermophilus* and *L. delbrueckii* subsp. *bulgaricus*) or that are added to yogurt in the form of culture adjuncts (e.g., lactobacilli and bifidobacteria) that are responsible for the desirable health benefits of yogurt.

Of course, yogurt has nutritional properties other than those derived from the culture organisms. A single 170 g (6 ounce) serving of plain, nonfat yogurt contains about 170 calories and supplies 18% of the Daily Value requirements for protein, 30% for calcium, and 20% for vitamin B_{12}. Still, these are the same nutrient levels one would get from milk (provided one had accounted for the milk solids normally added to the yogurt mix). Although there are reports that yogurt contains more vitamins than the milk from which it was made (due to microbial biosynthesis), these increases do not appear to be significant. Thus, if indeed yogurt has an enhanced nutritional

quality compared to milk, those differences must be due to the microorganisms found in yogurt. As noted earlier, there are many health claims ascribed to yogurt, and especially the probiotic bacteria added as nutritional adjuncts. Among these claims are that these bacteria are anti-cholesterolemic and anti-tumorigenic, enhance mineral absorption, promote gastrointestinal health, and reduce the incidence of enteric infections. There are also some rather surprising health benefits that yogurt may provide (Box 4-4). Despite volumes of research on these claims, however, few have been unequivocally established (Box 4-5.).

Perhaps the health benefit that is most generally accepted is the claim that yogurt organisms can reduce the symptoms associated with lactose intolerance. Lactose intolerance is a condition that is characterized by the inability of certain individuals to digest lactose. Its specific cause is due to the absence of the enzyme β-galactosidase, which is ordinarily produced and secreted by the cells that line the small intestine (Figure 4-4). In individuals expressing β-galactosidase, lactose is hydrolyzed and the glucose and galactose are absorbed across the epithelial cells and eventually enter into the blood stream (in the case of galactose, only after conversion to glucose in the liver). If β-galactosidase is not produced in sufficient levels, however, the lactose remains undigested and is not absorbed. Instead, it passes to the large intestine, where it either causes an increase in water adsorption into the colon (via osmotic forces) or is fermented by colonic anaerobes. The resulting symptoms can include diarrhea, gas, and bloating, leading many lactose intolerant individuals to omit milk and dairy products from their diet. Lactose intolerance has a genetic basis, affecting African, Asian, American Indian, and other non-Caucasian populations far more frequently than Caucasian groups. These individuals could typically tolerate lactose (i.e., milk) when young, but lose this ability during adulthood. As many as 50 million people in the United States may be lactose intolerant.

It has often been noted that lactose-intolerant individuals could consume yogurt without ill effect. It is now known that the bacteria in

Box 4–4.   Stressed out? Have a cup of yogurt and relax

It's exam time and you're feeling a bit stressed, maybe even depressed. How can you possibly relax so that you can study effectively? The solution is not have yet another cup of coffee, but to have a cup of yogurt (yes, yogurt) instead. At least, that's the suggestion made recently by a research group at the Consejo Superior de Investigaciones Científicas research institute in Spain (Marcos et al., 2004).

In this study, two groups of college students were fed, on a daily basis, either plain milk or milk fermented with *Streptococcus thermophilus* and *Lactobacillus delbrueckii* subsp. *bulgaricus* (i.e., yogurt bacteria), plus a probiotic organism, *Lactobacillus casei*. The study began three weeks prior to the start of an annual college testing period and continued for the three weeks during which the tests were administered. Consumption of other fermented milk products was not allowed, although no other dietary restrictions were imposed.

Various indicators of stress and anxiety were measured at the start (baseline) and at the end of the study. Anxiety self-report tests and analyses for lymphocytes, phagocytic activity, and serum cortisol levels were included. The latter analyses can be used to assess the effect of stress on the immune response of the subject. Although no significant differences were reported between the anxiety scores for the control (milk) and treatment groups (fermented milk), differences were observed for other measures. Specifically, the total lymphocyte numbers were higher in the treatment group, as was the number of CD56 cells (a specific subset or type of "killer cell" lymphocytes).

Because lymphocytes, and CD56 cells in particular, are part of the host defense system for fighting infections, the lower levels in the control group (compared to treatment subjects) may indicate that these subjects could be exposed to a greater risk of infection. Conversely, the yogurt-eating group may have a more resilient immune system, and, therefore, are better prepared to tolerate stress.

Recently, it was suggested that probiotic bacteria may also be effective at alleviating symptoms related to major depressive disorder or MDD (Logan and Katzman, 2005). This hypothesis is based, according to the authors, on several observations. First, there appears to be a relatively high correlation between the number of individuals who suffer from MDD and who also have intestinal disorders, atopic disease, endometriosis, malabsorption syndromes, small intestine bacterial overgrowth syndrome, and other chronic diseases. Second, the intestinal flora in these patients often contain lower levels of lactobacilli and bifidobacteria, and stress can lead to even greater long-term reductions in these organisms. Third, as the authors note, the intestinal tract is "a meeting place of nerves, microorganisms, and immune cells." Thus, the microflora, by virtue of their ability to produce various neurochemicals and to influence host production of cytokines and other bioactive compounds, may have intimate interactions with the immune and central nervous system.

Therefore, while more studies are certainly needed to establish the relationship between probiotic bacteria and stress and depression, students may, nonetheless, wish to consider the "chilling" effects of yogurt before the next big test.

References

Logan, A.C., and M. Katzman. 2005. Major depressive disorder: probiotics may be an adjuvant therapy. Med. Hypotheses. 64:533–538.

Marcos, A., J. Wärnberg, E. Nova, S. Gómez, A. Alvarez, R. Alvarez, J. A. Mateos, and J. M. Cobo. 2004. The effect of milk fermented by yogurt cultures plus *Lactobacillus casei* DN-114001 on the immune response of subjects under academic stress. Eur. J. Nutr. 43:381–389.

Box 4–5.   Sorting Through the Pros and Cons of Probiotics

Few research areas in the food and nutritional sciences have received as much worldwide attention and study as the subject of probiotics. Since the late 1980s, there has been a rapid rise in the amount of research being conducted on probiotic microorganisms (Figure 1). Included in the scientific literature are descriptions of the biochemical and physiological properties of probiotic bacteria, their taxonomical classifications, criteria for their selection, and data from *in vitro* and *in vivo* studies in which physiological effects on hosts have been tested. In addition, there is a plethora of information in the popular press, trade magazines, and especially the Internet, on probiotics.

The challenge for probiotic researchers has been to establish mechanisms and degrees of efficacy of probiotics to prevent or treat particular diseases or conditions. These efforts are complicated by many factors, including:(1) the complexity of microbial communities of the alimentary canal; (2) the variety of potential probiotic organisms to evaluate; (3) the diversity of possible diseases or conditions that might be influenced by probiotics; and (4) the need for plausible mechanisms. Furthermore, there is a need for validated "biomarkers"—biological characteristics that predict underlying health or disease and that can be objectively measured.

Finally, clinical studies, designed like those in the pharmaceutical industry, have not (until recently) been done. In an effort to reconcile these issues, several scientific organizations (e.g., International Scientific Association for Probiotics and Prebiotics, the Canadian Research and Development Centre for Probiotics, and Proeuhealth) have been formed to provide structure to the debate, assess the state of the science, coordinate research, and, ultimately, formulate recommendations.

Which organisms are probiotic?

In truth, the probiotic field is replete with mis-named, mis-characterized, and mis-represented organisms. In some cases, an organism was long ago assigned to a particular species and the name stuck, even though the particular species was no longer officially recognized. As a result, some probiotic-containing foods and formulations contain microorganisms whose value, as a probiotic, at least, is rather dubious. Moreover, the lack of appropriate regulations or standards (with regard to probiotics in foods) means that for cultured dairy products and other foods that profess to contain probiotic bacteria, there may or may not be organisms present at sufficient levels and viability (assuming they are alive), or that have the expected functional characteristics.

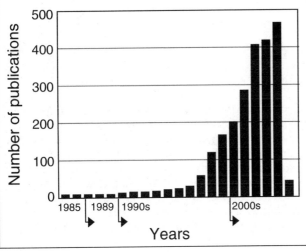

Figure 1. Publications on probiotics in the PubMed database. Citations were obtained using "probiotic" or "probiotics" as key words for each of the year(s) indicated. Data for 2005 are through February.

(Continued)

Box 4–5. Sorting Through the Pros and Cons of Probiotics *(Continued)*

A good example of the nomenclature confusion concerns the organism "Lactobacillus sporogenes." This organism was described in early editions of *Bergey's Manual* (prior to 1939), but has since been shown to be a misclassification and now "Lactobacillus sporogenes" is considered to be scientifically invalid (Sanders et al., 2003).

It is important to recognize, however, that just because a given organism is appropriately named does not ensure that the organism has probiotic activity. In other words, while solid evidence may exist to support the probiotic functions of a specific strain of *Bifidobacterium bifidus* or *Lactobacillus acidophilus*, that does not mean that all strains of those organisms meet the minimum relevant criteria for a probiotic or that they have even been evaluated in any way for probiotic activity (Table 1).

Table 1. Functional characteristics of probiotic bacteria[1].

Property	Survival	Colonization	Bio-functional	Safety
Acid tolerance	+			
Bile tolerance	+			
Stress tolerance	+			
Compatibility with starter culture	+			
Adherence factors		+		
Attachment factors		+		
Prebiotic metabolism		+		
High growth rate		+		
Competition factors		+		
Produce short-chain acids		+		
Immunomodulating			+	
Hypocholesteremic			+	
Antitumorogenic			+	
Enhance gut barrier function			+	
Lactose hydrolysis			+	
Antimicrobial activity			+	
Stimulate mucus production			+	
Avirulent				+
Noninflammatory				+
Absence of side-effects				+
Absence of gene transfer				+

[1]Note, the desirability of these properties for a given probiotic is target-dependent

In addition, the number and viability of cells present in the food vehicle or formulation may be inadequate to promote a probiotic effect. For example, 10^9 cells per gram are thought to be the minimum level that should be present in yogurt, although no regulatory requirement currently exists. Moreover, questions have been raised regarding whether bacteria in yogurt actually survive the route through the alimentary system. It was recently reported, for example, that volunteer subjects consuming yogurt daily for two weeks failed to shed yogurt bacteria in their feces at any time during the study (del Campo et al., 2005). Another complication concerns the claim that organisms other than bifidobacteria and lactobacilli may also have probiotic effects. These include organisms not ordinarily associated with growth in the intestinal tract, such as *Bacillus, Sporolactobacillus,* and *Brevibacillus* (Sanders et al., 2003). Whether or not these organisms can actually survive digestion and have an impact in the gastrointestinal tract remains uncertain. In addition, enterococci, *Escherichia coli,* and even yeasts (especially *Saccharomyces boulardii*) have been similarly promoted as probiotics and are also widely available. Finally,

Box 4–5. Sorting Through the Pros and Cons of Probiotics *(Continued)*

questions have been raised in recent years regarding the safety of lactobacilli and other lactic acid bacteria (despite their long time consumption in foods).

These safety concerns are based mainly on just a few sporadic cases in which probiotic organisms were associated with endocarditis, bacteremia, or other invasive infections or underlying conditions (Borriello et al., 2003). Most of these cases, however, involved immunocomprimised or at-risk individuals undergoing treatments that would expose them to any opportunistic microbial agent. Moreover, the genome sequences from several probiotic bacteria are now available (Chapter 2), and no functional virulence factors have yet been identified.

It is important to recognize that some bacteria used as probiotics are intrinsically resistant to antibiotics, and, in some cases, may even harbor antibiotic resistant genes on mobile genetic elements (Temmerman et al., 2003; von Wright, 2005). As noted above, *Bacillus cereus* may be used as a probiotic, despite the fact that some strains of this species cause foodborne disease. Thus, safety evaluations must be considered for any strain to be used a probiotic.

Making the Case for Probiotics

As noted above, literally thousands of research articles have been published on probiotics, with many suggesting that beneficial health effects occur as a result of consuming probiotics or probiotic-containing foods. However, these studies are frequently criticized for either lacking relevant controls or having relied on poor experimental design. Even when a particular study reports a positive result, it is frequently followed by another study in which such effects are absent.

Therefore, determining which claims are supported by solid scientific evidence and identifying the appropriate criteria to evaluate this evidence have become critical goals (Pathmakanthan et al., 2000; Sanders et al., 2004). Indeed, meta-analyses and other retrospective studies, in which published reports are critically evaluated, indicate that while some claims are weakly supported by experimental or clinical data, others appear to be more convincing (Tables 2 and 3).

Table 2. Probiotics: evidence for intestinal health benefits[1].

Reasonable evidence	Suggestive evidence	Potential use
Inflammatory bowel disease	Pouchitis[5]	Crohn's disease[3,5]
Chronic pouchitis[2]	Traveler's diarrhea[3,4]	Ulcerative colitis[3]
Ulcerative colitis[2]	Episodic diarrhea[4]	Pouchitis[3]
Intestinal infections		
Acute diarrhea[3]		
Relapsing *Clostridium difficile*[4]		
Antibiotic-associated diarrhea[4]		

[1]Adapted from Sartor, 2005
[2]Prevention of relapse
[3]Treatment
[4]Prevention
[5]Postoperative prevention

For example, based on a review of the literature on the effect of probiotic bacteria on diarrheal disease, serum cholesterol, the immune system, and cancer prevention, the only consistent finding was a positive effect of *Lactobacillus* GG as a treatment against rotavirus intestinal infections (de Roos and Katan, 2000). This result was supported by a meta-analysis that also concluded that *Lactobacillus* therapy could be used safely and effectively to treat diarrheal disease in children (Van Niel et al., 2002).

(Continued)

Box 4–5. Sorting Through the Pros and Cons of Probiotics *(Continued)*

Table 3. Probiotics: recommendations according to evidence-based medicine[1,2].

Recommendation Grade[3]	Evidence Level[3]	Claim, treatment, or benefit
Grade A	Level 1A	Treatment of infectious diarrhea in children
		Prevention of antibiotic-associated diarrhea
		Prevention of nosocomial and community acquired diarrhea in children
		Treatment of lactose malabsorption
Grade A	Level 1B	Prevention and maintenance of remission of pouchitis
		Prevention of post-operative infections
		Prevention and management of atopic diseases in children
Grade B	Level 2	Prevention of traveler's diarrhea
		Prevention of pancreatitis-associated sepsis
		Maintenance of remission of ulcerative colitis
		Lowering of blood cholesterol

[1]According to this approach, recommendations regarding the effectiveness of a treatment for particular diseases are assigned a "grade" that is based on the strength of the clinical evidence that supports the specific claim. The evidence is also assigned a strength "level" that is based on the experimental methodology and the manner in which the supporting studies were performed. Level 1A and 1B evidence are from multiple or single, respectively, randomized, placebo-controlled trails with low false-positive rates, adequate sample size, and appropriate methodology. Level 2 evidence is from randomized, placebo-controlled studies with high false-positive rates, inadequate sample size, or inappropriate methodology. Accordingly, Grade A recommendations are supported by evidence from two or more Level 1 studies without conflicting evidence from other Level 1 studies. Grade B recommendations are supported by evidence from two or more Level 1 studies with conflicting evidence from other Level 1 studies or by evidence from two or more Level 2 studies.
[2]Adapted from Gill and Guarner, 2004
[3]Grade recommendations and hierarchies of evidence as defined by Fennerty, 2003

The ability of probiotic organisms (lactobacilli and *Saccharomyces boulardii*) to prevent diarrheal disease was also suggested by a meta-analysis of nine trials, but treatment efficacy was not supported by the data (D'Souza et al., 2002). Another meta-analysis on the effect of probiotics on antibiotic-associated diarrhea similarly concluded that the evidence for beneficial effects was not definitive (Cremonini et al., 2002).

Objective Measures of Probiotic Activity

Determining whether a probiotic has actually had a biological effect is not always easy. Clearly, it is not difficult to measure changes in cholesterol or blood lipid levels or other relevant markers in clinical trials (whether the treatment is a probiotic or a pharmaceutical drug). However, it is an entirely different matter when the outcome marker takes longer to become evident or when the marker provides an indirect measure of efficacy.

Double-blind, randomized, and placebo-controlled studies are essential, considering that in many cases any probiotic effect is often based on the subjective judgment of human subjects. For example, a positive effect of probiotic feeding on digestion and intestinal well-being in otherwise healthy subjects may be difficult to substantiate. In contrast, objective measures, or biomarkers, can provide much more credible evidence. Biomarker measures may, in fact, show no differences between treatments, even when subjective scores appear to suggest that differences do exist. For many studies, biomarkers are mandatory, since they may be the only reliable, non-invasive way to measure the efficacy of probiotics. Biomarkers, which must first be validated for efficacy, may include immunological factors, GI tract populations, blood lipids, and fermentation metabolites. More than twenty colon cancer biomarkers were described in one recent report (Rafter, 2002).

Box 4–5. Sorting Through the Pros and Cons of Probiotics *(Continued)*

References

Borriello, S.P., W.P. Hammes, W. Holzapfel, P. Marteau, J. Schrezenmeir, M. Vaara, and V. Valtonen. 2003. Safety of probiotics that contain lactobacilli or bifidobacteria. Clin. Infect. Dis. 36:775-780.

Cremonini, F., S. Di Caro, E.C. Nista, F. Bartolozzi, G. Capelli, G. Gasbarrini, and A. Gasbarrini. 2002. Meta-analysis: the effect of probiotic administration on antibiotic-associated diarrhoea. Aliment. Pharmacol. Ther. 16:1461-1467.

del Campo, R., D. Bravo, R. Cantón, P. Ruiz-Garbajosa, R. García-Albiach, A. Montesi-Libois, F.-J. Yuste, V. Abraira, and F. Baquero. 2005. Scarce evidence of yogurt lactic acid bacteria in human feces after daily yogurt consumption by healthy volunteers. Appl. Environ. Microbiol. 71:547-549.

de Roos, N.M., and M.B. Katan. 2000. Effects of probiotic bacteria on diarrhea, lipid metabolism, and carcinogenesis: a review of papers published between 1988 and 1998. Am. J. Clin. Nutr. 71:405-411.

D'Souza, A.L., C. Rajkumar, J. Cooke, and C.J. Bulpitt. 2002. Probiotics in prevention of antibiotic diarrhoea: meta-analysis. Br. Med. J. 324:1361-1366.

Fennerty, M.B. 2003. Traditional therapies for irritable bowel syndrome: an evidence based appraisal. Rev. Gastroenterol. Disord. 3 (Supple. 2):S18-S24.

Gill, H.S., and F. Guarner. 2004. Probiotics and human health: a clinical perspective. Postgrad. Med. J. 80:516-526.

Pathmakanthan, S., S. Meance, and C.A. Edwards. 2000. Probiotics: a review of human studies to date and methodological approaches. Microb. Ecol. Health Dis. 12 (Supple. 2):10-30.

Rafter, J.J. 2002. Scientific basis of biomarkers and benefits of functional foods for reduction of disease risk: cancer. Br. J. Nutr. 88 (Supple. 2):S219-S224.

Sanders, M.E., T. Tompkins, J.T. Heimbach, and S. Kolida. 2004. Weight of evidence needed to substantiate a health effect for probiotics and prebiotics—regulatory considerations in Canada, E.U., and the U.S. Eur. J. Nutr. 44:303-310.

Sanders, M.E., L. Morelli, and T.A. Tompkins. 2003. Sporeformers as human probiotics: *Bacillus*, *Sporolacto-bacillus*, and *Brevibacillus*. Comp. Rev. Food Sci. Food Safety. 2:101-110. (www.ift.org/publications/crfsfs).

Sartor, R.B. 2005. Probiotic therapy of intestinal inflammation and infections. Curr. Opin. Gastroenterol. 21:44-50.

Temmerman, R., B. Pot, G. Huys, and J. Swings. 2003. Identification and antibiotic susceptibility of bacterial isolates from probiotic products. Int. J. Food Microbiol. 81:1-10.

Van Niel, C.W., C. Feudtner, M.M. Garrison, and D.A. Christakis. 2002. *Lactobacillus* therapy for acute infectious diarrhea in children: a meta-analysis. Pediatrics 109:678-684.

von Wright, A. 2005. Regulating the safety of probiotics—the European approach. Curr. Pharm. Design. 11:17-23.

Probiotic websites

www.proeuhealth.vtt.fi
www.isapp.net
www.crdc-probiotics/ca
www.usprobiotics.org

yogurt are largely responsible for this effect, since heat-treated yogurt is not as effective at alleviating lactose intolerance symptoms. It is interesting, however, that while yogurt can be tolerated, other cultured milk products such as sour cream and buttermilk cannot. The explanation for this observation is related directly to the different routes by which lactose is metabolized by the cultures used in the manufacture of these products.

The fermentation that occurs in sour cream and cultured buttermilk production is performed mainly by lactococci (described below), whereas *S. thermophilus* and *L. delbrueckii*

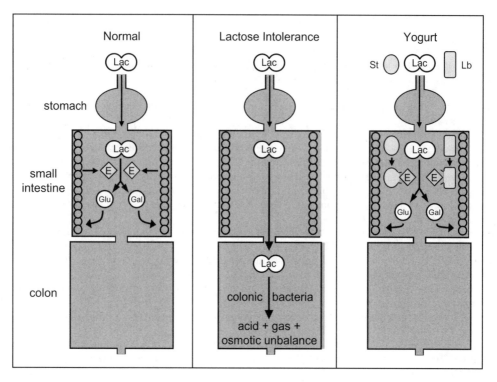

Figure 4–4. In normal individuals (left panel), ingested lactose (Lac) passes through the stomach. Hydrolysis occurs in the small intestine via endogenous β-galactosidase (E) produced by cells that coat the surface of the intestinal wall. The hydrolysis products glucose (Glu) and galactose (Gal) are subsequently adsorbed. In lactose-intolerant individuals (center panel), those intestinal cells do not produce β-galactosidase, and the lactose escapes hydrolysis and instead reaches the colon where it causes colonic distress, either by virtue of its metabolism by anaerobes or by osmotic disturbances. When yogurt is consumed (right panel), *Streptococcus thermophilus* (St) and *Lactobacillus bulgaricus* (Lb) serve as a source of exogenous β-galactosidase (after permeabilization in the small intestine) and restore normal lactose hydrolyzing ability to the host.

subsp. *bulgaricus* are responsible for the yogurt fermentation. Recall that the latter organisms hydrolyze lactose via β-galactosidase. When yogurt is consumed and these bacteria reach the small intestine, cell lysis occurs (due to bile salts), intracellular β-galactosidase is released, and lactose is hydrolyzed. Thus, the yogurt culture serves as an exogenous source of this enzyme, substituting for the β-galactosidase not produced by the host. In contrast, lactococci produce phospho-β-galactosidase. The substrate of this enzyme is lactose-6-phosphate (the product of the lactose PTS), which can only be formed by intact cells. This enzyme does not hydrolyze free lactose. Thus, when lactococci are ingested and lysed in the small intestine, there

will be little, if any, impact on lactose digestion, and lactose-intolerant individuals will, unfortunately, get no relief.

Frozen Yogurt and Other Yogurt Products

Frozen yogurt is a product for which no federal standard of identity exists. This has led to confusion and even mis-representations about what this product actually is. Several states now have their own definitions of frozen yogurt (mainly dealing with fat levels), and the FDA has been petitioned to implement standards of identity. The main reason why such measures are thought necessary is to protect the image of yogurt as a healthy, nutritious food—one low in fat and calories and containing health-promoting bacteria. Obviously, there are marketing reasons for including "yogurt" on the label of a product, whether or not that product actually contains yogurt, or what consumers think is yogurt. For example, yogurt-covered pretzels, yogurt-covered raisins, and yogurt-containing salad dressings are made using material that may have once been yogurt, but after processing and dilution with other ingredients, little of the original yogurt character may actually be present. Similarly, ice cream mix containing only a trace of yogurt could, in theory, be called frozen yogurt and gain the market advantages associated with yogurt. Certainly, and perhaps most importantly, the presence of live yogurt organisms should not be assumed for these products.

As noted, some states have written definitions for frozen yogurt. However, while there may be requirements for the presence of particular organisms, there are generally no such requirements for minimum numbers. In Minnesota, for example, frozen yogurts refer to a "Frozen dairy food made from a mix containing safe and suitable ingredients including, but not limited to, milk products. All or a part of the milk products must be cultured with a characterizing live bacterial culture that contains the lactic acid producing bacteria *Lactobacillus bulgaricus* and *Streptococus thermophilus* and may contain other lactic acid producing bacteria." Some states require that the frozen yogurt products have a minimum acidity, ranging from 0.3% to 0.5%.

Cultured Buttermilk

Buttermilk is the fluid remaining after cream is churned into butter. It is a thin, watery liquid that is rarely consumed as a fluid drink. Because it is rich in phospholipids (derived from the rupture of milk fat globules during churning), it has excellent functional properties and is an especially good source of natural emulsifiers. It is typically spray dried and used as an ingredient in processed food products. Cultured buttermilk, in contrast, is made from skim or low-fat milk that is fermented by suitable lactic acid bacteria. The only relation this product has to buttermilk is that butter granules or flakes are occasionally added to provide a buttery flavor and mouth feel. How, then, did this product come to be called buttermilk? In the traditional manufacture of butter, it was common practice to add a mixed, undefined lactic culture to cream prior to churning. The lactic acid would provide a pleasant tart flavor, and the cream-ripened butter would be better preserved. The resulting by-product, the buttermilk, would also be fermented.

Cultured Buttermilk Manufacture

Cultured buttermilk is usually made from low-fat milk, although non-fat and whole milk versions also exist. Non-fat milk solids are frequently added to give about 10% to 12% non-fat solids (Figure 4–5). Next, the milk is heated to 85°C to 88°C for thirty minutes. This not only pasteurizes the milk, but also satisfies other functional requirements, as described above for yogurt. The mix is then cooled to 21°C to 22°C and inoculated with a mesophilic starter culture, specific for cultured buttermilk.

The starter culture for buttermilk usually contains a combination of acid-producing bacteria and flavor-producing bacteria, in a ratio of about 5:1. Many culture suppliers now also offer body-forming (i.e., EPS-producing) strains. The acid producers include strains of *L. lactis* subsp. *lactis* and/or *L. lactis* subsp. *cremoris*. These bacteria are homolactic, and their function is simply to produce lactic acid and lower the pH. For manufacturers who prefer a decidedly tart, acidic product, the acid-producing strains are sufficient.

It is more common, however, to include flavor-producers in the starter culture. Among the flavor-producing bacteria used in buttermilk cultures are *L. lactis* subsp. *lactis* (diacetyl-producing strains), *Leuconostoc mesenteroides* subsp. *cremoris*, or *Leuc. lactis*. The latter organisms are heterofermentative,

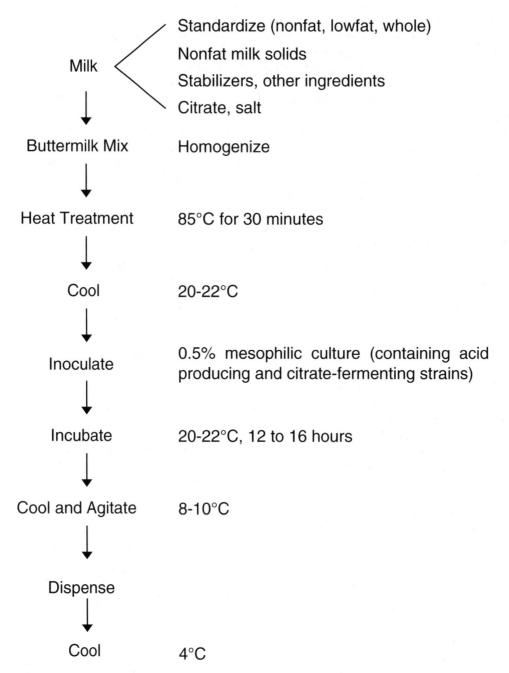

Figure 4–5.   Manufacture of cultured buttermilk.

producing lactic acid, as well as small amounts of acetic acid, ethanol, and carbon dioxide. These metabolic end-products contribute to the flavor and, in the case of the carbon dioxide, to the mouth feel of the product. Importantly, these bacteria also have the metabolic capacity to ferment citric acid and to produce diacetyl, a compound that has major impact on the flavor of cultured buttermilk. Diacetyl has a buttery aroma and flavor, and, according to buttermilk afficionados, imparts a delicate and characteristic flavor. The ability of the starter culture to perform the citrate fermentation, therefore, is a critical trait. How lactic acid bacteria convert citrate into diacetyl has been a subject of considerable interest, not only because of the practical importance of the pathway in fermented foods, but also because it posed a particular challenge to biochemists. For many years, two pathways, one enzymatic and the other, non-enzymatic, were thought possible.

In both pathways, the early steps are the same (Figure 4–6; also see Chapter 2). Citrate is first transported by a citrate permease (CitP) that is pH-dependent, with an optimum activity between pH 5.0 and 6.0. The intracellular citrate is then hydrolyzed by citrate lyase to

form acetate and oxaloacetate. The oxaloacetate is decarboxylated by oxaloacetate decarboxylase to give pyruvate and carbon dioxide, and the acetate is released into the medium. Normally, lactic acid bacteria reduce pyruvate to lactate, which is then excreted. However, pyruvate reduction requires NADH, which is ordinarily made during glycolysis, but which is not made during citrate fermentation. In the absence of reduced NADH, pyruvate would accumulate inside the cell, eventually reaching toxic or inhibitory levels. The excess pyruvate is instead oxidatively decarboxylated by thiamine pyrophosphate (TPP)-dependent pyruvate decarboxylase and acetaldehyde-TPP is formed (see Chapter 2 for details). The latter compound condenses with another pyruvate molecule to form α-acetolactate, a reaction

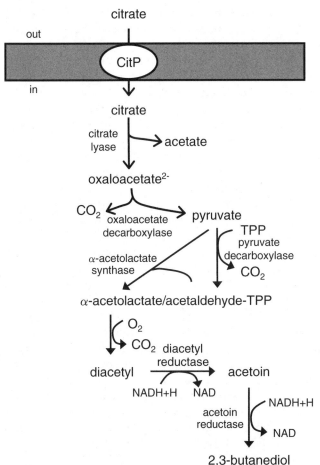

Figure 4–6. Citrate fermentation pathway in lactic acid bacteria. Adapted from Hutkins, 2001.

catalyzed by α-acetolactate synthase and requiring high concentrations of pyruvate. The α-acetolactate is unstable in the presence of oxygen and is finally decarboxylated, nonenzymatically, to form diacetyl. It is important to note that diacetyl is not necessarily the terminal end-product of the pathway. Further reduction of diacetyl can also occur, forming acetoin and 2,3-butanediol, compounds that contribute no flavor or aroma to the product.

Following the addition of the culture, the mix is incubated at 21°C to 22°C for twelve to sixteen hours. At the end of the fermentation, when the titratable acidity has reached 0.85% to 0.90% and the pH has decreased to about 4.5, the product is cooled to 2°C and agitated to break up the coagulum. Salt is usually added, and, if desired, butter granules or flakes are added. The finished product should be viscous and pourable. The product is pumped into containers and distributed.

Factors Affecting Diacetyl Formation in Cultured Buttermilk

Even if the culture contains citrate-forming strains, diacetyl formation does not always occur in amounts necessary to impart the desired flavor. Several reasons may account for reduced citrate fermentation and diacetyl synthesis. First, there may simply not be enough citrate in the milk. Although milk contains, on average, about 0.15% citrate, this amount varies, depending largely on the diet of the cow. Therefore, sodium citrate is frequently added to the milk to provide a consistent source of substrate. If the temperature during incubation is too high (>24°C), growth of the homolactic lactococci will be favored, too much acid will be produced, and the citrate-fermentors may be inhibited. Not only will the product have a "lacks flavor" defect, but it will instead have a harsh acid flavor.

In contrast, if the incubation temperature is too low (< 20°C), not enough acid is produced and the flavor will be flat or "lacks acid." Acid production is also necessary for diacetyl formation, since the citrate transport system is not activated unless the pH is below 5.5. In fact, maximum diacetyl synthesis occurs between pH 5.0 and 5.5. Finally, once diacetyl is made, low pH inhibits the reduction reactions that convert diacetyl to acetoin and 2,3-butanediol. One other factor that is critical for synthesis of diacetyl is oxygen, which can stimulate diacetyl formation by as much as thirty-fold. Several mechanisms are responsible for the oxygen effect. As noted above, pyruvate, the metabolic precursor of diacetyl, can serve as the substrate for several alternative enzyme reactions, including lactate dehydrogenase. In the presence of high atmospheric oxygen, lactate dehydrogenase activity is reduced, and the oxidative decarboxylation reaction responsible for diacetyl synthesis is enhanced. In addition, by oxidizing NADH, oxygen also slows the rate at which diacetyl is reduced to acetoin or 2,3-butanediol. For these reasons, it may be useful to stir air into the final product during the agitation step. In fact, maximum diacetyl production occurs about seventy-two hours after manufacture, so fresh product may not have as good a flavor as one that is slightly "aged". The flavor and texture defects that occur in cultured buttermilk are, in general, similar to those in yogurt. The most common flavor defect is "lacks flavor," caused by insufficient diacetyl formation due to reasons outlined above. Excess acidity is also objectionable to consumers. Texture defects include wheying off, too thin, or too viscous. The latter may be caused by body-forming bacteria that over-produce EPS. Milk-borne defects, such as yeasty flavor, unclean, and rancidity, are caused by poor quality milk, especially milk that had been contaminated with psychrotrophic microorganisms.

Sour Cream

Despite the apparent differences in the appearance and texture, sour cream is actually quite similar to cultured buttermilk in several respects (Figure 4-7). The sour cream culture, for example, is the same as that used for buttermilk. The incubation conditions and the flavor compounds produced by the culture are also similar for both products. There are, however, several notable differences. For sour cream manufacture, cream (containing varying levels of milkfat) is used instead of lowfat or skim milk. The cream is pasteurized, but not quite at the severe conditions used for buttermilk or yogurt. This is because for sour cream, denaturation of protein is not as crucial, since the milk fat will impart the desired creaminess, thickness, and body. The cream is homogenized, which also promotes a desirable heavy body.

Sour Cream Manufacture

In the United States the starting material for sour cream manufacture is cream at 18% to 20% milkfat. Lower-fat versions, such as sour half-and-half (10% milkfat), are produced in some regions, as are higher-fat products (containing as much as 50% milkfat). The latter are generally produced as an ingredient for dips and other sour cream products. There is no

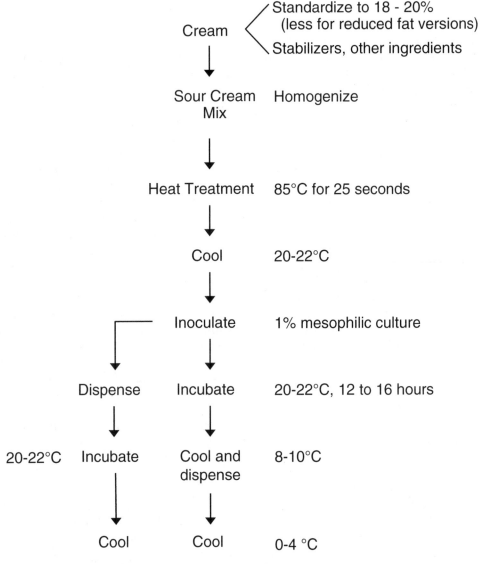

Figure 4–7. Manufacture of sour cream.

need to add additional nonfat dry milk, as for yogurt and buttermilk, since achieving a firm gel and thick body is not an issue for these high-solids products. For the same reason, the cream is pasteurized at more typical time and temperature conditions (85°C for twenty-five seconds). However, the cream must be homogenized (usually twice) to produce a smooth-textured product with good viscosity. After the mix is cooled, the sour cream culture is added. The culture, containing mesophilic acid-producing, flavor-producing, and body-forming strains, is often the same as that is used for cultured buttermilk. The mix is then either filled into cups and incubated or incubated directly in vats (analogous to the two styles of yogurt). Incubation is at 20°C to 25°C for ten to sixteen hours. When the pH reaches about 4.4 to 4.6 (about 0.7% to 0.9% lactic acid), the sour cream is cooled, either by moving the cup-fermented product into coolers or, in the case of vat-fermented product, by stirring the product in jacket-cooled vats. The product is then pumped into containers. Sour cream should have a similar flavor profile as cultured buttermilk, with lactic acid and diacetyl predominating. Body characteristics are especially important, and various gums and other stabilizing agents are frequently added to the mix. Some manufacturers even add a small amount of chymosin to provide additional firmness. When defects do occur, they are often due to poor quality ingredients and post-pasteurization contamination by acid-tolerant yeasts and molds and psychrotrophic bacteria.

Kefir

Kefir is a fermented dairy product that is described in many reference texts and has long been of academic interest to microbiologists, but was one for which few U.S. students (or consumers for that matter) had any first-hand knowledge. Although several brands of kefir are now available throughout the United States, it is still a mostly unknown product. This is despite the popularity of this product throughout a large part of the world. In the

Middle East, Eastern Europe, and Central Asia, especially Turkey, Russia, Ukraine, Poland, and the Czech Republic, kefir is one of the most widely consumed cultured dairy products, accounting for as much as 70% of all cultured milk products consumed. It is interesting that yogurt manufacturers have begun to introduce fluid or pourable yogurt-like products that are only slightly different than traditional kefir.

Kefir originated in the Caucasus Mountain in Russia. The traditional kefir manufacturing process, which is still widely practiced, relies on a mixed assortment of bacteria and yeast to initiate the fermentation (Figure 4–8). The kefir fermentation is unique among all other dairy fermentations in that the culture organisms are added to the milk in the form of insoluble particles called kefir grains (Box 4–6). Moreover, once the fermentation is complete, the kefir grains can be retrieved from the fermented milk by filtration and reused again and again.

The flavor of plain kefir is primarily due to lactic and acetic acids, diacetyl, and acetaldehyde, produced by homofermentative and heterofermentative lactic acid bacteria. However, because kefir grains also contain yeast, in addition to lactic acid bacteria, other end-products are formed that make the finished product quite different from other cultured dairy products. This is because ethanol is produced when the yeasts ferment lactose, such that kefir can contain as much as 2% ethanol (accounting, perhaps, for its appeal). In the United States, this much ethanol would trigger action from the regulatory agencies (i.e., Alcohol and Tobacco Tax and Trade Bureau). Thus, if yeast are present in the culture (most kefir products made in the United States claim yeasts on their labels), they must be low- or non-ethanol producers. Kefir grains can also contain acetic acid-producing bacteria, such as *Acetobacter aceti.*

Kefir Manufacture

In the United States, kefir is usually made with low-fat or whole milk (Figure 4–8). The milk is pasteurized at 85°C to 90°C for up to thirty minutes (like yogurt), then cooled to 20°C to

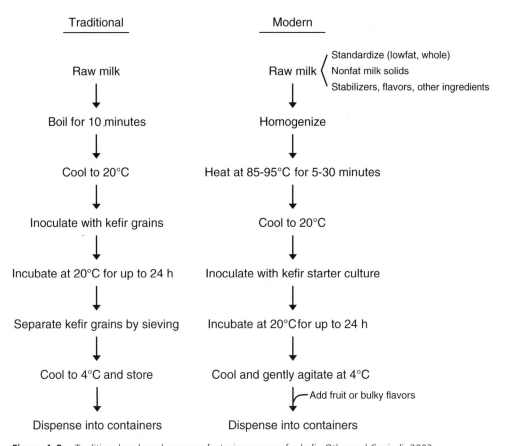

Figure 4–8. Traditional and modern manufacturing process for kefir. Otles and Cagindi. 2003.

22°C. Kefir grains are generally not used in the United States; instead, the milk is inoculated with a kefir starter culture containing *Lactobacillus kefiranofaciens* and *Lactobacillus kefir*. The mixture is incubated for sixteen to twenty-four hours or until a pH of 4.2 to 4.6 is reached. The coagulum is then gently stirred, flavoring ingredients (usually fruit) are added, and the product is then dispensed. Kefir should contain 0.8% to 1.0% lactic acid along with other heterofermentative end-products (acetic acid, ethanol, and CO_2). Acetaldehyde and diacetyl are also usually formed. The kefir will have a tart flavor and a smooth, viscous body. Production of CO_2 by the starter culture also provides effervescence and mouth feel.

Other Cultured Dairy Products

The emphasis of this chapter has been on those products produced and consumed in the United States. However, there are hundreds of other products produced around the world that, although manufactured via similar processes, have unique and interesting features (Table 4–4). Villi, for example, is widely consumed in Finland is known for its high viscosity and musty flavor and aroma. The ropy texture is due to capsular EPS production by *L. lactis* subsp. *lactis* and *L. lactis* subsp. *cremoris*, and the musty flavor is caused by growth of the fungal organism *Geotrichum candidum*. Another popular product is Koumiss, which was traditionally produced from

Box 4–6. The Microbial Diversity of Kefir Grains

To microbial ecologists, kefir grains represent an amazingly diverse ecosystem. A single grain can harbor more than a million cells, with perhaps dozens of different species present (Table 1). The grains are essentially a natural immobilized cell bioreactor, with the kefir grain serving as an inert support material for the cells.

Table 1. Bacteria, yeast, and mold isolated from kefir grains[1].

Bacteria	Yeasts and Mold
Lactobacillus brevis	*Candida kefir*[2]
Lactobacillus fermentum	*Candida maris*
Lactobacillus kefiranofaciens[3]	*Candida inconspicua*
Lactobacillus kefiri	*Candida lambica*
Lactobacillus parakefir	*Candida krusei*
Lactobacillus plantarum	*Saccharomyces cerevisiae*
Lactococcus lactis subsp. *cremoris*	*Kluyveromyces marxianus*
Lactococcus lactis subsp. *lactis*	*Geotrichum candidum*
Leuconostoc mesenteroides subsp. *cremoris*	*Zygosaccharomyces* sp.
Leuconostoc sp.	
Streptococcus thermophilus	
Acetobacter aceti	

[1]Adapted from various sources (in references)

[2]syn *Candida kefyr*

[3]Recently reclassified as *Lactobacillus kefiranofaciens* subsp. *kefirgranum* or subsp. *kefiranofaciens* (Vancanneyt et al., 2004)

Attachment to the grains is mediated via one of several different types of polysaccharide-like material (kefiran being the prototype) that are produced by the resident organisms. The kefir organisms do not appear to be randomly distributed; rather, some are contained mainly within the core of the grains, whereas others are located primarily in the exterior. Symbiotic and synergistic relationships likely exist between different organisms, but these relationships are difficult to establish in such a complex system (Leroi and Pidoux, 1993a and 1993b).

Although the precise chemical composition of kefir grains varies depending on the source, they generally contain about 80% to 90% water, 2% to 5% protein, and 8% to 10% carbohydrate (Abraham and De Antoni, 1999; Garrote et al., 2001). Most of the protein faction is in the form of casein, whereas most of the carbohydrate portion is polysaccharide. The irregular, cauliflower-shaped grains range in size from 0.2 cm to more than 2 cm in diameter. In their dry form, they are stable for many months. Although kefir grains are available commercially, many modern manufacturers now use pure lyophylized cultures containing many of the strains ordinarily found in the grains.

The predominant organisms in kefir grains are lactic acid bacteria, with *Lactobacillus* species accounting for up to 80% of the total (Garrote et al., 2001; Simova et al., 2002; Vancanneyt et al., 2004). Other studies have shown that homofermentative species, including *Lactobacillus kefirgranum* and *Lactobacillus kefiranofaciens,* were the most frequently isolated species in kefir grains, whereas heterofermentative *Lactobacillus kefir* and *Lactobacillus parakefir* were less common (Takizawa et al., 1998; Simova et al., 2002). Other lactic acid bacteria that have been identified include species of *Lactococcus, Streptococcus*, and *Leuconostoc*. Yeasts are also well-represented, and include lactose-fermenting (*Saccharomyces kefir*) and non-fermenting strains.

It was reported recently that kefir grains likely contain organisms that cannot easily be cultured and that are instead only viable within the confines of the grain environment (Witthuhn et al., 2005). Collectively, the kefir microflora has considerable metabolic diversity, which ex-

Box 4–6. The Microbial Diversity of Kefir Grains *(Continued)*

plains, in part, the incredible stability and viability of the grains, as well as the wide spectrum of products produced during the fermentation.

References

Abraham, A.G., and G.L. De Antoni. 1999. Characterization of kefir grains grown in cow's milk and in soya milk. J. Dairy Res. 66:327–333.

Leroi, F., and M. Pidoux. 1993a. Detection of interactions between yeasts and lactic acid bacteria isolated from sugary kefir grains. J. Applied Bacteriol. 74:48–53.

Leroi, F., and M. Pidoux. 1993b. Characterization of interactions between *Lactobacillus hilgardii* and *Saccharomyces florentinus* isolated from sugary kefir grains. J. Applied Bacteriol. 74:54–60.

Garrote, G.L., A.G. Abraham, and G.L. De Antoni. 2001. Chemical and microbiological characterization of kefir grains. J. Dairy Res. 68:639–652.

Simova, E., D. Beshkova, A. Angelov, T. Hristozova, G. Frengova, and Z. Spasov. 2002. Lactic acid bacteria and yeasts in kefir grains and kefir made from them. J. Ind. Microbiol. Biotechnol. 28:1–6.

Takizawa, S., S. Kojima, S. Tamura, S, Fujinaga, Y. Benno, and T. Nakase. 1998. The composition of the *Lactobacillus* flora in kefir grains. System. Appl. Microbiol. 21:121–127.

Vancanneyt, M., J. Mengaud, I. Cleenwerck, K. Vanhonacker, B. Hoste, P. Dawynndt, M.C. Degivry, D. Ringuet, D. Janssens, and J. Swings. 2004. Reclassification of *Lactobacillus kefirgranum* Takizawa *et al.* 1994 as *Lactobacillus kefiranofaciens* subsp. *kefirgranum* subsp. nov. and emended description of *L. kefiranofaciens* Fujisawa *et al.* 1988. Int. J. Syst. Evol. Microbiol. 54:551–556.

Witthuhn, R.C., T. Schoeman, and T.J. Britz. 2005. Characterisation of the microbial population at different stages of Kefir production and Kefir grain mass cultivation. Int. Dairy J. 15:383–389.

Table 4.4. Cultured dairy products from around the world.

Product	Origin	Culture Organisms	Unique Features
Villi	Finland	*Lactococcus* spp. *Leuconostoc* spp. *Geotrichum candidum*	Ropy texture Musty flavor
Skyr	Iceland	*Lactobacillus delbreckii* subsp. *bulgaricus* *Streptococcus thermophilus*	Concentrated, high protein content
Dahi	India	*Lactobacillus delbreckii* subsp. *bulgaricus* *Streptococcus thermophilus* *Lactobacillus* spp.	Yogurt-like
Koumiss	Russia	*Lactobacillus delbreckii* subsp. *bulgaricus* *Lactobacillus acidophilus* *Kluyveromyces* spp.	Mare's milk > 1% ethanol
Bulgarian Milk	Bulgaria	*Lactobacillus delbreckii* subsp. *bulgaricus*	High acid (>2% lactic acid)

mare's milk. Similar to kefir in that lactic acid and ethanol are both present, this product owes much of its popularity to its putative therapeutic properties. The yogurt-like products dahi and laban are among the most widely-consumed cultured dairy products in India and the Middle East, respectively.

Bibliography

Chaves, A.C.S.D., M. Fernandez, A.L.S. Lerayer, I. Mierau, M. Kleerebezem, and J. Hugenholtz. 2002. Metabolic engineering of acetaldehyde production by *Streptococcus thermophilus*. Appl. Environ. Microbiol. 68:5656–5662.

Hutkins, R.W. 2001. Metabolism of starter cultures, p. 207241. *In* E.H. Marth and J.L. Steele (ed.), *Applied Dairy Microbiology*. Marcel Dekker, Inc., New York, NY.

Lucey, J.A. 2002. Formation and physical properties of milk protein gels. J. Dairy Sci.5:281–294.

Mistry, V.V. 2001. Fermented milks and creams. p. 301–325. *In* E.H. Marth and J.L. Steele (ed.), *Applied Dairy Microbiology*. Marcel Dekker, Inc., New York, NY.

Otles, S. and O. Cagindi. 2003. Kefir: a probiotic dairy-consumption nutritional and therapeutic aspects. Pak. J. Nutr. 2:54–59.

Reid, G. 2001. Regulatory and clinical aspects of dairy probiotics. FAO/WHO Expert Consultation on Evaluation of Health and Nutritional Properties of Powder Milk with Live Lactic Acid Bacteria. (Online at ftp://ftp.fao.org/es/esn/food/Reid.pdf)

Tamine, A.Y., and H.C. Deeth. 1980. Yogurt: technology and biochemistry. J. Food Prot.43:939–977.

Sanders, M.E. 1999. Probiotics. Food Technol. 53:67–77.

Sodini, I., F. Remeuf, S. Haddad, and G. Corrieu. 2004. The relative effect of milk base, starter, and process on yogurt texture: a review. Crit. Rev. Food Sci. Nutr. 44:113–137.

Vedamuthu, E.R. 1991. The yogurt story—past, present and future. Part I. Dairy Food Environ. Sanit. 11:202–203. [Note, this is the first of a ten-part series of articles on yogurt by E.R. Vedamuthu that appeared in this journal through Vol. 12, No. 6, 1992]

5

Cheese

"We are still, however, far from having arrived at a complete elucidation of all the questions involved. It is particularly difficult to understand how various sorts of hard cheese, apparently containing the same microflora, should each have its own characteristic taste and smell. There can hardly be any doubt that these sorts of cheese in reality contain different species of bacteria, only we are unable to distinguish them by the methods hitherto available."

From The Lactic Acid Bacteria *by S. Orla-Jensen, 1918*

Introduction

Perhaps no other fermented food starts with such a simple raw material and ends up with products having such an incredible diversity of color, flavor, texture, and appearance as does cheese. It is even more remarkable that milk, pale in color and bland in flavor, can be transformed into literally hundreds of different types of flavorful, colorful cheeses by manipulating just a few critical steps. How so many cheeses evolved from this simple process undoubtedly involved part trial and error, part luck, and plenty of art and skill. It is fair to assume that, until very recently, most cheese makers had only scant knowledge of science, and microbiology in particular. Now, however, it is likely that few fermented foods require such a blend of science, technology, and craftsmanship as does the making of cheese.

On a volume basis, the cheese industry is the largest of all those involved in fermented foods manufacture. Of the 75 billion Kg (165 billion pounds) of milk produced in the United States in 2001, more than one-third was used in the manufacture of 3.7 billion Kg (8.1 billion pounds) of cheese. About a fourth of that cheese was used to make various types of processed cheese (discussed later).

On a per capita basis, cheese consumption in the United States has increased in the past twenty-five years from 8 Kg in 1980 to nearly 14 Kg (30.1 pounds) per person per year in 2003 (of U.S. made cheese). The most popular cheeses have been the American style (e.g., Cheddar, Colby) and Italian style (e.g., Mozzarella and pizza cheese) cheeses, accounting for 41.5% and 40.6%, respectively, of all cheeses consumed in the United States (Figure 5-1). In addition, another 0.75 Kg of imported cheese is consumed per person per year. Worldwide, Greece (26 Kg per person per year), France (24 Kg per person per year), and Italy (21 Kg per person per year) are the leading consumers of cheese, with other European countries not too far behind (Figure 5-2).

On the production side, American and Italian-types cheeses are, by far, the main cheeses produced in the United States. Although the American-type cheeses accounted for nearly 70% of all cheese produced in the United States in the 1960s, Italian-type cheeses and Mozzarella, in particular, are about to exceed that of American-type cheese, based on the current trend (Figure 5-3).

The U.S. cheese industry began in the mid-1800s, with factories opening first in New York (1851) and Wisconsin (1868). Prior to that time, cheese was mainly produced directly on farms and sold locally. By the late 1800s, about 4,000 cheese factories accounted for the nearly 100 million Kg (217 million pounds) of cheese. Most

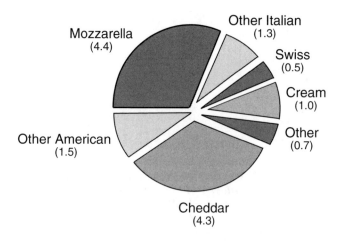

Figure 5–1. Per capita consumption (Kg per person per year) of different varieties of cheese in the United States in 2003. Adapted from USDA Economic Research Service statistics.

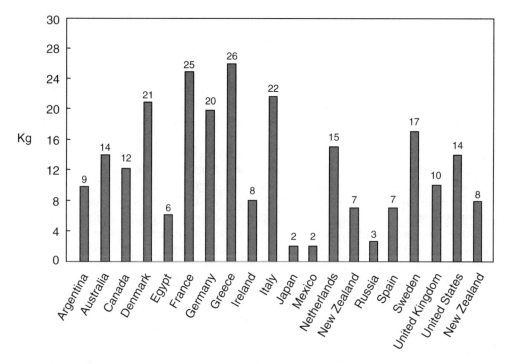

Figure 5–2. Worldwide per capita cheese consumption (Kg per person per year). Adapted from 2002 USDA and United Nations/FAO Agricultural Database statistics.

cheese factories were located in the dairy-producing states of Wisconsin, Minnesota, Pennsylvania, and New York. In the past twenty years, California has emerged as the leading producer of milk and the second leading (to Wisconsin) manufacturer of cheese. However, like other segments of the food industry, more and more product is made by fewer and fewer plants. In 2001, there were half as many production facilities making American and Italian style cheese as there were in 1980, while, at the same time, production capacity increased by 100%.

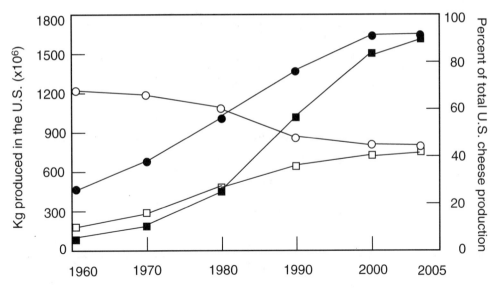

Figure 5–3. Production of American (●) and Italian (■) type cheeses in Kg and as a percent of total cheese production (open symbols) from 1960 through 2003. Adapted from USDA Economic Research Service statistics.

The basic cheese manufacturing principles will be outlined in this chapter, emphasizing the key variables at each step that account for the myriad types of cheeses that exist throughout the world. Manufacturing details for specific cheeses representing the main cheese groups or families will then be described. Along the way, an understanding of dairy chemistry and the milk milieu and the role of the starter culture will be examined. Finally, how flavor and texture development occurs during cheese ripening, current problems and challenges faced by the cheese industry, and other issues will be addressed.

Manufacturing Principles

Like so many fermented foods, the first cheese made by human beings was almost certainly a result of an accident. Some wandering nomad, as the legend goes, filled up a pouch made from the stomach of a calf or cow with a liter or two of fresh milk. After a few hours, the milk had turned into a solid-like material, and when our would-be cheese maker gave the container a bit of a shake, a watery-like fluid quickly sep-

arated from the creamy white curd. This moderately acidic, pleasant-tasting curd and whey mixture not only had a good flavor, but it also probably had a longer shelf-life than the fresh milk from which it was made. And despite the rather crude production scheme, the product made several thousands of years ago was not much different than many of the cheeses currently produced and consumed even today.

Just what happened to cause the milk to become transformed into a product with such a decidedly different appearance, texture, and flavor? To answer that question, it is first necessary to compare the composition of the starting material, milk, to that of the product, the finished cheese (Figure 5–4). Cow's milk consists of, in descending order (and in general concentrations), water (87%), lactose (5%), fat (3.5% to 4%), protein (3.2% to 3.4%), and minerals (<1%), mainly calcium. In contrast, a typical cheese, such as Cheddar cheese, contains 36% to 39% water, 30% to 32% fat, 26% to 28% protein, 2% to 2.5% salt, 1% mineral (mostly calcium), and <1% lactose.

The differences should be evident—cheese contains less water and more milk solids, in the

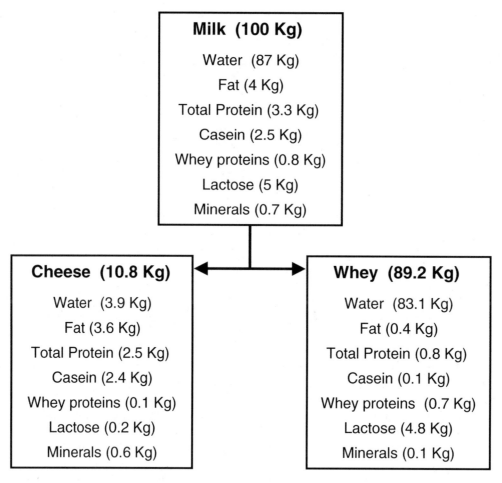

Figure 5–4. Partition of milk into cheese and whey.

form of fat and protein, than the milk from which it was made. Thus, cheese making can simply be viewed as a concentration process, in which the water portion, or whey, is removed and the solids are concentrated. In fact, as we shall see later, most of the steps involved in cheese making are performed for the singular purpose of removing water. Doing so not only concentrates and solidifies the solid matter, but also decreases microbial and enzymatic activities such that cheese is much better preserved than milk. As noted in the previous chapter, buttermilk, yogurt, sour cream, and other cultured dairy products are different from cheese in several respects, but chief among these differences

is that these products do not involve a water removal step.

Converting a liquid into a solid

Water removal and the concentration of protein and fat occur via a combination of biochemical, biological, and physical-chemical events. Many of these events happen at nearly the same time and often have complex effects on one another. For example, exposure of cheese curds to both high temperature and low pH enhances removal of water from the curd (a phenomenon known as syneresis). But if the temperature is too high, the microorganisms that produce the acid that lowers the pH will be inactivated, resulting in poor syneresis (and poor quality cheese). If, on the other hand, too much acid is produced, then significant mineral loss (specifically calcium) will occur, making the cheese crumbly (which may or may not be a good thing, depending on the cheese).

To further complicate the situation, removing water from the curd also removes the solutes dissolved in the curd. Thus, the amount of lactose in the curd is reduced as the curds become dryer, and less substrate will be available later for the culture. The point is that each process step has consequences (hopefully, intended) not only on other cheese making activities, but also on the overall properties and quality of the finished cheese.

The conversion of liquid milk into a solid mass of cheese is done via coagulation (or precipitation) of milk protein. Milk, as noted above, contains about 3.3% protein. Of the protein fraction, about 80% is casein (2.5% of the milk), and the remaining 20% are known collectively as whey proteins. For most cheeses, the protein portion consists almost entirely of casein. When milk coagulates and the coagulated material is separated, the soluble whey proteins are released into the water or whey fraction. The casein matrix not only contains some water (and whatever solutes are dissolved or suspended in the water phase), but also a large portion of the lipid fraction that was originally present in the milk, depending on how coagulation occurs.

There are three ways the initial coagulation step is accomplished. First, milk can be coagulated by acids produced by lactic acid bacteria, based on the same principle used for yogurt and other cultured dairy products (Chapter 4). When the milk pH reaches 4.6, casein is at its isoelectric point and its minimum solubility, and therefore it precipitates. However, unlike the process used for cultured products, the coagulum or gel is not left intact, but rather is then cut into die-sized curds. After separation from the whey and cooking and washing steps are completed, the acid-precipitated curds are comprised almost entirely of casein and water. In fact, pure casein and caseinate salts (made by neutralizing acid casein), both of which are of considerable commercial importance, are made via the acid precipitation method. In cheese making, of course, the curds are then further processed, resulting in products such as Cottage cheese and farmers' cheese. It is important to realize that casein coagulates at pH 4.6 whether acidification occurs via fermentation-generated acids or simply by addition of food-grade acids direct into the milk. In fact, the latter process is preferred by some producers of Cottage, cream, and other acid-coagulated cheeses, due in part to the ease of manufacture and the elimination of starter cultures as an ingredient. However, because theses products are not fermented (and, therefore, they must be labeled as acid-set), their manufacture will not be considered further in this text.

The second and most common way to effect coagulation is by the addition of the enzyme chymosin (or rennet). This enzyme hydrolyzes a specific peptide bond located between residues 105 (a methionine) and 106 (a phenylalanine) in κ casein. The hydrolysis of this bond is sufficient to cause a part of κ casein (the glycomacropeptide fraction) to dissociate from the casein micelle, exposing the anionic phosphates of β-casein. Thus, the remaining casein micelle becomes sensitive to calcium-mediated precipitation (Box 5-1). In contrast to acid-precipitated casein, the coagulated casein network formed by chymosin treatment traps nearly all of the milkfat within the curd. Most of the cheeses manufactured around the world rely on chymosin coagulation. It is worth emphasizing that even though chymosin, alone, is sufficient to coagulate milk, lactic starter cultures are also absolutely essential for successful manufacture of most hard cheeses. The lactic acid bacteria that comprise cheese cultures not only produce acid and reduce the pH, they also contribute to the relevant flavor, texture, and rheological properties of cheese, as described later.

Until relatively recently, most chymosin was obtained from its natural source, the stomachs of suckling calves after slaughter,

Box 5–1. Casein Chemistry and Chymosin Coagulation

Casein, the main protein in milk, consists of several subunits, including $\alpha_{s1}, \alpha_{s2}, \beta$, and κ. They are assembled as complexes called micelles that contain thousands of casein subunits, along with calcium phosphate, that are held together primarily via hydrophobic and electrostatic interactions. Although some researchers have suggested that casein is organized in the form of sub-micelles, how casein micelles actually exist in milk is uncertain (Lucey, 2002; Lucey et al., 2003). Moreover, it also interesting to understand how these micelles form into a gel when exposed to the enzyme chymosin. Over the past forty years, several models have been proposed that describe the casein micelle structure and how gel formation occurs.

One simplified model suggests that the hydrophobic α and β subunits are located within the interior core of the micelle, and are surrounded by κ casein, as shown below. Accordingly, a part of the kappa casein molecule (called the glycomacropeptide) protrudes from the micelle, providing steric hindrance between neighboring micelles (Figure 1). Since this section of the κ casein is negatively charged, electrostatic repulsion between casein micelles also prevents micelles from coming into contact with other micelles. However, once this barrier is removed (via the action of the enzyme chymosin; see below), glycomacropeptide is released and the micelles can interact with one other. There will still be a small negative charge on the micelles; however, soluble cationic calcium serves to neutralize the charge. This allows the micelles to come into contact with one another, resulting in formation of a gel. Importantly, this gel will entrap fat, forming the basis of cheese manufacture (Figure 2).

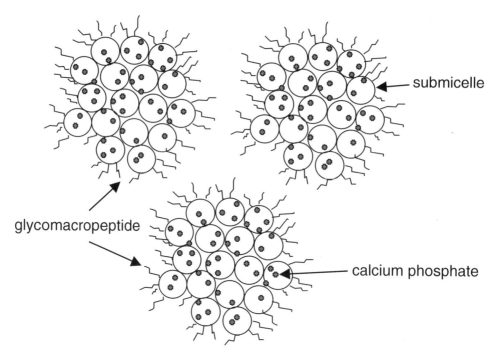

Figure 1. Structural model of the casein micelle, showing individual submicelles. Adapted from Walstra, 1999.

Box 5–1.　Casein Chemistry and Chymosin Coagulation *(Continued)*

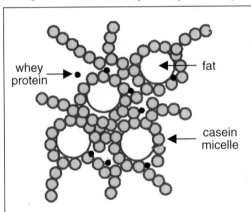

Figure 2.　Chymosin-coagulated cheese matrix model. Casein is envisioned as surrounding fat globules; small amounts of whey protein may also be present. Adapted from Hinrichs, 2004.

In effect, κ casein can be viewed as a stabilizing protein or physical barrier that protects the casein micelle from a spontaneous calcium-induced coagulation. Similarly, according to another recently proposed model (the Horne Model), casein molecules are "polymerized" via hydrophobic cross-linking and calcium phosphate bridges (Horne, 1998). Since κ casein does not bind calcium and cannot cross-link, it acts a chain-terminator (thereby limiting the size of the micelles).

Chymosin is an aspartic protease (meaning that it contains aspartic acid residues within the active site of the enzyme) synthesized by mucosal cells within the fore-stomach of young calves. Its physiological function in the suckling calf is presumably to transform casein from a soluble or dispersed form in the stomach to a solid state, where it can then be attacked more efficiently by other gastric enzymes. The peptides and amino acids can then be adsorbed across the intestinal wall, satisfying the nutritional requirements of the growing calf.

Chymosin performs a similar biochemical function in cheese manufacturing, albeit under quite different circumstances. When chymosin is added to milk, hydrolysis of κ casein occurs precisely at the peptide bond located between residues 105 (phenylalanine) and 106 (methionine). As noted above, the product of this hydrolysis reaction, glycomacropeptide, separates from the casein micelles. Because milk contains more than 1,000 mg/L calcium, about 10% of which is in the ionic form, calcium-mediated coagulation readily occurs, and the first step in cheese making has been accomplished.

It is also important to note that the casein subunits (and βcasein, in particular) can also be attacked by chymosin and other proteinases present in the chymosin mixture by milk-derived proteinases (particularly plasmin) and by proteinases produced by microorganisms (including the starter culture). The hydrolysis of casein by these proteinases may have serious consequences, both good and bad, during the manufacture of cheese, as will be discussed later in this chapter.

References

Horne, D.S. 1998. Casein interactions: casting light on the black boxes, the structure in dairy products. Int. Dairy J. 8:171–177.

Lucey, J.A. 2002. Formation and physical properties of milk protein gels. J. Dairy Sci. 85:281–294.

Lucey, J.A., M.E. Johnson, and D.S. Horne. 2003. Perspectives on the basis of the rheology and texture of properties of cheese. J. Dairy Sci. 86:2725–2743.

Walstra, P. 1999. Casein micelles: do they exist? Int. Dairy J. 9:189–192.

via a salt extraction process. The relatively high cost of this enzyme along with increasing demand and sporadic supply problems have long driven cheese ingredient suppliers to develop less expensive, alternative enzyme products that could perform the same function as chymosin. A genetically engineered form of chymosin was approved by the FDA in 1990, and several such products were later approved that have since captured much of the American chymosin market (Box 5-2).

Finally, it is possible to form a precipitate by a combination of moderate acid addition (pH 6.0), plus high heat (> 85°C). Whey proteins are denatured under these conditions, thus the precipitate that form consists not only of casein, but also whey proteins. Fat may also be retained. Even in solutions where casein is absent

---

Box 5–2. Making Calf Chymosin in Fermentors

Chymosin, the enzyme that causes milk to coagulate, is an essential ingredient in the cheese-making process. Until the 1990s, calf chymosin (the major source) had been the most expensive ingredient (other than milk) used in cheese manufacture, adding about $0.03 to each kg of cheese. This was because the chymosin supply depended on veal production, which was subject to considerable market variations, as well as on production and purification costs. As the cheese industry grew worldwide, but especially in the United States in the 1960s, '70s, and '80s, the substantial increase in demand for calf chymosin created supply problems and led to even higher prices. Thus, cheese makers sought other, less expensive sources of coagulant.

Among the early non-calf rennet substitutes were bovine and porcine pepsin. These enzymes coagulate milk; however, they have a number of undesirable features, most important being that cheese quality is not as good. In the 1960s, fungal enzymes (sometimes referred to as microbial rennets) derived from *Mucor miehei* and *Mucor pusilus* were isolated and commercialized. Although considerably less expensive than calf chymosin, these enzymes were also far from perfect. They had much more non-specific casein hydrolysis activity, resulting in yield loss and flavor and texture defects. Some were heat stable and residual activity could be detected in the whey, limiting the application of whey as an ingredient in other products. Still, price and other considerations (e.g., kosher and vegetarian status) led these products, and their improved, second-generation versions, to gain a substantial portion of the coagulant market.

The search for enzymes with properties more like calf chymosin shifted to an entirely new direction in the mid- to late 1980s when recombinant DNA technology was developed. During this time, several new biotechnology companies began projects to identify the gene coding for chymosin in calf abomasum mucosal cells and expressing the gene in suitable host cells (Marston et al., 1984; van den Berg et al., 1990; Dunn-Coleman et al., 1991). It was immediately recognized that the actual genetic material of interest was not going to be the chymosin gene in the form of DNA, but rather the mRNA transcribed from the DNA.

This is because it is the mRNA, not the DNA, that contains the actual coding regions in eukaryotic cells. Most eukaryotic genes contain coding regions (exons) along with introns, intragenic regions of non-coding DNA. The latter are excised from the mRNA following transcription, leaving behind the edited mRNA that is ultimately transcribed. In the case of the chymosin gene, there are eight introns and nine exons.

The problem of identifying the chymosin gene was even more complicated because the gene also contains a sequence that encodes for the pro-enzyme form of the chymosin protein. Prochymosin contains a 42 amino acid N terminal region whose function is to maintain the enzyme in a stable but inactive form thereby preventing the enzyme from hydrolyzing proteins within the producer cell itself. The chymosin genes also encode for a leader sequence that directs secretion of the protein across the membrane and out of the producing cell. This "pre" region is cleaved during the secretion step, and the "pro" portion is ultimately autohydrolyzed in the acid environment of the stomach, leaving active chymosin.

Box 5–2. Making Calf Chymosin in Fermentors *(Continued)*

As molecular biologists began this research, therefore, they first isolated total mRNA from the mucosal cells (Figure 1). Following a simple purification step, the enzyme, reverse transcriptase, was used to make complimentary DNA (cDNA) directly from the mRNA. This cDNA was then ligated into a plasmid vector and cloned into *E. coli* to make a library of *E. coli* clones, representing, in theory, all of the different mRNA species transcribed by the mucosal cells (including clones containing the preprochymosin mRNA). The library was screened, either using anti-chymosin antibodies or chymosin gene DNA probes, to identify clones capable of expressing the chymosin protein. Eventually, chymosin-expressing, recombinant *E. coli* clones were identified.

Yet another expression problem, however, quickly became apparent. When *E. coli* expresses certain heterologous genes, especially genes from eurkaryotic organisms, the gene products—in

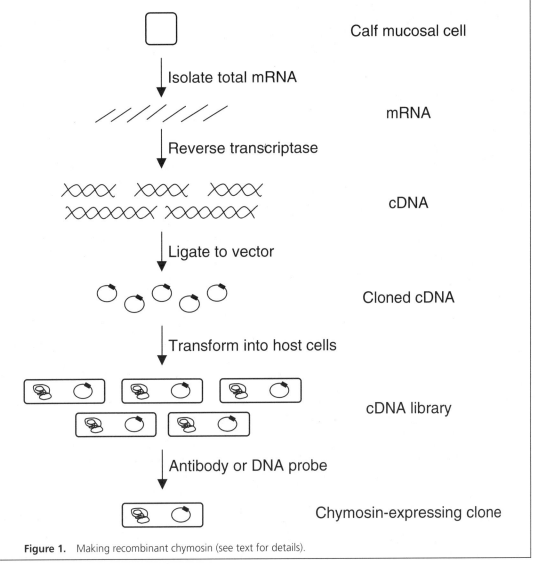

Figure 1. Making recombinant chymosin (see text for details).

(Continued)

Box 5–2. Making Calf Chymosin in Fermentors *(Continued)*

this case, chymosin—are not secreted but instead are packaged inside the cell in the form of inclusion bodies. Thus, to retrieve the chymosin, the cells had to be collected and disrupted, either mechanically or chemically, to release the chymosin-containing inclusion bodies. The latter, then, had to be solubilized to liberate active chymosin. Nonetheless, despite these awkward and expensive steps, genetically engineered chymosin won FDA approval and quickly gained acceptance in the marketplace (Flamm, 1991). Subsequently, expression systems for eukaryotic organisms, including *Kluyveromyces lactis* and *Aspergillus nidulans*, were developed that led to secreted product with high yield. Engineered chymosin is currently marketed worldwide by several major companies, including DSM Food Specialties (formerly Gist-Brocades), Chrs. Hansens, and Danisco. Among the production organisms currently used for these so-called recombinant chymosin products are *K. lactis* and *Aspergillus niger*.

Aside from the obvious cost advantage of the genetically engineered chymosins, other benefits were also realized. These products are derived from a fermentor, rather than calf stomachs, and are not subject to supply problems or price fluctuations. They can be classified as vegetarian and can more easily obtain kosher status. Importantly, the chymosin is relatively easy to purify and can be produced free of other proteases that are sometimes present in calf chymosin preparations. In fact, manufacturers claim that the engineered chymosin is 100% pure and has high specific activity. Thus, according to some experts, cheese made using these chymosin products has organoleptic properties as good as (if not better than) cheese made with calf chymosin. (There was also a contrary argument—that pure chymosin makes a less flavorful aged cheese, since non-specific proteolysis is absent.)

Currently, engineered chymosin sells for about half the cost of what calf chymosin sold for in the 1980s. Thus, it is not surprising that engineered chymosin has, for all practical purposes, entirely displaced calf chymosin in the United States. Calf chymosin has essentially become a niche product, with less than 10% of the coagulant market. However, in Europe, where many cheese manufacturers prefer or are required to produce cheese by traditional techniques, calf chymosin is more widely used. It is worth noting that unlike other foods produced via biotechnology, when engineered chymosin was introduced, there was little public concern or protest. Perhaps this was because the notion of a product produced by microorganisms in fermentors was more appealing to consumers than the image of calf stomach extracts. In addition, the FDA determined that no special labels would be required for cheese made with engineered chymosin, which they considered to be "not significantly different" from calf chymosin.

References

van den Berg J.A., K.J. van der Laken, A.J. van Ooyen, T.C. Renniers, K. Rietveld, A. Schaap, A.J. Brake, R.J. Bishop, K. Schultz, D. Moyer, M. Richman, and J.R. Shuster. 1990. *Kluyveromyces* as a host for heterologous gene expression: expression and secretion of prochymosin. Bio/Technology 8:135–139.

Dunn-Coleman, N.S., P. Bloebaum, R.M. Berka, E. Bodie, N. Robinson, G. Armstrong, M. Ward, M. Przetak, G.L. Carter, R. LaCost, L.J. Wilson, K.H. Kodama, E.F. Baliu, B. Bower, M. Lamsa, and H. Heinsohn. 1991. Commercial levels of chymosin production by *Aspergillus*. Bio/Technology 9:976–981.

Flamm, E.L. 1991. How FDA approved chymosin: a case history. Bio/Technology 9:349–351.

Marston, F.A.O., P.A. Lowe, M.T. Doel, J.M. Schoemaker, S. White, and S. Angal. 1984. Purification of calf prochymosin (prorennin) synthesized in *Escherichia coli*. Bio/Technology 2:800–804.

(i.e., whey), this process can result in enough precipitated whey protein to form a cheese. Examples of precipitated cheeses include Ricotta cheese, the Hispanic-style cheeses queso fresco and queso blanco, and Gjetost, a whey-derived cheese popular in Norway.

Squeezing out water

Once milk is transformed from a liquid into a solid (actually a gel), the next goal is to remove water. The first step involves increasing the surface area of the single large gel mass by cutting it into literally millions of smaller cubes of curd (e.g., 1 m^3 of a cheese gel cut into 1 cm^3 particles yields 10^6 such curd particles). Because the distance the water molecules must travel (from the interior of the gel to the outside environment) decreases as the curd size decreases, this step has the effect of substantially increasing the rate of syneresis. Then, when these curds are stirred, and then heated, the curds shrink and syneresis is further enhanced.

Acidification is another factor involved in concentrating milk solids. As discussed previously in Chapter 4, as the pH approaches the isoelectric point of casein (i.e., 4.6), water-holding capacity decreases and syneresis, or whey expulsion, increases. In cheese manufacture, as for cultured milk products, acidification is due to fermentation of lactose by the starter culture. As we will see later, the culture performs other critical functions; in particular, it is also responsible for desirable flavor and texture properties. But it is the ability to convert lactose to lactic acid and to reduce the cheese pH that makes the culture essential. Finally, it is possible to use physical force (i.e., pressure or gravity) to squeeze out even more water.

General Steps in Cheese Making

On a worldwide basis, there are probably thousands of different types of cheeses produced and consumed. As Charles De Gaulle, the former French president, famously lamented, there are hundreds of different cheeses made in France alone[1]. Anyone who has visited a fromageri in Paris or a formaggio in Milan (or perhaps the National Cheese Emporium in England; Box 5-3), can certainly appreciate the incredible variety of cheeses that are available. How could there be so many? Are the procedures for making cheese so complex as to allow for all the cheeses produced?

In reality, the basic manufacturing steps for all cheeses are surprisingly similar. However, it is the almost unlimited number of variables that exist at each of these steps that ultimately account for the myriad number of different cheeses. Described below are the various ingredients and manufacturing steps that are used in cheese making, with an emphasis on those variables that distinguish one type of cheese from another.

Milk

As shown in Figure 5-5, the first variable starts with the milk itself. Although most cheeses are made from cow's milk, many cheeses are made using milk from other sources. For example, Feta and Chevre are ordinarily made from goat milk, Roquefort and Romano are made from sheep milk, and Mozzarella is often made from the milk of water buffaloes. These milks have dramatically different gross compositions from bovine milk; in particular, they all contain more fat (water buffalo and goat milk contain nearly twice as much). However, even the composition of cow's milk varies according to the breed of cow, the nature of the feed consumed by the cow, and even when the milk was obtained (i.e., morning versus evening).

Moreover, in cheese making, not only does the gross composition affect cheese properties, but so does the specific composition of each of the milk constituents. The lipid portion of goat milk, for example, contains a higher percentage of volatile, short-chain fatty acids, such that rancid flavor notes are most evident when the triglycerides are hydrolyzed by lipases. The fat content also has a profound influence of other properties of cheese. Fat not only contributes to the body and texture of cheese, but it also serves as substrate for important flavor-generating reactions performed by microorganisms. Also, many of the flavor constituents derived from non-lipid substrates that form during cheese ripening are soluble in the lipid phase. For example, hydrophobic peptides derived from casein hydrolysis (many of which are bitter) are found in the fat portion of the cheese.

For some cheeses, the milk is standardized to give a fat content that is particular to a given cheese. In general, the minimum fat content for most cheeses is usually around 50% (on a dry basis). For example, Cheddar type cheeses are made with whole milk, containing 3.5 to 4.5% milkfat. However, other cheeses are made with

Box 5–3. The Monty Python Cheese Shop Sketch

More than forty different types of cheese are mentioned, making it, for cheese science students, at least, the most educational television sketch ever presented.

CUSTOMER: Good Morning.

OWNER: Good morning, sir. Welcome to the National Cheese Emporium!

CUSTOMER: Ah, thank you, my good man.

OWNER: What can I do for you, sir?

CUSTOMER: I want to buy some cheese.

OWNER: Certainly, sir. What would you like?

CUSTOMER: Well, eh, how about a little red Leicester, sir.

OWNER: I'm afraid we're fresh out of red Leicester, sir.

CUSTOMER: Oh, never mind, how are you on Tilsit?

OWNER: I'm afraid we never have that at the end of the week, sir, we get it fresh on Monday.

CUSTOMER: Tish tish. No matter. Well, stout yeoman, four ounces of Caerphilly, if you please.

OWNER: Ah! It's been on order, sir, for two weeks. Was expecting it this morning.

CUSTOMER: 'Tis not my lucky day, is it? Aah, Bel Paese?

OWNER: Sorry, sir.

CUSTOMER: Red Windsor?

OWNER: Normally, sir, yes. Today the van broke down.

CUSTOMER: Ah, Stilton?

OWNER: Sorry.

CUSTOMER: Ementhal? Gruyere?

OWNER: No.

CUSTOMER: Any Norwegian Jarlsburg, per chance?

OWNER: No.

CUSTOMER: Lipta?

OWNER: No.

CUSTOMER: Lancashire?

OWNER: No.

CUSTOMER: White Stilton?

OWNER: No.

CUSTOMER: Danish Brew?

OWNER: No.

CUSTOMER: Double Gloucester?

OWNER: No.

CUSTOMER: Cheshire?

OWNER: No.

CUSTOMER: Dorset Bluveny?

OWNER: No.

CUSTOMER: Brie, Roquefort, Pol le Veq, Port Salut, Savoy Aire, Saint Paulin, Carrier de lest, Bres Bleu, Bruson?

OWNER: No.

CUSTOMER: Camembert, perhaps?

OWNER: Ah! We have Camembert, yes sir.

CUSTOMER: (surprised) You do! Excellent.

OWNER: Yessir. It's . . .ah, . . .it's a bit runny . . .

CUSTOMER: Oh, I like it runny.

OWNER: Well, . . .It's very runny, actually, sir.

CUSTOMER: No matter. Fetch hither the fromage de la Belle France! Mmmwah!

OWNER: I . . .think it's a bit runnier than you'll like it, sir.

Box 5–3. The Monty Python Cheese Shop Sketch *(Continued)*

CUSTOMER: I don't care how blinking runny it is. Hand it over with all speed.
OWNER: Ooooooooooohhh...!
CUSTOMER: What now?
OWNER: The cat's eaten it.
CUSTOMER: Has he.
OWNER: She, sir.
CUSTOMER: Gouda?
OWNER: No.
CUSTOMER: Edam?
OWNER: No.
CUSTOMER: Case Ness?
OWNER: No.
CUSTOMER: Smoked Austrian?
OWNER: No.
CUSTOMER: Japanese Sage Darby?
OWNER: No, sir.
CUSTOMER: You ... do have some cheese, don't you?
OWNER: Of course, sir. It's a cheese shop, sir. We've got—
CUSTOMER: No no ... don't tell me. I'm keen to guess.
OWNER: Fair enough.
CUSTOMER: Uuuuuh, Wensleydale.
OWNER: Yes?
CUSTOMER: Ah, well. I'll have some of that!
OWNER: Oh! I thought you were talking to me, sir. Mister Wensleydale, that's my name.
CUSTOMER: Greek Feta?
OWNER: Uh, not as such.
CUSTOMER: Uh, Gorgonzola?
OWNER: No.
CUSTOMER: Parmesan?
OWNER: No.
CUSTOMER: Mozzarella?
OWNER: No.
CUSTOMER: Paper Cramer?
OWNER: No.
CUSTOMER: Danish Bimbo?
OWNER: No.
CUSTOMER: Czech sheep's milk?
OWNER: No.
CUSTOMER: Venezuelan Beaver Cheese?
OWNER: Not today, sir, no.
CUSTOMER: Aah, how about a Cheddar?
OWNER: Well, we don't get much call for it around here, sir.
CUSTOMER: Not much ca—It's the single most popular cheese in the world!
OWNER: No 'round here, sir.
CUSTOMER: And what IS the most popular cheese 'round hyah?
OWNER: 'Illchester, sir.
CUSTOMER: Is it.
OWNER: Oh, yes, it's staggeringly popular in this manusquire.
CUSTOMER: Is it.

(Continued)

Box 5–3. The Monty Python Cheese Shop Sketch *(Continued)*

OWNER: It's our number one best seller, sir!
CUSTOMER: I see. Uuh ... 'Illchester, eh?
OWNER: Right, sir.
CUSTOMER: All right. Okay. 'Have you got any?' he asked, expecting the answer 'no.'
OWNER: I'll have to look, sir ... nnnnnnnnnnnnnnnno.
CUSTOMER: It's not much of a cheese shop, is it?
OWNER: Finest in the district!
CUSTOMER: (annoyed) Explain the logic underlying that conclusion, please.
OWNER: Well, it's so clean, sir!
CUSTOMER: It's certainly uncontaminated by cheese ...
OWNER: You haven't asked me about Limburger, sir.
CUSTOMER: Would it be worth it?
OWNER: Could be ...
CUSTOMER: Have you got any Limburger?
OWNER: No.
CUSTOMER: Figures. Predictable, really I suppose. It was an act of purest optimism to have posed the question in the first place. Tell me:
OWNER: Yes sir?
CUSTOMER: Have you in fact any cheese here at all.
OWNER: Yes, sir.
CUSTOMER: Really?
OWNER: No. Not really, sir.
CUSTOMER: You haven't.
OWNER: No sir. Not a scrap. I was deliberately wasting your time, sir.
CUSTOMER: Well, I'm sorry, but I'm going to have to shoot you.
OWNER: Right-O, sir.

The Customer takes out a gun and shoots the owner.

CUSTOMER: What a senseless waste of human life.

milk adjusted to 3% fat (e.g., Swiss cheese) or less (part-skim Mozzarella). In part, this is because the body characteristics of certain hard cheeses, for example, Swiss and Parmesan, require less fat and a higher casein-to-fat ratio. In contrast, some cheeses are made from milk that is enriched with cream, such that the fat content of the cheese (so-called double cream cheese) will be 60% (or even 72% for triple cream cheese).

Another key variable involves the handling of the milk, and in particular, whether the milk has been heated or not. In the United States, all unripened cheese (aged less than sixty days) must be made from pasteurized milk, whereas only aged cheese (held for more than sixty days

at a temperature not less than 1.7°C) can be made from raw milk. In many other parts of the world, including France, Italy, and other major cheese-producing countries, there are no such pasteurization requirements, and even fresh cheeses can be made from raw milk (unless they are to be exported to the United States—then they must conform to U.S. requirements).

In reality, however, even cheese milk that is not pasteurized is often heat-treated to sub-pasteurization conditions (even for aged cheese). The reason for pasteurizing milk, regardless of how that milk is to be used, is to kill pathogenic and spoilage microorganisms. Given the concern about food safety in the United States, there has been a trend among

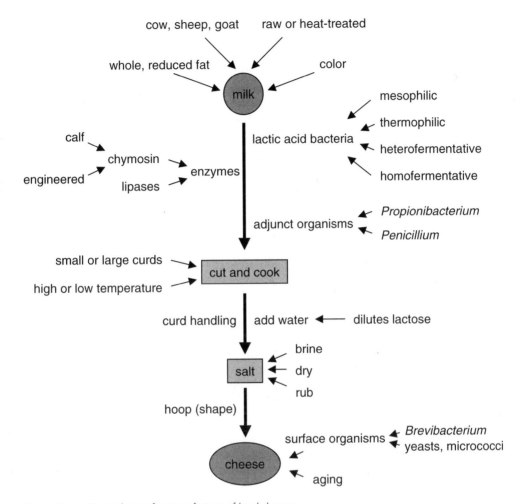

Figure 5–5.   General steps for manufacture of hard cheeses.

large U.S. manufacturers to use pasteurized milk for most cheeses, even those that are aged.

Pasteurization, however, not only kills pathogens and undesirable spoilage bacteria, but it also inactivates much of the endogenous microflora and enzymes ordinarily present in raw milk. Since the microflora and enzymes both contribute to the overall flavor and texture properties of the finished cheese, especially if the cheese is aged, quality differences between cheese made from raw or heat-treated milk can be significant. Thus, there is now a debate between those who believe that

pasteurization should be required and those who contend that pasteurization is detrimental to cheese quality (Box 5–4). It should be emphasized, however, that whether the milk is raw or pasteurized, it should still be of high microbiological quality, free of antibiotics, and within the standards specified by government regulations.

Finally, one of the most obvious ways to treat milk to distinguish one cheese from another is by the color. Milk, and the finished cheese, can be made more yellow by adding the natural coloring agent, annato, or made

Box 5–4.   Safety of Cheese Made From Raw Milk

In the United States, only cheese that is aged for more than sixty days can be made from raw milk; all other cheeses must be made from milk that has been pasteurized. The rationale for this policy, which was established in the 1930s, is based on early research that indicated any pathogenic bacteria present in the raw milk would die during a sixty-day ripening period. Factors believed to contribute to cell death included low pH, low water activity, low Eh, high salt-in-moisture levels, the presence of organic acids and competing microflora, and the absence of fermentable sugars. In other words, the collective effect of these inhospitable conditions were thought to create an environment that would result in the eventual lysis and death of any pathogenic bacteria that may have been present in the original cheese.

The sixty-day requirement was really nothing more than an approximation based on a rather limited number of studies. Moreover, these studies were done long before *Listeria monocytogenes, Escherichia coli* O157:H7, and other pathogenic organisms were recognized as potential contaminants of raw milk. That these pathogens could be inherently more resistant to environmental extremes or that they could induce stress resistance defense systems were possibilities that could hardly have been imagined.

It is now well-known that some of these pathogens can indeed tolerate all sorts of inhibitory conditions. For example, *L. monocytogenes* is particularly capable of surviving, even growing, at acid pH and in the presence of high salt concentrations. Furthermore, when exposed to various shocks, such as high, sub-lethal temperatures, this organism synthesizes a set of proteins that enable it to tolerate other subsequent stresses. These observations led researchers to challenge the conclusions made seventy years ago, i.e., that a sixty-day aging process would result in raw milk cheese that is free of pathogens. Recent research, in fact, now indicates that these "newly" recognized pathogens could survive aging and, if present at sufficiently high numbers in the raw milk, could be present in the aged cheese. For example, several publications from the Marth laboratory at the University of Wisconsin showed that *L. monocytogenes* could survive more than one year in Cheddar cheese and that levels greater than 10^7 cfu/g could be reached in Camembert cheese (summarized by Ryser, 1999).

In response to these findings, the FDA announced in 2002 that this policy would be reviewed and left open the possibility that all cheese, regardless of whether it was aged or not, would have to made from pasteurized milk. Although manufacturers of aged cheese were undoubtedly concerned about this announcement, the greatest furor of opposition came from gourmet-type consumer groups. Their position, as articulated on various Web sites, is that cheese made from raw milk is superior in flavor, texture, and overall sensory appeal to that of pasteurized milk, and that any perceived food safety risks are exaggerated or non-existent. Indeed, despite the research showing that pathogens might, at least in theory, survive sixty days of aging, the food safety record for aged cheese is extraordinarily good. There are few published accounts of aged raw milk cheese having been responsible for foodborne disease, and in those cases where the cheese was implicated, other extenuating circumstances were involved (e.g., post-aging contamination by workers or equipment).

Despite the excellent record for aged raw milk cheese, fresh raw or pasteurized milk cheese has been the cause of several serious foodborne disease outbreaks (Table 1). In the majority of these outbreaks, poor handling of the milk or cheese or poor sanitation was responsible. Post-pasteurization contamination of milk, in particular, is the most frequent cause of these outbreaks. However, there are some fresh or non-aged cheeses that, even when properly made, may still pose a risk to certain consumers. Those at risk include the very young and very old, pregnant women, and other immuno-compromised populations. Brie cheese, for example, is on the "do not eat" list for pregnant women and HIV-positive individuals, due to the potential (but infrequent) occurrence of *L. monocytogenes* in this cheese. The use of sub-pasteurization treatments, such as "thermization" (performed by heating milk to about 65°C for fifteen to twenty

Box 5–4. Safety of Cheese Made From Raw Milk *(Continued)*

seconds), may be somewhat effective against potential pathogens and provides a margin of safety at which some cheese makers may be comfortable.

Table 1. Safety record of cheese.

Year	Cheese	Cases	Organism	Cause
1970s	Brie		Enteroinvasive	Poor sanitation
		347	*Escherichi coli*	
1980s	Brie		Enterotoxigenic	Unknown
		169	*Escherichi coli*	
1980s	Mont d'Or	122	*Listeria monocytogenes*	Environment
1984	Cheddar	2,000	*Salmonella typhimurium*	Unknown
1985	Mexican-style	300	*Listeria monocytogenes*	Raw milk contamination; poor sanitation
1989	Mozzarella	164	*Salmonella javiana*	Poor sanitation
1995	Brie	20	*Listeria monocytogenes*	Raw milk contamination

Reference

Ryser, 1999. Incidence and behavior of *Listeria* in fermented dairy products, *In* E.H. Marth and E.T. Ryser (ed.) Listeria, *Listeriosis, and Food Safety, 2nd ed.* Marcel Dekker, New York.

more white by adding bleaching-type agents. Although the color has no effect on flavor or texture, manufacturers have learned that visual appeal can have a pronounced effect on acceptability and preference. In the United States, for example, consumers from the Midwest, in general, prefer orange-colored Cheddar cheese, whereas Northeasterners favor white Cheddar.

Starter cultures

The composition of the starter culture depends on the intentions of the cheese maker. If the cheese-making procedure includes a step where the curds will be exposed to high temperatures, such as during manufacture of Swiss, Parmesan, and Mozzarella cheese, then a thermophilic lactic acid bacterial culture able to withstand those temperature must be used. If a particular flavor compound is desired, such as diacetyl in Gouda cheese, then again, the culture must contain specific organisms capable of producing those flavor compounds. In fact, culture technology has advanced now to the point where each strain present in the cul-

ture can be selected on the basis of the specific performance attributes desired by the customer. Thus, the culture can contribute to considerable variation in the finished cheese, and even modest changes in the culture composition or amount can result in dramatic differences in the cheese.

As described in Chapters 2 and 3, cultures used for cheese fermentations consists of several genera and species of lactic acid bacteria. Aside from selecting a culture that gives the desired rate and extent of acid development and that produces the desired flavor and texture, the main distinction for culture selection is based on temperature. Mesophilic cultures, containing strains of *Lactococcus lactis* subsp. *lactis* and *Lactococcus lactis* subsp. *cremoris*, grow within a wide temperature range of about 10°C to 40°C. Growth rates at the extreme ends of this range are, however, quite low and temperatures much above the upper limit for growth are lethal. Moreover, the optimum temperature for growth of mesophilic cultures is about 28°C to 32°C (depending on the strain), which is generally very near the temperature range that the milk or cheese will be held during manufacture (i.e., when fermentation is expected to occur). The mesophilic cultures are the workhorses of the cheese industry, and are used for the majority of cheeses.

Thermophilic cultures, mainly *Lactobacillus helveticus* and *Streptococcus thermophilus*, are also widely used. They have a temperature optima around 42°C to 45°C and are able to grow at temperatures as high as 52°C. Like the mesophiles, however, temperatures just a few degrees higher than their maximum growth temperature may result in thermal inactivation. This means that when curds are heated, the temperature must not exceed that which the culture can tolerate. Otherwise, the culture will be inactivated or injured and the subsequent fermentation will either be slow or not occur. However, it should be noted that thermal effects on microorganisms obey first-order kinetics, such that inactivation or killing occurs at a logarithmic rate. Thus, even when the temperature reaches a lethal level, some cells will survive (depending on the exposure time) and be able to grow (albeit slowly), once a compatible temperature is established.

Most of the strains that comprise mesophilic and thermophilic cultures are homofermentative. However, heterofermentative lactic acid bacteria may be included in mesophilic cultures that are used for particular cheeses. Specific heterofermentative species include *Leuconostoc mesenteroides* subsp. *cremoris* and *Leuconostoc lactis*. Not coincidentally, these organisms also ferment citrate and produce diacetyl. Thus, cheeses such as Gouda and Edam have a buttery aroma, as well as a few small eyes, due to CO_2 formation. Finally, cultures may also contain non-lactic acid bacteria for Swiss-type and surface-ripened cheeses, as well as fungal spores for mold-ripened cheeses, as will be discussed later.

Until relatively recently, bulk cultures grown in whey or milk without pH control were the dominant form in which cultures were used (reviewed in Chapter 3). Culture inoculum for Cheddar type cheeses was usually about 1% (w/w), which gave an initial cell concentration of about 5×10^6 cells per ml of milk. Other cheeses are made using more culture (e.g., Mozzarella is normally made with a 2% starter) or less culture (e.g., Swiss is made with 0.5% culture). Although these traditional bulk cultures are still used, in the past twenty years, bulk culture tanks began to be equipped with external pH control systems. This simple technology made it possible for operators to partially neutralize the medium as the cells grew, and, therefore, maintained the pH at a level that was more to the liking of the starter culture organisms.

Specialized culture media was also introduced that provided "internal" pH control. With the widespread use of these pH controls, bulk culture systems, culture viability, and cell density (i.e., cells per gram) have been significantly enhanced. Thus, less culture is necessary, perhaps half as much on a weight or volume basis. In addition, highly concentrated, direct-to-vat cultures are also now widely used. These products may contain ten or more times as many cells as traditional bulk cultures. One 500 ml can, for example, is sufficient to inoculate 10,000 L of milk (i.e., 0.005%).

Despite the consistent nature of modern cultures, their activity and cell density may still be somewhat variable. Thus, it is important for cheese manufactures to adjust inoculum levels to satisfy production schedules and performance expectations. In the case of Cheddar cheese, if too much culture is added, acid development might occur too rapidly, resulting in early loss of calcium and demineralization of the cheese (discussed later). In addition, a large culture inoculum may lead to excessive production of proteolytic enzymes that can eventually affect cheese yield, flavor, and texture. In contrast, if not enough culture was added, fermentation and acid development is delayed, causing production schedule headaches as well as opportunities for spoilage or pathogenic organisms to grow.

It is important to note that, despite the availability of defined and consistent starter cultures, there are some cheeses made using more traditional starter cultures. Parmigiano Reggiano, the authentic version manufactured only in the Parma region of Italy, is made using whey as starter culture, a form of backslopping. Many of the Dutch-type cheeses (e.g., Gouda and Edam) are still made using undefined strains maintained simply as a mixed culture.

Whether a pure, mixed, or undefined starter culture is used, the milk for most cheeses made with a mesophilic culture is ordinarily brought to a temperature between 30°C and 40°C (higher for cheeses made using thermophilic cultures). If early culture growth is to be encouraged, the milk may be held for a period of time within this temperature range. However, despite a mostly favorable temperature, little culture activity and little acid development normally occurs during this early stage of the process. This is because the cells either are coming out of a somewhat dormant phase (in the case of frozen or lyophilized direct-vat-set cultures), or are adapting from a rich, almost ideal bulk culture medium to a milk medium that requires induction of at least some new biochemical pathways, such as those involving protein hydrolysis and amino acid use. Depending on the culture media and culture preparation conditions, it is possible for cultures to be rather active, with relative short lag phases. In general, most cultures still experience a lag phase (that may be quite extended, especially if a cooking step is also included), before active log phase growth occurs.

Although growth of the culture and production of fermentation end-products (other than lactic acid) usually does not occur until later in the cheese making process, delays due to culture inhibition can cause serious problems for the manufacturer. Large cheese factories that process a million or more Kg of milk per day require high throughput (i.e., the rate that milk is converted to finished cheese products). A bottleneck at the fermentation step may place upstream operations are on hold, disrupting production schedules and perhaps causing employees to work extra hours. Of course, if the fermentation is slow, sluggish, or fails altogether, cheese quality will be poor.

There are several possible causes of starter culture inhibition. Although state and federal laws require that milk be free of antibiotics, if residues were present they could inhibit the lactic culture. However, it is now very rare that milk would contain undetected antibiotics. Milk may also contain natural immunoglobulins that bind to culture bacteria, forming clumps that eventually settle in the vat. This is particularly a problem in Cottage cheese due to the long fermentation time. Another potential inhibitor is formed via the lactoperoxidase reaction. This reaction occurs when the enzyme lactoperoxidase oxidizes thiocyanate in the presence of hydrogen peroxide to form hypothiocyanate. This reactive compound can be inhibitory to certain starter culture lactic acid bacteria. Lactoperoxidase and thiocyanate are naturally present in milk, and hydrogen peroxide can be produced by the endogenous microflora. It is also possible to activate this reaction by inoculating raw milk with hydrogen peroxide-producing lactic acid bacteria in an effort to control psychrotrophic spoilage bacteria.

Chemical agents used to sanitize cheese vats can occasionally inhibit cultures if residues are not adequately rinsed. By far, however, the main cause of culture inhibition are bacteriophages, viruses that infect bacteria—in this case, lactic acid bacteria (Chapter 3). Their detrimental role in the cheese making process, and the means by which phage problems can be controlled, will be discussed later in this chapter.

Coagulation

In many cheese factories, chymosin is added to the milk immediately after or nearly at the same time as the culture is added. Some cheese makers allow a pre-ripening period to give the culture a brief opportunity to produce a small amount of acid and a slight lowering of the milk pH. Since chymosin is an acid protease (its optimum activity on κ-casein occurs at pH 5.5), it will be more active as milk pH decreases. The solubility of calcium also increases as the pH decreases. Thus, with pre-ripening, less chymosin can be used to give the same clot firmness. For similar reasons, it is also common to add calcium chloride to the milk to promote coagulation (and yield).

In any event, the amount of chymosin added, and the length of the setting period prior to cutting, depends on the cheese being made and the curd firmness desired. Usually, about 200 ml of single-strength chymosin per 1,000 kilograms of milk will give a suitable coagulation within about 30 minutes. For many large, automated manufacturers, the point at which the curd is sufficiently firm and ready for cutting is based strictly on the clock (but also on *a priori* knowledge of what times give the best cheese). Although specialized instruments are available for this purpose, curd firmness is more often than not determined on a subjective basis, i.e., when the operator deems it ready based on a simple cutting method.

Cutting and cooking

The coagulated mass is next cut using harp-like, wire knives that cut the curds into die-sized particles. The knives are constructed such that the curds can vary in size. Since this step is performed to enhance syneresis, the size, or more importantly, the surface area of the curd particles, has a major influence on the rate of water removal from the curd. Hard, low-moisture cheeses like Parmesan and Swiss are typically cut into kernel or wheat berry-sized curds, whereas soft, high-moisture cheeses are cut into die-sized pieces. The size of the curd also influences fat loss—more fat is retained in large curds than in small ones. Regardless of the size of the curd at cutting, the actual composition varies relatively little. The curd is constructed of a calcium-casein complex that contains entrapped fat and bacteria (including starter culture organisms), as well as water and water-soluble components including whey proteins, enzymes, vitamins, minerals, and lactose.

Syneresis begins as soon as the curds are cut, and increases during the ensuing minutes when the curds are gently stirred. The initial rate depends on the starting pH of the curd, because prior acid development greatly enhances syneresis. However, the cooking step is the primary means of enhancing syneresis. All other factors being equal, the higher the temperature

and the longer the curds are cooked and stirred, the dryer will be the finished cheese. Thus, the cooking step is one of the major variables that cheese manufacturers can manipulate to produce different types of cheese.

As noted previously, the more water removed from the curd, the less lactose will be available for fermentation. Cheese manufacturers can, therefore, influence acid production and cheese pH by modulating the cooking time and temperature conditions. Whatever the cooking temperature, however, the culture must be able to withstand that temperature, or else the cells will be attenuated (or worse yet, inactivated). Furthermore, when heat is applied it must be done gradually via a step-wise progression, since too rapid heating causes the exterior of the curds to harden and actually reduce syneresis. The cooking and stirring step has so much of an effect on the finished cheese that some experienced cheese makers can tell how dry the cheese will be simply by touch and the feel of the curd. Of course, some soft cheeses, like Brie, are not cooked at all, but rather are stirred at the setting temperature. Finally, although the actual heating step usually occurs via indirect heat transfer through jacketed vats, it is also possible to inject steam directly into the whey-curd mixture.

Curd handling

Perhaps the most influential step during the cheese making process involves the means by which the curd is handled during and after the cooking and stirring steps. For many cheeses, the whey is removed when the desired acidity is reached, when the curd has been cooked for a sufficient length of time, or when it is sufficiently firm or dry. There are several means by which the curd is separated and the whey is removed. In traditional Cheddar cheese manufacture, the curds are simply pushed to the sides of the cheese vat and the whey is drained down from the center, with screens in place at the drain end to prevent curd loss. Alternatively, the curds can be collected in cheese cloth and hoisted above the whey, as in traditional Swiss cheese manufacture. In more modern, large production factories, where cheese vats must be cleaned and re-filled, curds and whey are typically pumped to draining tables or Cheddaring machines, where whey separation occurs. Similarly, the curd-whey mixture can be added directly into perforated cheese hoops, where the whey drainage step is completed.

Although there are many ways to manipulate the curd to alter the properties of the cheese, one simple twist in the separation step is worth special mention. If, during the draining step, water is added back to the curds, then so-called "sweet" or low-acid cheeses will result. This is due to lactose dilution. Since the lactose concentrations in the curd and whey are ordinarily in equilibrium, when whey is removed and replaced by water, lactose will diffuse from curd to water, leaving the curd with markedly less lactose available for subsequent fermentation by the starter culture. Thus, the initial pH of Colby, Gouda, Havarti, and Edam cheeses are generally in the range of 5.2 to 5.4. If the added water is warm (i.e., about 35°C), cooking and syneresis will continue (as is the case for Gouda, Edam, and Havarti). However, if the water is cold (about 15°C), the moisture content of the cheese may increase (e.g., Colby). This washing step is also used for Mozzarella, in part for pH control, but more so to remove lactose and galactose from the curd (discussed later).

Once the curds are separated from the whey, several things begin to happen. First, the starter culture finds itself at a temperature conducive for growth, and soon the fermentation of lactose to lactic acid occurs. A subsequent decrease in curd pH and an increase in the titratable acidity (expressed as percent lactic acid) of the expressed whey is evident. The curds, almost immediately after whey is removed, mat or stick together. The matted curds, helped by piling slabs on top of one another, begin to stretch out and become plastic-like, a process known as Cheddaring. Alternatively, the dry curds can be stirred to facilitate whey removal and to lower the moisture in the finished cheese.

Salting

Salt is an essential ingredient that provides flavor, enhances syneresis, and contributes to the preservation of most cheeses. Even the simple step of salting, however, represents an important variable during cheese manufacture. Salt can be applied directly (i.e., in dry form) to the milled curds, as in the case of Cheddar, or salt can be rubbed onto the surface of hooped cheese, as in the case of some blue cheese varieties (e.g., Gorgonzola and Roquefort). Alternatively, some cheeses, such as Swiss, Mozzarella, and Parmesan, can be placed in brines. Obviously, when salting occurs via brining methods, the amount of salt that ends up in the cheese is a function of the diffusion rate into the cheese, as well as the geometry of the cheese block, the duration of brining, and brine strength. If the cheese is shaped or cut into small units and left in the brine, as with Feta cheese, salt concentrations can be very high (>3%). In contrast, large blocks or wheels of brined Swiss cheese typically contain less salt (<1%), especially in the interior sections. Brined cheeses that are then allowed to air dry develop natural rinds, due to surface dehydration.

Aging

The last step in the cheese making process has as much influence as any previous step with regard to the properties and qualities of the finished cheese. As noted in Chapter 1, it is a fine line that separates the production of a perfectly flavored, three-year old Cheddar cheese and a bitter, rancid, sour Cheddar cheese that is quickly rejected by any discerning consumer. Although the distinctly different properties of both of these two cheese are the result of microbial and enzymatic activities, there is one clear distinguishing factor. The key difference is that the gourmet cheese is produced when aging occurs under controlled conditions, whereas the rejected cheese occurs when control is absent or lost. As a general rule, any cheese that is intended for aging must be manufactured, from the very start, differently than an unaged cheese. The handling of the milk, the cheese pH, the moisture and salt content, and the water/salt ratio, in particular, all are important determinants that influence aged cheese quality.

In addition to its impact on the finished cheese, aging or ripening is also one of the most complex and most variable of all cheese making steps. This is due largely to the enzymes and microorganisms that are primarily responsible for flavor and texture changes that occur during ripening. The enzymes in cheese may occur naturally in the milk or be added directly in the form of rennet, chymosin, or lipase extracts. Enzymes are also derived from starter culture bacteria, adjunct organisms, or endogenous milkborne organisms. Furthermore, the availability of substrates and the pH and Eh conditions in the cheese influence the activity of these enzymes and the types and amounts of products that are formed. Similarly, microorganisms in cheese originate from the milk, the environment, and the starter culture. Although the temperature in aging rooms is generally low (usually around 3°C to 7°C, but sometimes much higher), and the cheese milieu is not particularly conducive for growth (as noted above), metabolism of the various substrates in cheese by intact organisms still occurs. A ripening cheese represents a rather vibrant ecosystem.

Types of Cheese

Given the hundreds of cheeses produced worldwide, it is obviously not possible to discuss each particular one. There are, however, several ways to categorize the many different types of cheese into manageable groups, based, for example, on their level of hardness (e.g., from soft to hard), moisture content, cooking temperature, or extent of aging. In the sections that follow, the manufacturing procedures for different cheeses and the role of microorganisms involved in their production will be reviewed, based on the primary properties of those cheeses and their distinguishing characteristics (Table 5-1). For the most part, only

the most well-known and widely consumed cheeses are discussed and only general procedures are described. The reader seeking detailed procedures is advised to consult other excellent sources for specific manufacturing details (see Bibliography).

Acid-coagulated cheeses

In the United States, the most popular of the acid-precipitated cheeses are Cottage cheese and cream cheese. And although per capita consumption of Cottage cheese (all varieties) has declined in the past twenty years by nearly 30% (despite modest increases in low-fat versions), cream cheese per capita consumption has increased by more than 100% (from less than 0.5 Kg to more than 1 Kg per person per year). This increase in cream cheese consumption is undoubtedly due to an equal increase in the popularity of bagels and cheesecakes. The availability of flavored, whipped, and low- and reduced- fat cream cheese products has also contributed to this increase. Other cheeses in this category, including bakers' cheese and farmers' cheese, have only a small share of the market.

These cheeses rely on the fermentation of lactose to lactic acid by a suitable starter culture, such that a pH of 4.6 or below is reached. In Cottage cheese manufacture, the starting material is simply skim milk. Often, nonfat dry milk is added to increase the throughput, or the amount of cheese produced per vat, and to improve body. The milk is always pasteurized, as required by law, since this is a fresh or non-aged product. A mesophilic lactic starter culture, containing strains of *Lactococcus*, is then added at a rate or amount that depends on the production schedule preferences of the manufacturer. For a fast make, as much as a 5% culture inoculum is added (that is, 5% of the total milk volume, by weight, is culture). The inoculated milk is mixed, then allowed to incubate quiescently in the vat at 30°C to 32°C, the optimum temperature for the culture. At this inoculum level and at this temperature, an active culture can coagulate the milk in five hours or less.

Table 5.1. Properties of major cheese groups.

Cheese	Starter Culture	Other organisms	Salt	Moisture	pH
Cheddar type					
Cheddar	mesophilic[1]		1.5	37	5.5
Cheshire	mesophilic		1.7	38	4.8
Colby	mesophilic		1.5	39	5.5
Dutch type					
Gouda	mesophilic	*Leuconostoc* sp.	2.0	41	5.8
Edam	mesophilic	*Leuconostoc* sp.	2.0	42	5.7
Cheese with eyes					
Emmenthal	thermophilic[2]	*Propionibacterium*	0.7	35	5.6
Gruyere	thermophilic	*Propionibacterium*	1.1	33	5.7
Grating type					
Parmesan	thermophilic		2.6	31	5.4
Romano	thermophilic		5.5	23	5.4
Pasta filata					
Mozzarella	thermophilic		1.2	53	5.2
Provolone	thermophilic		3.0	42	5.4
Blue mold					
Roquefort[3]	mesophilic	*Penicillium roqueforti*	3.5	40	6.4
Gorgonzola	mesophilic	*Penicillium roqueforti*	2.5	45	6.2
Stilton	mesophilic	*Penicillium roqueforti*	2.3	39	6.2
External mold					
Brie[4]	mesophilic	*Penicillium camemberti*	1.6	52	6.9
Camembert[4]	mesophilic	*Penicillium camemberti*	2.5	49	6.9
Surface ripened					
Havarti	mesophilic		1.9	43	6.4
Muenster	mesophilic	*Brevibacterium linens*	1.6	42	6.4
Limburger[5]	mesophilic	*Brevibacterium linens*	2.0	45	6.8
Brined					
Feta	mesophilic		3.0	53	4.5

[1]Mesophilic cultures = *Lactococcus lactis* subsp. *lactis* and *Lactococcus lactis* subsp. *cremoris*
[2]Thermophilic cultures = *Streptococcus thermophilus, Lactobacillus delbrueckii* subsp. *bulgaricus,* and/or *Lactobacillus helveticus*
[3]Citate-fermenting *Leuconostoc* or *Lactococcus* sp. may be added
[4]*Streptococcus thermophilus* may be added
[5]For Limburger and other surface-ripened cheeses, species of *Arthrobacter, Micrococcus,* and yeasts may also be present.
Data from Guinee and Fox, 2004; Marcos et al., 1981; and other sources

In contrast, if a 1% inoculum is added and the incubation temperature is set at 20°C to 22°C, well below the optimum for growth (essentially, room temperature), coagulation may take twelve to sixteen hours. Thus, a busy cheese maker could, under the first scenario, produce multiple batches of cheese each day out of the same vat. However, it is also possible, in the alternative procedure, to inoculate or set the milk at 5:00 p.m., go home, sleep, and return to work early the next morning and have the cheese ready for the next step. Of course, inoculum and temperature regimens in between the long-set and short-set make times are also possible. Finally, although not required for coagulation, a small amount of chymosin (1 to 2 ml per 1,000 Kg) is sometimes added to promote a more firm coagulum (usually for large curd cottage cheese).

Once a pH of about 4.7 is obtained, a soft curded gel mass is formed. The gel is then cut using a pair of cheese knives that resemble square or horizontal-shaped harps. One knife contains vertical wires, and the other horizontal wires. The distance between the wires determines the curd size, which in turn depends on the product being made (e.g., large, medium, or small curd Cottage cheese products are available). Following the appropriate passes through the gel, die-shaped cubes are formed.

Almost immediately, whey becomes apparent and the curds begin to separate. The curds after cutting are very soft and fragile, and must be handled carefully to avoid shattering or fracture. Not only do fractured curds result in product defects, but, importantly, the small, broken curds or "fines" are lost when the whey is drained, resulting in loss of yield. Thus, the cut curds are initially left undisturbed for about fifteen minutes, and then are stirred gently.

Stirring continues as the curds are heated by raising the temperature in the jacketed vats. The temperature must be raised slowly at first (whether from 20°C or from 32°C) to prevent the curd exterior from cooking too fast and drying out. If this phenomenon, known as case-hardening, occurs, water molecules in the interior of the curd cannot escape and are held within the cheese. Ultimately, the curd-whey mixture is heated to about 52°C to 56°C, usually within one and a half hours (but according to some procedures, as long as three hours), with constant stirring.

Heating not only accelerates the rate of syneresis, the driving out of water from the curd, but temperatures above 45°C also arrest the fermentation. In fact, inactivating the culture at this step is the only way to keep the pH from decreasing too low. If the coagulated cheese is not cut until the pH has already reached 4.6 or below, then by the time the cook temperature is sufficiently high enough to inactivate the culture, the curds may already be too acidic. Cooking also inactivates coliforms and other heat-sensitive microorganisms (especially psychrotrophs, such as *Pseudomonas*).

After cooking has been completed, the whey is drained and cold (4°C) water is applied to the curds to quickly reduce the temperature. The curd usually is washed two or three times. The water used for this washing step must be slightly acidic so that the precipitated casein is not re-solubilized. Also, the wash water is often chlorinated to ensure that spoilage microorganisms are not inadvertently added back to the curds. The water is drained, leaving behind what is called dry curd Cottage cheese. This product has a bland, acidic flavor (pH about 4.6 to 4.8), and contains mostly protein (20%) and water (nearly 80%). It is used mostly as an ingredient in lasagna, blintzes, and other prepared products. It is far more common to add a cream-based dressing to the dry curds, producing the familiar creamed Cottage cheese products of varying fat levels (generally ranging from 0% to 4%). The cream dressings typically contain gums and thickening agents, salt, emulsifiers, anti-mycotic preservatives (e.g., natamycin, sorbates, and other organic acids), and flavoring agents. Of the latter, diacetyl distillates are frequently added to impart buttery-like flavor notes. Because the dressing contains lactose (which can be fermented), it is important that the starter culture bacteria be inactivated during the cooking step.

The packaged dry or creamed Cottage cheese is a perishable product and has a shelf-life of only two to three weeks under refrigeration conditions. However, the addition of new generation preservatives, such as Microgard, and the application of modified atmosphere (using CO_2) and aseptic packaging, may increase shelf-life to as long as forty-five days. Cottage cheese is susceptible to spoilage for several reasons. The cheese vats are typically open (although most newer vats are enclosed), and exposure to air and environmental microorganisms can be significant. Because the final pH of creamed products can be as high as 5.4, and the water activity is also high (0.98 to 0.99), *Pseudomonas* and other psychrotrophic Gram negative bacteria can grow and produce fruity, rancid, and bitter off-flavors. Yeast and molds are the other main spoilage organisms, causing appearance as well as flavor and aroma defects. Flavor defects, such as high acidity and bitterness, are the result of excess growth by the starter culture. Other quality defects, including shattered, gummy, or soft curd, are usually caused by manufacturing flaws.

Manufacture of cream cheese (and its lower fat version, Neufchatel) is similar in principle to Cottage cheese, but the starting material, the manufacturing steps, and the finished product are quite different. First, cream cheese is made from pasteurized and homogenized milk containing as much as 12% fat. Most other cheeses are made using non-homogenized milk because homogenization results in a soft curd, which, in the case of cream cheese, is desirable. A mesophilic culture is added, followed by either a long or short set incubation. The acid-induced coagulum that forms at pH 4.7 is stirred (not cut) while the temperature is raised to as high as 73°C. The curds are then separated from the whey by special centrifuge-type devices or ultrafiltration systems. Although the resulting cheese material can be packaged as is, most cream cheese is mixed with other ingredients, including cream and gums, and then homogenized or mixed. The temperature is maintained above 72°C throughout the process. Packaging is done under aseptic conditions (i.e., in rooms under positive pressure with high efficiency air filtration systems in place), giving this product a long shelf-life (> 45 days).

Cheddar family

Cheddar cheese is the most popular cheese consumed in the United States, with per capita consumption in 2001 of nearly 6 kg (12.7 pounds) per person per year (including other Cheddar types). Approximately 1.2 billion Kg (2.7 billion pounds) are produced each year in the United States. Several cheeses, including Colby, Monterey Jack, and other washed or stirred curd cheeses, are closely related to Cheddar cheese and are collectively considered as American type cheeses. However, this designation can be confusing since American cheese is often considered synonymous with processed American cheese, a totally different product.

Cheddar cheese was first made in the village of Cheddar, England. It is different from most cheeses due to a specific curd handling technique called Cheddaring. During Cheddaring, the curds are separated from the whey and allowed to mat, after which the matted curd slabs are flipped and stacked. The Cheddaring practice, like many great discoveries, probably occurred as a result of an accident. Perhaps the cheese maker had drained the whey, and instead of stirring the curds prior to filling forms or hoops, he or she was delayed or distracted. The curds then matted, staying warmer than usual, and then began to stretch and become plastic. The cheese that was then produced had a unique texture and body that led to the adoption of the Cheddaring process. The manufacture of Cheddar cheese has evolved from a traditional, strictly batch operation to a more modern process that is more automated and mechanized.

The early steps in Cheddar cheese manufacture are rather straightforward. Whole milk, either pasteurized or raw, is brought to a temperature of 30°C to 32°C, and is inoculated with a mesophilic lactic culture containing strains of *L. lactis* subsp. *lactis* and *L. lactis* subsp. *cremoris*. The culture, or rather the specific strains present in that culture, is selected based on the desired properties expected in the finished cheese. Sometimes *S. thermophilus* strains are added to promote rapid acidification. If the cheese is intended for the process market, then speed is the main criteria, and fast-growing strains of *L. lactis* subsp. *lactis* are often used. If, however, the cheese is to be aged, then appropriate flavor development will likely drive selection of the culture (usually containing *L. lactis* subsp. *cremoris*). Two properties, however, are required: the strains must ferment lactose and they must be able to hydrolyze and use proteins as a nitrogen source. Of course, other properties, as well as the form of the culture (whether bulk set or direct-to-vat set, frozen or lyophilized, mixed or defined) are also important, and are discussed later in this chapter. After culture addition, chymosin is added, coagulation occurs, and the curd is cut using medium sized knives (giving curds about 0.6 to 0.8 cm in diameter). Following a short five- to ten-minute "healing" period to allow the curds to form, the curds are gently stirred. Heat is then applied gradually, about 0.3 degrees per minute, to a final curd-whey temperature of about 38°C or 39°C. As the curds are cooked, the stirring speed is increased to promote heat transfer, and stirring is maintained for an additional forty-five minutes. Although the starter culture bacteria are contained within the curd, as a general rule, little fermentation should occur during the cooking or stir-out steps. A decrease of only 0.1 to 0.2 pH units is normal. Although *L. lactis* subsp. *lactis* has a slightly higher temperature range for growth compared to *L. lactis* subsp. *cremoris*, both grow slowly, if at all, at the upper end of common cooking temperatures. If however, lower cooking temperatures are used, growth of the culture can occur, resulting in fermentation and lactic acid formation. Although fermentation during this step may be desirable for some cheeses, for Cheddar it generally is not. This is because acidification is accompanied by demineralization of casein as the calcium becomes solubilized. When the whey is drained, the remaining curds contain less calcium, resulting in cheese that holds fat poorly and is less elastic, with a short, brittle texture. The importance of calcium during later stages of the Cheddar process is discussed below.

Once most of the whey is drained, the curds are assembled on both sides of the vat, where they quickly mat or stick together. By the time the last of the whey is removed, the curds have turned into a cohesive mass. A knife is then used to cut the matted curds into approximately 20 cm x 80 cm slabs. Depending on the cheese maker's preferences, the slabs are then either rotated or flipped or flipped and stacked, a process known as Cheddaring. Because the temperature of the Cheddar slabs is maintained at 28°C to 32°C, the culture can begin to grow in earnest, increasing from about 10^6/g to 10^8/g of curd within the Cheddaring period. As the lactic culture grows, it performs two critical functions. First, it ferments lactose in the curd, in homolactic fashion, to lactic acid. Second, it hydrolyzes casein via a cell wall-anchored pro-

teinase, forming a variety of variously sized peptides. The consequences of both of these activities will be discussed below.

During the Cheddar process, several biological, chemical, and physical changes occur that give this cheese it's characteristic properties. However, without an active starter culture, none of these changes will occur. As the culture produces lactic acid and as pH becomes more acidic, calcium that was initially associated with the negatively-charged amino acid residues of casein (e.g., serine phosphate) is displaced by protons. In this state, the casein is more soluble, smooth, and elastic, as is evident as the Cheddar slabs stretch and become plastic-like. Eventually, however, if the pH of the cheese becomes too acidic (less than 4.95), the cheese will become short and crumbly. Fermentation during Cheddaring, what Cheddar masters refer to as dry acid, is much preferred over acid formation during the cooking step, or wet acid development. In addition to the changes in casein structure that occur as a result of acid formation, gravitational forces from stacking the Cheddar slabs on top of one another also affect cheese structure. This facilitates linearization and stretching of the casein and allows the curds to knit when pressed (see below).

The longer the cheese is Cheddared, the more the culture will grow and with it, the greater will be the acidity. The extent of Cheddaring, and with it the extent of acid formation, depends, therefore, on the desired properties of the finished cheese. Lactic acid levels of 0.4% to 0.6% are normally achieved by the end of Cheddaring (about one and a half to two hours after whey separation), resulting in a finished cheese pH of 5.0 to 5.2. In addition, longer Cheddaring times also result in more cell mass produced. Since the lactic acid bacteria in the culture produce enzymes that degrade milk proteins, higher concentrations in the curd at this stage likely results in greater proteolysis at later stages, i.e., during ripening.

Although the traditional Cheddar process is still practiced in the United States in small cheese factories, most mid-sized and large operations use different procedures and produce a somewhat different type of product. This is because large cheese factories require much greater throughput than can be obtained by traditional procedures. Thus, setting, coagulation, and cutting steps are performed in enclosed vats, and the cut curds are then pumped to another location for further processing. This allows the manufacturer to clean, sanitize, and re-fill the original vat.

In addition, modern operations have eliminated the labor-intensive flipping and stacking Cheddaring steps by employing alternative methods. Cheese makers learned that a similar texture could be obtained if the dry curds, after draining, were simply stirred in the vat. An additional variation of this process involves adding cold water to the curds. As noted above, this step results in less lactose in the curd, and it also increases the moisture. The resulting cheese, known as Colby, is a less acidic, high moisture cheese, compared to traditional Cheddar. In some modern cheese factories, curds are continuously conveyed to the top of Cheddaring towers. By the time the curds have descended from these vertical towers and emerge out at the bottom, they have achieved a similar level of Cheddaring as traditional curd.

After Cheddaring, or when the pH is about 5.2 to 5.4, the plastic, elongated Cheddar slabs are chopped up in a special milling device (called, appropriately, a Cheddar mill) to reduce the slabs to uniform, thumb-sized pieces. These Cheddar curds are bland, with a squeaky, rubbery texture. Next, salt is applied, in an amount ranging from 2% to 3%, and in a manner that permits even salt distribution (i.e., while the curds are continuously being stirred).

Salting is a critical step, since it has a profound effect on the quality of the finished cheese. This is because, in addition to providing flavor, salt has a major influence on controlling microbial and enzymatic activities in the ripening cheese. Although 2% salt might not be expected to have much of an effect on microorganisms or enzymes in food, one must consider that cheese consists of two phases, a fat phase and a water phase. It is in the water phase that the salt is dissolved, and likewise, that is where the microorganisms live. Since Cheddar cheese contains no more than 39% water, the relevant salt concentration, i.e., that with which the microorganisms must contend, will be more than 5% ($2/39 \times 100 = 5.1\%$). At this concentration, the growth of the starter culture bacteria and many other microorganisms and the activities of microbial and milk-derived enzymes are effectively controlled. This is not to imply that microbial and enzymatic activities are actually halted, because salt-tolerant organisms and enzymes remain active in the presence of high salt concentrations. Rather, the salt provides a means to check or contain those activities and to create a selective environment that aids in establishing a desired microflora.

The active salt concentration is often referred to by cheese manufacturers as the salt-in-moisture or S/M ratio. For example, a cheese with 2.4% salt at a moisture of 38% will have a S/M of 6.3. The S/M value is arguably the main determinant affecting cheese ripening. The higher the S/M level, the more inhibitory will that environment be to microorganisms and enzymes. In contrast, for cheeses having low S/M values (i.e., below 4.5), the microflora will not be effectively constrained or controlled and production of off-flavors and other spoilage defects may occur. However, cheese ripening rates, which are also a function of microbial and enzymatic activities, will also be inhibited if the S/M is too high. Thus, cheese manufacturers must carefully adjust salt concentrations and moisture levels (as well as pH) to achieve the desired ripened cheese properties (Figure 5-6). In general, an S/M between 4.5 and 5.0 is desirable.

The next step involves filling forms, hoops, or barrels with the salted curds. The size of these forms range from 9 Kg to as high as 290 Kg (20 pounds to 640 pounds). Obviously, the form also gives shape to the cheese, varying from rectangular blocks to barrels. Many of the cheese forms have collapsible ends, such that pressure can be applied to enhance the transformation of the curds into a solid mass and to squeeze out whey (through perforations in the

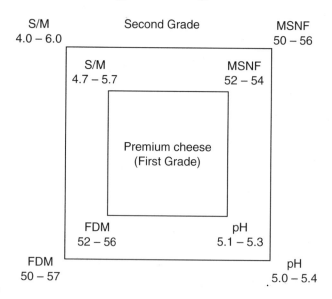

S/M
4.0 – 6.0

Second Grade

MSNF
50 – 56

S/M
4.7 – 5.7

MSNF
52 – 54

Premium cheese
(First Grade)

FDM
52 – 56

pH
5.1 – 5.3

FDM
50 – 57

pH
5.0 – 5.4

Figure 5–6. Relationships between S/M (salt to moisture ratio), MSNF (percent of moisture in the non-fat portion), FDB (percent of fat on a dry basis), and pH in Cheddar cheese at fourteen days after manufacture. To obtain a Premium cheese (or First Grade, according to New Zealand standards), the more narrow compositional specifications are usually necessary (compared to Grade B cheese which has broader compositional specifications). Adapted from Lawrence et al., 2004.

hoops). Usually, about 20 psi is applied for twelve to sixteen hours. The cheese is then removed from the forms, and, for the 9 Kg and 18 Kg blocks, placed in Cryovac bags, vacuum is applied, and the bags are sealed. The cheese then is moved into cold rooms, ranging in temperature from 4°C to 12°C for storage or aging.

Cheese with Eyes

Swiss cheese, the most famous of the eye-containing cheeses, is considered by experienced cheese makers to be the easiest cheese to make but the hardest cheese to make well. This is because the manufacture of high quality Swiss cheese, and proper eye development, in particular, depends on two rather independent processes. First, Swiss cheese requires excellent curd handling technique, such that conditions are correct for eyes to form. Second, there must be precise control over the microorganisms that are involved in the fermentation and produce the gas that ultimately results in eye formation. It is possible, easy in fact, to produce a cheese that tastes like Swiss cheese, but that has either small eyes, large eyes, irregularly-shaped eyes, too many eyes, too few eyes, or no eyes. The goal is to produce a cheese with the correct flavor and body char-

acteristics, *and* that has the right amount of uniformly distributed, evenly-shaped round eyes with the desired size. The trick is to manage the early cheese-making steps such that the curd has just the right texture or elasticity necessary to accommodate the carbon dioxide that is produced much later in the process.

The manufacture of Swiss cheese seems easy because the steps involved in curd making appear rather simple and straightforward. However, care must be taken from the outset. First, the milk is usually standardized to 3% fat. If too much or not enough fat is present in the milk, the body will be too soft or too hard, leading to poor eye development. Second, the milk is given a modest heat treatment, about 56°C for fifteen seconds, which is below that used for pasteurization but which still is effective at inactivating many of the milk-borne bacteria that could otherwise cause problems during the ripening period. Finally, the quality of the cheese also depends heavily on the performance of the bacteria used in the mixed species starter culture, as well as the ratio of those organisms.

To achieve the texture and body characteristics necessary for proper eye development, Swiss cheese must have the correct pH and moisture content and retain (or lose) the right amount of calcium. The curds are cut to about rice-sized particles, and then are cooked to a much higher temperature than for Cheddar-type cheeses. Typically, the temperature will be raised over a forty-minute period from 32°C to 35°C at setting to as high as 55°C, and then is held at that temperature during stir-out for another forty-five minutes or until the curd is sufficiently dry. Thermophilic cultures that are tolerant of high temperature must be used (see below), although there are some manufacturers who cook to a lower temperature and add mesophilic *L. lactis* subsp. *lactis*.

It is important to note that reducing the moisture in the curd will also result in less lactose in the curd. As noted above, controlling the amount of lactose in the curd is an effective way to limit the amount of acid that will ultimately be produced during fermentation, a consideration that will be further discussed later. High cooking temperatures also slow acid development, remove water, and prevent solubilization of calcium phosphate into the whey. The whey is drained at the end of cooking, while the pH is still high (about 6.3). This keeps phosphate in the curd, which promotes buffering and keeps the pH from becoming too low later during the fermentation.

In the traditional process, when the curds are sufficiently firm, they are collected or dipped into cheese cloth, which is then placed into a large round form. In a modified version used by many U.S. manufacturers, the curds are allowed to settle under the whey, sometimes with weights applied, then the whey is drained, the matted curd is cut into block-sized sections, and placed into forms. In the larger operations, curd-whey mixtures are pumped into draining vats and then the curd is filled into large forms. For all of the processes, the cheese blocks are then pressed and held for up sixteen hours at near ambient temperature.

Unlike Cheddar cheese, in which the fermentation occurs in the vat or on draining tables, for Swiss cheese there has been no opportunity, up this point, for fermentation. Rather, the fermentation occurs after the cheese is out of the vat and filled into forms. In addition, the fermentation takes a much longer time, as much as twenty-four hours. The cheese, however, will still be warm following the cooking step, so the internal temperature may be as high as 35°C for several hours. The actual fermentation is quite different from that which occurs during Cheddar cheese manufacture. In Cheddar cheese, the culture contains mesophilic lactococci, and although more than one species may be present (e.g., *L. lactis* subsp. *lactis* and *L. lactis* subsp. *cremoris*), they are obviously closely related and serve an almost identical function. In contrast, the Swiss cheese starter culture contains three different organisms, from three different genera: *Streptococcus thermophilus*, *Lactobacillus helveticus*, and *Propionibacterium freudenreichii* subsp. *shermanii*. All three are essential, and all are responsible for the fermentation pattern that is unique to Swiss cheese.

The culture that is initially added to the milk contains different proportions of each organism. The *S. thermophilus* generally outnumbers the *L. helveticus* by as much as ten to one. Therefore, at the outset, growth of *S. thermophilus* occurs first, in part because it is present at a higher concentration, but also because its simple physiological requirements are more easily met, especially compared to the more fastidious *L. helveticus*. As noted in Chapter 4, *S. thermophilus* and *L. helveticus* (or *Lactobacillus delbrueckii* subsp. *bulgaricus*) have a synergistic relationship, such that growth of one organism promotes growth of the other. In Swiss cheese, *S. thermophilus* growth is stimulated by amino acids and peptides released from casein via *L. helveticus* proteinases. Growth of the latter does not commence until the pH and Eh within the cheese are sufficiently reduced by *S. thermophilus*, a period that may take as long as twelve hours.

Another factor that influences the outcome of the Swiss cheese fermentation relates directly to the metabolic properties of these two organisms. In fact, one of the most interesting peculiarities of all lactic acid bacteria occurs when *S. thermophilus* ferments lactose in Swiss cheese (actually, the same phenomena also occurs in Mozzarella cheese, yogurt, or anytime *S. thermophilus* grows in milk). This organism actively transports lactose from the extracellular environment across the cell membrane and into the intracellular cytoplasm. The enzyme, β-galactosidase, then immediately hydrolyzes the accumulated lactose to glucose and galactose. Glucose is phosphorylated to glucose-6-phosphate, which then feeds into the glycolytic pathway and is rapidly metabolized to lactic acid. However, most strains of *S. thermophilus* do not express the enzymes necessary for galactose phosphorylation and metabolism (enzymes that comprise the Leloir pathway), and instead secrete or efflux the galactose back into the extracellular medium. In the case of Swiss cheese, galactose will appear in the curd almost as fast as lactose is consumed. Researchers have learned that the excretion of galactose not only coincides with

lactose consumption, but that the efflux reaction actually provides a driving force for lactose uptake (described in Chapter 2).

Thus, in the first several hours of the fermentations, *S. thermophilus* grows and ferments lactose and excretes galactose (Figure 5-7). *L. helveticus* begins active growth after about eight to twelve hours and competes with *S. thermophilus* for the remaining lactose, which is subsequently consumed. However, one of the main roles of *L. helveticus* is to then ferment the galactose left behind by *S. thermophilus*, such that after about eighteen to twenty-four hours all of the carbohydrate in the curd has been fermented. If the curd contained just the right amount of lactose at the start of the fermentation (i.e., after cooking), and all of the lactose and its constituent monosaccharides were fermented to completion, then the final pH should be very near 5.2 ± 0.1. If there is too much lac-

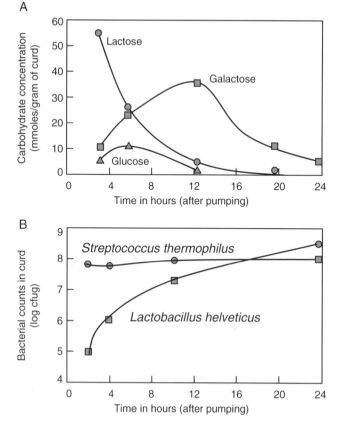

Figure 5–7. Carbohydrate utilization by *Streptococcus thermophilus* and *Lactobacillus helveticus* during the Swiss cheese fermentation. The upper pane (A) shows carbohydrate concentrations during twenty-four hours after the curds have been pumped from the vat. The lower panel (B) shows growth of the culture organisms during the same period. Adapted from Giles, et al., 1983 (*N.Z. J. Dairy Sci. Technol. 18*:117-123) and Turner et al., 1983 (*N.Z. J. Dairy Sci. Technol. 18*:109-115).

tose or too little lactose at the start, due to insufficient or excessive cooking, then the pH can end up being less than 5.0 or above 5.4. Both of these situations could result in poor quality cheese, as described below.

After the primary fermentation period, the cheese blocks or wheels are placed into brines containing 20% salt for as long as three days. The blocks may be flipped and additional salt may be applied. Then the blocks are removed and allowed to air dry in coolers at 10°C to 15°C for five to ten days. As noted for Cheddar cheese, salt provides flavor and influences the activities of microorganisms and enzymes present in the cheese, although the average concentration in Swiss cheese is much lower ($< 1\%$). In the case of traditional Swiss cheese and other brined cheeses, salt also helps form a natural rind, due to dehydration at the surface of the cheese. The hard rind provides an excellent natural protective barrier or casing. Brining, in contrast to direct or dry salt methods, also creates a salt gradient, with the concentration decreasing from the surface toward the interior of the cheese. This also affects the development of the microflora.

The next step is perhaps the most crucial. The dried blocks or wheels are moved into warm rooms where the temperature is maintained at 20°C to 25°C. It is during the ensuing three to four weeks that growth of *Propionibacterium freudenreichii* subsp. *shermanii* occurs. Although this bacterium is morphologically, physiologically, and genetically distant from the lactic acid bacteria, it is similar in that it prefers an anaerobic atmosphere and has a fermentative metabolism. Propionibacteria are neutrophiles and are salt-sensitive, so low pH and high salt conditions are inhibitory. They are added to the milk as part of the starter culture, but at a much lower rate—the inoculated milk contains only about 10^2 to 10^3 cells per ml. And although these bacteria are moderately resistant to high temperature, growth does not occur until the cheese is moved into the warm room. When these bacteria are grown on lactose or other fermentable carbohydrates, large amounts of propionic and acetic acids and lesser amounts of carbon dioxide are produced. However, as explained above, by the time Swiss cheese is placed in the warm room, there should not be any carbohydrate still in the curd and available for fermentation. So what does *P. freudenreichii* subsp. *shermanii* use as a substrate? In fact, this organism is capable of metabolizing the lactic acid produced by the lactic starter culture via the propionate pathway (Figure 5–8). This pathway yields propionic acid, acetic acid, carbon dioxide and ATP. If, however, any lactose is still available, then more of these products will be formed. Moreover, excess production of carbon dioxide, in particular, has serious consequences for the final product.

As noted above, *P. freudenreichii* subsp. *shermanii* is initially present at relatively low levels in the cheese. As it grows in the warm room, small micro-colonies within the curd matrix are formed. Likewise, the fermentation end products are evolved in that same vicinity and diffuse out into the neighboring region within the curd. Although the acids are readily dissolved, the carbon dioxide molecules will diffuse through the curd only until they reach weak spots, where they will then collect along with other CO_2 molecules made by other micro-colonies. Eventually, enough CO_2 molecules will have accumulated, and *voila*—an eye is formed. However, this entire sequence of events depends on several hard-to-control variables.

First, CO_2 formation must be slow and steady. If too much CO_2 is produced all at once or the curd is too firm, the gas pressure can exceed the ability of the curd to sustain the gas and large, even exploded holes are formed. If the body of the curd is too soft and the weak spots too numerous, then many small eyes will form. The hard rind produced as a result of the brining and drying steps also serves an important role; without a rind, the CO_2 could theoretically escape clear out of the cheese. Of course, CO_2-impermeable bags provide an easy remedy for this problem (thus was born rindless Swiss, as described below). Obviously, time is a critical factor as well—too much or not enough incubation time will result in less than perfect eye development. Experienced cheese

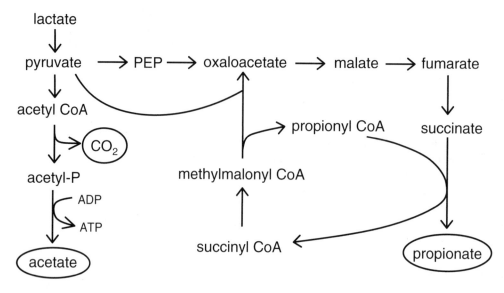

Figure 5–8. The propionic acid pathway of propionibacteria. Only the key intermediate compounds are shown. Adapted from Hutkins, 2001.

makers have adopted various professional tricks, from tapping the cheese and listening for just the right echo, to visually examining the expansion of the wheel or block.

Finally, when the cheese leaves the warm room, it is moved into a cooler at 2°C to 5°C for aging, which can vary from three months to two or more years. In addition to the acid products (i.e., propionic and acetic acids), *P. freudenreichii* subsp. *shermanii* is also responsible for producing other important end-products. In particular, *P. freudenreichii* subsp. *shermanii* produces specialized peptidases that release proline, which has a sweet-like flavor. Various other amino acids and peptides are also metabolized, generating nutty flavors that are characteristic of Swiss cheese.

Variations of the traditional Swiss or Em-menthaler cheese (made in Switzerland) include Gruyere (France), Jarlsberg (Norway), and Samsoe (Denmark). All of these have a natural rind and varying levels of eye formation. In the United States, rindless versions are more common, with the cheese blocks wrapped in CO_2-impermeable plastic wrapping.

Mozzarella and Pasta Filata Cheese

As a recently as a generation ago, Mozzarella was still considered a mostly ethnic cheese, used primarily as an ingredient in Italian cuisine. The popularity of this cheese—it is now equal to that of Cheddar cheese among American consumers (Figure 5-3), is due to one product: pizza. Of the more than 1.2 billion Kg (2.7 billion pounds) of Mozzarella and related cheeses consumed each year, about 70% is used by the food service industry as an ingredient on pizza. And while Cheddar production has grown by about 60% since 1980, over that same period, Mozzarella production has increased by 400%. In fact, in the last twenty years, many cheese factories that once made Cheddar converted their operation to Mozzarella manufacturing. And despite the apparent differences in flavor and appearance between Cheddar and Mozzarella cheese, both share several common manufacturing steps.

In the United States, all Mozzarella cheese is made from pasteurized milk. In large part, this is because Mozzarella is considered a fresh cheese that is essentially never ripened. Thus, legal requirements dictate that the milk be pasteurized. Although some Mozzarella cheese is made from whole milk, most is made from reduced fat or partially skimmed milk. Special considerations for the manufacture of reduced fat Mozzarella and other low- or reduced fat cheeses will be described later. Although almost all Mozzarella cheese made in the United States comes from cow's milk, it is worth noting that some traditional Italian Mozzarella is made from the milk of water buffaloes.

After the milk is pasteurized and standardized, a thermophilic culture is added. Although the organisms used for Mozzarella, *Streptococcus thermophilus* and *Lactobacillus helveticus,* are the same as those used for Swiss cheese, the specific strains and ratios are likely quite different. Strain selection for Mozzarella cheese is based, like all cultures, on the desired properties of the particular cheese. This is especially true for Mozzarella cheese, where the strains may have a profound effect on the properties of the finished cheese (Box 5-5).

Box 5–5. The Science of Making Pizza

Given that the U.S. pizza industry is a $32 billion industry, that 350 slices of pizza are eaten every second of every day, and that Mozzarella and pizza-type cheeses are the main ingredients on a pizza, it is not surprising that Mozzarella producers are under considerable pressure to produce a cheese with the functional properties desired by the pizza manufacturer.

Of course, pizza restaurants are the main users of Mozzarella, with several companies dominating the industry (Dominos, Pizza Hut, Godfathers, and Papa Johns). When considering the cheese, each pizzeria seems to have their own particular preference. Some pizza chains desire cheese that stretches a particular distance when a slice is moved from plate to mouth. Others want a cheese that remains white and resists browning even when the pizza is baked quickly at extremely high temperatures. Thus, Mozzarella production is now often customized to the exact needs of the customer, such that even before the vat is filled with milk, the manufacturer is ready to produce a cheese meeting not only compositional requirements, but functional specifications as well.

Because of these considerations, there is now much interest in understanding and predicting how a given cheese will perform during pizza manufacture and how manufacturing conditions can be manipulated to produce a cheese that has specific functional properties (Kindstedt, 1993). Unlike most other cheeses, Mozzarella, at least when used on pizza, is prized less for its flavor and more for its physical and functional attributes. Mozzarella, after all is a bland, slightly acidic cheese, with no aged cheese flavor. However, it has a unusual ability to stretch, retain fat, melt evenly, and provide a chewy mouth feel. These properties, however, are not necessarily automatic, and are influenced by the culture, the coagulant, the manufacturing conditions, and post-manufacturing handling and storage. Ultimately, these factors affect pH, loss of calcium, proteolysis, and the overall composition of the finished cheese.

The metabolic activities of the thermophilic starter culture bacteria have a major impact on the finished product. After the curds are cooked and the whey is drained, the curds are either Cheddared or dry stirred. During this time, *Streptococcus thermophilus* and *Lactobacillus helveticus* (or *Lactobacillus delbrueckii* subsp. *bulgaricus)* ferment lactose, producing lactic acid and causing the curd pH to decrease. However, both organisms metabolize only the glucose portion of lactose; the galactose is released back into the curd.

When the pH reaches 5.2, the curds head off to the cooker-stretcher (operating at 85°C), which heats the curd to a temperature near 60°C. This process inactivates enzymes (although some residual chymosin might remain) and the starter culture (or at least reduces the cell concentration several fold). Thus, there is essentially no opportunity left for this galactose (and any remaining lactose) to ferment. Instead, it will remain intact and be present in the finished cheese.

Box 5–5.   The Science of Making Pizza *(Continued)*

When the cheese is exposed to high baking temperatures (which dry the cheese surface and reduce the water activity), these reducing sugars react with amino acids, via the non-enzymatic Maillard reaction, to from brown pigments. Although moderate browning may be desirable, excessive browning or blistering is undesirable. Because most pizza manufacturers prefer white, non-browning cheese, steps must be taken to reduce the galactose concentration in the cheese. This can be done either mechanically, by simply washing the curds, or biologically, by using cultures that can ferment galactose. The latter may consist of mesophilic *Lactococcus lactis* subsp. *lactis* that have no problem fermenting galactose, but their use requires other changes in the manufacturing process (i.e., lower cooking temperature). Selected galactose-fermenting strains of *S. thermophilus* and *L. helveticus* can also be used, and although such cultures are available from starter culture suppliers, their availability appears limited (Mukherjee and Hutkins, 1994). Plus, even galactose-fermenting strains do not ferment galactose as long as lactose is still available. It is also technically possible to engineer such strains, but this strategy has not yet been adopted.

The culture, and specifically, the proteolytic activity of *L. helveticus*, also affects other important properties. The ability of Mozzarella cheese to stretch is largely a function of protein structure. If, however, the culture is proteolytic (or residual coagulant activity is present in the curd after cooking), protein hydrolysis will occur, and the shortened casein strands will lose their ability to stretch. This may or may not be a good thing, depending on the preference of the pizza manufacturer (Dave et al., 2003).

If a short, non-stretching cheese is desired, then greater proteolysis may be encouraged. However, manipulating any one variable to affect a given property will likely affect some other property. As noted previously, extensive proteolysis makes the cheese "soft" and "gummy," and less amenable to shredding. In addition, it is the protein network that traps fat in Mozzarella. During baking, the fat melts and liquefies, but is retained by the protein matrix. If proteolysis has occurred and the protein network is disrupted, the fat is no longer contained and leaks out. On a pizza, the fat forms unsightly pools of melted oil, a defect known as oiling-off. Finally, since proteolysis increases the free amino acid concentration, the more proteolytic the culture (or the coagulant), the greater the rate of the Maillard reaction and the potential for browning problems.

Ordinarily, the amount of time between manufacture of Mozzarella and its use on a pizza is only a few weeks. Thus, proteolysis in Mozzarella is nowhere near as extensive as in an aged cheese. Still, even moderate proteolysis can affect cheese functionality. Many Mozzarella producers, therefore, freeze the cheese within a few days of manufacture, giving it just enough aging time to develop appropriate properties. Some manufacturers have even developed processes in which the cheese is frozen within hours of manufacture.

Manufacturing variables have also been modified to affect other functional properties. For example, most Mozzarella is eventually used in a shredded or diced form, so producing a cheese that is easy to shred, without gumming up the equipment, is important. These modifications have led some to question whether the cheese produced is actually Mozzarella, as defined by the standards of identity. Nonetheless, it seems likely that the demand for Mozzarella (or Mozzarella-type) cheese with particular functional characteristics, combined with significant economic pressures, will continue to be a driving force for innovative research within the Mozzarella and starter culture industries.

References

Mukherjee, K.K., and R.W. Hutkins. 1994. Isolation of galactose fermenting thermophilic cultures and their use in the manufacture of low-browning Mozzarella cheese. J. Dairy Sci. 77:2839–2849.

Dave, R.I., D.J. McMahon, C.J. Oberg, and J.R. Broadbent. 2003. Influence of coagulant level on proteolysis and functionality of mozzarella cheeses made using direct acidification. J. Dairy Sci. 86:114–126.

Kindstedt, P.S. 1993. Effect of manufacturing factors, composition, and proteolysis on the functional characteristics of mozzarella cheese. Crit. Rev. Food Sci. Nutr. 33:167–187.

After rennet addition, coagulation, and cutting, the curd-whey mixture is pumped into draining tables, where the curds are gently stirred and then cooked to 56°C. The whey is drained and the curds are either allowed to mat, as far Cheddar, or are dry-stirred (as for stirred-curd Cheddar). Frequently, the curds will be washed after draining to reduce lactose levels. Mozzarella is a brine-salted cheese; however, salt may also be added directly to the curds prior to the cooking-stretching step to minimize the lengthy brine-salting time ordinarily required.

The fermentation of lactose in Mozzarella is similar to that which occurs in Swiss cheese. In both fermentations, galactose appears in the curd as a result of the inability of *S. thermophilus* to efficiently metabolize both of the monosaccharide moieties of lactose. In Swiss cheese, there is plenty of time for the companion strain (i.e., *L. helveticus*) to eventually ferment the galactose and remaining lactose. However, the situation is quite different for Mozzarella. This is because when the pH during Cheddaring or dry-stirring reaches 5.2, the curds are moved to a cooker-stretcher device that exposes the curds to temperatures as high as 85°C. Although the curd temperature may not reach such a high temperature (usually the curd is about 54°C to 57°C), this treatment is still sufficient to inactivate many of the starter culture bacteria present in the cheese. Thus, any galactose or lactose still in the curd at the cooking-stretching step will likely remain in the cheese for the duration of its manufacture and storage. The consequences of this are discussed in Box 5-5.

In addition to its effect on the culture and residual enzymes, the cooking-stretching step also has an important impact on the physical properties of the cheese. Prior to this step, Mozzarella curds are not that much different than other rennet set curds. However, the cheese exiting the cooking-stretching machines has properties unlike any other cheese. The cooking-stretching machines use augers to knead, stretch, and convey the curds in an upward manner (about a 30° incline), all the while exposing the curds to a hot water-steam

mixture at 85°C to 90°C. In the United States and Italy, almost all Mozzarella is stretched in high through-put continuous cooker-stretcher devices. Only a relatively few very small manufacturers rely on traditional techniques—manually kneading the curd in hot water, an arduous and time-consuming task.

Although what actually happens to the casein matrix during this step is not clearly understood, it appears that the casein network becomes linearized, giving the fibrous property characteristic of Mozzarella cheese and desired by most consumers. It is critical that the cooking-stretching step occurs when the curd pH reaches 5.2, otherwise the cheese will be short-textured and will stretch poorly. Finally, the molten, plastic, and fluid curds are then dropped into a hopper and filled into molds of varying loaf sizes and shapes. The cheese sets up quickly in the molds and is then dropped into a cold salt brine. After brining (about sixteen hours, but much less if dry salt had been added to the curd), the cheese is dried and packaged. In Italy and in small U.S. factories that employ traditional manufacturing practices, Mozzarella can be formed into small round shapes, which are then packaged in water or dilute brines.

Mozzarella cheese, as noted above, is not aged; however, even within just a few days or weeks, important functional properties can change. One of the first changes that occurs is the equilibration of calcium, that which exists as part of the casein complex and that in the form of calcium phosphate. Importantly, the loss of calcium from the casein complex improves the melting and stretching properties of the cheese.

The other main change in the functional properties of Mozzarella is due to protein hydrolysis. Proteolysis can occur as a result of residual proteinases released by the lactic starter culture, as well as by residual coagulant (which appears to be the main source of proteinase activity). Although flavor changes due to protein hydrolysis are usually minor, texture properties can be significant. As the long casein strands are hydrolyzed, the cheese stretches less. Also, the hydrolyzed casein is less able to hold or contain the fat. The latter may result in a defect known as "oiling off" that is readily apparent when the cheese is used in pizza and other cooked products. Also, since most Mozzarella cheese is marketed in a shredded form, excessive proteolysis makes the cheese less shreddable. For these reasons, fresh Mozzarella is best used within two to three weeks. Alternatively, the cheese can be frozen shortly after manufacture or after shredding (the preferred form for most of the large pizza manufacturers), and although this will prevent proteolysis, freezing is not without its own set of effects.

Mozzarella is but one of several so-called pasta filata (Italian for *stretched curd*) types of cheese. Another well-known variety is Provolone cheese. This cheese is made essentially as for Mozzarella except that a lipase preparation is added to the milk to promote lipid hydrolysis and release of free fatty acids. Lipases are either fungal-derived or are from animal tissues. A slightly rancid but pleasant flavor, described as piquant, develops during a short aging period. Provolone is often shaped into pear-shaped balls (*prova* is Italian for *ball*), and is frequently smoked.

Hard Italian Cheese

Given the popularity of Mozzarella cheese and Italian cuisine, in general, it is not surprising that production of other Italian cheeses, such as Parmesan, Romano, and other hard grating types, have increased by more than 120% in the last twenty years. Although Parmesan is the prototype of the hard Italian grating type cheeses, there are many variations of this cheese, including Romano, Asiago, and Grana.

As recently as 2002, the Food and Drug Administration's official Standard of Identity for Parmesan was changed to accommodate newer methods of manufacturing, such that less aging was required. In contrast, Parmesan made in Italy must conform to rigid manufacturing procedures if it is to be called Parmigiano Reggiano. The milk, for example, must be obtained from cows raised in a specific region

(the Po Valley) and fed a specified diet. Only raw milk can be used, and it must be standardized in a specific manner (which includes a natural creaming step). Calf rennet is used for coagulation. The cheese is ultimately aged for at least twelve months, but usually longer (two to three years). Despite these considerable challenges, annual production of Parmigiano Reggiano is still rather high (about 100 million kg); however, even in Italy it is expensive (often more than $18 per Kg). Recently, the European Union ruled that only cheese made following these prescribed procedures could be called Parmesan. Other famous European cheeses, including Roquefort and Feta, have received similar EU protections (the so called Designation of Protected Origin or DPO).

In the United States, where manufacturers are not bound by EU rules, Parmesan is made from cow milk standardized to about 2.4% to 2.8% fat. The milk is inoculated with a thermophilic culture and rennet is added. The curds are cut to give small curds, similar to Swiss, cooked to 45 to 48°C, and stirred for up to forty-five minutes. Some acid development during the stir-out step occurs, and when the pH reaches about 5.9, the whey is drained. The curds are placed into forms and pressed for twelve to sixteen hours, during which time the thermophilic culture ferments the lactose. Next, the cheese is brined for two weeks and dried for up to six weeks (to obtain a moisture content of 32%). Aging is essential for Parmesan cheese, and until recently, ten months of aging was required to satisfy the U.S. Standards of Identity (and to produce good flavor).

In 1999, Kraft successfully petitioned the FDA to permit cheese aged for a minimum of six months, instead of the nine months, to be called Parmesan, provided the original characteristics and properties were still maintained. Apparently, the use of enzyme technologies, combined with a slight increase in ripening temperatures, can produce a cheese that, while not intended to be the quality equivalent of a traditional, well aged Parmesan, still has a clean and acceptable flavor that is quite suitable as a grating cheese. One of the more popular variations of Parmesan, Romano, is made in a similar manner as Parmesan, but it contains lipase and, at least in Italy, is made from sheep's milk.

Dutch-type Cheeses

Among the most pleasant-tasting, colorful, and microbiologically complex cheeses are the so-called Dutch-type cheeses, of which Edam and Gouda are the most well-known. In fact, there are comparable cheeses produced throughout the world, using similar manufacturing procedures. Although eyes are usually present in these cheeses, there are fewer of them and they are much smaller than in Swiss cheese. The texture and flavor is also completely different from Swiss—these cheeses are softer, with a sweet, mild, buttery flavor. They are also easily distinguished from other cheeses by the characteristic red wax that covers the round wheels.

Although the basic manufacturing procedures for these cheeses are similar to those already described for other cheeses, there are several unique steps. The starter culture contains, in addition to acid-forming lactococci, one or more flavor-forming lactic acid bacteria. The latter include various species of *Leuconostoc*, including *Leuconostoc mesenteroides* subsp. *cremoris* and *Leuconostoc lactis*. Not only are these organisms heterofermentative, producing lactic acid, acetic acid, ethanol, and carbon dioxide, but they are also capable of fermenting citrate. Fermentation of citrate by these bacteria results in formation of diacetyl (see Chapter 2), which imparts the buttery, creamy flavor that characterizes these cheeses. In addition, the citrate fermentation pathway generates carbon dioxide, accounting for the eyes that are often present.

After culture and rennet are added to the milk (held at about 30°C to 32°C), the curds are then cut and stirred. Next, a unique manufacturing step occurs. A portion of the whey is drained, and then hot water is added to raise the temperature to 37°C to 38°C and to very near the original volume. Stirring continues until the desired curd firmness is achieved, and the remaining whey-water mixture is drained.

This washing step reduces the lactose concentration in the curd, making less available for the lactic culture. However, by using hot water, the moisture content is not increased. Ultimately, the cheese will be sweeter, and less acidic, with a final pH between 5.4 and 5.6.

Surface-ripened by Bacteria

Among the cheeses that are the most aromatic and flavorful, Limburger, Muenster, Brick, and other surface-ripened cheeses are right at the top of the list. A well-ripened Limburger can fill the room with the strong sulfury volatile compounds that are produced by *Brevibacterium linens*. This organism (and others that are generally present; see below) is neither added to the milk nor curds, but rather is applied to the cheese after its manufacture. As profound as is the aroma and flavor of this cheese, its orange-red appearance is nearly as dramatic. In fact, the natural pigment produced by *B. linens* that accounts for the color of these cheeses has recently found use as a coloring agent in other food products.

The reader is probably wondering why the Muenster or Brick cheese common to U.S. consumers lacks the punch described above, despite having the expected appearance. This is because the early manufacturers of these cheeses, who typically had emigrated from Germany and other European countries where these cheese were popular, realized that U.S. consumers preferred more mild-flavored cheeses. Thus, the manufacturing process had to be modified to accommodate the particular preferences of the new American customers. Thus was born the mild-flavored Muenster and Brick cheeses that are so widely consumed today. Only Limburger has retained the traditional flavor and appearance properties of the original. Still, the preference for the mild continues even today—Limburger production in the United States in 2001 was less than 50% of that in 1980, and Brick and Muenster continues to outsell Limburger by more than 100-fold.

Limburger and related cheeses are initially made according to rather standard procedures.

Whole milk is inoculated with a mesophilic lactic culture (*L. lactis* subsp. *lactis*) and chymosin is added. The coagulated milk is cut and stirred, but the curds are cooked to more moderate temperatures, usually between 30°C and 34°C. The curds are collected into open-ended, brick-shaped forms, which are turned every three to five hours to promote whey drainage. After about twelve hours, the cheese is brined or dry salted at the surface and held in warm (20°C) and humid (90% relative humidity) rooms.

Traditional manufacturers may rely on the endogenous flora that is present on the shelves in the ripening room to initiate surface ripening. In addition to *B. linens,* micrococci, yeast, and other organisms are likely present and will participate in the ripening process by raising the pH and producing flavor and color compounds. The presence of *Corynebacterium, Arthrobacter,* and other coryneform bacteria in these cheeses is common, although their role in flavor and pigment production is not yet established. Alternatively, a pure *B. linens* culture can be applied. Under these warm, humid, and aerobic conditions, ripening does not take long, and there is considerable surface growth, color formation, and flavor development within just a couple of weeks. The cheese is packaged and moved into a 10°C cooler for another month or two. These cheeses can easily become over-ripened, so they must be kept at low temperature. The packaging materials usually include parchment, wax paper, and foil to minimize oxygen availability (and to contain volatile aroma).

The strong aroma and flavor of Limburger and related cheeses is due to production of several volatile compounds. In defense of these cheeses, it is fair to say that the bark is worse than the bite, in that the flavor is nowhere near as strong as the aroma might portend. In fact, the use of surface smears in cheese making has fast become popular in the United States, especially among small, farmstead or artisan-type cheese makers. Thus, domestic versions of Tilsiter, Raclette, Beaufort, Gruyere, and other similarly-produced cheeses are now more widely available. In July, 2005, a Wisconsin-made Beaufort style cheese received the best in show award at the American Cheese Society competition, besting nearly 750 other cheeses.

Mold-ripened Cheese

Aside from the fact that the two main mold-ripened cheeses, the blue-type and the Brie-type, are both fungal fermentations, they share few common properties. The blue mold cheeses contain visible mold growth throughout, at the surface, and within the curd. The mold responsible, *Penicillium roqueforti,* produces blue-green mycelia, in addition to a myriad of enzymes that ultimately generate typical blue cheese flavors. In contrast, the Brie-type cheeses are made using *Penicillium camemberti,* which produces white mycelia, and grows only at the surface. Blue cheese is acidic, salty, brittle and crumbly, while the Brie-type cheeses are satiny smooth, soft, and creamy, with a near neutral pH.

Blue-mold Ripened Cheese

Although blue mold-ripened cheeses are made throughout the world, three specific types have achieved a significant measure of fame to warrant their own name (and have DPO status). Roquefort, perhaps the most well-known of all blue cheeses, must be made according to a strict set of manufacturing requirements. For example, the milk must come from specially-bred sheep that have grazed in the Causses region of France. It is neither pasteurized, standardized, nor homogenized, and the cheese must ultimately be aged in caves within that same region. In Italy, the manufacture of Gorgonzola cheese is similarly restricted to the Po Valley region of Northern Italy and its manufacture is also subject to specified procedures. This cheese is made from cow's milk, but is otherwise very similar to Roquefort. Gorgonzola, however, is usually not quite as strongly flavored as Roquefort. Finally, the English representative to the blue-mold family is called Stilton. It also is made from cow's milk and only in three counties of central England.

Other blue mold-ripened cheeses are ordinarily referred to simply as blue cheese, although regional varieties, often times of excellent singular quality, exist worldwide.

In general, the manufacture of blue mold-ripened cheese requires several specialized steps. In large part, the goal is to ensure that the aerobic mold can grow well within the interior of the cheese, where the Eh is ordinarily low. Although the blue mold organism, *P. roqueforti*, is capable of growing at high CO_2 and low O_2 levels, air incorporation is still an important feature of the process. The coagulated milk is cut, when very firm, into larger than normal-sized curds (as much as 2.5 cm). As noted earlier, the larger the curds at cutting, the higher the moisture will be in the cheese. However, an additional effect is to create a more open or porous texture such that diffusion of air is increased.

Frequently, the mesophilic lactic starter culture, in addition to containing *L. lactis*, will also include heterofermentative *Leuconostoc* or citrate-fermenting lactic acid bacteria. These bacteria produce CO_2 that contributes to the open texture that enhances oxygen diffusion and gas exchange. Spores of *P. roquefortii* can be added to the curds or to the milk, along with the lactic starter culture, prior to setting. Later, once the cheese is hooped and the lactose fermentation is complete, the cheese wheels are brine- or dry-salted. At this point, a set of large bore needles (diameter about 0.24 cm) may be used to pierce the cheese and provide a means for air to penetrate the inner portions. Later, when the cheese is cut vertically, one can see that mold growth followed the spike lines where oxygen was available.

Next, the cheese is aged for several weeks in a warm (10°C to 12°C), humid (90% to 95% relative humidity) environment that promotes growth of the *P. roqueforti* throughout the interior of the cheese. Lower ripening temperatures (4°C to 8°C) and longer times may also be used. Since the only way to arrest further mold growth is to cut off its supply of oxygen, the ripened cheeses are then wrapped in oxygen-impermeable foil. Although some consumers like blue cheese flavor, they may not be fond of the moldy appearance; thus, some manufacturers may limit growth of *P. roqueforti*, giving a somewhat whiter cheese with only a few streaks of blue. Even consumers who enjoy blue veined cheese may object to extensive growth at the surface, so the wheels are often cleaned prior to final packaging. Pimaricin may also be applied to control surface growth.

Although other organisms, including yeast and bacteria, are often present in these cheeses, especially at the surface, the flavor compounds characteristic of blue cheese clearly are generated primarily via mold growth and metabolism. Fungi, in general, are prolific producers of proteases, peptidases, and lipases, and *P. roqueforti* is no exception. Thus, release and subsequent metabolism of protein hydrolysis and lipolysis products are important in blue cheese flavor. Of particular importance are ammonia and amines, derived from amino acid metabolism, and methyl ketones, derived from free fatty acids.

Free fatty acids, themselves, may also contribute to cheese flavor, but their metabolism, via β-oxidation pathways to 2-heptanone, 2-nonanone, and other ketones, are primarily responsible for the flavor of blue cheese. As much as 20% of the triglycerides in the milk may be hydrolyzed. When blue cheese is made from raw milk, natural milk lipases may also contribute to formation of free fatty acids. These lipases are loosely associated with the surface of the casein micelles and are dislodged by agitation. In addition, lipases' substrates (triglycerides) are exposed when the fat globule is disrupted. Therefore, some manufacturers add a portion of homogenized raw cream to the cheese milk to accelerate flavor development.

Finally, it is important to note that mold growth during blue cheese ripening is accompanied not just by flavor development, but also by a marked increase in pH. This occurs because *P. roqueforti*, not having lactose on which to grow (all of the lactose is fermented by the starter culture), instead uses lactic acid as an energy source. Therefore, consumption of lactic acid causes the pH to rise from about 4.6 to above 6.0. Production of ammonia and other amines also contributes to the increase in pH. This increase in pH, in turn, promotes flavor production because many of the fungal decarboxylases and other enzymes involved in methyl ketone formation have neutral pH optima. However, the rise of pH to near neutrality may also have food safety and preservation consequences in that acid-sensitive organisms may be able to grow once the pH reaches non-inhibitory levels.

White-mold Ripened Cheese

The white mold-ripened cheeses, of which Camembert and Brie are the most well-known, are primarily made in France, where they are also among the most popular. These cheeses vary only slightly; Camembert is made principally in the Normandy region of France, whereas most Brie is produced in Melun and Meaux, just outside of Paris. Brie wheels are usually a bit larger, with bacteria on the surface contributing to flavor development.

Although similar versions of both Brie and Camembert are made in Germany, Switzerland, and other European countries, as well as South America and even the United States, manufacturing conditions are not that different. In general, whole cow's milk is used, and double or even triple cream versions exist in which the milk is supplemented with cream (not for the faint of heart). Since these cheeses are aged for as little as a few weeks, in the United States, the milk must be pasteurized (Box 5-4). In France, however, raw milk is most frequently used. The milk is inoculated with a mesophilic starter culture, which is allowed to grow and produce acid before the chymosin is added. Spores of *P. camemberti* can also be added to the milk (or applied later to the surface of the hooped cheese). A related species, *Penicillium caseicolum* is also used, because it produces a whiter mycelia.

As with blue cheese, the coagulum is firm when cut into large curds particles. The curds are gently stirred, but not cooked. Rather, they are almost immediately filled into forms. These steps decrease syneresis and result in cheese with high moisture levels. The hooped cheeses are flipped several times over an eighteen-hour period, during which time the lactose is fermented and the pH decreases to as low as 4.7. As with blue cheese, the cheese is then brined or dry-salted, and, if not already added to the milk, spores are applied, but only onto the surface.

The cheese wheels are then moved into ripening rooms (similar to those used for blue cheese, e.g., 10°C, high humidity). The cheese sits on shelves designed to promote contact with air, and are turned periodically. It takes only a week or two for a white mycelium mat to form across the entire surface. However, the extent of mold growth and subsequent flavor development depends on the target market, so the ripening time may be several weeks longer. In cheese made from raw milk, a diverse microflora, consisting of yeasts, brevibacteria, enterococci and staphylococci, and coliforms, may also emerge during extended ripening. Thus, not only will there be more flavor and texture development (see below) as ripening continues, but the surface mat may become less white and more pigmented. In the United States, a more mild flavored, white-matted product is preferred, so ripening times are usually short. To control growth of mold, the cheese is wrapped and either held for more ripening or else stored at <4°C.

Not only do the Brie-type cheeses have a different appearance from the blue mold cheeses, but they also have a quite different flavor and texture. Despite these differences, the progression of flavor and texture development is similar. The proteinases and peptidases produced by *P. camemberti* are similar to those produced by *P. roqueforti*. Subsequent production of ammonia, methanethiol, and other sulfur compounds are derived from amino acid metabolism and are characteristic of Brie-type cheese. The scent of ammonia can be striking in a well-aged Brie. Lipolysis of triglycerides and fatty acid metabolism by *P. camemberti* are also important, and methyl ketones can be abundant.

It is interesting that whereas blue cheese is crumbly and brittle, Brie-type cheeses are soft and creamy. The creamy, even fluid texture of these cheeses is now thought to occur as a result of protein hydrolysis, as well as the increase in pH due to ammonia. In particular, not only are fungal proteases important in texture development, but α_{s1} casein hydrolysis by chymosin and the natural milk protease, plasmin, is also involved.

For many of the cheeses discussed in this chapter, the geometry of the forms or hoops that give shape to the cheese have not been described in much detail. In many cases, whether the cheese is collected and shaped into rectangular blocks or round forms or whether the hoops or forms contain 5 Kg or 200 Kg has only modest bearing on the properties of the cheese. This is most definitely not the case for Brie-type cheese, where shape has a major impact on flavor and texture development. Since ripening depends on the fungi growing exclusively on the surface, the rate of flavor and texture development within the interior of the cheese is necessarily a function of the diffusion rate of enzymes and enzyme reactants and products from the exterior.

Although Camembert wheels usually have a smaller diameter (about 10 cm to 12 cm) than Brie (about 30 cm to 32 cm), their heights are essentially the same (about 3.5 cm) and the geometric center of these cheeses is never more than 2 cm from the surface. Thus, the enzymes and reactants (e.g., ammonia) produced at the surface have only a short distance to travel before reaching the center. If Brie cheese were to be made in thicker wheels, like those used for blue cheese (e.g., 15 cm to 20 cm diameter), it would obviously take a long time for the reactants to diffuse into the center region. During that period, the exterior portions would have been exposed to those enzymes far too long. The cheese might be perfectly ripened in the center, but well over-ripened at the surface.

Finally, as with *P. roqueforti,* growth of *P. camemberti* in the manufacture of Brie-type cheeses depends on lactic acid as an energy source. The subsequent rise in pH (from 4.6 to as high as pH 7.0 at the surface) is similarly due to lactate consumption and ammonia produc-

tion. This return to a neutral pH also contributes to the diversity of microorganisms that grow on the surface of these cheeses (see above) and, more importantly, poses a special food safety risk due to the possible presence of *Listeria monocytogenes* (see Box 5-4).

Pickled Cheese

High-salt, high-acid cheeses were likely among the first cheeses intentionally produced by humans. They would have had a much longer shelf-life than other similarly produced soft cheeses that had lower salt content. These cheeses have long been popular in Greece, Turkey, Egypt, Israel, and other Middle Eastern countries, as well as the Balkan region of Eastern Europe. In the past twenty years, production and consumption of these cheeses has spread throughout Europe, the United Kingdom, the United States, New Zealand, and Australia. The most popular cheese in this category is undoubtedly Feta cheese, which has its origins in Greece. Other similar types exist and vary based mostly on salting method (whether applied before or after curd formation). Although they are often packaged dry in the United States, it is more common for retail Feta-type cheeses to be packaged in tubs containing salt brine.

The general manufacturing procedures for Feta-type cheese start with a pasteurization step, since the cheese is usually consumed fresh (aged for as little as a few weeks). In Greece and other European countries, raw milk is often used. The milk itself can be obtained from cows, sheep, or goats. In many modern facilities, ultrafiltered milk is used. The milk is inoculated with a mesophilic starter containing *L. lactis* subsp. *lactis* and *L. lactis* subsp. *cremoris.* The culture is added at the rate of 2% and allowed to ripen in the milk for as long as two hours prior to addition of chymosin. The large inoculum and long ripening time results in considerable acid development and a decrease in pH even before chymosin is added.

A rennet paste, a crude preparation containing a mixture of enzymes, is used in traditional

Feta manufacture. Other proteases as well as lipases are among the enzymes present in this paste. The latter can also be added separately, in the form of a lipase extract obtained from animal or microbial sources. The coagulation time can be an hour or longer. The coagulum will be very firm at cutting. The large curds (2 cm) are stirred for only a short time before being dipped and filled into hoops, which are flipped and turned for several hours. Fermentation occurs quickly and the pH will reach 4.7 or lower within eight hours. The cheese is then either placed into a 18% salt brine for twelve to twenty-four hours, or dry salted for one to three days. The salted cheeses are then placed in a 10% brine and held until packaging and distribution.

The Feta-type cheeses have a very short, crumbly texture, due, in part, to low pH (<4.7) and high salt. As the pH approaches the isoelectric point of casein (pH 4.6), less water is retained. Also, high brine concentrations cause the casein network to shrink, releasing even more water from the cheese. Both of these fac-

tors make Feta crumbly. The main flavors in Feta are due to lactic acid and salt, but, depending on the presence of lipases (exogenous or from milk), there can also be considerable hydrolysis of triglycerides and formation of free fatty acids. It is the release of short, volatile fatty acids, including acetic, propionic, butyric, and valeric acids, that account for the rancid-like flavor notes characteristic of these cheeses.

Processed and Cold Pack Cheese

Although cheese is the main ingredient, processed cheese (or what many consumers mistakenly call American cheese) is not a fermented food. However, because processed cheese is so popular in the United States and there is so much confusion regarding the differences between natural cheese and processed cheese (Box 5–6), it is worthwhile to describe the manufacture of processed cheese.

Briefly, processed cheese is made by adding emulsifying salts to natural cheese, along with water and other dairy and non-dairy ingredi-

Box 5–6. "American" Cheese

Ask any American kid (even college kids) what kind of cheese he or she prefers and the likely answer will be "American" or perhaps even a popular brand name. What they are really referring to, in most cases, at least, is one of several types of processed cheese products. In fact, more than half of the cheese now purchased at retail (i.e., from supermarkets) is processed cheese. In contrast, the term "American" has no legal meaning, but is simply an adjective that describes those natural cheeses belonging to the Cheddar-style family.

The processed cheese category consists of three general types of products, each with a specific standard of identity that describes in detail the types of ingredients that are allowed, how it is to be manufactured, and the gross composition of the finished product. Pasteurized processed cheese is made from a single type of cheese and is labeled accordingly. Thus, pasteurized processed American cheese is made from Cheddar or other American-type cheese. It contains a maximum of 43% moisture and a minimum of 27% milkfat. Pasteurized processed cheese food may contain additional optional ingredients, and more water and less fat are allowed (44% and 23%, respectively). Pasteurized processed cheese spread contains even more water (44% to 60%), less fat (20%), and stabilizers are allowed. Many of the loaf-type products fit this description.

Not only do these products provide consumers with convenience and functionality, they can be considered an excellent example of value-added food technology. Cheese is, after all, a commodity whose dollar value is rather modest. Aging can certainly increase its value, but wholesale block and barrel prices for Cheddar cheese in the United States during 2004 averaged only about $1.50 per pound. Considering that it may cost the cheese manufacturer as much as $1.30 just for the milk that goes into a pound of cheese, earning a reasonable profit making commodity-style cheese is a real challenge.

ents. The mixture is then agitated while being heated to about 70°C or higher. The emulsification that occurs in processed cheese is different than other true food emulsions. In the case of process cheese, sodium or potassium polyphosphate and citrate emulsifying salts raise the pH and displace calcium ions from the casein complex, resulting in more soluble sodium casein. The latter contains both lipophilic and hydrophilic regions that then forms an emulsion-like mixture with the lipid portion of the cheese. After heating, the cooled mixture is then formed into slices or loaves or filled into jars or cans. The finished products have excellent functionality and convenience features. Due to the heat treatment and other inherent conditions (pH, water activity, antimycotic agents), these products also have long shelf-lives, even at ambient temperature.

The quality of the cheese used as the starting material is among the factors that influence processed cheese manufacture. A typical blend consists of 15% aged and 85% young (or current) cheese. The aged cheese provides flavor; the young cheese provides elasticity, body, and emulsifying properties. Too much of the former may cause a soft, soupy body, too much of the latter may lead to hard body and brown pigment formation. Cold pack cheese, in contrast, is made by mixing or grinding different types of natural cheese in the absence of heat. Various optional ingredients, including color, spices, and other flavoring agents, as well as antimycotic agents, can also be added. These products have excellent flavor, but typically lack the functionality of process cheeses.

Cheese Ripening

Freshly made cheese has essentially none of the flavor, aroma, rheological, or appearance properties of aged or ripened cheese. Rather, the metamorphosis from a bland, pale, rubbery mass of protein and fat into a flavorful, textured fusion of complex substances takes time and requires patience. Although efforts to reduce aging time and accelerate the ripening process have been somewhat successful (see below), for the most part, cheese ripening is a sequential process, with each step relying on a preceding step. In other words, a particular flavor compound may be present in a cheese only as a result of several preceding metabolic steps.

For example, methyl ketones, which contribute to the characteristic flavor of blue cheese, are synthesized by *P. roqueforti* from free fatty acids. Thus, methyl ketone formation depends on release of these acids from triglycerides via lipolytic enzymes produced by microorganisms or naturally present in the milk. Similarly, hydrogen sulfide, which is an important flavor note in aged Cheddar cheese (or may also be a defect; see below), is derived from sulfur-containing amino acids that form via protein and peptide hydrolysis.

Of course, not only is ripening a sequential process, but it also is subject to potential chaos and disarray, such that the cheese may ripen poorly or unexpectedly. In some cases, over-production of an otherwise desirable compound occurs, resulting in serious flavor defects. For example, hydrogen sulfide, at concentrations in the ppb range, imparts a pleasant aroma in Cheddar cheese, but when present at ppm levels, the cheese is nearly inedible. In contrast, while blue cheese may contain 90 ppm to 100 ppm of methyl ketones, a small amount (1 ppm to 2 ppm) is perfectly fine in Cheddar cheese. Controlling the ripening process, then, is key to successful and consistent production of aged and well-ripened cheese.

Many factors contribute to the ripening process, including live microorganisms, dead microorganisms, enzymes, and chemical and physical reactions. Manipulating and controlling these activities and reactions depends on characteristics intrinsic to the cheese, such as moisture, pH, salt, and Eh, as well as those extrinsic factors that are influenced by the cheese manufacturer. The latter include the source and handling of the milk, the temperature and humidity of the ripening room or environment, and other manipulations performed by the manufacturer. It is the collective result of these events that dictate the properties of the ripened cheese.

As noted above, cheese contains mostly water, protein, and fat as it leaves the vat and moves into the ripening coolers. Although some protein hydrolysis certainly occurs during the lactic fermentation by starter culture bacteria and by the action of milk proteases and the coagulant, α_{s1}, α_{s2}, and β-caseins are still intact. Similarly, the triglycerides in milk are also mostly unaffected by the early cheese-making steps. However, within just a few days, enzymes begin to attack these substrates and initiate the ripening process.

Proteolysis in cheese

It was long argued that milkfat was the primary constituent responsible for cheese flavor. While it is certainly true that many cheese flavors are either evolved from the lipid fraction or are soluble in the lipid phase, it is now generally accepted that, for most cheeses (the main exception being the blue mold-type cheeses), it is the protein fraction that makes the more important overall contribution (Table 5-2). Cheeses made under controlled conditions in which proteolysis does not occur develop neither the flavor nor texture of a normal cheese. The recognition that proteolysis has such a profound influence on cheese flavor and ripening has led to a detailed understanding of many of the specific steps involved in the degradation of milk proteins and the metabolism of the hydrolysis products (Figure 5-9).

For the most part, the relevant organisms responsible for these activities are lactic acid bacteria, either those added as part of the starter culture or those present as part of the ordinary microflora. Lactic acid bacteria cannot synthesize amino acids from ammonia and instead require pre-formed amino acids for protein biosynthesis and cell growth. However, milk contains only a small amount of free amino acids (<300 mg/L), which are quickly assimilated. Thus, growth of lactic acid bacteria in milk depends on the ability of these cells to metabolize proteins in milk. The means by which large casein molecules are broken down by these bacteria into their component amino acids (and beyond) consists of four main steps: (1) hydrolysis of casein; (2) transport of casein-derived amino acids, peptides, and oligopeptides into the cytoplasm of starter and non-starter lactic acid bacteria; (3) intracellular hydrolysis of accumulated peptides; and (4) metabolism of amino acids.

In Cheddar-type cheeses, starter culture lactic acid bacteria, notably, *L. lactis* subsp. *lactis*, produce a cell-wall anchored proteinase that hydrolyzes specific casein subunits (Box 5-1), producing as many as 100 or more different peptides varying in length from four to thirty amino acid residues. Although proteinases are produced by several strains of *L. lactis* subsp. *lactis* and other lactic acid bacteria, the enzymes are structurally quite similar (in terms of their DNA and amino acids sequences).

Table 5.2. Sources of flavor compounds in cheese[1].

Protein (casein)	Carbohydrate	Lipid
peptides	lactate	fatty acids
amino acids	acetate	keto acids
sulfur compounds	pyruvate	esters
ammonia and amines	ethanol	methyl ketones
pyruvate	diacetyl	lactones
acetate	acetoin	
aldehydes	2,3-butanediol	
alcohols	acetaldehyde	
keto acids		

[1]Adapted from McSweeney, 2004 and Singh et al., 2003

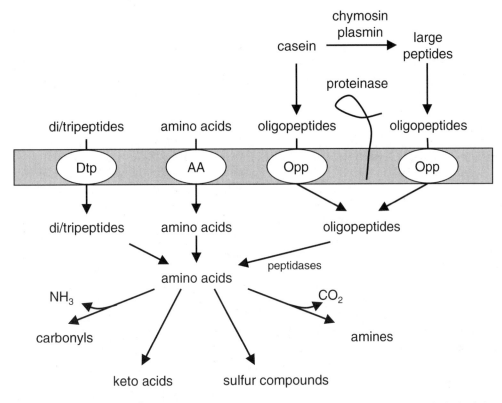

Figure 5–9. Proteolysis during cheese ripening. The proteolytic system in lactococci starts with the hydrolysis of casein by a cell envelope-associated proteinase (PrtP). The main products are oligopeptides, which are then transported across the cell membrane by the oligopeptide transport system (Opp). Any free amino acids and di- and tripeptides in the milk are similarly transported by amino acid (AA) and di- and tripeptide transporters (DtpT, DtpP), respectively. Once inside the cell, the oligopeptides are hydrolyzed by various peptidases, including PepA, PepC, PepN, and PepX. The di- and tripeptides are also hydrolyzed. The free amino acids are then metabolized or used for protein biosynthesis (not shown). Adapted from McSweeney, 2004, and Hutkins, 2001.

However, the specific casein substrates and hydrolysis products of the different PrtP enzymes can vary considerably.

For example, some proteinases hydrolyze α_{S1}-, β-, and κ-caseins, whereas others have preference for β-casein and relatively little activity for α_{S1}- and κ-caseins. It should be noted that residual chymosin and milk proteinases (mainly plasmin) may also contribute to the peptide pool, especially during the early stages of cheese ripening. However, it is the starter and non-starter lactic acid bacteria that are responsible for most of the subsequent casein hydrolysis. In fact, when cheese is made with

starter strains that do not produce cell wall proteinase, casein degradation occurs to a very limited extent (and cheese flavor and texture do not develop). Likewise, cheese made in the absence of non-starter lactic acid bacteria is similarly bland.

Hydrolysis of the casein-derived peptides was previously thought to occur by extracellular peptidases produced by starter or non-starter lactic acid bacteria. It now is generally accepted that peptide hydrolysis does not occur outside the cells, but rather the peptides are transported via various peptide transport systems and are subsequently hydrolyzed by intracellular peptidases. The main transporter, called the oligopeptide transport system (OPP), is present in most starter lactic acid bacteria and transports oligopeptides across the cytoplasmic membrane and into the cell. Mutants defective in OPP, like mutants defective in proteinase production, grow poorly in milk because they are unable to use casein as a nitrogen source. Starter lactococci also have amino acid as well as di- and tri-peptide transport systems; however, they are somewhat less important since most of the peptides present in cheese are larger than three amino acid residues.

Once inside the cells, the peptides can be hydrolyzed by a variety of peptidases, depending on their size and amino acid content. Aminopeptidases hydrolyze peptides at the amino end (i.e., the peptide bond located between the amino terminal amino acid residue and the penultimate residue), whereas carboxypeptides act at the carboxy end. Endopeptidases also exist that hydrolyze peptide bonds within a peptide. Lactococcal peptidases also have specificity with regard to the amino acids and their position within a given peptide bond. For example, aminopeptidase P (PepP) from *L. lactis* subsp. *lactis* hydrolyzes peptides containing proline at the penultimate residue.

Protein hydrolysis also has a major impact on texture. The solubility of coagulated casein is limited, and is even less at low pH, making young cheese plastic and rubbery. However, as cheese ages and casein is degraded, the peptides are more soluble, and elasticity increases. Eventually, however, when the casein and casein-derived peptides are more extensively degraded, the cheese develops a crumbly, short texture.

As noted above, cheese flavor and texture development is largely a function of the proteolytic capacity of the bacteria present in the ripening cheese. However, for the most part, the starter culture bacteria are not maintained very long after the first few weeks of aging. That is not to say that their impact is not significant. In fact, as the starter bacteria lyse (due to salt, low water activity, lack of fermentable substrates, and other factors), their entire array of intracellular enzymes are released into the curd matrix. Thus, the various peptidases important for peptide degradation are free to hydrolyze available substrates. Likewise, enzymes that act directly on amino acids, triglycerides, and other substrates are also released, as are the substrates themselves.

Finally, non-starter lactic acid bacteria, as well as other indigenous bacteria, play an important (perhaps the single most important) role in cheese ripening. In particular, several *Lactobacillus* spp., including *Lactobacillus casei*, *Lactobacillus paracasei*, and *Lactobacillus plantarum,* increase in number (from 10^2 per gram to more than 10^7) during aging, and produce peptidases and other enzymes that generate aged cheese flavor. Of course, some of these adventitious organisms can also produce off-flavors. Heterofermentative lactobacilli are especially of concern due to their ability to produce acetic acid and CO_2 from residual sugars.

Current Issues in Cheese Technology

Given the important role of microorganisms in the manufacture of cheese, it should come as no surprise that among the most important issues faced by the cheese industry, most are microbiological in nature (not withstanding yield, costs, and other economic issues). The key to making any high quality fermented food, including cheese, is to control the activities of the microbial starter culture and other organisms. This means ensuring that they grow when they are supposed to, that agents that would otherwise interfere with their growth are controlled, and that they produce the correct amount and types of enzymes, flavors, and other products.

Although consumer trends often influence and raise specific sorts of problems, several long-term issues continue to challenge cheese technologists. These questions include: (1) how to improve and accelerate cheese flavor development and reduce bitterness, especially in low-fat cheese; (2) how to control bacteriophages that infect, inhibit, and sometimes inactivate starter cultures; (3) how to enhance shelf-life and prevent growth of spoilage microorganisms and ensure that cheese is free of potential pathogenic microorganisms; and (4) how to increase the value of whey using microorganisms.

Bitterness and accelerated ripening

For most cheeses, production of good aged cheese flavor and texture requires that there be significant hydrolysis of casein. At the same time, the main flavor defect in cheese is bitterness, which is also caused by casein hydrolysis and subsequent formation of bitter peptides. Thus, aging cheese is like walking a tightrope—

it requires a fine sense of preparation, balance, and proportion. A single misstep may lead to an unacceptable end. Accelerated ripening—obtaining good cheese flavor and texture in half or even a third of the time—is, to continue the analogy, like running across the tightrope. There is tremendous economic incentive, however, to decrease aging time, because the aging process adds about one cent per Kg per month to the cost of the cheese. Reduced fat cheese poses another set of flavor problems (Box 5-7). Not surprisingly, considerable research has been aimed at developing accelerated cheese aging programs.

Box 5–7. Reduced Fat Cheese—Taking the Fat (But Not the Flavor) Out of Cheese

Cheese, like other dairy products, is heavily regulated by the federal government, and standards of identity exist for most of the cheeses consumed in the United States (Figure 1). These standards (contained within the Code of Federal Regulations, Title 21, Part 133) not only describe the ingredients allowed in the manufacture of various cheeses, they specifically spell out the fat and compositional requirements for them. Importantly, the minimum amount of fat (on a dry or solids basis) and maximum amount of water are specified. Until relatively recently, Cheddar cheese that contained less than 50% fat on a dry basis (or in the dry matter or FDM) could not be called cheese. Similar standards exist for other cheeses, such that most contain at least 45% fat and derive more than 75% of their calories from fat.

Because milkfat was viewed, for hundreds of years, as a precious and desirable commodity, these standards were established to protect consumers and make sure they received full value for their money. Products labeled as cheese, but containing less than the required amount, were considered inferior and in violation of the law. This public perception of milkfat, however,

Reduced fat = 25% reduction in total fat per reference amount when compared to an appropriate reference food

Light = food that has 50% less fat per reference amount when compared to an appropriate reference food, when 50% or more of its calories are from fat, or in the case of food containing less than 50% of its calories from fat, less than one-third less calories from fat or less than 50% fat per reference amount when compared to an appropriate reference food.

Nonfat = Less than 0.5% fat per reference amount when compared to an appropriate reference food

Lowfat = Three grams or less fat per reference amount.

Figure 1. Reduced Fat Cheese Standards (Implemented 1996).

(Continued)

Box 5–7.   Reduced Fat Cheese—Taking the Fat (But Not the Flavor) Out of Cheese *(Continued)*

changed in the 1960s and 70s, as consumers in the United States began to reduce their consumption of animal fats and to seek out lower-fat products. Sales of low-fat fluid milk products began to increase, and a market for other low-fat dairy products became evident. Although the cheese industry was well aware of the change in consumer attitudes and had been working on lower-fat versions, these new products could not be called cheese, due to existing standards. It was not until the 1990s that the U.S. government modified the definitions of cheese so that low-fat versions could still be labeled as cheese. These new standards for cheese follow the reduced fat, low-fat, and nonfat definitions used for other foods (see below).

The resolution of the labeling problem, however, did not lead to immediate consumer acceptance of reduced fat cheese. Making cheese with low-fat milk is technically no different than conventional cheese manufacture, but the products that were initially introduced suffered from serious flavor and body defects. This situation has slowly begun to change, and as of the year 2000, about two-thirds of consumers had purchased reduced or non-fat cheese. However, many of the problems that plagued the initial products have yet to be resolved and the market for these products is still relatively modest.

Making reduced fat cheese with flavor and texture properties comparable to full fat products, it turns out, is not easy, for two main reasons. First, when milkfat is taken out of cheese, it is essentially replaced by casein and water. Thus, the body becomes more firm and less smooth and pliable. Second, a higher moisture content means that the salt-in-moisture ratio decreases, resulting in an undesirable increase in microbial and enzymatic activities. All other things being equal, the cheese ages poorly, with potential for significant production of bitterness and other off-flavors. More moisture also results in more lactose in the curd, which may lead to excessive acid production. Not only does this affect flavor, but as the pH decreases, calcium is lost, which may cause other body defects.

Manufacturers have adopted several strategies to produce more acceptable reduced fat cheeses. First, gums and other fat mimics can be added to give the body, texture, and mouth feel of full-fat cheese. To control acidity, the curds can be washed to remove excess lactose. Alternatively, researchers at the University of Wisconsin developed a process, based on enhancing the buffering capacity in the curd (by retaining calcium phosphate), that results in higher acid levels but moderate pH (Johnson et al., 2001). This also resulted in a lower level of calcium bound to casein and a softer body. An experimental centrifugation method in which the fat is removed after manufacture and aging was recently described (Nelson and Barbano, 2004).

Despite these improvements, flavor, either the lack thereof or the presence of undesirable flavors, remains a problem. According to some researchers, milkfat not only serves as an important

The main approach of most aging strategies is to promote proteolytic and other enzymatic activities. As described above and in Figure 5–9, the first step in the use of casein by the lactic starter culture is hydrolysis of casein by the cell wall anchored proteinase (PrtP). Of the hundred or more peptides that are produced, most contribute little to overall cheese flavor. However, several of these peptides have a bitter flavor. If the starter culture cannot degrade these bitter peptides, the cheese will be bitter. Starter culture strains that produce PrtP (or related proteinases), but that are otherwise unable to degrade the bitter peptides so produced, are referred to as bitter strains.

In general, bitter peptides are hydrophobic and contain proline, and are hydrolyzed only by specific intracellular aminopeptidases, including PepN and PepX. Bitter strains lack the ability to produce these enzymes. It is interesting that most commercial strains of *L. lactis* subsp. *lactis*, the most widely used organism in the cheese industry, are bitter. In contrast, most of the *L. lactis* subsp. *cremoris* strains are non-bitter. However, *L. lactis* subsp. *lactis* grows faster and, at least in unripened cheese (e.g.,

Box 5–7. Reduced Fat Cheese—Taking the Fat (But Not the Flavor) Out of Cheese *(Continued)*

flavor-generating substrate (i.e., fatty acids), it also performs as a solvent in cheese, since many of the important Cheddar cheese flavor compounds are found in the lipid phase (Midje et al., 2000). Although others contend that cheese flavor is associated with the water-soluble fraction (McGugan et al., 1979; Nelson and Barbano, 2004), it is clear that the quality of reduced fat cheese could be improved by increasing flavor intensity.

One popular approach to improve the flavor of reduced fat cheeses (and conventional cheese, as well) has been to add selected flavor-producing lactic acid bacteria to the milk, in the form of adjunct cultures (Fenelon et al., 2002; Midje et al., 2000). As described below (Box 8), adjunct cultures are added to the milk in much the same way as the starter culture. However, their function is to produce enzymes and flavor compounds, rather than lactic acid. Adjunct cultures typically contain strains of *Lactococcus* and *Lactobacillus* that produce aminopeptidases that specifically hydrolyze bitter peptides. Although bitter peptides are often present in full-fat cheese, they can be even more of a problem in reduced fat cheese, in part because they partition more in the non-lipid or serum phase, where they are more readily detected. Also, the lower salt-in-moisture ratio typical of reduced fat cheese may result in higher cell numbers and greater proteinase production and activity by the starter culture bacteria, which ultimately leads to bitter peptide accumulation. The use of non-bitter strains may, however, significantly reduce bitterness problems in these cheeses.

References

Fenelon, M.A., T.P. Beresford, and T.P. Guinee. 2002. Comparison of different bacterial culture systems for the production of reduced-fat Cheddar cheese. Int. J. Dairy Technol. 55:194–203.

Johnson, M.E., C.M. Chen, and J.J. Jaeggi. 2001. Effect of rennet coagulation time on composition, yield, and quality of reduced-fat Cheddar cheese. J. Dairy Sci. 84:1027–1033.

McGugan W.A., D.B. Emmons, and E. Larmond. 1979. Influence of volatile and nonvolatile fractions on intensity of Cheddar cheese flavor. J. Dairy Sci. 62:398–403.

Midje, D.L., E.D. Bastian, H.A. Morris, F.B. Martin, T. Bridgeman, and Z.M. Vickers. 2000. Flavor enhancement of reduced fat Cheddar cheese using an integrated culturing system. J. Agric. Food Chem. 48: 1630–1636.

Nelson, B.K., and D.M. Barbano. 2004. Reduced-fat Cheddar cheese manufactured using a novel fat removal process. J. Dairy Sci. 87:841–853.

cheese destined for the process kettle), its inability to degrade bitter peptides is not an issue.

Even cheese made with non-bitter strains, if aged long enough, can eventually accumulate enough bitter peptides to develop bitter flavor. Bitter peptide production is further increased at higher ripening temperatures (i.e., >12°C), which is perhaps the most common way to accelerate cheese ripening. Although some bitterness is unavoidable, the addition of adjunct cultures, or strains capable of producing aminopeptidases, has been promoted as a means of accelerating the aging process (Box 5–8). Adjunct cultures, which typically contain selected strains of *Lactobacillus* and *Lactococcus,* can be added directly as part of the starter culture. However, because excess acid production is to be avoided, variants that are unable to ferment lactose are preferred.

The observation that some strains of lactococci and streptococci were autolytic led to another innovative means for accelerating the ripening process. As described previously, ripening depends largely on the presence of enzymes released by bacteria as they lyse during the aging process. Ordinarily, cell lysis occurs only when the cells are no longer able to maintain cell wall integrity, due to the lack of energy sources and the generally inhospitable conditions (i.e., low pH, high salt, etc.) present within the cheese. In contrast, if lysis and enzyme release were to occur during the early

Box 5–8. Adding Flavor With Adjunct Cultures

The development of cheese flavor and texture is a complicated process, involving numerous enzyme-catalyzed reactions. The enzymes involved in these reactions originate from three main sources: the coagulant, the milk, and bacteria. The latter are either normally present in milk due to unavoidable contamination or are deliberately added. And although it is certainly possible to make cheese without bacteria (i.e., aseptic cheese), the finished product would contain few, if any of the flavor or texture characteristics one might expect. This would be especially true if the cheese were aged, since aged flavor and texture would simply not develop.

While the coagulant and natural milk enzymes certainly contribute to cheese flavor and texture development, their roles are considered to be rather limited, whereas the presence of bacteria is essential (El Soda et al., 2000). Moreover, it is not just the lactic starter culture bacteria that are responsible for the desirable changes that occur during cheese ripening—bacteria present as part of the natural milk flora also serve as a vital reservoir of enzymes that are necessary for proper cheese maturation. If, however, the milk is pasteurized or if it is produced and handled under strict hygienic conditions, the natural microflora will be rather limited, both in number and diversity. The finished cheese will ripen slowly and, in some cases, never achieve the desired quality expectations.

The realization that the natural milk flora is important for cheese ripening led investigators first to identify the relevant organisms, and then to consider the possibility of adding them directly to the milk to boost or enhance flavor development (Reiter et al., 1967). These studies revealed that the bacteria that were most frequently associated with good cheese flavor were lactic acid bacteria, but not the same species or strains used as the starter culture. Therefore, as a group, they are referred to as non-starter lactic acid bacteria (NSLAB). Cheese flavor is enhanced when the correct NSLAB, which generally consist of species of *Lactococcus* and *Lactobacillus*, are added to cheese milk. It also appears that the presence of these NSLAB may out-compete "wild" NSLAB that might otherwise cause flavor defects. They are now commercially available in the form of adjunct cultures, and are widely used for accelerated ripening programs as well as for reduced fat cheese, which often suffer from flavor problems (Box 5–7).

When adjunct cultures were first considered for modifying cheese flavor, cells attenuated by freeze or heat shocking, physical-chemical treatment, or genetic means were used (El Soda et al., 2000). This was done to make the cells nonviable and to minimize subsequent acid production and metabolism. However, as noted above, adjunct cultures are now commercially available and are added directly to the cheese milk, albeit at much lower inoculation rates compared to the starter culture (Fenelon et al., 2002).

Among the species currently used commercially are strains of *Lactobacillus helveticus*, *Lactobacillus casei*, and *Lactobacillus paracasei*. In particular, the specific NSLAB used as adjunct cultures are selected based largely on their ability to produce aminopeptidases and to lyse dur-

stages of aging, the ripening process could be shortened.

Bacteriophages

Bacteriophages (or phages, for short) are arguably the number one problem in cheese production. They are certainly the main reason why starter culture activity is sometimes inhibited. Infection of starter cultures by phages may result in slow or sluggish fermentations; occasionally the fermentation may fail completely. Phage

infections ultimately can lead to prolonged make times and inferior or poor quality cheese, both of which cause substantial economic hardships on the manufacturer. Although phages were once thought to be primarily a problem for cheeses made using mesophilic lactococcal cultures, it is now recognized that thermophilic cultures (especially *S. thermophilus*) are also susceptible to attack by phages.

As discussed in Chapter 3, classification of lactic phages is based mainly on morphological criteria (i.e., head shape and size and tail

Box 5–8. Adding Flavor With Adjunct Cultures *(Continued)*

ing cheese aging. The aminopeptidases released by cell lysis are then able to degrade the bitter peptides that accumulate in cheese as a result of partial casein hydrolysis by chymosin and starter culture proteinases. Thus, the cheese becomes less bitter, and the released amino acids can be metabolized to yield desirable end products.

The latter point deserves emphasis, because de-bittering is not the only function of culture adjuncts. There appears to be an association, for example, between strains with high glutamate dehydrogenase activity (which forms α-ketoglutarate from glutamate) and formation of desirable cheese flavor and aroma (Tanous et al., 2002). Similarly, lactic acid bacteria that metabolize glutathione release cysteine into the cheese, which serves as precursor for sulfur-containing flavor compounds.

Recently, the use of non-lactic acid bacteria as cultures adjuncts has been proposed, based largely on the ability of these organisms to metabolize amino acids and generate various flavor compounds (Weimer et al., 1999). Specifically, strains of *Brevibacterium linens* (the organism used for Limburger and other surface-ripened cheese) have been reported to enhance flavor development via production of sulfur-containing volatiles, such as methanethiol, and fatty acids, such as isovaleric (Ganesan et al., 2004).

References

El Soda, M., S.A. Madkor, and P.S. Tong. 2000. Adjunct cultures: recent developments and potential significance to the cheese industry. J. Dairy Sci. 83:609–619.

Fenelon, M.A., T.P. Beresford, and T.P. Guinee. 2002. Comparison of different bacterial culture systems for the production of reduced-fat Cheddar cheese. Int. J. Dairy Technol. 55:194–203.

Ganesan, B., K. Seefeldt, and B.C. Weimer. 2004. Fatty acid production from amino acids and α-keto acids by *Brevibacterium linens* BL2. Appl. Environ. Microbiol. 70:6385–6393.

McGugan W.A., D.B. Emmons, and E. Larmond. 1979. Influence of volatile and nonvolatile fractions on intensity of Cheddar cheese flavor. J. Dairy Sci. 62:398–403.

Midje, D.L., E.D. Bastian, H.A. Morris, F.B. Martin, T. Bridgeman, and Z.M. Vickers. 2000. Flavor enhancement of reduced fat Cheddar cheese using an integrated culturing system. J. Agric. Food Chem. 48:1630–1636.

Reiter, B., T.F. Fryer, A. Pickering, H.R. Chapman, R.C. Lawrence, and M.E. Sharpe. 1967. The effect of microbial flora on the flavor and free fatty acid composition of Cheddar cheese. J. Dairy Res. 34:257–272.

Tanous, C., A. Kieronczyk, S. Helinck, E. Chambellon, and M. Yvon. 2002. Glutamate dehydrogenase activity: a major criterion for the selection of flavour-producing lactic acid bacteria. Antonie van Leeuwenhoek 82:271–278.

Weimer, B., K. Seefeldt, and B. Dias. 1999. Sulfur metabolism in bacteria associated with cheese. Antonie van Leeuwenhoek 76:247–261.

length). Importantly, they vary widely with respect to host range and virulence. Some phages have a narrow host range, infecting only a few host strains, whereas others are able to attack many different strains. Although temperate phages that lysogenize their hosts may occasionally become virulent, lytic phages are, by far, the more serious.

Following infection, lytic phages may replicate and release as many as 100 or more new infective phage particles from a single host cell within thirty to forty minutes. Even if the initial phage concentration in the milk is low, phage replication may be so rapid that enough phages are produced to significantly inhibit or decimate the starter culture before that culture does its job. In the example shown in Table 5–3, a single phage with a latent period of thirty minutes and a burst size of fifty will outnumber even a fast-growing culture in less than three hours. In a cheese operation, that means the curd pH may never reach the desired level, leaving the manufacturer with a "dead" vat and a huge headache. Phages that are more virulent

Table 5.3. The phage problem during industrial cheese fermentations[1].

Time	Phage absent		Phage present		
	Cells/ml	pH	Cells/ml	Phage/ml	pH
0:00	1×10^6	6.6	1×10^6	1	6.6
0:30	2×10^6	6.5	$2 \times 10^6 - 1$	50	6.5
1:00	4×10^6	6.3	$4 \times 10^6 - 50$	2,500	6.3
1:30	8×10^6	6.1	$8 \times 10^6 - 2,500$	125,000	6.2
2:00	1.6×10^7	5.8	$1.6 \times 10^7 - 125,000$	6.25×10^6	6.0
2:30	3.2×10^7	5.5	$2.5 \times 10^7 - 6.25 \times 10^6$ $= 1.9 \times 10^7$	3.1×10^8	5.7
3:00	6.4×10^7	5.2	$1.9 \times 10^7 - 3.1 \times 10^8$ $= 0$	3.1×10^8	5.6

[1]This simulated fermentation profile is based on the following assumptions:
Culture generation time = 0.5 hours
Phage replication time = 0.5 hours
Final desired pH after 3 hours = 5.2
Average phage burst size = 50

than the one cited in the example above, or that are present initially at much higher levels, may inactivate the culture even faster. Conversely, if the culture itself is more resistant to the phages present in the plant, or if the phage levels are low when its host is introduced into the cheese environment, then the effects of that phage may be much less severe.

Ever since the phage problem was first recognized (seventy years ago!), manufacturers and scientists have developed strategies to prevent or at least minimize the incidence and impact of phage infections. In fact, phages and the problems they cause cheese manufacturers have provided the driving force not only for much of the scientific research on dairy starter cultures, but also for the development of a highly sophisticated starter culture industry. Research on bacteriophages has advanced to a remarkable extent in the past two decades due to advances in molecular biology and biotechnology by groups in Australia, New Zealand, the Netherlands, United Kingdom, the United States and Switzerland.

Several control strategies have evolved from these phage research programs, and are now part of most cheese manufacturing operations. However, before describing these approaches, it is important to note that the first requirement of any phage control plan is to ensure that the starter culture propagation environment, as well as the cheese production areas, be well sanitized and that phage entry points are as restrictive as possible. Phages are sensitive to hypochlorite and heat, and can, therefore, be inactivated from equipment and culture media. Phages move about via whey aerosols, air, and personnel, so these represent important control points. The bottom line is that, unless a cheese manufacturer practices excellent sanitation, has appropriate air handling systems in place (e.g., HEPA filters, positive pressure in culture rooms, etc.), segregates whey processing to separate areas, and limits personnel traffic, more high-tech phage control measures will have little chance of being effective.

Even when good sanitation programs are in place, phages cannot be eliminated from the cheese manufacturing environment. After all, wherever there are growing lactic acid bacteria, there will be phages that attack them. Given that reality, the culture industry offers a number of products designed to minimize or prevent phage problems. These include: (1) phage inhibitory media; (2) phage-resistant cultures; and (3) culture rotation programs.

The bulk culture tank is the first place where phage propagation can occur. Traditionally, milk or whey was used as the culture medium, and although good growth of lactic starter bacteria could be obtained, phages were also well-suited to this simple medium. Phage-inhibitory media, in contrast, are also milk- or whey-based media, but, in addition to some added nutrients and growth factors, they also contain phosphate and citrate salts. These salts bind calcium, which is required by phages for adsorption to host cells (the first step in the infection process). There are some phages, however, that are unaffected by the lack of calcium and replicate fine in phage-inhibitory media. Thus, despite its widespread use, this approach has only a modest overall effect on phage control. Moreover, this medium only affects phage propagation within the bulk tank, and not later, during cheese manufacture.

Dairy microbiologists have long searched for lactococci and other lactic acid bacteria that naturally resist bacteriophages. Although such strains exist, in many cases, these isolates were often found to lack other relevant characteristics necessary for cheese production. For example, they either grew too slowly in milk, produced poor-quality cheese, or eventually became phage sensitive. An alternative approach that has been more successful has been to isolate spontaneous phage-resistant mutants. These mutants or variants can be obtained by repeated exposure of cheese production strains to the predominant phage present in a given environment. To ensure that fermentative properties have not been altered, the mutants must be re-evaluated for acid production rates while exposed to phages and under conditions mimicking those that occur during cheese manufacture (the so-called Heap-Lawrence procedure, named after two New Zealand researchers).

The most powerful and effective means of developing phage-resistant cultures has been to exploit the innate ability of certain strains to resist lytic phage infections. Several mechanisms have been identified that are responsible for the phage resistance phenotype in lactic acid bacteria. In general, phage-resistant strains block infection at one of several points in the lytic cycle (Figure 5–10). In some strains, phage adsorption fails to occur, due either to modifi-

cation of cell wall attachment sites or interference by polysaccharides. Once adsorption occurs, phages ordinarily inject DNA into the host cell to initiate phage DNA replication. This step is inhibited in some cells, although the mechanism is not clear. Following injection of phage DNA, the host cell can attack that DNA via expression of restriction enzymes that hydrolyze DNA at specific restriction sites.

Finally, if the phage has survived all of these host defense systems, its replication can still be inhibited via a mechanism known as abortive infection. Although some phages may still be released, and the infected cell is killed, the burst size is significantly reduced, and the impact on the remaining cells is minimized.

The discovery in the 1980s and 1990s of the mechanisms described above coincided with the observation that genes coding for most of these traits in lactococci were located on plasmids. Thus, it was possible to mobilize the relevant plasmid from one lactic acid bacterium to another, and to confer a phage resistance phenotype into the recipient strain. This strategy was first used, in 1986, to develop a phage-resistant cheese-making strain of *L. lactis* subsp. *lactis*. The plasmid, pTRK2030, actually encoded for several phage-resistant mechanisms (at the time, the researchers were aware of only a restriction system). Because conjugation, a natural means of genetic exchange, was used to effect gene transfer, the phage-resistant transconjugant strain required no regulatory approval. Once introduced into a cheese plant, it performed extremely well.

The application of this and other natural phage resistance mechanisms to improve starter cultures quickly followed. If one plasmid is good, two (or more) reasoned the Klaenhammer group, would be better. Thus, a stacking strategy was developed in which a single strain contained different plasmids, each coding for unique phage resistance properties. Alternatively, the different plasmids can be separately introduced into a single strain, yielding multiple isogenic strains harboring a unique phage resistant plasmid. These strains can then be used in a rotation program (see below) to

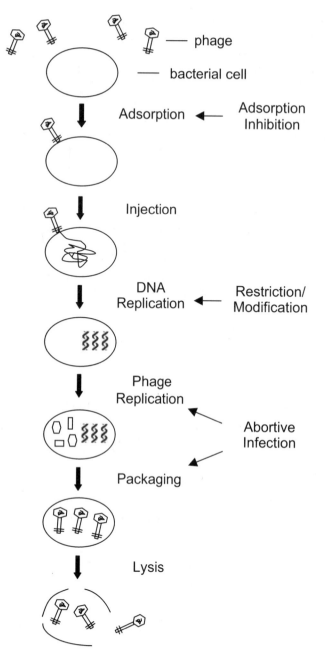

Figure 5–10. Lytic cycle of lactic bacteriophages and steps at which phage resistance mechanisms function.

prevent any one phage from reaching high enough levels to cause problems.

As noted above, it is inevitable that a phage capable of infecting a given starter culture strain will eventually emerge in a cheese plant. The infecting phage may be naturally occurring, or more likely, is the result of DNA modification, mutation, or a gene acquisition event. There is, after all, no other phage-host relationship that is more dynamic than that which occurs in a large cheese plant. Every day, there may be as many as 10^{18} bacteria growing in such a plant, all serving as potential hosts for phages. Thus, staying a step or two or three ahead of the phage is no easy matter, and engineering more durable and longer-lasting phage resistance into lactic acid bacteria is a major challenge. The availability of genome information (for both the host cells and their phages) in the hands of clever microbiologists has made it possible to deploy powerful strategies for development of new phage-resistant strains (Box 5-9).

Despite the opportunity for phages to find and infect host cells, the only scenario that would result in a particular phage actually affecting a fermentation would be if the phage were initially present at reasonably high levels. Phages can only replicate, and therefore increase in number and reach danger levels, when they infect a suitable host. If the host cell is not present, the phage will not last long in that environment, especially when good

Box 5–9. How to Outsmart a Virus

It is remarkable that the main biological problem that confronts the modern and sophisticated fermented dairy foods industry is viruses, nature's simplest life form. Given the discovery in the 1980s that some lactic acid starter culture bacteria are armed with natural anti-viral defense systems, one might think that controlling these viruses, or bacteriophages, would be a relatively easy task. However, this is clearly not the case, as "new" bacteriophages continue to emerge, much to the frustration of cheese manufacturers.

The inhibition of milk fermentations by bacteriophages was first described by New Zealand researchers in 1935 (Whitehead and Cox). Although phage infections of starter culture bacteria were initially only occasionally reported, such occurrences became more common as the size of the cheese industry grew. As previously noted, many innate phage-resistant mechanisms, and the genes responsible for those mechanisms, have been identified in lactococci and other lactic acid bacteria and subsequently used to construct new phage-resistant strains. However, the conferred resistance has generally been effective only for the short-term, because new virulent phages invariably appear within a relatively short time. Given the parasitic nature of bacteriophages (i.e., they can only replicate on suitable hosts) and the incredible opportunity for phage-host interactions, it is really no surprise that wherever lactic acid bacteria live and grow, phages capable of infecting those bacteria will eventually arise. This realization has inspired researchers, especially the Todd Klaenhammer group at North Carolina State University, to develop novel and quite clever strategies for combating these phages and protecting starter cultures from phage infections. These "intelligent" approaches (also referred to as "artificial," since they do not exist naturally) rely on molecular trickery to either prevent phage adsorption or to short-circuit the phage lytic cycle and thereby block phage replication.

Pips and Pers

The first stage of the phage infection process occurs when a phage recognizes and then attaches to a specific site on the surface of the host cell. One such receptor site, referred to as a phage infection protein, or *pip*, has been identified in *Lactococcus lactis* subsp. *lactis* C2 (Geller et al., 1993; Garbutt et al., 1997). Infection by phage c2 does not occur in cells lacking an intact copy of *pip*. Since mutations in the *pip* gene do not appear to have any other phenotypic effect on the cell, engineered *pip*-defective cells would be insensitive to phages that depend on Pip for binding. Recently, similar results were obtained for another class of lactococcal phages that do not recognize *pip*, but instead bind to membrane proteins called receptor-binding proteins (Dupont et al., 2004).

Replication of the phage genome occurs soon after a phage has infected a host cell. Replication begins at a unique location within the phage genome, the so-called origin of replication (or

(Continued)

Box 5–9. How to Outsmart a Virus *(Continued)*

ori site), which is characterized by several direct and inverted repeated sequences. The *ori* region, therefore, serves as the recognition site for the DNA replication machinery. If, however, there were multiple "decoy" copies of the *ori* region within the cell, then the DNA replication system would, in theory, be diluted out and phage genome replication would be reduced. Indeed, this approach, called phage-encoded resistance, or *Per*, was shown by Hill et al. (1990) to work in *Lactococcus lactis*. In this case, a phage genome fragment (containing the *ori* region) from phage φ50 was introduced into a phage-sensitive strain via a multi-copy plasmid vector. The resulting transformant had a phage-resistant phenotype, and, as predicted, the greater the expression of the plasmid, the more phage resistant was the cell.

Suicide Traps

For a given phage to inhibit a fermentation, that phage must first infect a host cell, replicate inside that cell and make more phage particles, and then lyse the cell to release the fully assembled and infectious phage. If the lytic cycle is somehow blocked prematurely, then viable phage particles will not be formed or released. Although these naturally occurring abortive infection systems (Abi) are widespread in lactic acid bacteria, it is also possible to contrive or engineer such a system by genetic means. This strategy, envisioned by O'Sullivan et al. (1996) and Djordjevic et al. (1997), involved construction of a plasmid that contained lactococcal-derived genes encoding for a restriction/modification system (*Lla*IR) positioned downstream from a promoter region obtained from a lactococcal phage (φ31). When strains harboring this plasmid were attacked by phage, the restriction enzyme encoded by *Lla*IR was induced and expressed, which then lysed host DNA, leading to cell death. Thus, the single infected cell dies, but phage propagation is blocked. Or, as the authors eloquently stated, "infected cells . . . undergo programmed cell death in an altruistic fashion," sparing the remaining population from a more severe phage infection (Djordjevic et al., 1997).

Antisense mRNA

When DNA is transcribed into mRNA, the relevant transcription signals are present on only one of the two strands of DNA, the coding or sense strand. The non-coding or antisense strand is not transcribed. If, however, the transcription elements (consisting of a promoter and terminator regions) are positioned around the antisense strand, the latter will also be transcribed. Assuming enough of the antisense RNA transcripts are produced, they will hybridize with the sense mRNA, preventing translation and protein synthesis. Phage replication can be reduced by targeting genes encoding for essential phage proteins, although in lactococci, this approach has been met with mixed results (Kim et al., 1992; Walker and Klaenhammer, 1998). However, by combining antisense technology with a Per approach, it was possible to obtain "explosive" amplification of the number of antisense *ori* transcripts and achieve a more significant reduction in phage proliferation (McGrath et al., 2001; Walker and Klaenhammer, 2000).

Phages strike back

As noted above, the phage-host relationship is extremely dynamic. There may easily be more than 10^{15} cells in a large cheese plant and just as many viable phages. Even if the frequency of spontaneous mutations within the phage population is low ($< 10^{-9}$), many "new" phages will certainly appear on a regular basis. Some of these mutations will confer an ability to infect a previously resistant cell. However, given the nature of the phage infection, DNA packaging, and cell burst processes, co-mingling of host and phage DNA is also likely to occur via recombination-type events.

Box 5–9. How to Outsmart a Virus *(Continued)*

Thus, lytic phages that have acquired host or prophage DNA will emerge within phage populations. In some cases, it turns out, the acquired DNA provides the phage with the means necessary to counter or circumvent the resistance of the host cell. For example, in one well-known study in which a once resistant *L. lactis* commercial starter culture strain (LMA12-4) harboring a phage resistance plasmid (pTR2030) had become phage-sensitive, it was shown that the phage (φ50)had acquired a methylase gene that was carried by the host cell (Sanders et al., 1986; Alatossava and Klaenhammer, 1991). Since methylases modify host DNA so that it is not degraded by host restriction enzymes, the presence of this methylase gene in phage φ50 enables this phage to self-methylate its DNA and thereby evade restriction by the host cell.

References

Alatossava, T., and T.R. Klaenhammer. 1991. Molecular characterization of three small isometric-headed bacteriophages which vary in their sensitivity to the lactococcal phage resistance plasmid pTR2030. Appl. Environ. Microbiol. 57:1346-1353.

Djordjevic, G.M., D.J. O'Sullivan, S.A. Walker, M.A. Conkling, and T.R. Klaenhammer. 1997. Triggered-suicide system designed as a defense against bacteriophage. J. Bacteriol. 179:6741-6748.

Dupont, K., T. Janzen, F.K. Vogensen, J. Josephsen, and B. Stuer-Lauridsen. 2004. Identification of *Lactococcus lactis* genes required for bacteriophage adsorption. Appl. Environ. Microbiol. 70:5825-5832.

Geller, B.L., R.G. Ivey, J.E. Trempy, and B. Hettinger-Smith. 1993. Cloning of a chromosomal gene required for phage infection of *Lactococcus lactis* subsp. *lactis* C2. J. Bacteriol. 175:5510-5519.

Garbutt, K.C., J. Kraus, and B.L. Geller. 1997. Bacteriophage resistance in *Lactococcus lactis* engineered by replacement of a gene for a bacteriophage receptor. J. Dairy Sci. 80:1512-1519.

Hill, C., L.A. Miller, and T.R Klaenhammer. 1990. Cloning, expression, and sequence determination of a bacteriophage fragment encoding bacteriophage resistance in *Lactococcus lactis*. J. Bacteriol. 172:6419-6426.

Kim, S.G., and C.A. Batt. Antisense mRNA-mediated bacteriophage resistance in *Lactococcus lactis*. Appl. Environ. Microbiol. 57:1109-1113.

O'Sullivan, D.J., S.A. Walker, and T.R Klaenhammer. 1996. Development of an expression system using a lytic phage to trigger explosive plasmid amplification and gene expression. Biotechnology 14:82–87.

McGrath, S., G.F. Fitzgerald, and D. van Sinderen. 2001. Improvement and optimization of two engineered phage resistance mechanisms in *Lactococcus lactis*. Appl. Environ. Microbiol. 67:608–616.

Sanders, M.E., P.J. Leonhard, W.D. Sing, and T.R. Klaenhammer. 1986. Conjugal strategy for construction of fast acid-producing, bacteriophage-resistant lactic streptococci for use in dairy fermentations. Appl. Environ. Microbiol. 52:1001-1007.

Walker, S.A., and T.R Klaenhammer. 1998. Molecular characterization of a phage-inducible middle promoter and its transcriptional activator from the lactococcal bacteriophage φ31. J. Bacteriol. 180:921-931.

Walker, S.A., and T.R Klaenhammer. 2000. An explosive antisense RNA strategy for inhibition of a lactococcal bacteriophage. Appl. Environ. Microbiol. 66:310–319.

Whitehead, H.R., and G.A. Cox. 1935. The occurrence of bacteriophage in starter cultures of lactic streptococci. N. Z. J. Sci. Technol. 16:319-320.

sanitation is practiced. This rationale forms the basis of the culture rotation programs that are widely used in the cheese industry. A given culture, which may contain only one or two, but as many as six different strains, will be used for a day or two, then followed by a different culture containing strains with different phage sensitivity patterns. Thus, phages that infect one or more of the strains present in the initial culture may have begun to become established in the plant, but absent their host cells (which were rotated out), these phages will quickly die out and decrease in number. Rotation programs can be supplemented by various phage monitoring procedures in which phage titers (i.e., phage concentrations) are enumerated or estimated. This allows cheese manufacturers to continue using the same culture (therefore, minimizing product variation) and only make a switch when the phage levels reach a particular threshold.

Microbial Defects, Preservation, and Food Safety

Given that cheese is made in a mostly open, non-sterile environment, using non-sterile raw materials, and is exposed to or held at non-lethal, non-inhibitory temperatures, it is not surprising that all sorts of microorganisms can gain entry in cheese. Although the combined effects of moderately low pH, moderately high salt concentrations, and low Eh provide a barrier against some microorganisms, there are others, including spoilage organisms, and in rare cases, pathogens, that can grow in the cheese environment. Thus, microbial defects are not uncommon in cheese and can be responsible for significant economic loss. Of course, chemical and physical defects (e.g., oxidation, crystal formation, discoloration, mechanical openings) can also be responsible for decreased shelf-life or consumer rejection, however, they generally are less serious.

Microorganisms can cause several general types of spoilage in cheese. These include appearance defects, texture defects, and flavor and aroma defects. Appearance defects are those that may cause the consumer to reject a product based strictly on sight. Perhaps the best example is cheese contaminated with mold growth. Although there are many different fungi capable of growing on cheese, *Penicillium* spp. are the most common. Fungi are obligate aerobes, so growth occurs almost exclusively on the surface, where oxygen is available. However, mold growth on packaged cheese, even vacuum-packaged products, can also occur, provided there is enough oxygen present. Mold growth is more common in block cheese, and even though it can be trimmed prior to cutting and packaging, trimming is time-consuming and results in loss of

yield. Since some fungi that grow on cheese belong to species known to produce mycotoxins, there has been concern that mycotoxins could have diffused in the cheese and that trimming visible mold from the surface may not be sufficient to render the product safe. However, numerous studies have revealed that mycotoxins are not produced on cheese, even when the fungi is theoretically capable of synthesizing toxins. Presumably, the cheese environment restricts or inhibits expression of toxin biosynthesis pathways.

Appearance spoilage also occurs due to formation of undesirable pigments. The most common and best-studied example is the pink defect (or pink ring) that occurs around the exterior of Parmesan-type cheeses. The defect is worse after prolonged aging, and if the cheese is subsequently grated, a pink-to-brown color may occur throughout the grated cheese. Several factors contribute to the formation of the pink color, including oxygen and the presence of reducing sugars and free amino acids (i.e., reactant of the Maillard browning reaction). The starter culture, and the *L. helveticus* strains in particular, appears to be the main culprit, by virtue of its ability to produce oxidized pigments from the amino acid, tyrosine.

Finally holes or slits may form as a result of gas production. Several organisms are capable of producing gas (mainly carbon dioxide) in cheese, including yeast, coliforms, propionibacteria, and clostridia. The clostridia, especially *Clostridium tyrobutyricum*, are more common in milk obtained from silage-fed cows and can produce enough CO_2 to cause formation of large holes (>3 cm in diameter). However, the heterofermentative lactic acid bacteria are probably the most common cause of the slit defect. Relevant species include *Lactobacillus fermentum* and *Lactobacillus curvatus*. These bacteria are either naturally present in the milk or are normal contaminants in a cheese plant; thus, they may be particularly difficult to control, especially in raw milk cheese. Residual lactose or galactose in the cheese provides substrate for these anaerobic bacteria, whose growth is enhanced within the interior of large cheese blocks or barrels where the cheese is slow to cool and where low Eh conditions prevail.

The most common texture defects include slime, crystal or sandy mouth feel, and poor body. Polysaccharides that impart a slimy texture to the cheese can be produced by psychrotrophic Gram negative bacteria such as *Pseudomonas* and *Alcaligenes*, as well as by yeast and lactic acid bacteria. Crystals that form on cheese are usually comprised of calcium lactate and are mainly a physical-chemical problem, caused by low solubility of calcium lactate salts at the cheese surface. However, pediococci, lactobacilli, and other non-starter lactic acid bacteria may contribute to this problem by converting L(+) lactate to its isomer, D(−) lactate, which is less soluble. Tyrosine-containing crystals can also form in dry cheeses (e.g., Parmesan and Romano).

Flavor defects can be caused not only by contaminating organisms, but also by the starter culture itself. If the culture is given the opportunity, it may produce too much lactic acid, causing an acid defect. As described earlier, certain peptides generated by starter culture proteinases are bitter-tasting, and unless further metabolized, cause a bitter defect. Heterofermentative, non-starter lactic acid bacteria produce ethanol and acetic acid in addition to CO_2. Acetic acid imparts a highly objectionable vinegar-like sour flavor to the cheese. These bacteria may also hydrolyze triglycerides, releasing free fatty acids that contribute fruity flavors into the cheese. Production of hydrogen sulfide and other sulfur-containing compounds, ammonia, and other volatiles by bacteria and fungi, are also common flavor defects.

Minimizing the entry of spoilage organisms into the cheese milk and subsequently into the cheese is one of the most important steps in preventing spoilage. In most cases, the cheese will only be as good as the milk from which it was made. The use of high quality milk, application of good sanitation practices, and proper cheese manufacture are essential. Pasteurization of milk, even for aged cheese, is now common, not only to inactivate pathogens (see below), but also to kill potential spoilage or-

ganisms. Some specific spoilage problems can be addressed by the use of chemical preservatives, either added to the milk or applied to the cheese. For example, mold growth can be inhibited by sorbic acid (usually in the form of potassium sorbate salts) or the antibiotic natamycin (also known as pimaricin). In some European countries, but not in the United States, lysozyme and sodium nitrate can be added to milk to control clostridia.

Although spoilage of cheese by microorganisms is a constant concern, the presence of pathogens in cheese occurs only infrequently, and rarely does foodborne disease result (Box 5-4). However, due to the serious public health consequences, as well as potential economic loss due to recalls and liability, attention to food safety is an absolute requirement. Just a single positive test for *L. monocytogenes* or *E. coli* O157:H7 may initiate a product recall. Thus, the goal is not just to prevent the growth of pathogens, but also to reduce or eliminate their very presence in cheese. In the United States, this means more and more cheese is being made from pasteurized milk, and exposure to environmental sources of contamination is minimized (e.g., via the use of enclosed vats). Most cheese manufactures have also adopted Hazard Analysis Critical Control Points (HACCP) plans that are designed to anticipate and prevent food safety problems.

Another type of foodborne disease that occasionally occurs in cheese is caused by the presence of the biogenic amines histamine and tyramine, but is unrelated to foodborne pathogens. In sensitive individuals, these amines (mainly histamine) can cause headaches, nausea, and cramps and other gastrointestinal symptoms. These compounds are produced from the amino acids histidine and tyrosine, respectively, via specific amino acid decarboxylases. Bacteria that produce these enzymes and form these amines include *Lactobacillus buchneri* and other lactobacilli, lactococci, and enterococci. Aged cheese, particularly aged Cheddar, Gouda, and Swiss, are more likely to contain biogenic amines due to the higher concentration of free amino acid substrates that accumulate during more extensive proteolysis. Because the bacteria responsible for amine production are ordinarily present as raw milk contaminants, pasteurization effectively minimizes this problem.

Whey Utilization

As shown earlier in this chapter (Figure 5-4), water, in the form of whey, is released when milk is converted into cheese. In fact, the whey accounts for 90% of the original milk volume. The dilute nature of the whey (about 92% to 94% water) and low protein concentration (<1%) have historically contributed to the perception (at least in the United States) that whey has little economic value. However, in other parts of the world, whey is more widely used, especially in the manufacture of whey-derived cheese. For example, in Norway, the goat whey cheeses Gjetost and Mysost are among the most popular of all cheeses (accounting for about 20% of cheese consumption), and Ricotta, another whey-based cheese, is popular not only in Italy, but throughout Europe and the United States. Whey cheeses are also produced in Greece (Manouri, Anthotyros), Portugal (Requejño), and the Czech Republic (Urda).

Whey cheese manufacture is based on the principle that whey proteins will precipitate when slightly acidified (pH 6) and then heated to high temperatures (>70°C). Although whey alone is ordinarily used as the starting material, milk can also be added. Much of the Ricotta produced in the United States, for example, contains a portion of skim or whole milk. The finished cheese, in this case actually consists of a casein-whey co-precipitate. Whey cheeses are not fermented, and since the lactose content of whey may be as high as 75% (on a solids basis), these cheeses will also contain 3% to 4% lactose and are generally sweet. Although Ricotta is a very soft cheese with a white, creamy color, other whey cheeses, such as Gjestot, can be very hard and a light brown color. The tan-to-brown color results from lactose caramelization and Maillard browning reactions that occur during high temperature heating.

In the United States, where the market for whey cheeses has been extremely small, most cheese manufacturers instead used whey as animal feed or fertilizer or, more likely, simply discarded it altogether. In the past thirty years this practice has changed for several reasons. First, whey is now considered to be environmental problem. It has a high biological oxygen demand (BOD), and, given that nearly 25 million kg are generated each year, discharging this much high BOD material down the drain or into the environment exceeds that which sewage treatment plants can handle.

Second, the dairy and food processing industries have developed products and applications in which economic value can be derived from whey and whey components. Many of these applications involve separation technologies (e.g., ultrafiltration and reverse osmosis) in which the more valuable whey protein fractions are obtained. What remains from these processes is a lactose-rich material (whey permeate) that can be used directly, or further purified, as food- or chemical grade lactose.

Importantly, it is technologically possible to use whey, whey permeate, or whey-derived lactose as feedstocks for various industrial fermentations, including the production of organic acids, ethanol, and other small molecules. However, given their current market value and the capital and processing costs that would be necessary, it does not make economic sense at the present time to produce these commodity chemicals from whey fermentations.

Bibliography

Coffey, A., and R.P. Ross. 2002. Bacteriophage-resistant systems in dairy starter strains: molecular analysis to application. Antonie van Leeuwenhoek. 82:303–321.

Fox, P.F., and P.L.H. McSweeney. 2004. Cheese: an overview, p. 1—18. *In* P.F. Fox, P.L.H. McSweeney, T.M. Cogan, and T.P. Guinee (ed.). *Cheese: Chemistry, Physics and Microbiology, Third Edition, Volume 1: General Aspects*. Elsevier Ltd., London.

Guinee, T.P., and P.F. Fox, 2004. Salt in cheese: physical, chemical and biological aspects, p. 205—259. *In* P.F. Fox, P.L.H. McSweeney, T.M. Cogan, and T.P.

Guinee (ed.). *Cheese: Chemistry, Physics and Microbiology, Third Edition, Volume 1: General Aspects*. Elsevier Ltd., London.

Hutkins, R.W. 2001. Metabolism of starter cultures, p. 207—241. *In* E.H. Marth and J.L. Steele (ed.), *Applied Dairy Microbiology*. Marcel Dekker, Inc., New York, NY.

Kosikowski, F.V., and V.V. Mistry. 1997. *Cheese and Fermented Milk Foods. Volume 1: origins and principles. 3rd ed.* F.V. Kosikowski, Great Falls, Virginia.

Kosikowski, F.V., and V.V. Mistry. 1997. *Cheese and Fermented Milk Foods. Volume 2: Procedures and analysis. 3rd ed.* F.V. Kosikowski, Great Falls, Virginia.

Lawrence, R.C., J. Gilles, L.K. Creamer, V.L. Crow, H.A. Heap, C.G. Honoré, K.A. Lawrence, and P.K. Samal. 2004. Cheddar cheese and related dry-salted cheese varieties, p. 71-102. *In* P.F. Fox, P.L.H. McSweeney, T.M. Cogan, and T.P. Guinee (ed.). *Cheese: Chemistry, Physics and Microbiology, Third Edition, Volume 2: General Aspects*. Elsevier Ltd., London.

Lucey, J.A., M.E. Johnson, and D.S. Horne. 2003. Perspectives on the basis of the rheology and texture properties of cheese. J. Dairy Sci. 86: 2725-2743.

Marcos, A., M. Alcala, F. Leon, J. Fernandez-Salguero, and M.A. Estaban. 1981. Water activity and chemical composition of cheese. J. Dairy Sci. 64: 622-626.

McSweeney, P.L.H. 2004. Biochemistry of cheese ripening. Int. J. Dairy Technol. 57:127-144.

McSweeney, P.L.H. 2004. Biochemistry of cheese ripening: Introduction and Overview, 347—360. *In* P.F. Fox, P.L.H. McSweeney, T.M. Cogan, and T.P. Guinee (ed.). *Cheese: Chemistry, Physics and Microbiology, Third Edition, Volume 1: General Aspects*. Elsevier Ltd., London.

Pintado, M.E., A.C. Macedo, and F.X. Malcata. 2001. Review: technology, chemistry and microbiology of whey cheeses. Food Sci. Tech. Int. 7:105-116.

Singh, T.K., M.A. Drake, and K.R. Cadwallader. 2003. Flavor of cheese: a chemical and sensory perspective. Comp. Rev. Food Sci. Food Safety. 2: 139-162.

Footnotes

[1]"How can you govern a country which has 246 varieties of cheese?" Charles De Gaulle, as quoted in *Les Mots du General*, 1962.

6

Meat Fermentation

"Stomach of goat, crushed
sheep balls, soft full
pearls of pig eyes,
snout gristle, fresh earth,
worn iron of trotter, slate
of Zaragoza, dried cat heart,
cock claws. She grinds
them with one hand and
with other fists
mountain thyme, basil,
paprika, and knobs of garlic.
And if a tooth of stink thistle
pulls blood from the round
blue marbled hand
all the better for
this ruby of Pamplona,
this bright jewel of Vich,
this stained crown
of Solsona, this
salami."
  From "Salami" *by Philip Levine, poet*

Introduction

If the origins of fermentation and the making and eating of fermented foods was based on food preservation, then meat serves as one of the best examples of the application of this technology. When, many thousands of years ago, wild or domesticated animals were slaughtered for food, unless some sort of low temperature storage was available, either the entire animal had to be cooked and consumed within a relatively short amount of time or the meat would spoil. Even if these early consumers were less discerning than the more sophisticated consumers of today, rotten meat still must have had limited appeal. Of course, it must have been far more troubling to eat meat that lacked obvious signs of spoilage only to suffer hours or days later from a serious illness of one form or another. Comminuted, or chopped or ground, meat products must have been even more prone to spoilage.

It probably did not take long, therefore, for some aspiring food scientists to develop ways to preserve meat. Drying was undoubtedly among the first of these technologies proven to be successful. It was probably observed that smoking and salting were also effective. If, however, a small amount of sugar was added to a meat mixture, and if the salt level was not too high and other conditions were just right, then an entirely different type of product could be produced, one that had a tart but appealing flavor and lasted a long time, especially if combined with drying. Moreover, if the salt also contained sodium or potassium nitrate, a highly desirable cured meat color and flavor was also achieved.

History and Evolution of the Fermented Meats Industry

Like other fermented foods, recorded references to fermented meats date back thousands of years. The manufacture of these products likely originated in southern Europe and areas surrounding the Mediterranean Sea during the Roman era, although there were probably Asian counterparts that appeared around the same time. Even though it is not absolutely clear from the historical records whether these early sausage products were actually fermented, it is difficult to imagine, given the circumstances, that there wasn't some sort of natural fermentation occurring.

The manufacture of these early fermented meat products most certainly occurred as a result of accidents, and repeating these successes on a consistent basis must have been very difficult. We know that the key to produce these meat products via a natural fermentation relies on creating conditions that select for the proper organisms (discussed below). However, the technique of backslopping—taking a portion of a good finished product and adding it to the starting raw material—was likely adopted to increase the probability of success, despite the fact that the rationale behind this step was not understood. This technology is still practiced throughout the world, even though, as we shall see later, there are now far more reliable methods for ensuring a consistent fermentation. In any event, the technology of fermented meats production remained an art for many years. In fact, sausage manufacturing was, like the manufacture of other fermented foods, performed by craftsmen, monks, and other trained specialists.

The actual number and types of products that evolved were many, and like other fermented foods, depended largely on geography. Certainly, preservation was the likely driving force for many of the processing practices that were adopted. In warmer areas, such as those in the Middle East and around the Mediterranean, spices were often added, and a drying step was common. Thus, dry, peppery, and spicy sausages, such as Genoa salami and pepperoni, evolved in Italy. In colder, more Northern areas, where sausage technology is more recent, spices were rarely added, and instead products were usually smoked or sometimes cooked following fermentation. Most of the moist and semi-dry German-style sausages, such as Thuringer, Lebanon bologna, and cervelat, are of this variety.

The drying, cooking, and smoking steps, the addition of salt and spices, as well as the incorporation of nitrate salts, not only added flavor and appeal, but also would have given these products a long shelf-life, even at ambient temperature. That these products are enclosed within casings also contributes to their preservation, since post-processing microbial contaminants are excluded. In fact, the preservation of fermented meats serves as a perfect example of what food scientists now refer to as the hurdle or barrier concept of food preservation. Preservation is achieved not by any one specific process, but rather by a combination of processes, such that fermentation is combined with drying, salting, smoking, cooking, and antimicrobial chemicals to ensure safety and to enhance preservation (discussed later in more detail).

The fermented meats industry, compared to other fermented foods industries, is relatively new. In fact, it was not until well into the twentieth century that producers of fermented meats began to develop and apply modern production technologies. Previously, sausage manufacture was mostly confined to craftsmen and at-home producers who made products on a batch-by-batch basis. Product quality, consistency, and especially safety were not always achieved. Equipment for grinding, mixing, and stuffing procedures, as well as fermentation chambers, only became available in the past sixty to seventy years. Moreover, pure curing agents, synthetic casings, and starter cultures have only been available since the 1960s. It is no coincidence that it was during this period that the industry increased in size and that rapid, high throughput manufacturing methods capable of producing products of consistent quality and safety developed.

Today the variety of fermented meat products available around the world is nearly equal to that of cheese. In Spain, for example, there are at least fifty different types of fermented sausages, and in Germany there are more than 350. Not only is this variation due to the meat source (i.e., beef, pork, goat, sheep, etc.), but the cut of meat, the amount and coarseness of the fat, and the casing material used to form the shape all have a profound influence on the finished product. The level of dryness and whether or not smoking is applied or mold growth permitted to occur are especially important factors and form the basis of fermented meats classification (Table 6-1).

It is worth emphasizing, especially for readers less familiar with sausage nomenclature, that many of the sausage products available in the marketplace are simply cured, comminuted products, meaning they contain curing salts and undergo many of the same processing steps as do fermented sausages, but they are not fermented. Thus, frankfurters, bologna, and breakfast sausages are not fermented and will not be discussed in this chapter. In contrast, there are fermented meat products that are made from whole, intact meat materials, such as the country hams popular in selected regions

Table 6.1. Examples of fermented meats and sausages.

Product	Meat	Areas Produced	Features
Moist sausages $(a_w > 0.94)$			
Lebanon bologna	Beef	North America	Low pH (<4.9) Smoked
Mortadella	Pork	Italy, France, USA	Low pH Un-smoked Cooked
Semi-dry sausages $(a_w = 0.90 - 0.95)$			
Summer sausage	Beef/Pork	USA	Moderate pH ($4.9 - 5.0$) Smoked
Thuringer cervelat	Pork	Italy, France, USA	Low pH Un-smoked Cooked
Dry sausages $(a_w < 0.90)$			
Pepperoni	Beef/Pork	Italy, North America	Low pH Un-smoked Spiced
Salami	Pork	Europe, USA, Mexico	Low a_w (<0.85) Un-smoked Cooked
Chorizo	Pork/Beef	Italy	Highly spiced Low acid
Whole meats			
Country-cured hams	Pork	North America	Heavily salted
Parma hams	Pork	Italy	Heavily salted

in the United States and Europe. These products will be discussed separately.

Although fermented meats are produced and consumed worldwide, they are most popular in Europe (Table 6–2). More than 600 million kg of fermented sausage are consumed in Germany, Spain, France, and Italy. About 3% to 5% of all meat consumed in these countries is in the form of fermented sausages.

Meat Composition

Fresh meat is a nutrient-rich medium and is, in fact, one of the best for supporting growth of microorganisms. Skeletal bovine muscle contains nearly 20% of high quality protein, about 2% to 3% lipid, and a small amount of carbohydrate, non-protein nitrogen, and inorganic material. The balance, about 75%, is water, so the water activity (a_w), not surprisingly, is nearly 0.99. The pH of the fresh tissue, before rigor, is 6.8 to 7.0, but decreases to about 5.6 to 5.8 following rigor, due to post-mortem glycolysis by endogenous enzymes present within muscle cells.

These values may vary somewhat, depending on the animal and its status before slaughter, but regardless of the source of meat (e.g., pork usually has a slightly higher pH), these conditions are just about perfect for most bacteria, including spoilage organisms and pathogens. Consequently, raw meat is highly perishable. And although the interior of intact meat tissue is sterile, the exterior surface harbors a wide array of aerobic and facultative bacteria. In its comminuted form, i.e., chopped or ground, meat

spoilage can be even more rapid, because the surface microflora become homogeneously distributed throughout the meat. And even though oxygen may also be incorporated into the mixture, enhancing growth of aerobes, anaerobic pockets can also develop within the interior, providing suitable environments for anaerobic bacteria. Included among the latter are *Clostridium botulinum* and other related species capable of causing serious, sometimes fatal food poisoning disease. Since fermented sausages are typically made using raw, comminuted meat, controlling the resident flora, especially pathogens, is of utmost importance. If control is not maintained, the results can be disastrous.

Fermentation Principles

In contrast to the lactic acid fermentation that occurs in milk, the meat fermentation has been, until recently, considerably less well studied and understood. In fact, the use of pure, defined starter cultures in the fermented meats industry is a relatively recent development (begun only in the 1950s and '60s). Before the use of meat starter cultures, the most common way to start the fermentation practice was, as noted above, backslopping.

Backslopping works for several reasons. First, backslopping ordinarily selects for those bacteria that are well suited for growth in the sausage environment. Strains that are slow to scavenge for carbohydrates, are inhibited by fermentation acids, or are sensitive to salt or nitrite are not maintained. Instead, they are displaced by more competitive bacteria that have particular metabolic and physiological advantages in that environment. Thus, even a prolific acid-forming strain does have much of a chance if it is not also tolerant of salt and nitrite and able to grow in a low oxygen environment.

Second, the bacterial population that becomes established during repeated transfers and fermentations is heterogenous in nature, consisting of multiple species and strains. If, for example, one such strain were to suddenly die or otherwise be lost due to the presence of bacteriophages or an inhibitory agent, then the

Table 6.2. Consumption of fermented meats.

Country	Kg/person/year
Canada[1]	0.9
Finland[1]	3.1
France[2]	3.1
Germany[2]	7.1
Italy[2]	4.5
Spain[2]	2.7
United Kingdom[2]	0.1
United States[1]	0.3

[1]Estimated from 2001 WHO statistics
[2]Estimated from Fisher and Palmer, 1995

remaining strains, acting as back-up, would still be able to complete the fermentation.

Finally, backslopping is effective due simply to the size of the inoculum, which is usually around 5%, but which can be as high as 20% of the total mass. Such a large inoculum provides reasonable assurance that the desired organisms will overwhelm the background flora and that undesirable interlopers will have little chance of competing.

Despite backslopping's long history and wide use, there are several drawbacks associated with it. Such products often have inconsistent quality and fermentations can be unreliable and difficult to control. In large production facilities in Europe, and especially in the United States, short and consistent fermentation times, standardized production schedules, and consistent product quality have become essential. A manufacturer producing a single, small batch of product can tolerate delays, and may even come to appreciate differences in product quality. However, inconsistent product quality and production delays due to sluggish fermentations are unacceptable for large manufacturers who have substantial employee payrolls and tight production schedules. Above all, the entire backslopping process can be considered microbiologically risky, since any deviation from the norm (i.e., slow or delayed fermentation) may permit growth of *Staphylococcus aureus*, *Listeria monocytogenes*, *C. botulinum*, or other pathogens of public health significance.

Although obvious, the following point cannot be over-emphasized: sausage is made from raw meat that may well contain pathogenic organisms. If a cooking step is not included, fermentation represents the primary means of preservation and the main barrier against pathogens. Since the actual fermentation can take a long time (from twelve to thirty-six hours and in some cases, even longer), a slow or failed fermentation may not be discovered right away, permitting growth of pathogenic bacteria. If the bacteria are able to reach high levels, even subsequent acid production (or even a heating step) may not be sufficient to inactivate these pathogens.

It is important to mention, as noted above, that it is entirely possible to produce fermented sausages without any type of culture at all. This practice is not uncommon, and is still practiced in many parts of Europe. One can simply prepare the sausage mixture and wait for the natural lactic flora to take over. This method is successful because the formulation and fermentation conditions (i.e., salt, nitrate, low temperature, and an anaerobic environment) provide sufficient selection of desirable lactic acid bacteria. However, for obvious food safety reasons, none of the large, modern operations in the United States rely on natural fermentations. Finally, there is one other way to produce sausages that have a tangy or acid-like flavor without a starter culture and without relying on the indigenous microflora. The meat mixture can be directly acidified using a food-grade acidulant (e.g., glucono-δ-lactone), which results in a product with sensory properties that mimic (somewhat) those of fermented sausage.

Meat Starter Cultures

Once microbiologists began to study and identify the microorganisms present in fermented sausages in the 1940s, it became clear that lactic acid bacteria were the primary organisms responsible for the fermentation. This conclusion was based on the fact that the predominant organisms isolated from naturally fermented sausages were species of *Lactobacillus*. When the isolates were propagated and re-inoculated into fresh meat, a well-fermented sausage could be produced with all the expected characteristics.

Patents, based on using these bacteria as meat starter cultures, were assigned in Europe and the United States. However, application of this technology was initially unsuccessful. This was because the *Lactobacillus* strains that had been identified and used successfully in trial situations were difficult to mass produce in a form convenient for sausage manufacturers. For any organism to function well as a starter culture, it must not only satisfy the performance criteria, but it must also be present in high numbers and be viable at the time of use. In the case of the *Lactobacillus* cultures, cell viability following lyophilization (or freeze drying, the main form of starter culture preservation) was poor, leading to slow and unacceptable fermentation rates.

Demand for a culture that could be used for fast, consistent, large-scale production of fermented sausage eventually led to the discovery of other lactic acid bacteria that not only had the relevant performance characteristics, but also the durability required for commercial applications. One organism, classified as *Pediococcus cerevisae*, was found to have these properties. Even though pediococci are not normally found in fermented sausage (they are, however, involved in vegetable fermentations), this organism was introduced in the United States in the late 1950s as the first meat starter culture. Strains of this species, which were later re-classified as either *Pediococcus acidilactici* or *Pediococcus pentosaceus,* are still widely used today.

In addition, improvements in starter culture technology and the development of frozen concentrated cultures (see Chapter 3) enhanced culture viability such that *Lactobacillus* starter cultures are also now available. Initially, the *Lactobacillus* strains that were isolated from fermented sausage and used as starter cultures were classified as *Lactobacillus plantarum*. Other strains of *Lactobacillus* that had different physiological properties and that performed well in sausage manufacture were subsequently isolated. Thus, the closely related species, *Lactobacillus sake* and *Lactobacillus curvatus,* are also used in starter culture preparations.

Although the different *Pediococcus* and *Lactobacillus* strains used as starter cultures perform the same basic role—to ferment sugars and produce organic acid—they vary with respect to several important physiological and biochemical properties (Table 6-3). These differences influence how they are used as starter cultures. First, different species have different temperature optima and different thermal tol-

erances. For example, *L. sake* and *L. curvatus* are considered psychrotrophic, meaning they are capable of growth, albeit slowly, at temperatures as low as 4°C. Thus, they are suitable for fermentations conducted at cool ambient temperatures.

As discussed in Box 6-1, many of the European sausages are fermented at low temperatures (20°C) to control the fermentation rate and to provide sufficient time for nitrate-reduction and subsequent color and flavor development. In contrast, *P. acidilactici* has a growth optimum near 40°C and is preferred by manufacturers interested in fast, high temperature fermentations. Metabolic differences also exist between these organisms. Whereas all pediococci and lactobacilli ferment glucose, other sugars are fermented only by specific strains. Pediococci, for example, do not ferment lactose. Sausage formulations must, therefore, account for the metabolic capacity of the culture.

There are occasions when starter culture metabolism can lead to problems and defects in the finished product. All pediococci are considered to be homofermentative, producing only lactic acid from glucose. However, under some circumstances, such as during sugar limitation, heterofermentative products such as ethanol, carbon dioxide, and acetic acid can be formed. Similarly, small amounts of these end products can also be produced by homofermentative *L. plantarum* during sugar limitation or aerobic growth. Of the heterofermentative end products that are formed, acetic acid is especially undesirable in fermented sausage because it imparts a sour, vinegar-like flavor rather than the tart, tangy flavor contributed by lactic acid.

The ability of some strains to produce peroxides also is a serious problem. Depending on the starter culture strain and the level of oxygen in the environment, hydrogen peroxide can be formed directly in the fermenting sausage. Hydrogen peroxide can react with heme proteins in the muscle tissue to form undesirable, green pigments. In addition, hydrogen peroxide and peroxide radicals promote lipid oxidation, a serious flavor defect.

Table 6.3. Properties of bacteria used in meat start cultures[1].

Organism	Minimum Temperature	Temperature Optimum	Acid from Glucose[2]	Nitrate Reductase	Primary Function
Lactobacillus sakei	4°C	32°C — 35°C	+	−	acid
Lactobacillus curvatus	4°C	32°C — 35°C	+	−	acid
Lactobacillus plantarum	10°C	42°C	+	−[3]	acid
Pediococcus pentosaceus	15°C	28°C — 32°C	+	−	acid
Pediococcus acidilactici	15°C	40°C	+	−	acid
Kocuria varians[4]	10°C	25°C — 37°C	−	+	flavor, aroma
Staphylococcus carnosus	10°C	30°C — 40°C	−	+	flavor, aroma
Staphylococcus xylosus	10°C	25°C — 35°C	−	+	flavor, aroma

[1]Adapted from *Bergey's Manual of Systematic Bacteriology, Volume 2, 8th Ed.*
[2]Under anaerobic conditions
[3]Some strains reduce nitrate under low glucose, high pH conditions
[4]Formerly *Micrococcus varians*

Box 6–1. Micrococcaceae, Nitrate, and "Old World" Sausage

In the United States, most fermented sausage starter cultures contain only lactic acid bacteria. The fermentation temperatures are high and fermentation times are short. In contrast, for many European products (and a few in the United States), the microbial flora contain species of *Staphylococcus, Micrococcus*, and *Kocuria* in the Family Micrococcaceae (Hammes and Hertel, 1998; Tang and Gillevet, 2003). In Europe, fermentation temperatures are low and fermentation times are long.

Not coincidentally, nitrite salts are, by far, the most common curing agents used in fermented sausage manufacture in the United States, whereas in Europe, nitrate salts are more common. These differences reflect not only the different manufacturing requirements, but also the desired quality attributes relative to "new world" and "old world" sausage-making technologies and styles.

To achieve the expected color, flavor, and antimicrobial property peculiar to all cured meats (whether fermented or not), either nitrate or nitrite must be added as a curing agent. However, it is the nitrite form, and not the nitrate, that actually reacts with the meat pigments and provides the curing effect. If nitrate is used as the curing agent, it must first be converted to nitrite (Figure 1). This conversion is simply a reduction reaction catalyzed by the enzyme nitrate reductase.

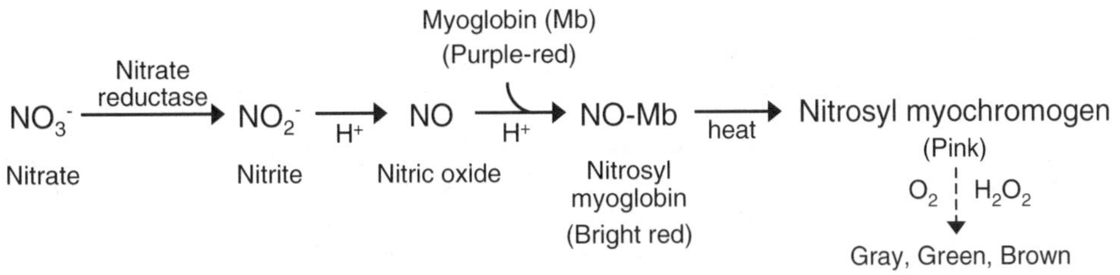

Figure 1. Nitrate and nitrite reactions in meat. Nitrate is first reduced to nitrite by nitrate reductase, an enzyme produced by strains of *Micrococcus* and *Staphylococcus*. Nitrite is further reduced to nitric acid, which then reacts with myoglobin at low pH to form nitrosyl myoglobin. When heated, nitrosyl myochromogen (also called nitrosyl hemochromogen) is formed, giving a pink color. The latter can be oxidized to form an undesirable gray, green, or brown appearance.

(Continued)

Box 6–1. Micrococcaceae, Nitrate, and "Old World" Sausage *(Continued)*

In general, bacteria that produce nitrate reductase are respiring aerobes that use nitrate as an electron acceptor during growth under anaerobic or low oxygen conditions. Among the organisms having high nitrate reductase activity are species of *Staphylococcus, Micrococcus,* and *Kocuria.* Thus, if the sausage formulation contains nitrate, then its conversion to nitrite will depend on the presence of nitrate reductase-producing strains of these bacteria, either those naturally present in the raw meat or added in the form of a starter culture. In the latter case, it might seem rather odd to add staphylococci to food, since some coagulase-positive staphylococci (e.g., *Staphylococcus aureus*) are pathogenic and produce exo-toxins. However, the strains of *Staphylococcus carnosus* and *Staphylococcus xylosus* used for sausage fermentations are coagulase-negative and non-toxigenic. Still, despite this caveat, the use of these strains in sausage fermentations is not acceptable in some areas. This also points out why it is important to accurately identify strains of *Staphylococcus* that might end up in a starter culture (Morot-Bizot et al., 2004).

If nitrate must be converted to nitrite to produce cured sausage, then the obvious question to ask is why add nitrate in the first place? Why depend on the nitrate-reducing bacteria to perform this important function? In other words, why wouldn't all sausage manufacturers simply add nitrite as the curing agent, as is widely done in the United States? To answer these questions, one must consider what other functions the micrococci and other nitrate-reducing bacteria might perform in fermented sausage, other than to reduce nitrate. It turns out there are several.

The actual nitrate-to-nitrite reaction occurs rather slowly during the sausage fermentation because the prevailing conditions (low temperature and low pH) are less than optimal for nitrate reductase activity. It can take several weeks for the conversion of nitrate to nitrite. This affords the micrococci, kocuriae, and staphylococci the opportunity to secrete lipases and other enzymes that ultimately generate flavor precursors in the meat. If, instead of ripening the sausage at 20°C, it was fermented at 38°C to 40°C, the lactic acid bacteria would produce acid too quickly, and the non-lactic acid bacteria would be inhibited and unable to reduce nitrate or otherwise influence the properties of the finished product. It is also argued that slow conversion of nitrate to nitrite enhances color development (although why this should be the case is not understood).

Among the flavor compounds produced by staphylococci, micrococci and kocuriae are metabolic end products resulting from protein and fatty acid metabolism. Most strains are lipolytic and many are also proteoyltic. In addition, amino acid metabolism may also generate flavor and aroma compounds. For example, it was recently reported that metabolism of leucine by *Staphylococcus xylosus* resulted in formation of various metabolites, such as 3-methylbutanol and 2-methylpropanol, that contribute to the flavor properties of fermented meats (Beck et al., 2004).

References

Beck, H.C., A.M. Hansen, and F.R. Lauritsen. 2004. Catabolism of leucine to branched-chain fatty acids in *Staphylococcus xylosus.* J. Appl. Microbiol. 96:1185–1193.

Hammes, W.P., and C. Hertel. 1998. New developments in meat starter cultures. Meat Sci. 49 (Supple. 1): S125–S138.

Morot-Bizot, S.C., R. Talon, and S. Leroy. 2004. Development of a multiplex PCR for the identification of *Staphylococcus* genus and four staphylococcal species isolated from food. J. Appl. Microbiol. 97:10871094.

Tang, J.S., and P.M. Gillevet. 2003. Reclassification of ATCC 9341 from *Micrococcus luteus* to *Kocuria rhizophila.* Int. J. Syst. Evol. Microbiol. 53:995–997.

Although some strains may also produce cata-lase or catalase-like enzymes that degrade hy-drogen peroxide, the production of these en-zymes cannot always be counted on to inactivate accumulated peroxides.

Protective Properties of Cultures

It has long been suggested that lactic acid starter cultures were responsible for preserva-tion effects beyond their acidification and pH-lowering effects. During growth in meat, they scavenge sugars and other nutrients faster than competitors, and they lower the oxidation-reduction potential (Eh) of the environment such that growth of aerobic organisms is inhib-ited. In the 1980s it was discovered that some of the strains used as meat starter cultures may provide additional preservation effects via pro-duction of antagonistic agents that inhibited other bacteria. Although some strains could produce inhibitory levels of hydrogen perox-

ide, experimental evidence indicated that a substance other than hydrogen peroxide was responsible for the inhibitory effects. It was subsequently discovered that these bacteria were capable of producing a class of sub-stances known as bacteriocins.

Bacteriocins are proteinaceous substances with bactericidal activity, usually against bacte-ria that are closely related to the producer or-ganism. Among the bacteria used as starter cul-tures for fermented sausage, several species have been shown to produce bacteriocins, in-cluding strains of *P. acidilactici, L. plantarum,* and *L. sakei.* Moreover, several of these bacte-riocins—in particular the plantaricins and sakacins—have a somewhat broad spectrum of activity, inhibiting *L. monocytogenes, S. aureus, Clostridium* sp., and other Gram positive spoilage organisms. Greater application of these so-called bioactive or protective cultures is likely to increase (Box 6–2), perhaps leading to their use in traditionally non-fermented

Box 6–2. Improving the Safety and Preservation of Fermented Meats Using Bioprotective Cultures

It has long been recognized that lactic acid bacteria have inhibitory activity against other bacteria, above and beyond that due to the lactic acid they produce. Some lactic acid bacteria, for example, also produce acetic acid, hydrogen peroxide, and diacetyl, all of which can inhibit potential spoilage or pathogenic organisms. However, the antimicrobial compounds that have attracted the most attention are the bacteriocins. These agents are now being used, in a variety of delivery for-mats, in a wide range of foods, including fermented meats. Although fermented meats, by virtue of their intrinsic properties (organic acids, low pH, low a_w, smoke, spices), are not prone to microbial spoilage or infrequently serve as vehicles for food borne pathogens, the addition of a bacteriocin to these products provides manufacturers with one more barrier and an extra margin of safety.

Bacteriocins are defined as proteinaceous compounds produced by bacteria that inhibit other closely related bacteria. They are distinguished from antibiotics in that bacteriocins are ribosomally-synthesized (rather than via *de novo* synthesis from precursor molecules), they have a more nar-row antimicrobial spectrum of activity, and their mode of action is mediated primarily via distur-bances of the cytoplasmic membrane. Although the bacteriocins relevant to this discussion are produced by lactic acid bacteria, nearly every bacterial genus reportedly contains species capable of producing bacteriocins. In fact, so many bacteriocins have been described in the literature (a PubMed search on the term bacteriocin led to more than 2,762 published article hits since 1980, with more than 700 published since 2000), their classification has become a serious challenge.

In 1993, a four-class system, based on genetic, chemical, and physiological properties of lactic acid bacteria-produced bacteriocins, was described (Klaenhammer, 1993). Class I bac-teriocins are referred to as lantibiotics, because they contain the amino acid lanthionine. Nisin, the most well-studied bacteriocin, belongs to this class. Class II bacteriocins consist of several small *Listeria*-active non-lanthionine peptides that are heat stable, making them good

(Continued)

Box 6–2. Improving the Safety and Preservation of Fermented Meats Using Bioprotective Cultures *(Continued)*

antimicrobials for cooked sausage products. The large, heat-sensitive bacteriocins belong to Class III, and the complex, lipid- or carbohydrate-containing bacteriocins belong to Class IV (although whether this class actually exists is questionable). Most of the bacteriocins used in food applications belong to Class I or II.

Table 1.   Application of bacteriocins in fermented sausages.

Product	Producer(s)	Target(s)	Reference
Dry sausage	*Pediococcus acidilactici* PAC 1.0	*Listeria*	Foegeding et al., 1992
Salami	*Lactobacillus plantarum* MCS1	*Listeria*	Campanini et al., 1993
Dry sausage	*Lactobacillus sakei* CTC 494	*Listeria*	Hugas et al., 1995
Fermented sausage	*Lactobacillus sakei* CTC 494 *Lactobacillus curvatus* LTH 1174	*Listeria*	Hugas et al., 1996
Fermented sausage	*Lactobacillus sakei* LB 706 *Lactobacillus curvatus* LTH 1174	*Listeria*	Hugas et al., 1996
Spanish-style dry sausage	*Enterococcus faecium* CCM 4231 *Enterococcus faecium* RZS C13	*Listeria*	Callewaert et al., 2000
Dry sausage	*Staphylococcus xylosus* DD-34 *Pediococcus acidilactici* PA-2	*E. coli* *Listeria*	Lahti et al., 2001
Dry sausage	*Lactobacillus bavaricus* MI-401 *Lactobacillus curvatus* MIII	*E. coli* *Listeria*	Lahti et al., 2001
Merguez sausage	*Lactococcus lactis* M	*Listeria*	Benkerroum et al., 2003
Cacciatore sausage	*Enterococcus casseliflavus* IM 416K1	*Listeria*	Sabia et al., 2003
North European-type dry sausage	*Lactobacillus rhamnosus* LC-705 *Lactobacillus rhamnosus* E-97800	*Listeria*	Työppönen et al., 2003
Dry sausage	*Lactobacillus plantarum*	*Listeria*	Työppönen et al., 2003

There are two main reasons why bacteriocins are used in meat products: (1) to prevent growth of spoilage organisms and the production of spoilage end products, and (2) to inhibit or kill pathogens. Lactic acid bacteria, enterococci, *Brochothrix*, and clostridia are among the spoilage organisms inhibited by specific bacteriocins. However, control of pathogens and *Listeria monocytogenes*, in particular, is the main motivation for the use of bacteriocins in fermented meat products (*E. coli* O157:H7 is Gram negative and is largely unaffected by bacteriocins). The United States has a zero-tolerance policy for this organism in these products, so any measure that would reduce the frequency of a positive test result for *Listeria* would be very useful. It should also be noted that these same concerns about *Listeria* and other pathogens have led manufacturers of non-fermented, ready-to-eat meat products (e.g., hot dogs, luncheon meats) to consider the use of bacteriocins and bacteriocin-producing cultures, but under conditions in which fermentation does not occur or is minimized (Vermeiren et al., 2004).

Many of the bacteriocins produced by lactic acid bacteria are inhibitory or bacteriocidal to *L. monocytogenes*, including nisins A and Z (Class I) and pediocin PA-1 and sakacin A (Class II). The latter are produced by strains of *Pediococcus acidilactici* and *Lacobacillus sakei*, respectively. Because both *P. acidilactici* and *L. sakei* are ordinarily used as starter cultures for fermented

Box 6–2.   Improving the Safety and Preservation of Fermented Meats Using Bioprotective Cultures *(Continued)*

meats, if the appropriate bacteriocin-producing strains were included in the culture blend, bacteriocins could be produced *in situ* during the fermentation, as first demonstrated by Foegeding et al (1992). Many other examples have since been described (Table 1). In most of these experimental trials, however, the level of *Listeria* inhibition is modest, usually only one to two logs.

Although including bacteriocin-producing strains in a starter culture is a convenient way to introduce bacteriocins into a product, there is no assurance that the bacteriocin will always be produced on a consistent basis and at levels necessary to achieve the desired effect. Thus, other approaches have been considered to ensure that sufficient amounts are actually delivered. For example, a pure bacteriocin can either be added directly to the sausage batter or applied to the surface in the form of a dip or spray. It is also possible to incorporate bacteriocins into packaging films. However, nisin is the only bacteriocin that has been purified and granted "generally recognized as safe" (GRAS) status (and only for specific applications). Purification processes and clearing regulatory hurdles are expensive activities, which explains why other bacteriocins have not yet been commercialized.

An alternate strategy is to use the non-purified, pasteurized, and concentrated fermentation material obtained following growth of a bacteriocin-producing strain in a food-grade medium. Products containing such a mixture, in either a dehydrated or paste form, contain active bacteriocin along with the organic acids that were also produced. Importantly, these products (which are commercially available as shelf-life extenders), do not necessarily require regulatory approval, and are simply labeled as "cultured corn syrup" or "cultured whey."

References

Benkerroum, N., A. Daoudi, and M. Kamal. 2003. Behavior of *Listeria monocytogenes* in raw sausages (merguez) in presence of a bacteriocin-producing lactococcal strain as a protective culture. Meat Sci. 63:479–484.

Callewaert, R., M. Hugas, and L. De Vuyst. 2000. Competitiveness and bacteriocin production of enterococci in the production of Spanish-style dry fermented sausages. Int. J. Food Microbiol. 57:33–42.

Campanini, M., I. Pedrazzoni, S. Barbuti, and P. Baldini. 1993. Behavior of *Listeria monocytogenes* during the maturation of naturally and artificially contaminated salami: effect of lactic acid bacteria starter cultures. Int. J. Food Microbiol. 20:169–175.

Chen, H., and D.G. Hoover. 2003. Bacteriocins and their food applications. Comp. Rev. Food Sci. Food Safety. 2:82–100. (Available at www.ift.org/publications/crfsts.)

Foegeding, P.M., A.B. Thomas, D.H. Pilkington, and T.R. Klaenhammer. 1992. Enhanced control of *Listeria monocytogenes* by in situ-produced pediocin during dry fermented sausage production. Appl. Environ. Microbiol. 58:884–890.

Hugas, M., M. Garriga, M.T. Aymerich, and J.M. Monfort. 1995. Inhibition of *Listeria* in dry fermented sausages by the bacteriocinogenic *Lactobacillus sakei* CTC494. J. Appl. Bacteriol. 79:322–330.

Hugas, M., B. Neumeyer, F. Pages, M. Garriga, and W.P. Hammes. 1996. Antimicrobial activity of bacteriocin-producing cultures in meat products. 2. Comparison of bacteriocin-producing lactobacillli on *Listeria* growth in fermented sausages. Fleichwirtschaft 76:649–652.

Klaenhammer, T.R. 1993. Genetics of bacteriocins produced by lactic acid bacteria. FEMS Microbiol. Rev. 12:39–85.

Lahti, E., T. Johansson, T. Honkanen-Buzalski, P. Hill, and E. Nurmi. 2001. Survival and detection of *Escherichia coli* O157:H7 and *Listeria monocytogenes* during the production of dry sausage using two different sausage cultures. Food Microbiol. 18:75–85.

Sabia, C., S. de Niederhausern, P. Messi, G. Manicardi, and M. Bondi. 2003. Bacteriocin-producing *Enterococcus casseliflavus* IM 416K1, a natural antagonist for control of *Listeria monocytogenes* in sausage ("cacciatore"). Int. J. Food Microbiol. 87:173–179.

Työppönen, S.; A. Markkula, E. Petäjä, M.L. Suihko, and T. Mattila-Sandholm. 2003. Survival of *Listeria monocytogenes* in North European type dry sausages fermented by bioprotective meat starter cultures. Food Control 14:181–185.

Vermeiren, L., F. Devlieghere, and J. Debevere. 2004. Evaluation of meat born lactic acid bacteria as protective cultures for the biopreservation of cooked meat products. Int. J. Food Microbiol. 96:149–164.

sausages. For example, various research laboratories (including the author's laboratory) have shown that *P. acidilactici* can be incorporated into frankfurter mixtures, and even in the absence of fermentation, can reduce the *L. monocytogenes* population.

Micrococcaceae Cultures

Most meat starter cultures available in the United States contain species belonging to two genera of lactic acid bacteria, *Lactobacillus* and *Pediococcus*. In Europe, a quite different type of starter culture has been used. Most of the cultures used for European or European-style fermented sausages contain not only lactic acid bacteria, but also totally unrelated organisms belonging to the family *Micrococcaceae*. These include species of coagulase-negative *Staphylococcus, Micrococcus*, and *Kocuria*. In fact, when lactic acid bacteria starter cultures were first introduced in the United States nearly fifty years ago for sausage manufacture, the first European cultures contained only *Micrococcus*. These micrococci cultures are still available.

Whereas the main function of the lactic starter culture—to produce lactic acid and lower the pH—is generally considered to be essential, the inclusion of *Micrococcaceae* in meat starter cultures, while important, is strictly optional. These bacteria are not fermentative and they produce no acid end products. Moreover, although they are metabolically active, they hardly even grow in the sausage. Rather, these bacteria are included in starter cultures to convert nitrate to nitrite via expression of the enzyme nitrate reductase. Along the way, they help form flavor and enhance color (Box 6-1). Is it important to note that because they are mesophilic, the fermentation temperature must fall within the range of 18°C to 25°C. Thus, the lactic acid bacteria present in the culture must also be capable of growth within this range. If fermentation were to occur at higher temperatures (e.g., 32°C to 40°C), the rapid acid development would inhibit growth of the micrococci and the benefits they provide would not be achieved.

In summary, then, there are at least five functions performed by meat starter cultures (Table 6-4). The culture must: (1) produce lactic acid and lower the pH; (2) produce desirable flavors; (3) out-compete spoilage and pathogenic microorganisms for substrates and nutrients; (4) lower the Eh, since *Salmonella*, *S. aureus*, and other pathogens grow better aerobically; and (5) in the case of the *Micrococcaceae* cultures, enhance flavor and color development via reduction of nitrate. Although a number of factors account for the excellent safety record of fermented meat products (Box 6-3), the role of the culture in producing safe, high-quality products cannot be over-emphasized (see below).

Principles of Fermented Sausage Manufacture

There are actually only a few general steps involved in fermented sausage manufacture. First, the ingredients are selected, weighed, mixed, and stuffed into casings. Second, the stuffed sausages are held under conditions necessary to promote a fermentation. Third, the sausage is subjected to one or more post-fermentation steps whose purpose is to affect flavor, texture, and preservation properties. These latter steps can range in duration from as little as one week in the case of moist or semi-dry sausages to more than two months for very dry, strongly flavored sausages such as Italian salamis.

Table 6.4. Desirable properties of meat starter cultures[1].

Bacteria	Fungi
Non-pathogenic	Non-pathogenic
No toxins produced	No toxins produced
Grows well in meat	Competitive at the surface
Stable	Firm surface mycelium
Produces good flavor	Proteolytic and lipolytic
Nitrate/nitrite resistant	Moldy aroma
Salt-tolerant	
Bioprotective	
Easy to identify	

[1]Adapted from Hammes and Knauf, 1994

Box 6–3. Pathogens, Toxins, and the Safety of Fermented Sausage

Given that fermented sausages are made from raw meat and many are never heat processed or cooked, it is not unreasonable to question the microbiological safety of these products. The emergence of several serious foodborne pathogens in the 1980s, such as enterohemorrhagic *Escherichia coli* (EHEC) O157:H7 and *Listeria monocytogenes*, has raised additional concerns, since these organisms are frequently present in raw meat, have high mortality rates, and are more hardy than other food borne pathogens.

Of course, if not properly manufactured, even traditional foodborne pathogens, such as *Staphylococcus aureus*, *Clostridium botulinum*, and *Salmonella*, can present significant food safety risks in fermented sausages. Finally, there are several food safety hazards other than those caused by bacterial pathogens. Parasites, fungi, viruses, and other agents can potentially contaminate meat and sausage products. In addition, an unusual but not uncommon form of food poisoning is caused by the presence of biogenic amines that are produced by bacteria ordinarily present in fermented sausage.

Despite these concerns, manufacturers of fermented sausages have long considered these products to be safe, and they have generally withstood the test of time. Since 1994, there have been few food poisoning outbreaks associated with fermented meat products (Table 1). Their safety is undoubtedly due to the presence of multiple antimicrobial barriers or hurdles that exist in fermented sausages (Table 2). After all, for a given organism to survive and grow in these products, it would have to overcome each of these individual barriers. That is, at minimum, it would have to be nitrite-resistant, acid-resistant, salt-resistant, and osmotolerant. Moreover, when barriers are combined, there is a synergistic effect, such that the net antimicrobial effect of multiple barriers is greater than would be predicted based on their singular effects. For example, a strain of *Salmonella* may be fully capable of surviving a pH of 5.2, provided that all other conditions are optimal for growth. But if another hurdle is put in place (e.g., the water activity is reduced from 0.99 to 0.97), then pH 5.2 may be sufficient to inhibit this organism.

Table 1. Ten-year safety record for fermented meats.

Year	Country	Product	Cases	Organism
1994	USA	Dry-cured salami	23	*Escherichia coli* O157:H7
1995	Australia	Mettwurst	23	*Escherichia coli* O111:H⁻
1995	Germany	Teewurst	>300	*Escherichia coli* O157:H⁻
1998	Canada	Genoa salami	39	*Escherichia coli* O157:H7
1999	Canada	Sausage	155	*Escherichia coli* O157:H7
2001	USA	Fermented beaver	3	*Clostridium botulinum*

Table 2. Antimicrobial barriers in fermented meats.

Property	Level, range, or function
pH	4.5–5.5
a_w	0.7–0.9
salt	2–4%
Eh	Reduced
Acids	1–2%
Competition	Exclusion
Nitrite	125 ppm
Casing	Exclusion

(Continued)

The problem, however, is now more complicated, because it has been recognized recently that some pathogens appear to be tolerant even to multiple barriers. Some strains of *E. coli* O157:H7, for example, are much more tolerant to low pH and organic acids than are normal *E. coli* strains. Furthermore, *L. monocytogenes* is resistant to low pH, low water activity (a_w), high salt, and nitrite, and can even grow at refrigeration temperatures. And although there have been few food poisoning outbreaks caused by these organisms in fermented sausage, challenge experiments indicate that *E. coli* O157:H7, *L. monocytogenes*, and other pathogens can theoretically survive typical sausage manufacturing procedures. Thus, even the mere presence of these organisms in ready-to-eat meat products would likely initiate a recall. Indeed, the U.S. Department of Agriculture has established rules that now require a minimum five-log reduction of pathogenic organisms during the manufacture of uncooked, ready-to-eat fermented sausage. Manufacturers are also required to develop and implement HACCP plans that describe the relevant intervention steps necessary to produce safe products.

Of course, the simplest solution to many of these potential problem organisms and the most effective way to ensure a greater than five-log reduction would be to include a heating step somewhere during the process. However, many manufacturers are unwilling to accept the changes in the sensory properties caused by heating, and instead rely on ensuring that natural barriers are sufficient. It is also possible to include additional intervention measures into the product or process. For example, meat starter cultures capable of producing bacteriocins and other antagonistic agents that are inhibitory to pathogens are now available (discussed previously).

As noted above, bacterial pathogens are not the only food safety issue of concern to the fermented meats industry. Parasites and viruses can also be present in raw meat, and toxigenic fungi can contaminate sausage during fermentation and ripening. The parasite most commonly associated with meat, and pork in particular, is the tapeworm *Trichinella spiralis*, the causative agent of trichinosis. Although certified *Trichinella*-free pork is available in the United States (and in Europe), it is also possible to inactivate this nematode by a freezing treatment, as per USDA guidelines. Otherwise, a cooking step is required to destroy the cysts.

In contrast to parasites, viruses have not been considered as a serious food safety problem in fermented meats, and this is still largely true. Viruses found in meat are not usually pathogenic to humans, they do not replicate, and they are mostly sensitive to the acidification and drying steps used in sausage manufacture. Although not a human pathogen, the spread of the foot-and-mouth virus in the United Kingdom and Europe during 2001 had devastating economic and social effects on all segments of the meat processing industry, even though it posed no threat to humans.

This situation was different from that which occurred in the 1980s and 1990s when "mad cow disease" (or bovine spongiform encephalopathy) appeared in the United Kingdom. In the latter case, the causative agent was an unusual type of infectious protein called a prion that occurred in beef and that was fully capable of causing an extremely rare but fatal disease. The prion is not destroyed by fermentation, heat, or other food processing techniques, so the only means of control is to prevent the transmission of the prion during animal production.

Several different genera of fungi, including species of *Penicillium* and *Aspergillus*, are frequent contaminants of fermented meats. For many products, including both whole fermented meats (i.e., hams), as well as fermented sausages, their growth is encouraged, due to their ability to produce flavor-generating enzymes (as discussed previously). However, some strains isolated from mold-fermented products are capable of producing mycotoxins. Moreover, inoculation of ham and sausage with toxigenic fungi and incubation under optimized conditions results in toxin formation. Despite these findings, the presence of these toxins in mold-fermented meat products appears to occur rarely, if at all, a situation not unlike that for fungal-ripened cheese and other mold-fermented products.

Still, there is much concern about the potential for mycotoxin production in fermented meats, and there is now a trend to use defined, nontoxigenic strains rather than the wild or

Box 6–3. Pathogens, Toxins, and the Safety of Fermented Sausage *(Continued)*

house strains that have commonly been used. For example, *Penicillium nalgiovense,* a fungal meat starter culture, has many desirable properties, but since some strains produce mycotoxins, only strains demonstrated to be non-toxin producers are in commercial use. Undesirable mold growth on sausage can also be controlled by antimycotic agents such as sorbic acid and, if permitted, the antibioic pimaricin. Smoke, which is usually applied to high moisture, but not dry products, also contributes antimycotic constituents.

Finally, another biologically-active group of microbial end products found in a wide variety of fermented foods, including fermented meats, are referred to as biogenic amines (Suzzi and Gardini, 2003). These compounds are formed via decarboxylation of amino acids by various bacteria that are commonly found in fermented meats, including lactic acid bacteria, enterococci, *Enterobacteriaceae*, and *Micrococcaceae*. In some individuals, ingestion of biogenic amines results in a particular food poisoning syndrome marked by headache, nausea, and dilation of blood vessels.

In fermented meats, the most common biogenic amine capable of causing food poisoning symptoms is tyramine, derived from the amino acid tyrosine. In general, dry fermented sausages contain about 50 to 300 mg tyramine per kg (Table 3). Histamine, derived from histidine, may also be present, but usually at ten-fold less concentration. However, only when concentrations are very high (>1,000 mg per Kg of dry weight) do these products pose a health risk in most normal individuals.

Table 3. Biogenic amines in European fermented dry sausages[1,2].

Product (Country)	Tyramine	Histamine	Cadaverine	Putrescine
Soppressata (Italy)	178	22	61	99
Salsiccia (Italy)	77	0	7	20
Sobrasada (Spain)	332	9	13	65
Fuet (Spain)	191	2	19	72
Salchichón (Spain)	281	7	12	103
Chorizo (Spain)	282	18	20	60

[1]Adapted from Suzzi et al., 2003
[2]Concentrations given in mg/Kg

Other biogenic amines may also be formed in meat at appreciable levels, including putrescine and cadaverine, but these compounds do not generally elicit symptoms described above. Since formation of biogenic amines requires the presence of free amino acids, the amount produced depends on the extent of protein hydrolysis that had occurred in the food. Thus, the longer the meat is aged or fermented, the higher will be the concentration of amino acid substrates and the more likely it is that the product will contain biogenic amines (assuming the relevant decarboxylating enzymes are also present).

Because some lactic acid bacteria have the ability to produce amino acid decarboxylases, starter cultures strains should be screened to eliminate such strains from use. Moreover, it may be possible to use starter cultures that have inhibitory activity against potential bioamine producing bacteria, either by virtue of their ability to produce acids rapidly, produce bacteriocins, or outcompete them for nutrients. Finally, some staphylococci and lactic acid bacteria produce amine oxidases, enzymes that cause oxidative deamination of amines, and therefore, act to detoxify biogenic amines.

References

Suzzi, G., and F. Gardini. 2003. Biogenic amines in dry fermented sausages: a review. Int. J. Food Microbiol. 88:41–54.

Ingredients

It is entirely possible to manufacture a fermented sausage with just a handful of ingredients. In fact, only five ingredients are essential: meat, sugar, salt, culture, and a curing agent. As discussed above, fermented meats can indeed be made without adding a culture, but most large-scale manufacturers would not dream of making product without a culture. Likewise, fermented sausages also can be manufactured in the absence of the curing agent. However, these agents, either in the form of nitrite or nitrate, perform such important microbiological and organoleptic functions that they are nearly universally used. That being said, there is a small (but growing) market among organic foods proponents for reduced or even nitrite-free meat products. Even if the organoleptic properties provided by nitrite could be provided by other agents (a big if), removing nitrite from cured meat products would expose these products to a potentially serious food safety threat. Still, provided that other barriers are in place, especially low pH and low temperature, theoretically safe products can be produced.

Meat

Of the ingredients listed above, the main ingredient is obviously the meat, which contributes not only the protein and the bulk of the product matrix, but also the fat, which provides much of the flavor. The fat-containing cuts usually are chopped or ground separately from the leaner portions to impart a desired appearance and flavor. The grind also affects texture and accordingly determines the type of product. For example, some sausages (e.g., Plockworst) have large visible fat particles, whereas others (e.g., cervelat) are ground to a fineness such that the fat particles are so small as to be indistinguishable from the sausage matrix.

The fat and lean portions may even be derived from different animals. Beef fat contains more unsaturated lipids than pork fat, and is more susceptible to oxidation reactions that may result in undesirable rancid flavors. Thus, many sausage products, such as the popular U.S. product summer sausage, are typically made with mixtures of beef and pork. Cuts are also important—shanks, chucks, and bull meat have binding properties that are especially important in sausage manufacture. Obviously, there is a trend to use less expensive cuts, but high quality meats are often still used.

Sugar

The next essential ingredient is the sugar or carbohydrate. Although glucose, in the form of the polymer glycogen, is initially present in muscle tissue of slaughtered animals, glycogen stores are quickly depleted during the postmortem period. Thus, fresh meat contains little fermentable sugar, and addition of sugar is necessary. Glucose is most common in the United States, and is added to about 1% to 2% of the total batter weight. Since the amount of acid produced by the lactic culture is directly related to the amount of available glucose, the sugar concentration in the batter can be adjusted, in general, to give a particular final pH. Also, higher sugar levels promote faster fermentations, which are preferred in the United States. In contrast, many European fermented sausage manufacturers prefer less tanginess and more diverse flavor development. Achieving these characteristics require slower fermentation rates, thus less rapidly fermentable sugar is added (as little as 0.1% to 0.2%).

Salt

Salt is an essential ingredient in all types of sausage products (fermented or not). Salt, added in concentrations of 2.4% to 3%, performs several critical functions. First, it is responsible for extracting and solubilizing the muscle proteins, which are ordinarily in an insoluble form. Once extracted and solubilized, the proteins form a "sticky" film around the meat particles, creating an emulsion-type structure. Second, salt provides flavor. Finally, salt is the primary means, at least initially, for controlling the microflora. Although salt at a concentration of even 3% might not appear to be sufficiently high enough to inhibit many organisms, the actual concentration within the aqueous phase is considerably higher.

Culture

Most commercial cultures for sausage are supplied in either a frozen or lyophilized form. Frozen cultures, which are more common in the United States, are supplied as thick slurries in peel-back or flip-top cans ranging in size from 20 ml to 250 ml. Cell densities typically range from 10^8 to 10^9 cells per ml. A typical 70-ml can is sufficient for about 150 kg of sausage batter. These cultures are shipped frozen under dry ice and users are instructed to store the cans at $-40°C$ or below. The cans should be thawed in cold water prior to use.

Proper handling of frozen cultures is absolutely necessary to maintain culture viability and to ensure that culture performance (i.e., rapid fermentation) is not impaired. Lyophilized cultures, which are less commonly used, have the advantage of not requiring low-temperature storage. They are stable at refrigeration or even ambient temperatures. As free-flowing powders, they are easily measured and distributed into the batter. These cultures, which can contain up to 10^{10} cells per gram, are usually more expensive, however, than frozen cultures.

Curing agents

Finally, with few exceptions, fermented meat products include nitrite or nitrate as curing agents. These are added as either the sodium or potassium salt. Although nitrite salts are now used far more frequently, until the 1970s, nitrate salts were more common. For reasons discussed previously, some sausage manufacturers, still prefer nitrate. In any case, nitrite is added at a maximum of 156 ppm for dry and semi-dry sausages. Since a single ppm translates to 1 gram per 1,000 kg, only 156 grams (about one-third of a pound) are all that can be added to a 1,000-kg batch of sausage (more than 2,200 pounds). Despite this relatively

small amount, nitrite performs a number of important microbiological and organoleptic functions. In fact, without nitrite (or nitrate), fermented meat products would be far less popular and considerably less safe to eat.

Nitrite is mainly added to sausages (and not just those that are fermented) because of its effectiveness as an antimicrobial agent. In particular, nitrite inhibits the out-growth of *C. botulinum* spores (Box 6-4), making it one of the most powerful anti-botulinum agents available to the processed meats industry. Although fermented meats contain combinations of organic acids and salt, both effective antimicrobials, neither provide a sufficient degree of inhibition against this organism. It should be emphasized, however, that nitrite alone, at the levels currently used, does not entirely inhibit *C. botulinum*. Rather, it is the combined effects of nitrite along with organic acids, low pH, and low a_w that effectively control the growth of this organism during the manufacture and storage of fermented sausage.

The other reasons for adding nitrite to fermented meats are related to the organoleptic properties this agent imparts. Nitrite fixes color, acts as an antioxidant and prevents a warmed-over flavor, and imparts a desirable cured meat flavor.

Spices, flavoring, and other ingredients

A wide variety of spices, seasonings, and other flavoring agents are often added to fermented sausages. These include pepper (black and red), paprika, garlic, mustard, mace, and cardamom. These flavorings can be added in their natural form or as extracts. Levels vary, depending on the nature of the product and consumer preferences. In general, moist, smoked sausages that are popular in Germany and northern Europe are only slightly spiced, whereas the dried, non-smoked products consumed in southern Europe (e.g., Italy and Spain) and other Mediterranean regions are more heavily spiced (Table 6-1). Among the optional ingredients commonly added to fermented meats, ascorbate and erythorbate are

Box 6–4. Nitrite as an Anti-botulinum Agent

Using nitrate and nitrite salts as ingredients in fermented and non-fermented sausage is a rather recent practice, adopted only within the last century. It is likely, however, that sausage products have long contained nitrate salts, due to the contamination of salt with potassium nitrate, in the form of saltpeter (Cammack et al., 1999).

The development of the desirable cured meat color and flavor by nitrate, and more specifically, the nitrate reduction product, nitrite, must have certainly led to widespread use of saltpeter in meat processing. However, the discovery that nitrite also had inhibitory activity against *Clostridium botulinum* eventually led to the routine addition of nitrite salts in these products. The physiological mechanisms by which this organism is inhibited by nitrite, however, have yet to be established, although several possible explanations have been proposed (reviewed in Cammack et al., 1999 and Grever and Ruiter, 2001).

As noted previously (Box 6-1), when nitrite (NO_3^-) is added to foods, it is quickly converted to nitric oxide (NO) via dissimilatory nitrite reductases produced by the indigenous flora. It is also possible that chemical reduction of nitrite to nitric oxide can also occur. Nitric oxide is a reactive species, especially in the presence of metal ions. When nitric oxide-metal ion complexes are formed, the biological and chemical availability of those metals is decreased. Thus, microorganisms would be unable to acquire iron and other essential minerals from the medium.

In muscle systems, iron is present in the form of the heme-containing proteins, myoglobin, oxymyoglobin (oxidized), and metmyoglobin (reduced). When these proteins react with nitric oxide, the resulting nitrosyl myoglobin products give nitrite-cured meats their characteristic color. The strong binding of heme iron (which is enhanced by heating) is considered to be one of the factors responsible for nitrite-mediated inhibition of *Clostridium botulinum* and other bacteria.

Not only does nitric oxide react with heme iron, but it also binds to other iron- and iron-sulfur-containing proteins and enzymes. In particular, clostridia depend on ferridoxin and the enzymes pyruvate-ferridoxin oxidoreductase and hydrogenase for pyruvate metabolism and ATP generation (Figure 1). The nitrosyl-protein complexes that are formed in the presence of nitric oxide block pyruvate metabolism and subsequent ATP synthesis. The suggested net result, therefore, is the inhibition of outgrowth by germinated spores.

perhaps the most important. Both provide similar functional roles as nitrite in that they inhibit autooxidation and increase color and flavor intensity.

Manufacture of Fermented Sausage

Cutting and Mixing

Once the ingredients have been collected and weighed, the process is remarkably simple (Figure 6-1). The meat portions (lean and fat) are usually ground separately in a silent cutter—a rotating, bowl-shaped device that chops and mixes the sausage batter—or a similar grinding device to produce varying degrees of coarse-ness. The meat, along with all of the remaining ingredients, are then combined in a silent cutter. Mixing should be done to minimize or exclude oxygen, which not only can interfere with color and flavor development, but also is less conducive to growth of the starter culture. Above all, mixing, as well as all the grinding and weighing, must be performed at low temperatures for both quality and safety. The USDA requires that the temperature in the processing room be 40°C or below. In fact, in the United States, all meat processors must have a Hazard Analysis Critical Control Points (HAACP) plan, with the temperature during processing listed as a critical control point (Box 6-5).

Box 6–4. Nitrite as an Anti-botulinum Agent *(Continued)*

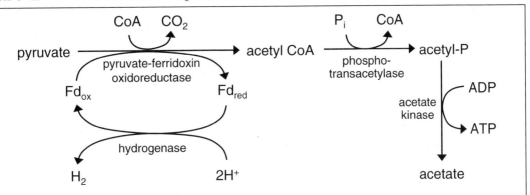

Figure 1. Metabolism of pyruvate to acetate in clostridia. The "phosphoroclastic" conversion of pyruvate to acetate is mediated by several iron-containing proteins. Pyruvate-ferridoxin oxidoreductase oxidizes pyruvate, using ferridoxin (Fd) as the electron acceptor. Reduced ferridoxin is then re-oxidized by a third iron-containing protein, hydrogenase. Protons serve as the electron acceptor, and hydrogen gas is formed. Phospho-transacetylase and acetate kinase catalyze the final two steps, with acetate and one mole of ATP formed as end products. Importantly, both pyruvate-ferridoxin oxidoreductase and ferridoxin are able to form complexes with nitric oxide. This would effectively block this pathway, depriving cells of ATP, as well as increasing the concentration of pyruvate to potentially toxic levels. Adapted from White, 2000.

References

Cammack, R., C.L. Joannou, X.-Y. Cui, C.T. Martinez, S.R. Maraj, and M.N. Hughes. 1999. Nitrtie and nitrosyl compounds in food preservation. Biochim. Biophys.Acta. 1411:475–488.

Grever, A.B.G., and A. Ruiter. 2001. Prevention of *Clostridium* outgrowth in heated and hermetically sealed meat by nitrite—a review. Eur. Food Res.Technol. 213:165–169.

White, D. 2000. *The physiology and biochemistry of prokaryotes, second edition*. Oxford University Press, Inc., New York.

Stuffing

After the batter has been sufficiently mixed, it is moved to a stuffer, a device that pumps the mix into casings. The casings are essentially long tubes that give the product its characteristic shape. The diameter of the casings can vary from less than 1.5 cm to more than 9 cm. As the tubes are filled, they are tied off or cut to give desired section lengths, again depending on the product being made. Lengths can vary from 5 cm to 100 cm. Shape and diameter size are important, not only because they are specific for a given product, but more importantly, because

they influence the rate of drying, cooking, smoking, and ultimately the flavor and texture of the finished product. Casings must be permeable to both moisture and smoke.

Two general types of casing materials are used, natural and synthetic. These terms are somewhat nebulous, since the latter are often made from natural sources. Traditionally, casings were made of animal intestines. Although there is still a significant market for products made using natural casings, especially in Europe, synthetic casings have several advantages are now widely used. They are usually made using cellulose or collagen.

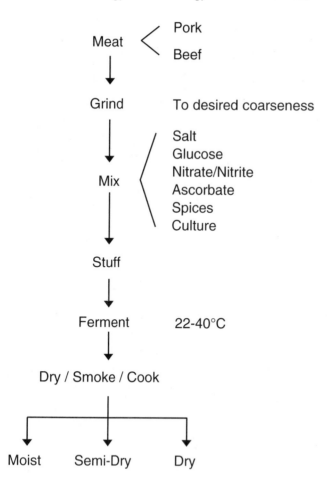

Figure 6–1. Manufacture of fermented sausage.

Fermentation

Once the sausage batter is packed into casings, the material is moved into specially designed ripening chambers where the fermentation occurs. These facilities, often referred to as the green room or smoke house or simply the house have controls for maintaining temperature, humidity, and air movement. Moreover, control systems in modern facilities are fully programmable so conditions can be ramped or adjusted depending on how the fermentation is proceeding or on the particular specifications of the manufacturers. Thermocouples and pH probes inserted directly into product samples can feed the appropriate information into a computer to provide constant monitoring, record-keeping, and feed-back control. Thus, the entire fermentation can proceed in the absence of a full-time operator.

Fermentation parameters vary depending on the culture and the desired product qualities. In general, lower incubation temperatures require longer fermentation times. For example, at the low temperature range of 21°C to 24°C (70°F to 75°F), which is very near ambient, fermentation can take as long as two to three days. At 29°C to 32°C (85°F to 90°F), twelve to sixteen hours of fermentation will be required. In the United States, where faster overall production times are preferred, the incubation temperature can be as high as 37°C to 40°C (98°F to 102°F) for as little as twelve to eighteen hours.

Box 6–5. Ensuring Meat Safety Through HACCP

In July 1996, the Pathogen Reduction/Hazard Analysis and Critical Control Point system became regulatory law for all USDA-regulated facilities. This system, abbreviated as HACCP, was designed to apply scientific principles to process control in food production.

The processing steps for manufacture of a representative fermented sausage product are illustrated in a flow diagram (Figure 1). Through careful analysis all hazards that are considered reasonably likely to occur are then identified (Table 1). Control measures designed to prevent hazards are applied at specific points in the manufacturing process; these are called critical control points.

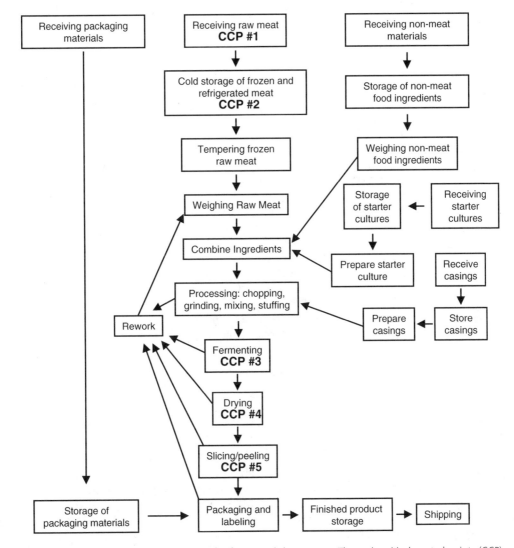

Figure 1. Generic process flow diagram for fermented dry sausage. The main critical control points (CCP) are shown.

(Continued)

Box 6–5. Ensuring Meat Safety Through HACCP *(Continued)*

In the fermentation step, for example, the pH must attain 5.3 or below within six hours. Control of this step may have the greatest impact on assuring the safety of the final product. Critical limits are established and then continuously monitored and recorded to verify that the process is under control.

Corrective actions are developed for each critical control point and documented in the HACCP plan. These corrective actions are applied whenever deviation from a critical limit occurs; specific and exacting records are kept. A separate HACCP program is designed and documented for each manufacturing process used in an establishment.

The HACCP method accompanied by proper prerequisite programs such as Good Manufacturing Practices and Sanitation Standard Operating Procedures can be an effective means of assuring the safety and quality of food products.

Table 1. Identification of critical control points during the manufacture of fermented semi-dry sausage.

Critical Control Point	Hazard	Reasoning	Control Measure
Receiving Raw Meat	*Salmonella* *E. coli* 0157:H7 *Listeria monocytogenes*	Reasonably likely to be present in raw meat ingredients	Supplier certification to guarantee laboratory analysis indicating absence of *Salmonella* and *E. coli* 0157:H7
Cold Storage	*Salmonella* *E. coli* 0157:H7	*Salmonella* and *E. coli* 0157:H7 may grow if temperature is not maintained below minimum growth requirement	Refrigeration temperature not to exceed 40°F; freezer temperature not to exceed 30°F
Fermentation	*Staphylococcus aureus*	Potential for growth and toxin production due to fermentation failure	Must attain pH of 5.3 or below within 6 hours
Drying	*Salmonella* *Staphylococcus aureus*	Potential for growth and toxin production due to drying failure	Attain established moisture to protein ratio (1.6:1 for pepperoni and 1.9:1 for sausage)
Slicing/peeling	*Listeria monocytogenes*	Potential for contamination from environment or handling	Sanitizer effective against *L. monocytogenes* applied to food contact surfaces every 4 hours

Since individual culture strains may have different temperature optima and tolerances, selection of cultures that perform under the selected conditions is critical. It is also important, for some applications, that the fermentation rate not proceed too fast. This is particularly relevant when flavor- or color-producing organisms (e.g., *Micrococcus*) are used, because they may be inhibited by fast lactic acid producers. However, if the fermentation is too slow, there may be sufficient opportunity for pathogens or spoilage organisms to grow. Ultimately, the pH at the end of fermentation should be less than 5.1, which, by itself, will not make the product shelf-stable, but which does provide a reasonable protective barrier against most foodborne pathogens. In addition, fermented sausages at pH 5.0 to 5.1 will have only a slight tart flavor that may actually be indistinguishable from non-fermented products.

Alternatively, depending on the culture, the incubation conditions, and the amount of substrate (i.e., fermentable sugar) provided, much higher acidities can be achieved. Typically, summer sausage has a pH around 4.9 but can be as low as 4.7. When the pH reaches 4.8 or less, a definite "tangy" flavor becomes apparent.

Fermentation rooms are also equipped with air movement devices and humidistats to maintain the relative humidity (RH) at desired levels. Since the RH in the atmosphere surrounding a food directly influences the water activity of that food (where $a_w \times 100 = RH$), as RH is decreased, the more moisture is lost to the atmosphere and the lower will be the a_w in the sausage. Typically, the RH should be about 10% less than the $a_w \times 100$ of the finished product.

Cooking, drying, and smoking

Several different treatments and combinations of treatments can be applied at end of the fermentation. These include cooking, drying, and smoking. In the United States, fermented sausages are often cooked after fermentation, whereas, in Europe and elsewhere, raw sausages are the norm, and post-fermentation heating steps are rarely applied.

In general, properly made dry, fermented, uncooked sausages, like salami and pepperoni, are still considered to be shelf-stable and ready-to-eat. Cooking, however, does provide important advantages. Cooking not only inactivates the culture and stops the fermentation, but it also kills pathogenic microorganisms that may have been present in the raw meat. Thus, the cooking step, at least in the United States, may be an important component of HACCP programs (Box 6-5) used by sausage manufacturers, since it serves as a terminal process step providing manufacturers (and consumers) with reasonable assurance of product safety.

Furthermore, in the United States, any pork-containing sausage must, by federal regulations, be cooked to destroy *Trichinae*, a nematode that infects swine and that can cause the disease Trichinella in humans. The USDA has developed various time-temperature regimens that are effective (e.g., 58.3°C), and which can easily be implemented as part of the cooking process. Exceptions are allowed if the pork is certified as *Trichinae*-free or if the pork is frozen according to the USDA guidelines for freeze-inactivation of the Trichinae cysts.

If a cooking step is included, the product can be moved to a separate chamber or, as is more common, the fermentation chamber itself is equipped with heating capability. The temperature is slowly raised until the desired internal temperature (usually 60°C to 62°C) is achieved. Following cooking—or fermentation, if there is no cooking step—the product can be smoked and/or dried. In general, Mediterranean sausages (e.g., Genoa and Milano) are dried, but not smoked, and northern sausages (e.g., German) are smoked, but not dried. In contrast, some semi-dry products, such as summer sausage, are slightly smoked.

If the product is to be dried, the chamber environment is set at about 7°C to 13°C (45°F to 55°F) and 70% to 72% RH. Good air movement is necessary to rapidly remove water vapor and any condensate that collects at the surface. The rate of drying is also critical. If the RH is too low and the temperature too high, drying will initially be rapid. However, at these high drying rates, the surface will become dehydrated and form a hard, water-impermeable skin. This phenomenon, called case hardening, results in slower drying and poor product quality, since water molecules are unable to diffuse through the hardened surface and are trapped within the sausage interior.

Drying times depend on product specifications and desired quality characteristics. Obviously, large diameter products require longer drying times than small diameter products. The longer the drying, the more water is lost, and the lower will be the final water activity. Semi-dry products are typically dried to remove about 20% to 30% of the original water and to give a final moisture of about 45% to 50%. The a_w of semi-dry products ranges from 0.90 to 0.94. Such products may be stable at ambient temperature, with a shelf-life of about thirty days. Examples include Thuringer, cervelat, Lebanon bologna, summer sausage, and semi-dry salami. Dry fermented sausages will lose about 35% water, giving a final moisture of 35% and an a_w between 0.85 and 0.91. Dry products can be shelf-stable for seventy-five days or more. Pepperoni and salami, especially the Italian, Mediterranean, and eastern European varieties, are the most common examples.

Most modern facilities are equipped with programmable systems that allow operators to set fermentation and cooking parameters, as well as air movement, smoking, and drying conditions. Recording devices provide a record of these conditions during the manufacturing process. Thus, once the sausages are loaded into the ripening chamber, they can be left on their own, more or less, until the process is completely finished.

Mold-ripening

Many of the European-style sausages are ripened by mold. These products are particularly popular in Hungary and Romania, as well as throughout the Mediterranean region. For many of these products, fungal growth can be extensive, with the mycelia covering the entire surface. Mold-fermented sausages are not nearly as common in the United States. However, there are some whole meat products, in particular, country-cured hams, that are fermented by yeasts and fungi and that are very popular in various regions of the United States. Although technically not fermented, aged meat owes its improved flavor and texture properties, in part, to growth of surface fungi, in particular, *Thamnidium elegans*.

The role of fungi in fermented whole meats and sausages is similar to that of other mold-fermented products. That is, the fungi are not involved in the primary lactic acid fermentation, but rather their function is to enhance flavor and texture properties. This is accomplished by production and secretion of proteinases and lipases and their diffusion into the meat, generating flavor and aroma products or their precursors. Ammonia, released from protein and amino acid metabolism, and ketones and other rancid and oxidized flavor notes, derived from fatty acid metabolism, are among the end products that accumulate in mold-ripened sausages. In this respect, the fungi produce organoleptic properties not unlike that found in mold-ripened cheese.

The manufacture of mold-ripened sausages is similar to that of normal fermented sausage, in that ingredients are mixed and stuffed into casings. Inoculation of sausages or whole meats with fungi can occur either naturally, via the fungal spores present in the manufacturing environment, or by dipping or spraying the product with a defined spore suspension. Salt applied to the surface of country-cured and Parma hams reduces the water activity and provides a selective environment for resident yeasts and fungal organisms. Yeast and fungal starter cultures, which are becoming more widely available for fermented sausage and ham manufacture, are generally comprised of strains of *Debaryomyces hansenii, Penicillium nalgiovense, Penicillium camemberti*, and *Penicillium chrysogenum*. The main problems with naturally-occurring fungi are (1) they may yield products with inconsistent quality; and (2) perhaps more importantly, they may also produce mycotoxins. Thus, the use of pure fungal starter cultures, selected on the basis of their functional properties as well as their inability to produce mycotoxins, have obvious advantages, since greater quality control and product safety can be achieved (Table 6–4).

Flavor of Fermented Meats

Like other fermented foods made with a lactic acid starter culture, the main flavor compounds are acids, principally lactic and acetic, derived via metabolism of sugars. In many of the U.S.-produced products that are cured with nitrite, fermented at high temperatures for a short time, and cooked following fermentation, there is only a brief opportunity for development of other flavors. In contrast, a much more complex array of flavor compounds is produced in sausages in which nitrate is used as the curing agent, that are fermented slowly, and are uncooked.

Flavor development may be especially enhanced if micrococci are included in the starter culture. The ripening process for these sausage products is not unlike that for aged cheese, in that microbial as well as endogenous enzymes act on proteins and fats in the raw material, generating hydrolysis products that contribute to the flavor and texture of the finished product. Importantly, the source of the microbial enzymes may be the lactic or micrococcal starter culture organisms or naturally-occurring bacteria, yeasts, and molds present in the raw meat material.

Defects and Spoilage of Fermented Meats

Defects of fermented meats can occur before, during, or after manufacturing. Like all fermented foods, the production of high quality products depends largely on the microbiological quality of the raw ingredients. Perhaps this is even more so for fermented sausage, since the starting material, meat, is raw and cannot be heat-processed to inactivate spoilage or other undesirable microorganisms. Thus, any organisms present in the raw meat will be present in the sausage batter and may even survive fermentation.

Of course, fermentation acids kill or inhibit many organisms, and, if the sausage is cooked after fermentation, most of the remaining organisms will be killed. However, spoilage products may have already been produced. For example, psychrotrophic bacteria, such as *Brochothrix thermosphacta* and various *Pseudomonas* spp., may produce ammonia and other volatile off-odors and flavors in the meat during storage even before fermentation. Lipases and other enzymes may also be produced by these organisms prior to fermentation, resulting in rancid, "cheesy," or bitter end-products.

The other main group of spoilage organisms are lactic acid bacteria. These bacteria are part of the natural meat microflora, and are quite tolerant of the barriers that ordinarily control microbial activity in fermented sausage (i.e., low pH, low a_w, low Eh, nitrite, salt, etc.). Under appropriate conditions, lactic acid bacteria can cause flavor, color, and tactile defects, and, as discussed in Box 6–3, some strains are also responsible for foodborne disease, due to their production of biogenic amines. Of course, the causative organisms are not necessarily endogenous to the product, because the starter culture itself is comprised of lactic acid bacteria, including *L. sake, L. curvatus, L. plantarum, P. acidilactici,* and other species capable, in theory, of producing these defects. Therefore, it is important that screening and selection of strains for starter cultures be based, in part, on their ability (or inability) to produce undesirable end-products.

Among the spoilage products formed by lactic acid bacteria, hydrogen peroxide is probably the most problematic. It is produced primarily by lactobacilli, but only by specific strains, and only under specific conditions. Oxygen is required for production of hydrogen peroxide by lactic acid bacteria; it serves as a reactant in the hydrogen peroxide-generating reactions and also induces expression of the enzymes involved in these reactions. Thus, minimizing exposure to air during mixing and subsequent steps is critical.

Once formed, hydrogen peroxide participates in several undesirable reactions. First, it reacts with heme-containing pigments, especially nitrosyl myochromogen, formed as a result of curing. When the heme iron is oxidized by peroxide, the desirable pink color is lost and an undesirable green pigment is formed (Box 6–1). Second, hydrogen peroxide can form hydroxyl radicals (e.g., O_2^-) that serve as initiators of lipid oxidation reactions.

One way to limit the formation of hydrogen peroxide in fermented meats is to include catalase-producing strains in the starter culture. Although some lactic acid bacteria produce a small amount of catalase or a pseudo-catalase, micrococci (present in some cultures) produce much greater levels of this enzyme. Another less common defect caused by lactic acid bacteria is slime formation. In particular, strains of *L. sakei* have been shown to produce exo-polysaccharides (i.e., slime) in vacuum-packaged meat products, and have occasionally been implicated in this form of spoilage in fermented sausages.

In general, microbial spoilage can best be prevented by keeping psychrotrophic bacteria out of the raw meat and keeping other potential spoilage organisms out of the finished fermented product via comprehensive sanitation programs. Few organisms should be present in cooked fermented sausages, provided post-processing contamination does not occur. Vacuum or modified atmosphere packaging with carbon dioxide inhibits aerobic psychrotrophs, but not facultative and anaerobic spoilage organisms, such as *B. thermosphacta* and lactic acid bacteria. Antimicrobial agents, including organic acids and bacteriocins, may be effective against these bacteria, as described previously.

References

Barbuti, S., and G. Parolari. 2002. Validation of manufacturing process to control pathogenic bacteria in typical dry fermented products. Meat Sci. 62:323–329.

Cook, P.E. 1995. Fungal ripened meats and meat products, p. 110–129. *In* G. Campbell-Platt and P.E. Cook (ed.), *Fermented Meats*. Blackie Academic and Professional (Chapman and Hall).

Fernández, M., J.A. Ordáñez, J.M. Bruna, B. Herranz, and L. de la Hoz. 2000. Accelerated ripening of dry fermented sausages. Trends Food Sci. Technol. 11:201–209.

Fisher, S., and M. Palmer. 1995. Fermented meat production and consumption in the European Union, p. 217–233. *In* G. Campbell-Platt and P.E. Cook (ed.), *Fermented Meats*. Blackie Academic and Professional (Chapman and Hall).

Hammes, W.P., A. Bantleon, and S. Min. 1990. Lactic acid bacteria in meat fermentation. FEMS Microbiol. Rev. 87:165–174.

Hammes, W.P., and C. Hertel. 1998. New developments in meat starter cultures. Meat Sci. 49, No. Suppl. 1, S125–S138.

Hammes, W.P., and H.J. Knauff. 1994. Starters in the processing of meats. Meat Sci. 36:155–168.

Incze, K. 1998. Dry fermented sausages. Meat Sci. 49, No. Suppl. 1, S169–S177.

Jessen, B. 1995. Starter cultures for meat fermentations, p. 130–159. *In* G. Campbell-Platt and P.E. Cook (ed.), *Fermented Meats*. Blackie Academic and Professional (Chapman and Hall).

Lücke, F.-K. 1994. Fermented meat products. Food Res. Int. 27:299–307.

Lücke, F.-K. 1998. Fermented Sausages, p. 441–483. *In* B.J.B. Wood (ed.), *Microbiology of Fermented Foods, Volume 2*. Blackie Academic and Professional (Chapman and Hall).

Suzzi, G., and F. Gardini. 2003. Biogenic amines in dry fermented sausages: a review. Int. J. Food Microbiol. 88:41–54.

7

Fermented Vegetables

"Oh, when I think of my Lena
I think of a girl who could cook
Some sweet sauerkraut that would swim in your mouth
Like the fishes that swim in the brook"
    From "Bring Back My Lena to Me," *by Irving Berlin, 1910*

Introduction

Wherever vegetables are grown and con-sumed, it is almost certain that fermented ver-sions exist. Moreover, despite the wide diver-sity of vegetables produced around the world, the principles involved in the manufacture of fermented vegetables are very near the same. Like other fermented foods, readily observed variations certainly exist, and are based on aes-thetic preferences, as well as the types of raw materials available in particular regions.

For example, cabbage is widely used as a fer-mentation substrate, but the actual cultivar that is used varies depending on culture and geography. Thus, in Germany, mild-tasting, white European cabbage is transformed via fer-mentation into sauerkraut, whereas in South Korea, Chinese cabbage is mixed with pep-pers, radishes, other vegetables, garlic, and spices, and fermented to make a much spicier version called kimchi.

The manufacture of fermented vegetables most likely evolved from simply dry-salting or brining vegetables. Salting vegetables was a com-mon means of food preservation and was prac-ticed for thousands of years in Europe, the Mid-dle East, and Asia, and for several centuries in the Americas. In general, salt or brine was added to the fresh raw material as a preservation aid, and

then the mixture was packed into suitable con-tainers and stored at an ambient temperature. If the salt concentration was not too high, this practice would have established ideal conditions for growth of naturally-occurring lactic acid bac-teria. The ensuing fermentation would have not only enhanced preservation, but it would have also created highly desirable flavor and aroma characteristics.

It is believed that fermented vegetable tech-nology actually began more than 2,000 years ago. In Asia and the Far East, fermented products were made from cabbage, turnips, radishes, car-rots, and other vegetables endemic to the local areas. This technology was exported to Europe sometime in the 1500s, with regional vegeta-bles, such as European round cabbages, serving as the starting materials. Eventually, European settlers to the New World brought with them cabbages and procedures for the manufacture of sauerkraut. It is also likely that other fer-mented vegetables, pickles and olives in particu-lar, were produced and consumed in the Middle East, at least since biblical times.

Fermented vegetables have long been a staple of Middle-East, Western and Far East diets, not only because of their enhanced preservation and desirable flavor and texture properties, but also because these products had important nu-tritional properties. For example, sauerkraut has

233

long been known to have anti-scurvy properties, due to the high vitamin C content of cabbage. Sauerkraut was an essential food for navies and seafarers, who had little access to the fresh vegetables and fruits that normally would have served as a source of vitamin C. Cabbage also contains high concentrations of thiocyanates and other sulfur-containing compounds that may have antimicrobial activity. In the Far East, and Korea in particular, kimchi, has become the most popular of all fermented vegetables. Again, kimchi has unique and desirable flavor and sensory attributes, but its popularity is also due to its perceived nutritional properties (see below).

Products and Consumption

In the United States, there are essentially only three fermented vegetable products that are produced and consumed on a large scale basis. These include sauerkraut, pickles, and olives. The raw materials for these products—cabbage, cucumbers, and olives—are high moisture foods, with little protein or fat (except for olives), and just enough fermentable carbohydrate to support a fermentation (Table 7-1).

Other fermented vegetables, such as peppers, cauliflower, and green tomatoes, are also produced, but these are not nearly as popular (at least in the West). There are also many acidified or pickled vegetable products that are made by adding mixtures of vinegar, salt, and flavoring materials to fresh vegetables. In fact, most of the pickle products consumed in the United States are not fermented, but rather are simply "pickled" by packing fresh cucumbers in vinegar or salt brines (discussed later). Likewise, most of the olives consumed in the United States are similarly produced.

Although more than 800 million Kg of fermented vegetables are produced annually in the United States, on a per capita basis, Americans are not particularly heavy consumers of fermented vegetables. In the United States, per capita consumption of sauerkraut for the past ten years has averaged about 0.6 Kg. Another 0.5 Kg of olives are also consumed. Although per capita U.S. consumption of pickles is somewhat higher, at about 4 Kg, less than half of those pickles are of the fermented variety. In contrast, Germans eat about 1.8 Kg of sauerkraut per person per year, and Syrians eat nearly 6 Kg of fermented olives. Remarkably, Koreans consume more than 43 Kg per year of kimchi (or 120 g per day!).

Production Principles

In a general sense, fermented vegetable technology is based on the same principles as other lactic acid fermentations, in that sugars are converted to acids, and the finished product takes on new and different characteristics. In reality, however, the actual production of fermented vegetables occurs quite differently. For example, whereas cheese, cultured dairy products, and fermented meats are usually produced using starter cultures, the fermented vegetable industry still relies on natural lactic microflora to carry out the fermentation.

Compared to the relatively few strains used for dairy and meat fermentations, the lactic acid bacteria that are ultimately responsible for vegetable fermentations are quite diverse. Several genera are usually involved, including both heterofermentative and homofermentative species. In addition, although the plant-based substrates (i.e., cabbage, cucumbers, olives) ordinarily contain the relevant lactic acid bacteria necessary

Table 7.1. Composition of substrates used in vegetable fermentations.

Vegetable	H_2O	Carbohydrate	Protein	Fat
Cabbage	92% – 94%	5% – 6%	1%	—
Cucumbers	95%	2% – 3 %	1%	—
Olives	78% – 80%	2% – 4%	1% – 2%	12% – 14%

to perform a lactic fermentation, they also harbor a complex microflora consisting of other less desirable organisms. In fact, the resident lactic acid bacteria population represents only a small faction of the total microflora present in the starting material. And unlike dairy fermentations, where pasteurization can substantially reduce the indigenous microflora present in raw milk, no such heating step can be used to produce fermented vegetables.

Although chemical pasteurization procedures have been developed for some products and can effectively reduce the resident flora (see below), these applications, for the most part, are not widely employed. Therefore, the essential requirement for a successful fermentation is to create environmental conditions that are conducive for the lactic acid bacteria, but inhibit or otherwise restrict the non-lactic flora.

The microflora of fresh vegetables

Plant material, including edible vegetables, serves as the natural habitat for a wide variety of microorganisms (Table 7-2). The endogenous or epiphytic flora consist of yeast, fungi, and both Gram positive and Gram negative bacteria. The plant environment is exposed to the air and the surfaces of plant tissue have a high Eh. Thus, aerobic organisms, such as *Pseudomonas, Flavobacterium, Bacillus,* and various mold species, would be expected to dominate freshly harvested material, as is indeed the case.

However, facultative anaerobes, including *Enterobacter, Escherichia coli, Klebsiella,* and other enteric bacteria, as well as sporeforming clostridia, are also part of the resident flora. Various yeasts, including *Candida, Saccharomyces, Hansenula, Pichia,* and *Rhodotorula,* may also be present. Lactic acid bacteria, mainly species belonging to the genera *Lactobacillus, Pediococcus, Streptococcus,* and *Enterococcus,* are ordinarily present, but at surprisingly low numbers. In fact, whereas the total population of *Pseudomonas, Flavobacterium, Escherichia,* and *Bacillus* may well reach levels as high as 10^7 cells per gram, lactic acid bacteria are nor-

Table 7.2. Representative microflora of vegetables.

Organisms	Log CFU/g
Aerobic bacteria	4 — 6
Pseudomonas	
Flavobacterium	
Micrococcus	
Staphylococcus	
Bacillus	
Lactic acid bacteria	0.7 — 4
Lactobacillus	
Pediococcus	
Streptococcus	
Tetragenococcus	
Leuconostoc	
Enteric bacteria	3 — 3.5
Enterococcus	
Enterobacter	
Klebsiella	
Escherichia	
Yeasts and Mold	0.3 — 4.6
Fusarium	
Ascochyta	
Aspergillus	
Penicillium	
Rhodotorula	

Adapted from Nout and Rombouts, 1992

mally present at only about 10^3 cells per gram. Thus, the lactic acid bacteria are outnumbered by non-lactic competitors by a thousand times or more, putting them at a serious disadvantage.

Given the diversity of microorganisms initially present in the raw material and the numerical disparity between the lactic and non-lactic bacteria, it would seem that rather severe measures must be adopted to establish the selective environment necessary for a successful lactic acid fermentation. Actually, selection is based on only a few simple factors: salt, temperature, and anaerobiosis. Thus, under appropriate conditions, most non-lactic acid bacteria will grow slowly, if at all. In contrast, lactic acid bacteria will generally be unaffected (but not totally,) and will instead grow and produce acidic end products. The acids, along with CO_2 that may also be produced, creates an even more stringent environment for would-be competitors. Within just a few hours, lactic acid bacteria will begin to grow, a lactic acid fermentation will commence, and the number of competing organisms will decline.

The lactic acid fermentation that occurs during most vegetable fermentations depends not on any single organism, but rather on a consortium of bacteria representing several different genera and species (Table 7-3). That is, a given organism (or group of organisms) initiates growth and becomes established for a particular period of time. Then, due to accumulation of toxic end products or to other inhibitory factors, growth of that organism will begin to slow down or cease. Eventually, the initial microbial population gives way to other species that are less sensitive to those inhibitory factors. Microbial ecologists refer to these sorts of processes as a succession. This is one reason why vegetable fermentations are ordinarily conducted without starter cultures, since duplicating a natural succession of organisms likely would not be achieved on a consistent basis.

Manufacture of Sauerkraut

Few fermented foods are produced in such a seemingly simple process as is sauerkraut (Figure 7-1). Only two ingredients, cabbage and salt, are necessary, and once these ingredients are properly mixed, there is little that the manufacturer needs to do until the fermentation is completed. The simplicity of the process is reflected by the U.S. Standards, which states that sauerkraut is the "product of characteristic acid flavor, obtained by the full fermentation,

chiefly lactic, of properly prepared and shredded cabbage in the presence of not less than 2 percent nor more than 3 percent of salt." After fermentation, sauerkraut should contain not less than 1.5% acid (expressed as lactic acid).

The manufacture of sauerkraut starts with the selection of the raw substrate material. Although various cabbage cultivars exist, white cabbage is typically used because it has a mild, slightly sweet flavor and contains 5% or more fermentable sugars (mostly equimolar amounts of glucose and fructose, with a small amount of sucrose). Cabbage used to make sauerkraut should be fully mature, and should contain few outer leaves. Some manufacturers allow the cabbage heads to wilt for a day or two.

Shredding and salting

Once the outer leaves and any spoiled leaves are removed, the cabbage heads are washed and the core is drilled out. The cabbage (along with the core) is shredded (according to the manufacturers specifications) to make a slaw. The shredded leaves are then weighed and conveyed directly to tanks or are deposited first into tubs or carts and then transferred into tanks. Salt can be added as the slaw is conveyed or it can be added to the slaw when it arrives in the tanks. In either case, both the amount of salt added and the means by which mixing and distribution occur are critical.

Usually, between 2% and 2.5% salt is added (by weight), although 2.25% is generally considered to be the optimum. Problems are al-

Table 7.3. Main lactic acid bacteria involved in vegetable fermentations.

Sauerkraut	Kimchi	Pickles	Olives
Leuconostoc mesenteroides	*Leuconostoc mesenteroides*	*Leuconostoc mesenteroides*	*Leuconostoc mesenteroides*
Leuconostoc fallax	*Leuconostoc kimchii*	*Lactobacillus plantarum*	*Lactobacillus plantarum*
Lactobacillus plantarum	*Leuconostoc gelidum*	*Lactobacillus brevis*	*Lactobacillus brevis*
Lactobacillus brevis	*Leuconostoc inhae*	*Pediococcus pentosaceus*	
Pediococcus pentosaceus	*Leuconostoc citreum*		
	Lactobacillus plantarum		
	Lactobacillus brevis		
	Lactococcus lactis		
	Weissella kimchii		

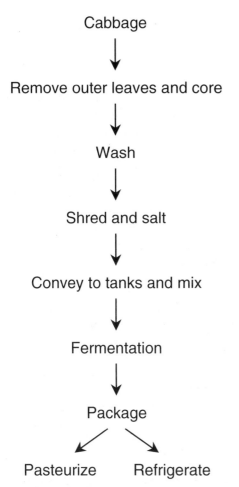

Cabbage

↓

Remove outer leaves and core

↓

Wash

↓

Shred and salt

↓

Convey to tanks and mix

↓

Fermentation

↓

Package

↙ ↘

Pasteurize   Refrigerate

Figure 7–1. Manufacture of sauerkraut. Adapted from Harris, 1998.

most certain to occur if too much or not enough salt is added or if the salt is not uniformly distributed, because salt performs several essential functions during the sauerkraut fermentation. Very soon after the salt is mixed with the shredded cabbage, water begins to diffuse out from the interior of the plant tissue to the exterior medium, due to simple osmosis. The brine that forms also contains sugars and other dissolved nutrients that diffuse out with the water. Thus, it is this water phase that ultimately serves as the location for most of the microbial activity.

Next, salt (dissolved in the brine) provides the selective conditions that discourage growth

of most of the non-lactic microorganisms that would otherwise compete with the lactic microflora. Although salt at a concentration of only 2.25% is, by itself, not ordinarily sufficient to inhibit all of the indigenous, non-lactic bacteria, it is enough to provide the lactic acid bacteria with a substantial growth advantage. Furthermore, combined with other environmental factors, the selective effects of this relatively moderate salt concentration can be increased appreciably (discussed below). Moreover, once the pH has been decreased by the production of organic acids, the combination of salt plus acid contributes significantly to the long preservation properties of the finished product. Finally, salt imparts a desirable flavor to the product and helps to maintain a crisp texture by preventing softening of the tissues.

Mixing

The shredded and salted cabbage is then placed into tanks and mixed well to distribute the salt. As noted above, mixing is an important step, because localized regions within the rather heterogenous material may contain more or less than the 2.25% salt that was added to the bulk mixture. Within those pockets, therefore, it is entirely possible that the salt concentration may vary considerably, perhaps by as much as 0.1%. This may result in either too little or too much inhibitory control over the organisms that reside in that microenvironment. If spoilage organisms were able to grow, their products (e.g., slime, pigments, off-flavors) could accumulate and, when the sauerkraut is mixed prior to packaging, contaminate the entire batch of product. It is worth noting that high salt levels can promote spoilage just as readily as low salt levels. For example, the "pink" defect (discussed below) is caused by growth of salt-tolerant yeasts that ordinarily would be suppressed by lactic acid bacteria whose growth is impaired at high salt levels.

The sauerkraut fermentation was traditionally performed in wooden barrels. Wood-stave tanks are still used; however, concrete vats are now common. The latter are lined with fiberglass or plastic, and can hold as much as 50,000 Kg. The cabbage is covered with a plastic, tarp-like material, large enough to drape over the sides of the tank. Water (or brine) is then placed on top to weigh down the cabbage and to drive out and exclude air. This also reduces exposure to air-borne organisms, foreign matter, and insects. The weight further enhances formation of a brine, which soon completely covers the shredded cabbage.

Fermentation

The sauerkraut fermentation has long been the subject of interest among food microbiologists as well as microbial ecologists. In fact, many of the biochemical and microbiological details of the sauerkraut fermentation were described as long ago as the 1930s. This interest has undoubtedly been due, in large part, to the very nature of the fermentation process, in that it involves several different naturally-occurring microorganisms acting as part of a complex ecosystem. Recent reports suggest that bacteriophages may also play an important role in the microbial ecology of the sauerkraut fermentation (Box 7-1).

The manufacture of sauerkraut and many other fermented vegetables depends on a succession of organisms that are naturally present in the raw material. Some appear early on in the fermentation, perform a particular function, and then, for all practical purposes, disappear from the product. Other organisms, in contrast, emerge later in the fermentation and then remain at moderate to high levels throughout the duration of the fermentation and post-fermentation process. However, growth of those organisms that occur late in the process depends on those organisms that had grown earlier and that had established the correct environmental conditions.

Microbial activity begins as soon as a brine has formed. Initially, the atmosphere is aerobic, with redox potentials (or Eh values) of over 200 mV. However, the combined effects of physical exclusion of air and residual respiration and oxygen consumption by plant cells quickly reduce the Eh and make the environment anaerobic. Thus, pseudomonads, fungi, and other obligate aerobic microorganisms that may initially be present at high levels, have little opportunity for growth. Some of these organisms are also salt-sensitive, further reducing their ability to grow in this environment. Still, at the temperatures used during the sauerkraut fermentation (20°C to 25°C), many other indigenous salt-tolerant, mesophilic, facultative organisms might be expected to grow, including *Enterobacter*, *E. coli*, *Erwinia*, and other coliforms. Instead, these organisms persist for only a short time, perhaps as little as a few hours, due to competition by lactic acid bacteria and the inhibitory effects of the acids produced by these bacteria.

The lactic fermentation in sauerkraut occurs in a series of overlapping stages or sequences. These stages and the succession of microorganisms associated with each stage have been very well studied. Remarkably, the fermentation almost always follows the exact same pattern (Figure 7-2).

The first stage, variously referred to as the initiation or heterolactic or gaseous phase, is marked by growth of *Leuconostoc mesenteroides*. This organism is salt-tolerant and has a relatively short lag phase and high growth rate at low temperatures (15°C to 18°C). Importantly, it metabolizes sugars via the heterofermentative pathway, yielding lactic and acetic acids, CO_2, and ethanol. The acidic environment (0.6% to 0.8%, as lactic acid) created by growth of *L. mesenteroides* not only inhibits non-lactic competitors, but it also favors other lactic acid bacteria. The production of CO_2 also contributes to making the environment even more anaerobic (as low as -200 mV), which again favors the more anaerobic lactic acid bacteria. Eventually, however, as the acid concentration approaches 1.0%, *L. mesenteroides* is, itself, inhibited, and within four to six days, this organism is barely detectable.

In the next stage or primary, homolactic, or non-gaseous phase, the decrease in the

Box 7–1. Bacteriophages Get Into the Mix

To microbial ecologists, there are few model systems better suited for the study of microbial interactions than that which occurs during the manufacture of sauerkraut and other fermented vegetables. Consider what transpires in the course of a typical vegetable fermentation. There are, for example, major shifts in the environment, from aerobic to anaerobic, and from a neutral to acidic pH. The availability of nutrients, and fermentable carbohydrates, in particular, decreases during the fermentation. At the same time, there is a succession of microorganisms, such that some species, present at the outset, are displaced by other microbial communities, and cannot even be detected at later stages.

In addition to these environmental and extrinsic factors, bacteriophages have recently been recognized as having a major influence on the microbial ecology and diversity of fermented vegetables (Lu et al., 2003 and Yoon et al., 2002). Phages, it now appears from these studies, are widespread in commercial sauerkraut fermentations and may directly influence microbiological succession during the fermentation. Over a two-year period, these investigators isolated more than twenty-six distinct phage types that were capable of infecting lactic acid bacteria, including species of *Leuconostoc, Lactobacillus,* and *Weissella* sp.

Importantly, the appearance of some members of the natural microflora was correlated with the appearance of their homologous phages. That is, phages capable of infecting *Leuconostoc* were only isolated during the first few days of the fermentation, when the flora are dominated by *Leuconostoc*. Phages that infect *Lactobacillus*, in contrast, were not found until later in the fermentation, when *Lactobacillus plantarum* and other species had become established. Thus, these phage infection events coincided with the main population shift that occurs during the sauerkraut fermentation, from mainly heterofermentative species to mainly homofermentative species.

Further characterization of these phages revealed that some were stable in low pH environments (pH 3.5) and were capable of persisting in the tanks for as long as sixty days. It appears from this work, therefore, that the well-studied phenomenon of succession that occurs during the sauerkraut fermentation may be mediated not just by the changing environment, but also by the emergence of bacteriophages.

There is another very practical reason why bacteriophages are important in vegetable fermentations. Although many manufacturers still rely on natural fermentations to produce these products, starter cultures are now being used more frequently in pickle production, and applications for their use in the manufacture of other fermented vegetables is expected to increase. However, like most industrial fermentation processes in which pure culture strains are used on a repeated basis, infective bacteriophages will invariably emerge. If the phage population in a pickle production environment, for example, was able to reach some critical level, infection and lysis of starter culture strains may occur to the point at which the fermentation fails.

Because cucumbers (and other raw materials used in fermented vegetable manufacture) cannot be heated, nor are these fermentations conducted under aseptic conditions, it is not possible to exclude phages from the production environment. Thus, the continuous use of starter cultures in large-scale manufacture of pickles and other fermented vegetables may provide ideal conditions for phage proliferation. Although there are no reports of phage problems in this industry, it has been suggested that development of phage control systems may still be warranted.

References

Lu, Z., F. Breidt, V. Plengvidhya, and H.P. Fleming. 2003. Bacteriophage ecology in commercial sauerkraut fermentations. Appl. Environ. Microbiol. 69:3192–3202.

Yoon, S.S., R. Barrangou-Poueys, F. Breidt, Jr., T.R. Klaenhammer, and H.P. Fleming. 2002. Isolation and characterization of bacteriophages from fermenting sauerkraut. Appl. Environ. Microbiol. 68:973–976.

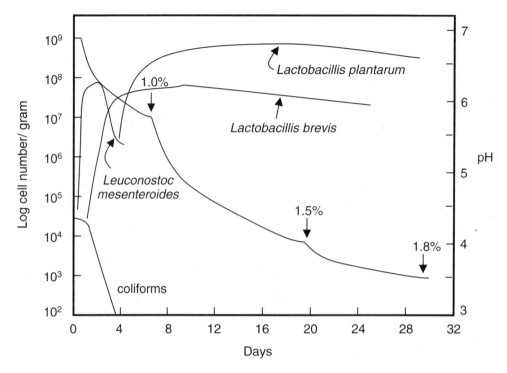

Figure 7–2. Fermentation succession. Idealized model for successive growth of lactic acid bacteria during the sauerkraut fermentation. The approximate acidities (as lactic) at varying pHs are indicated. Adapted from multiple sources.

Leuconostoc population coincides with the succession of several other lactic acid bacteria, most notably *Lactobacillus plantarum* and, to a lesser extent, *Lactobacillus brevis*. Although *L. plantarum* is a facultative heterofermentor (meaning it has the metabolic capacity to ferment different sugars via homo- or heterofermentative pathways) and *L. brevis* is an obligate heterofermentor, both organisms are strong acid producers, nearly doubling the acid content to about 1.4% to 1.6%. They are also quite stable in this acidic environment and dominate the fermentation during this period (especially *L. plantarum*). It is not unusual, however, for other lactic acid bacteria, including *Pediococcus pentosaceus*, *Pediococcus acidilactici*, *Lactobacillus cuvatus*, and *Enterococcus sp.*, to be present during the primary fermentation. Finally, as the acidity approaches 1.6% and the pH decreases below 4.0, only the acid-tolerant *L. plantarum* is able to grow. The entire process can take up to one to two months, and the fermentation is generally considered complete when the acidity is at about 1.7%, with a pH of 3.4 to 3.6.

End products

Although lactic acid is the major compound produced during the fermentation, other metabolic end products are also formed. Importantly, many of these products contribute to the overall flavor of sauerkraut. In particular, end products produced by *Leuconostoc* sp. and other heterofermentative lactic acid bacteria are essential for good-tasting sauerkraut. As much as 0.3% acetic acid and 0.5% ethanol can be present in the finished sauerkraut. In addition, these bacteria may also synthesize small amounts of diacetyl, acetaldehyde, and other volatile flavor compounds. Finally, the CO_2 that accumulated during the initiation stage of the fermentation provides carbonation and enhances mouth feel.

Mannitol is another end product that accumulates during the sauerkraut fermentation. It is formed directly from fructose by heterofermentative leuconostocs and lactobacilli via the NADH-dependent enzyme, mannitol dehydrogenase. As shown in Figure 7–3, regeneration of NAD in the heterofermentative pathway ordinarily occurs twice, first when acetyl CoA is reduced to acetaldehyde and then when acetaldehyde is reduced to ethanol. However,

fructose, when available in excess, can serve as an alternative electron acceptor. This allows the cell to use acetyl phosphate as a phosphoryl group donor in the acetate-generating reaction catalyzed by acetate kinase. As a consequence, one additional molecule of ATP is synthesized by substrate level phosphorylation.

More than 100 mM mannitol can be formed in sauerkraut by this pathway (with a commensurate amount of ATP made for the cell). Eventually, some of this mannitol may be consumed by lactobacilli that emerge later in the fermentation. Although most of the mannitol is produced directly from fructose, the appearance of mannitol even when fructose has been

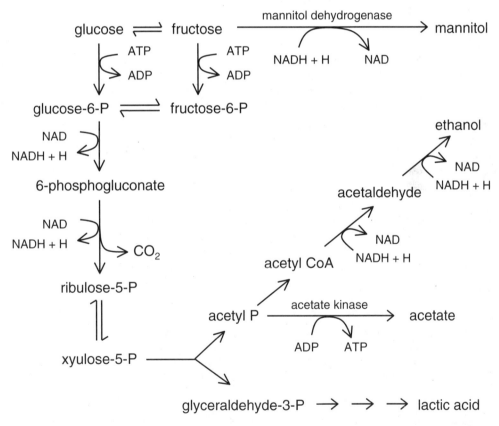

Figure 7–3. Mannitol formation in *Leuconostoc*. Phosphorylated glucose and fructose are metabolized by the heterofermentative (i.e., phosphoketolase) pathway. The reactions leading to lactic acid (from glyceraldehyde-3-P) are not shown. Only the relevant enzymes are indicated; enzymes for all other reactions are given in Chapter 2. P=Phosphate. Adapted from Salou et al., 1994 Appl. Environ. Microbiol. 60:1459-1466.

depleted suggests that some might be formed indirectly from glucose (following its conversion to fructose via glucose isomerase).

Packaging and processing

In the United States, commercial products are usually thermally processed, much like other high-acid foods at about 75°C, prior to packaging in cans or jars. Such products are essentially commercially sterile and are stable at room temperature. There is also a market for non-pasteurized, refrigerated sauerkraut that is packaged in glass jars or sealed plastic bags (polybags). These products also have a long shelf-life, provided antimycotic agents, such as benzoate and sulfite salts, are added and the product is kept cold.

Spoilage and defects

Although chemical or physical reactions are occasionally responsible for sauerkraut spoilage, most defects are caused by microorganisms (Table 7-4). Defects are more likely to occur if the production conditions are not properly controlled, leading to deviations in the fermentation pattern.

The two most common factors that influence the fermentation and that may lead to quality defects are temperature and salt. If, for example, the temperature is too high (>30°C) or if too much salt is added (>3%), *L. mesenteroides* will be in-

hibited, heterofermentative end-products will be absent, and the flavor will be harsh. Worse yet, when prompt acid formation is delayed by high salt conditions (or pockets of high salt, due to uneven mixing), growth of salt-tolerant yeasts may occur. Growth of *Rhodotorula* sp. is a particular problem, due to the undesirable pink pigment this yeast produces.

If, in contrast, the temperature is too low (<10°C) or too little salt is added (<2%), various Gram negative bacteria, including *Enterobacter*, *Flavobacterium*, and *Pseudomonas*, may grow. Some of these bacteria are capable of producing pectinolytic enzymes that cause a "soft kraut" defect, resulting from pectin hydrolysis.

Finally, some strains of *L. mesenteroides* produce dextrans and other polysaccharides that give a slimy or ropy texture to the product. Other lactic acid bacteria, including *L. plantarum*, may produce capsular materials that cause similar texture defects.

Kimchi

As noted previously, the Korean version of sauerkraut is called *kimchi*. Although the manufacture of kimchi and sauerkraut are very similar, the characteristics of the finished products are quite different. The main difference between sauerkraut and kimchi is that the latter product contains ingredients other than simply cabbage and salt. Kimchi, for example, is usually made from cabbage, but other vegeta-

Table 7.4. Microbial defects in fermented vegetables.

Product	Defect	Causative organisms
Sauerkraut	Pinking	*Rhodotorula*
	Ropy/slimy	*Leuconostoc, Lactobacillus*
	Softening	*Enterobacter, Flavobacterium, Pseudomonas*
Pickles	Bloating/floating	*Bacillus, Aeromonas, Achromobacter, Aerobacter, Fusarium, Penicillium*
	Softening	*Fusarium, Penicillium, Ascochyta*
Olives	Gassy/floater/fish eyes	*Enterobacter, Citrobacter, Klebsiella, Escherichia, Aeromonas, Hansenula, Saccharomyces*
	Softening	*Bacillus, Aeromonas, Achromobacter, Aerobacter, Fusarium, Penicillium*
	Zapatera/malodorous	*Clostridium, Propionibacteria*

bles, including radishes and cucumbers, can also be used, alone or mixed with cabbage. Other vegetables, spices, and flavoring agents are also commonly added to kimchi, depending on the particular type of kimchi being produced. Garlic, green onion, ginger, and red peppers are among the typical ingredients, but fish, shrimp, fruits, and nuts, can also be added. Kimchi, therefore, has a much more complex flavor and texture profile. However, kimchi has also been suggested to have unique nutritional properties, conferred in part by the raw materials, but also by the fermentative microorganisms and their end products (Box 7–2).

Box 7–2. Health Properties of Kimchi

In Korea, kimchi is arguably the most popular of all fermented foods. As many as 100 different types are produced using various raw materials and processes. On an annual basis, about 360 million kg are produced commercially, with a large amount also made directly in homes. Per capita consumption is more than 100 grams per day, accounting for about 12% of the total food intake. In Korea, it is common to eat kimchi at every meal, all year round.

The manufacturing steps involved in kimchi production are nearly the same as those used for its Western counterpart, sauerkraut (Figure 1). The main difference is that kimchi contains other ingredients that impart considerably more flavor and texture properties. While the popularity of

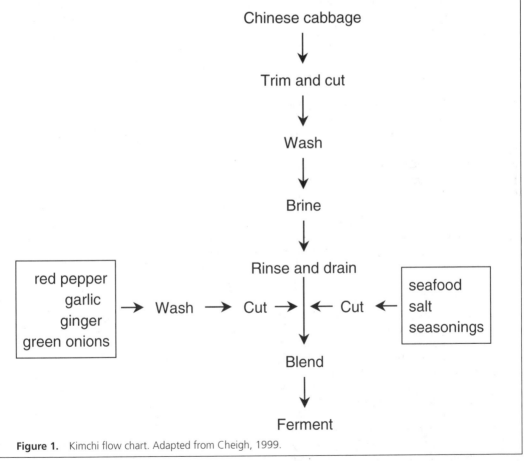

Figure 1. Kimchi flow chart. Adapted from Cheigh, 1999.

(Continued)

Box 7–2.   Health Properties of Kimchi *(Continued)*

kimchi can certainly be ascribed to the combination of these desirable sensory characteristics, there are also nutritional reasons that contribute to the its widespread consumption. For example, kimchi contains appreciable amounts of ascorbic acid (vitamin C), B vitamins, calcium, iron, potassium, dietary fiber, and naturally occurring antioxidants. It is also important to note that kimchi is most often eaten in its uncooked or raw state, and that a live microflora is ordinarily present.

Kimchi has been recognized as having other important nutritional and health-promoting properties and has even been regarded as a functional food (i.e., those with health benefits beyond their ordinary nutritional constituents). It is thought to reduce the risks of stomach, colon, and liver cancers, due to the putative antimutagenic and anticarcinogenic constituents present in kimchi. Although convincing evidence is still needed to support most of the claims, epidemiological studies indicate that consumption of kimchi may be associated with reduced incidence of various disease states among the Korean population. For example, a significant decrease in the risk of developing gastric cancer was correlated with increased intake of kimchi made from low-nitrate vegetables, such as Chinese cabbage (Kim et al., 2002). The same study, however, also showed that consumption of kimchi made from high-nitrate vegetables actually increased the risk.

If, indeed, kimchi has health-promoting properties beyond those present in the raw material, then those effects must be due either to the kimchi microflora directly, to products produced by the microflora during the fermentation, or to transformation of kimchi constituents. Various lactic acid bacteria isolated from kimchi have been reported to have antimutagenic activity, based on Ames tests. For example, *Lactobacillus plantarum* KLAB21 was found to produce extracellular glycoproteins that had antimutagenic activity against aflatoxin B1 and several chemical mutagens (Rhee and Park, 2001). A methanol extract from kimchi similarly reduced induced cytotoxicity in C3H/10T1/2 cells (Cheigh, 1999). In another recent study, four *Leuconostoc mesenteroides* strains isolated from kimchi were able to deplete nitrite, a potential carcinogen, from broth cultures, suggesting that this property would contribute to the safety of kimchi (Oh et. al., 2004). An active chemopreventative component was also recently identified in kimchi extracts (Park et. al., 2003). This plant-derived compound, β-sitosterol, inhibited proliferation of human cancer cells by decreasing DNA synthesis and cell signaling pathways. What role the fermentation plays in this process is not clear, however. Finally, it was recently suggested by researchers in Korea that kimchi somehow provided protection against the viral agent that causes severe acute respiratory syndrome (SARS). This claim was based largely on the observation that while SARS was widespread in China, Hong King, and Taiwan in 2003, there were few suspected cases in South Korea.

Kimchi is made from a wide variety of raw materials, including several that may have biological activity (e.g., garlic, radishes, peppers, ginger, and onion). The lactic acid microflora is also complex and variable. Thus, identifying the actual components that may be responsible for the suggested health benefits of kimchi, and then justifying the nutritional claims, remains a significant challenge.

References

Cheigh, H. 1999. Production, characteristics and health functions of kimchi. *In* Proceedings of the International Symposium on the Quality of Fresh and Fermented Vegetables, p. 405—419, J.M. Lee, K.C. Gross, A.E. Watada, and S.K. Lee (ed). Acta Hort. 483. ISHS.

Kim, H.J., W.K. Chang, M.K. Kim, S.S. Lee, B.Y. Choi. 2002. Dietary factors and gastric cancer in Korea: a case-control study. Int. J. Cancer 97:531–535.

Oh, C.K., M.C. Oh, and S.H. Kim. 2004. The depletion of sodium nitrite by lactic acid bacteria isolated from kimchi. J. Med. Food 7:38–44.

Park, K.-Y., E.-J. Cho, S.-H. Rhee, K.-O. Jung, S.-Y. Yi, and B.Y. Jhun. 2003. Kimchi and an active component, β-sitosterol, reduce oncogenic H-Ras^{v12}-induced DNA synthesis. J. Med. Food. 6:151–156.

Rhee, C.-H., and H.-D. Park. 2001. Three glycoproteins with antimutagenic activity identified in *Lactobacillus plantarum* KLAB21. Appl. Environ. Microbiol. 67:3445–3449.

Principles of Pickle Production

In a very general sense, pickles refer to any vegetable (or fruit) that is preserved by salt or acid. Certainly, the vegetable most often associated with pickles is the cucumber. Currently, about half of the total cucumber crop (1 billion Kg) in the United States is used for pickles. The acid found most often in pickled products is lactic acid, derived from a lactic fermentation. As previously noted, however, not all pickles are fermented. In fact, less than half of the pickles consumed in the United States undergo a lactic acid fermentation. Rather, acetic acid can be added directly as the pickling acid, omitting the fermentation step. The pickle slice on the top of a fast-food hamburger is probably not the fermented type.

Pickles and pickling technology have a long and rich history. The cucumber was brought from India to the Middle East about 4,000 years ago, and cured versions were eaten at least 3,000 years ago. Cleopatra endorsed pickle consumption, claiming they were responsible (in part) for her beauty. The ancient Roman historian and natural scientist Pliny the Elder and his contemporaries, Roman emperors Julius Caesar and Tiberius, were all fond of pickles. Columbus introduced pickles to the Americas, eventually leading to the origins of a pickle industry on the Lower East Side of New York City. American founding fathers George Washington, John Adams, and Thomas Jefferson reportedly derived inspiration from the pickle.

In the United States, pickles are generally divided into three different groups, based on their means of manufacture. Fresh-packed pickles are simply cucumbers that are packed in jars, covered with vinegar and other flavorings, then pasteurized by heat. They have a long shelf-life, even at room temperature. Fresh packed pickles are crisp, mildly acidic, and are the most popular. Refrigerated pickles are also made by packing cucumbers jars with vinegar and various flavorings, but they are not heated. Instead these pickles are refrigerated, giving them a crisp, crunchy texture and bright green color. Although a slight fermentation may occur, refrigerated pickles have a shorter shelf-life than fresh-packed pickles. Sodium benzoate is usually added as a preservative. The manufacture of both fresh-packed and refrigerated-style pickles is fast and easy and require few steps (Figure 7-4).

The only pickles that are fully fermented are those referred to as salt-stock or genuine pickles. They may also be referred to as processed, although this may be somewhat confusing since non-fermented pickles can be made into relishes and other processed pickle products. Fermented pickles are nearly as popular as fresh-packed pickles. They have a distinctly different flavor and texture compared to fresh-packed or refrigerated pickles, and take much longer to make. Fermented or processed pickles also have a very long shelf-life, about two years. Details regarding their manufacture and the fermentation process are described below.

It should be noted that within these three groups, pickles can be further distinguished based on the types of spices, herbs and flavoring agents used, the size or type of cucumber (gherkins and midgets), and the form or shape of the pickle (i.e., whole, spear, slice, etc.). For example, dill pickles, the most popular, refer to any type of pickle to which dill weed (either the seed or oil) is added. If the dill pickles are labeled as genuine dill, it means the pickles are of the processed type and are dill-flavored. Otherwise, dill pickles are usually made from fresh-packed or refrigerated pickles. Other common flavored pickle types include sweet, bread-and-butter, kosher (style), and garlic.

Manufacture of fermented pickles

The actual process steps used for the manufacture of fermented pickles are similar to those used for making sauerkraut. Both rely on salt, oxygen exclusion, and anaerobiosis to provide the appropriate environmental conditions necessary to select for growth of naturally-occurring lactic acid bacteria. There are, however, several differences between pickle and sauerkraut fermentations. First, salt concentrations are higher than those used for sauerkraut, resulting in the development of a less diverse microflora. In

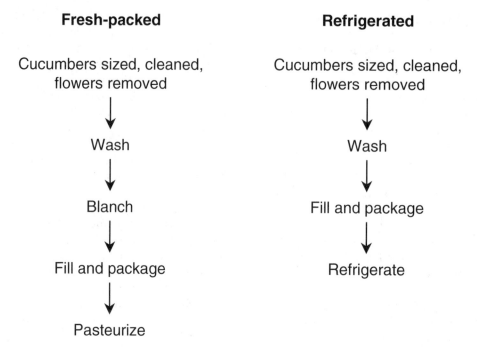

Figure 7–4. Non-fermented pickles and their manufacture. Adapted from Harris, 1998.

addition, a brine, rather than dry salt, is used for pickle fermentations. Finally, the pickle fermentation process, unlike sauerkraut, is amenable to the use of pure starter cultures and a more controlled fermentation. Indeed, such cultures are now available and some (but not many) pickle manufacturers have adopted controlled fermentation processes (Box 7-3).

The manufacture of fermented pickles starts with selection and sorting of cucumbers (Figure 7-5). Only small or immature cucumbers, harvested when they are green and firm, are used for pickles. They are then washed, sorted, and transferred to tanks, and a brine solution is added.

The brine typically contains at least 5% salt (or about 20° salometer, where 100° salometer = 26% salt). Because the cucumbers-to-brine ratio is nearly 1:1, the actual salt concentration is actually less. For so-called salt stock pickles, which may be held in bulk for long periods, the initial brine may contain 7% to 8% salt, which is followed by the addition of more salt to raise the total salt concentration to above 12%. For gen-

uine dill-type pickles, the brine concentration is usually between 7.5% and 8.5%. Dill weed is also added, usually in the seed or oil form.

Care must be taken when weighing down the pickles, because the buoyancy of the cucumbers may cause those at the top to become damaged. The large tanks used by large pickle manufacturers are usually located outdoors, and temperatures may, therefore, vary between 15°C to 30°C. The lower the temperature, the longer it takes to complete the fermentation. Thus, in Michigan (the largest northern producer of pickles), fermentation may require up to two months, whereas in North Carolina (the main southern producer), fermentations may be complete in three weeks. At the end of the fermentation, the pH will be about 3.5, with acidities between 0.6% and 1.2% (as lactic).

Pickle fermentation

As noted above, the high salt concentrations used in pickle manufacturing cause the fermentation to proceed quite differently from

Box 7–3. Starter Cultures and Fermented Vegetables

The modern manufacture of most fermented vegetables, in contrast to cheese, sausage, and other fermented food products, still relies on a natural fermentation. In large part, this is because vegetable fermentations occur as a succession, and duplicating this process with "controlled fermentations" using starter cultures has not been a viable option. Also, vegetable fermentations are often conducted in less than aseptic conditions, so adding a culture to a raw material comprised of a complex, well-populated background flora is unlikely to be very effective. Finally, paying for cultures to perform a step that ordinarily costs the manufacturer nothing makes little economic sense.

Yet, for all of the same reasons that eventually drove other industries to adopt pure starter culture technology (i.e., consistency, control, safety, and convenience), the fermented vegetable industries have indeed developed manufacturing procedures that depend on starter cultures, rather than the natural flora, to perform the fermentation. Research on pure culture technology for fermented vegetables actually began nearly fifty years ago, and cultures were developed in the 1960s (Daeschel and Fleming, 1987). Despite the availability of these cultures, however, they have not been widely used.

More recently, other factors have provided additional motivation for the use of starter cultures. One particularly relevant issue relates to the large volumes of high salt brines that are generated by pickle and olive fermentations. These salt solutions create significant environmental problems. The most obvious way to reduce the discharge of this material into the environment is to use less salt. However, since salt provides the major means for controlling the microflora, conducting a natural fermentation at low-salt concentrations is unlikely to be satisfactory. In contrast, less salt could be used if the background flora was controlled by other means (see below), and a starter culture was used instead to dominate the environment and to carry out the fermentation.

The first requirement for performing a controlled fermentation is to remove and/or inactivate the endogenous microflora. This can be done with chemical agents that kill organisms at the surface. Specifically, the raw product is washed first with dilute chlorine solutions, then with acetic acid. In reality, chemical pasteurization is possible only for cucumbers and olives and not for shredded cabbage due to surface area considerations. Although olives can tolerate a modest heat treatment, cabbage and cucumbers suffer severe texture defects if heated. Finally, a nitrogen purge drives out air and creates anaerobic conditions, and an acetate buffered brine is then added, followed by the starter culture.

The organisms that are currently available or are being considered for use as starter cultures include many of the same species ordinarily isolated from vegetable fermentations, e.g., *Lactobacillus plantarum*, *Leuconostoc mesenteroides*, *Pediococcus acidilactici*, and *Lactobacillus brevis*. Strain selection, however, is necessary and must be based on the specific application and desired characteristics of the particular fermented food (Table 1). For example, strains to be used as starter cultures for fermented olives must resist the antimicrobial phenolic compounds ordinarily present in olives.

As noted above, another inherent problem in controlled fermentation technology is the difficulty in establishing a microbial succession, such that a heterofermentative phase always precedes the homofermentative phase. That is, how can *L. mesenteroides* and *L. plantarum* both be added at the outset of a sauerkraut fermentation, yet have conditions controlled such that growth of *L. plantarum* is delayed until the later stages of the fermentation? Researchers at USDA and North Carolina State University in Raleigh, North Carolina, addressed this problem by developing a clever bio-controlled fermentation process (Breidt et al., 1995; Harris et al., 1992). In this model system, the starter culture consisted of an *L. mesenteroides* strain that resisted the bacteriocin nisin. As hypothesized, in the presence of nisin (either added directly or produced by a companion nisin-producing strain of *Lactococcus lactis*), the *L. mesenteroides* strain grew

(Continued)

Box 7–3. Starter Cultures and Fermented Vegetables *(Continued)*

fine, but growth of nisin-sensitive *L. plantarum* was inhibited. Thus, it was possible to prolong the heterofermentative (and flavor-generating) phase of the fermentation, while delaying the homofermentative phase.

Ultimately, the use of starter cultures for the manufacture of fermented vegetables is likely to increase as the size of the production facilities and the demand for speed, efficiency, and throughput both increase. In addition, the starter culture industry is now able to develop strains that have specific physiological properties, satisfy specific performance characteristics, are stable during storage, and are easy to use.

Table 1. Properties of starter cultures for fermented vegetables.

Minimum nutritional requirements
Able to grow at low temperatures
Able to ferment diverse carbohydrate substrates
Able to compete against wide array of organisms
Able to produce desirable flavor
Rapid growth and acid production
Tolerant to acids and low pH
Tolerant to salt
Tolerant to antimicrobial phenolics
Resistant to bacteriophage
Non-pectinolytic
Unable to produce dextrans or other polysaccharides
Unable to produce biogenic amines
Minimum loss of viability during storage

References

Breidt, F., K.A. Crowley, and H.P. Fleming. 1995. Controlling cabbage fermentations with nisin and nisin-resistant *Leuconostoc mesenteroides*. Food Microbiol. 12:109–116.

Daeschel, M.A., and H.P. Fleming. 1987. Achieving pure culture cucumber fermentations: a review, p. 141–148. *In* G. Pierce (ed.), Developments in industrial microbiology. Society for Industrial Microbiology, Arlington, Va.

Harris, L.J., H.P. Fleming, and T.R. Klaenhammer. 1992. Novel paired starter culture system for sauerkraut, consisting of a nisin-resistant *Leuconostoc mesenteroides* strain and a nisin-producing *Lactococcus lactis* strain. Appl. Environ. Microbiol. 58:1484–1489.

that in sauerkraut. Only those pickles made using brines at less than 5% salt will allow for growth of *L. mesenteroides*. Although heterofermentative fermentations may promote more diverse flavor development, the formation of CO_2 is undesirable, because it may lead to bloater or floater defects (see below). Moreover, low salt brines may also permit growth of unwanted members of the natural flora, including coliforms, *Bacillus*, *Pseudomonas,* and *Flavobacterium*. At salt concentrations between 5% and 8%, growth of *Leuconostoc* is inhibited and instead the fermentation is initiated by *Pediococcus* sp. and *L. plantarum*.

Pickle fermentation brines typically contain high concentrations of salt and organic acids and have a pH less than 4.5. These conditions are especially inhibitory to coliforms, pseudomonads, bacilli, clostridia, and other non-lactic acid bacteria that would otherwise cause flavor and texture problems. This environment, in fact, is hard even on lactic acid bacteria. However, the latter have evolved sophisticated physiological systems that enable them to survive under very uncomfortable circumstances (Box 7-4).

After fermentation, salt stock pickles can be held indefinitely in the brine. However, these pickles cannot be eaten directly, but rather

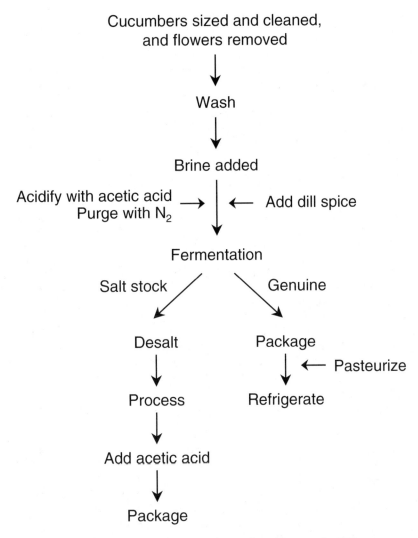

Figure 7–5. Manufacture of fermented pickles. Adapted from Harris, 1998.

must be de-salted by transfer to water. After several changes (a process called refreshing), the salt concentration is reduced to about 4%. They are then used primarily for relishes and other processed pickle products.

Defects

Among the microbial defects that occur in pickles, the most common are bloaters and floaters (Table 7–4). The defect is caused by excessive gas pressure that subsequently results in internal cavity formation within the pickles. The CO_2 gas is mainly produced by heterofermentative lactic acid bacteria (some of which may produce CO_2 via the malolactic fermentation), although coliforms and yeasts may also be responsible. Floaters and bloaters can still be used for some processed products (i.e., relish), however, they cannot be used for the whole or sliced pickle market. The most common way to control or minimize this defect is to remove dissolved CO_2 by flushing or purging the brine with nitrogen gas.

Box 7–4. In a Pickle: How Lactic Acid Bacteria Deal with Acids and Salts

The key requirement to ensure the success of vegetable fermentations is to create a restrictive, if not inhospitable, environment, such that most indigenous microorganisms are inhibited or otherwise unable to grow. Typically, this is initially accomplished by adding salt, excluding oxygen, and maintaining a somewhat cool temperature. As lactic acid bacteria grow and produce organic acids and CO_2, the ensuing decreases in pH and Eh provide additional hurdles, especially for salt-sensitive, neutrophilic, aerobic organisms. These conditions, however, may not only affect resident enteric bacteria, pseudomonads, clostridia, fungi, and other undesirable microorganisms, but they can also impose significant problems for the lactic acid bacteria whose growth is to be encouraged.

In addition, there may be other ionic compounds present in the vegetable juice brine that interfere with growth of lactic acid bacteria. For example, acetate, lactate, and other buffer salts are often added to brines, especially when pure cultures are used to initiate the fermentation. These acids and salts may have significant effects on cell metabolism and growth (Lu et al., 2002). Thus, how plant-associated lactic acid bacteria cope with these challenges may be of practical importance.

Lactic acid bacteria are prolific producers of lactic acid (no surprise there), and can tolerate high lactic acid concentrations (>0.1 M) and low pH (<3.5), much more so than most of their competitors. At least several physiological strategies have been identified that enable these bacteria to tolerate high acid, low pH conditions.

First, lactobacilli and other lactic acid bacteria can generate large pH gradients across the cell membrane, such that even when the medium pH is low (e.g., 4.0), the cytoplasmic pH (the relevant pH for the cell's metabolic machinery) is always higher (e.g, 5.0). For example, in one study (McDonald et al., 1990), *Lactobacillus plantarum* and *Leuconostoc mesenteroides* maintained pH gradients of nearly 1.0 or higher, over a medium pH range of 3.0 to 6.0 (acidified with HCl). In the presence of lactate or acetate, however, somewhat lower pH gradients were maintained. This is because organic acids diffuse across the cytoplasmic membrane at low pH (or when their pKa nears the pH), resulting in acidification of the intracellular medium. For some bacteria, e.g., *L. mesenteroides*, the pH gradient collapses at low pH (the so-called critical pH). At this point, the cell is in real trouble, as enzymes, nucleic acid replication, ATP generation, and other essential functions are inhibited. In general, these results reflect, and are consistent with the observed lower acid tolerance of *L. mesenteroides,* as compared to *L. plantarum.*

If maintenance of a pH gradient is important for acid tolerance, then the next question to ask is how such a gradient can be made. That is, how can the protons that accumulate inside the cell and cause a decrease in intracellular pH be extruded from the cytoplasm? Although there are actually several mechanisms for the cell to maintain pH homeostasis, one specific system, the proton-translocating F_0F_1-ATPase (H^+-ATPase) is most important. This multi-subunit, integral membrane-associated enzyme pumps protons from the inside to the outside using ATP hydrolysis as the energy source (Figure 1A). This enzyme is widely conserved in bacteria (in fact, throughout nature), but the specific properties of enzymes from different species show considerable variation. Thus, the H^+-ATPases from lactobacilli have a low pH optima, accounting, in large part, for the ability of these bacteria to tolerate low pH relative to less tolerant lactic acid bacteria.

Although the H^+-ATPase system is the primary means by which lactic acid bacteria maintain pH homeostasis, other systems also exist (Figure 1A). For example, deamination of the amino acid arginine releases ammonia, which raises the pH. Decarboxylation of malic acid, which is commonly present in fermented vegetables, also increases the pH by conversion of a dicarboxylic acid to a monocarboxylic acid. In fact, when lactobacilli and other lactic acid bacteria are exposed to low pH, a wide array of genes are induced (Van de Guchte et al., 2002). Collectively, this adaptation to low pH is referred to as the acid tolerance response. Some of the induced genes code for proteins involved in the machinery used by the cell to deal with other

Box 7—4. In a Pickle: How Lactic Acid Bacteria Deal with Acids and Salts *(Continued)*

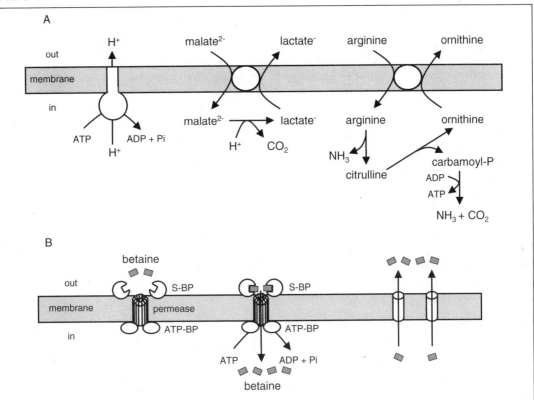

Figure 1. pH and osmotic homeostasis in *Lactobacillus plantarum*. Panel A shows three main systems whose function is to maintain pH homeostasis in *L. plantarum*. The F_0F_1-ATPase (left) is a primary proton pump that extrudes protons from the inside to the outside, using ATP as the energy source. In contrast, the malate and arginine systems rely on product efflux to drive uptake (no energy is required). In the malolactate system (center), proton consumption de-acidifies the medium and raises the pH. In the arginine diiminase system (right), medium pH is raised by virtue of the two molecules of NH_3 that are released per mole of arginine.

Panel B shows the QacT system responsible for osmotic homeostasis in *L. plantarum*. The components of this putative opuABCD-encoded system (left) include membrane-associated substrate-binding proteins (S-BP), a betaine permease, and ATP-binding proteins (ATP-BP). When the osmotic pressure is high, betaine is bound by the S-BP and taken up by the permease (center). Transport is driven by an ATPase following ATP-binding. If the osmotic pressure is reduced, accumulated betaine is effluxed via membrane channels (right).

physical or chemical stresses. Thus, the acid tolerance response may not only protect the cell against low pH, but also heat and oxidative stress.

In vegetable fermentations, the other important stresses encountered by lactic acid bacteria are high salt concentrations and high osmotic pressures. Salt concentrations in sauerkraut brines are around 0.4 M, giving an osmolality of about 0.8 Osm. Pickle and olive brines may contain more than 1.0 M salt, or osmolalities above 2 Osm. Salt is an extremely effective antimicrobial agent due to its ability to draw water from the cytoplasm, thereby causing the cell to become dehydrated, lose turgor pressure, and eventually plasmolyze (and die!). The ability of lactic acid bacteria to tolerate high salt conditions varies (just as it did for acid tolerance), depending on the organism. Given that *L. plantarum* usually dominates high salt fermentations, it should

(Continued)

Box 7–4. In a Pickle: How Lactic Acid Bacteria Deal with Acids and Salts *(Continued)*

be no surprise that this organism has also evolved physiological and genetic mechanisms that make it salt- and osmotolerant.

Like acid tolerance, the salt tolerance system depends on the activity of membrane pumps. However, in the latter case, the pumps are actually transport systems that take up a special class of molecules called compatible solutes. By accumulating these non-toxic solutes inside the cytoplasm to high concentrations, cell water is retained and osmotic homeostasis is maintained.

Compatible solutes are also referred to as osmoprotectants because they not only maintain osmotic balance, they also protect enzymes, proteins, and other macromolecules from dehydration and misfolding. Among the osmoprotectants accumulated by *L. plantarum*, the quaternary amine, glycine betaine (or simply betaine) is the most effective. Potassium ion, glutamate, and proline are also accumulated, but to lower concentrations, at least in lactobacilli. In *L. plantarum*, betaine is preferentially transported by the quaternary ammonium compound or QacT transport system (Glaasker et al., 1998). This transporter is a high affinity, ATP-dependent system whose activity is stimulated by high osmotic pressure (Figure 1B). In contrast, at low osmotic pressure, the efflux reaction is activated, and pre-accumulated betaine is released back into the medium. Analysis of the *L. plantarum* genome sequence indicates QacT may be encoded by the *opuABCD* operon, which is widely distributed in other Gram positive bacteria (Kleerebezem et al., 2003). The natural source of betaine, it is worth noting, is plant material, so perhaps it is no coincidence that *L. plantarum* is unable to synthesize betaine and instead relies on a transport system to acquire it from the environment (Glaasker et al., 1996).

References

Glaasker, E., W.N. Konings, and B. Poolman. 1996. Osmotic regulation of intracellular solute pools in *Lactobacillus plantarum*. J. Bacteriol. 178:575–582.

Glaasker, E., E.H. Heuberger, W.N. Konings, and B. Poolman. 1998. Mechanism of osmotic adaptation of the quaternary ammonium compound transporter (QacT) of *Lactobacillus plantarum*. J. Bacteriol. 180:5540–5546.

Kleerebezem, M., J. Boekhorst, R. van Kranenburg, D. Molenaar, O.P. Kuipers, R. Leer, R. Tarchini, S.A. Peters, H.M. Sandbrink, M.W. Fiers, W, Stiekema, R.M. Lankhorst, P.A. Bron, S.M. Hoffer, M.N. Groot, R. Kerkhoven, M. de Vries, B. Ursing, W.M. de Vos, and R.J. Siezen. 2003. Complete genome sequence of *Lactobacillus plantarum* WCFS1. Proc. Natl. Acad. Sci. U.S.A. 100:1990–1995.

Lu, Z., H.P. Fleming, R.F. McFeeters, and S.S. Yoon. 2002. Effects of anions and cations on sugar utilization in cucumber juice fermentation. J. Food Sci. 67:1155–1161.

McDonald, L.C., H.P. Fleming, and H.M. Hassan. 1990. Acid tolerance of *Leuconostoc mesenteroides* and *Lactobacillus plantarum*. Appl. Environ. Microbiol. 56:2120–2124.

Van de Guchte, M., P. Serror, C. Chervaux, T. Smokvina, S.D. Ehrlich, and E. Maguin. 2002. Stress response in lactic acid bacteria. Antonie van Leeuwenhoek 82:187–216.

Destruction and softening of the pickle surface tissue is another serious defect. When this happens, the pickle loses its crispness and crunch and becomes slippery and soft. These pickles are neither edible nor can anything be done to salvage them. The defect is caused by pectinolytic enzymes produced by microorganisms that are part of the natural cucumber microflora. The organisms responsible include mostly filamentous fungi, especially species of *Penicillium*, *Fusarium*, *Alternaria*, *Aschyta*, and *Cladosporium*. Pectins are complex heteropolysaccharides that serve as the main structural component of plant cell walls. Their hydrolysis requires the concerted action of three different enzymes—pectin methylesterase, polygalacturonase, and polygalacturonate lyase. Although *Penicillium* and other fungi are capable of secreting these enzymes, they may also be produced by various yeasts, as well as by the cucumber flowers. Bacteria, however, do not appear to be a major source of pectin-degrading enzymes.

The fungi responsible for the softening defect gain entry into the fermentation tank via their association with cucumber flowers. Thus, excluding these constituents from the fermentation may reduce or minimize this defect. Other preventative measures include maintaining sufficient salt concentrations, acidity, anaerobiosis, and temperature. Another method used to control adventitious microorganisms and to ensure a prompt fermentation (especially when the salt concentration is at the low end) is to partially acidify the cucumbers with acetic acid to about pH 4.5 to 5.0. This practice of chemical pasteurization is also used when starter cultures and controlled fermentation methods are used to produce pickles (Box 7-3).

Olives: Products and Markets

Olives refer not only to the usually salty, acidic product known as table olives, but to the fruit from which they are made. The main use of raw olives, in fact, is as olive oil—more than 90% of the total worldwide olive production is used for oil and only 7% to 10% are consumed as table olives.

Olive trees are native to the Middle East and, due to their hardy nature (some trees live as long as 1,000 years), olives have long been a major agricultural crop throughout the region. Olives and olive oil are among the most frequently mentioned foods in the Bible. Olive production subsequently spread from the Middle East across the Mediterranean, to Greece, Italy, France, Spain, and Northern Africa. Olives were not introduced to the Americas until the 18th century. Currently, four countries—Italy, Spain, Greece, and Turkey—are responsible for 75% of the total worldwide olive production (between 9 billion Kg and 15 billion Kg. In contrast, the United States accounts for less than 1% of total world olive production, with

essentially all 90 million Kg (200 million pounds) being produced in California (2003 statistics).

Currently, about 90% to 95% of the olives grown in California are used in the manufacture of table olives (making California among the leaders in table olive production). Although fermented olives are common in Europe and other olive-producing regions, most of the table olives produced and consumed in the United States are not fermented. In fact, more than 70% of the U.S. olive market consists of olives that are simply brined and canned (hence, this type is referred to as California-style olives).

There is also a small (but dedicated) market for fresh or raw tree-ripened olives. The famous Provence-style olives of France are of this type. More than thirty cultivars are grown worldwide; however, fermented olives are usually produced using one of eight main varieties: Manzanillo, Gordal, Picholine, Rubra, Mission, Sevillano, Ascolano, and Barouni. These cultivars differ widely with respect to their composition, size, texture, color, and flavor. Olive properties also change during ripening; most olives are picked when they are still green, straw-yellow, or cherry-red, even if later they acquire a much darker appearance. Ultimately, their selection for table olive production is based on the style of the olive produced (discussed below).

Composition

Phenol and polyphenol compounds are common to all olives. Some of these phenol compounds contribute to the color of the olive. Others have antimicrobial activity against a wide variety of microorganisms (including lactic acid bacteria). The phenol fraction of olives (and olive oil) also may have positive dietary and nutritional benefits, due to their antioxidant activities (Box 7-5). The most important and abundant phenol is oleuropein (part of a class of compounds called secoiridoids).

Structurally, oleuropein is a glucoside ester of 3,4-dihydroxytyrosol and elenic acid (Figure

7-6). Glucosidic phenols, and oleuropein in particular, are important in olives owing to the pronounced bitter flavor they impart. Some olive varieties can contain as much as 14% oleuropein (on a dry basis) during the early stages of growth, although most contain no more than 3% to 6%. As the olives mature, the oleuropein concentration decreases, and at maturation, about 1 mg to 2 mg per g of pulp is present. Still, between the remaining oleuropein and related derivatives, the olives are too bitter to consume. Thus, manufacturing processes for most olives, including some fermented varieties, include sodium hydroxide-treatment steps to remove the bitter oleuropein fractions.

Olives, of course contain a significant amount of oil (12% to 30%, depending on cultivar). The fermentable carbohydrate concentration of ripe olives generally ranges from 2% to 5%; most of this carbohydrate is glucose. However, when the olives are washed or treated to remove the bitter components, sugars are also lost.

Manufacture of fermented olives

There are three main styles or types of table olives, based on their method of production (Figure 7-7). Spanish-style (or green Spanish-style) olives are treated with sodium hydroxide (lye) and fermented. Greek-style or naturally-black, ripe-style olives are not treated with lye, but are fermented. The fermentation for both types is mediated by the natural microflora, much like that for other fermented vegetables (discussed below). The third type of olive is the ripe black- or green-style. They are lye-treated, but are not fermented. They may also undergo

Box 7–5. Olives and the Mediterranean Diet

The Mediterranean diet has long been promoted as a model for healthy eating (Willet et al., 1995). This diet takes its name from the foods normally consumed by Greek, Italian, Spanish, and other populations that reside around the Mediterranean Sea. The diet advocates high consumption of fruits and vegetables, grains and pasta, legumes, olive oil, and fish; moderate intake of wine and dairy products; and low intake of meat products. The epidemiological association of this diet with reduced risks of cancer and heart disease and total mortality is highly significant (Keys, 1995 and Trichopoulou et al., 2003).

Perhaps the most prominent foods associated with the Mediterranean diet, and among those thought to be responsible for many of the health benefits, are olives and olive oil. Studies on olive and olive oil components indicate that the phenolic fraction is particularly important (Visioli et al., 2002). As previously noted, phenolic compounds confer bitterness to olives (e.g., oleuropein) and are involved in color development, and some have antimicrobial activity. It now appears that some of these same phenolic compounds also have pharmacological activity and may be responsible, in part, for the health-promoting properties of the Mediterranean diet (D'Angelo et al., 2001).

The main biological activities associated with oleuropein, hydroxytyrosol, tryosal, verbascoside, and other phenolic compounds found in olives and olive oil are related to their antioxidant properties (Soler-Rivas et al., 2000). They have been found to inhibit low-density lipoprotein (LDL) oxidation and accumulation of oxidized end products. Oxidation of lipoprotein and other lipids is considered to be an important step in initiating coronary heart disease. Other cardioactive effects associated with these phenols include anti-arrhythmic and cardioprotective activities. Olive phenols also protect human cells against oxidative stress and injury. In addition, oxidative damage to DNA may be prevented by the free radical scavenging activity of olive phenols.

As noted earlier, most table olives are treated with lye as part of the de-bittering process (the exception being Greek-style, naturally black olives). Although this process hydrolyzes much of the bitter oleuropein, olives still contain appreciable amounts of the hydrolysis products,

Box 7–5. Olives and the Mediterranean Diet *(Continued)*

elenoic acid glucoside and hydroxytyrosol, as well as other so-called biophenols. One recent study (Romero et al., 2004), for example, reported total polyphenol concentrations of 2.4 to 11.0 mM in juice from Spanish- and Greek-style olives. Some of the Spanish-style olives contained up to 1 g per kg or 0.1%, which is even more than that found in virgin olive oil. The presence of these compounds is now considered so important to human health that researchers in Greece have encouraged the table olive industry to consider modifying processing conditions to increase polyphenol concentrations (Blekas et al., 2002).

References

Blekas, G., C. Vassilakis, C. Harizanis, M. Tsimidou, and D.G. Boskou. 2002. Biophenols in table olives. J. Agric. Food Chem. 50:3688–3692.

D'Angelo, S., C. Manna, V. Migliardi, O. Mazzoni, P. Morrica, G. Capasso, G. Pontoni, P. Galletti, and V. Zappia. 2001. Pharmacokinetics and metabolism of hydroxytyrosol, a natural antioxidant from olive oil. Drug Metab. Dispos. 29:1492–1498.

Keys, A. 1995. Mediterranean diet and public health: personal reflections. Am. J. Clin. Nutr. 61:1321S–1323S.

Romero, C., M. Brenes, K. Yousfi, P. García, A. García, and A. Garrido. 2004. Effect of cultivar and processing method on the contents of polyphenols in table olives. J. Agri. Food Chem. 52:479–484.

Soler-Rivas, C., J.C. Espín, and H.J. Wichers. 2000. Oleuropein and related compounds. J. Sci. Food Agric. 80:1013-1023.

Trichopoulou, A., T. Costacou, C. Bamia, and D. Trichopoulos. 2003. Adherence to a Mediterranean diet and survival in a Greek population. N.E.J. Med. 348:2599–2608.

Visioli, F., A. Poli, and C. Galli. 2002. Antioxidant and other biological activities of phenols from olives and olive oil. Med. Res. Rev. 22:65–75.

Willett, W.C., F. Sacks, A. Trichopoulou, G. Drescher, A. Ferro-Luzzi, E. Helsing, and D. Trichopoulos. 1995. Mediterranean diet pyramid: a cultural model for healthy eating. Am. J. Clin. Nutr. 61 (Suppl. 6): 1402S–1406S.

a special aeration treatment that promotes oxidation of pigments and conversion of a green color to black. This is the type referred to as California-style olives.

Spanish-style

Spanish-style olives are harvested when the skin color is green or straw-yellow. They are then treated with a lye solution for four to twelve hours at 15°C to 20°C to de-bitterize the olives via hydrolysis of oleuropein. The lye concentration may range from 0.5% to 3.5%, depending on the size and type of olive. Once the lye has penetrated to just outside the pit (about two-thirds of the way from the skin to the center), the olives are washed in one to three rinse cycles of water to remove the lye. The pH of the olives after washing should be less than 8.0. Although there

should be little residual lye remaining with the olives, a slight amount of bitterness may still be present, which is characteristic of these olives. Following the washing, the olives are moved to tanks or barrels and a brine of varying salt concentrations is added. For some olives, 10% to 15% salt brines are used (giving an actual concentration of 6% to 9%), whereas others start with lower salt brines (5% to 6%), and salt is added later to give comparable final concentrations. Glucose may be added to restore sugars lost during lye treatment and washing steps. The brined olives are subsequently held at 22°C to 26°C.

Although the endogenous microbial population is reduced by the lye and washing treatments, the olive production environment still contains a wide assortment of microorganisms. There is, in fact, an opportunity for growth of

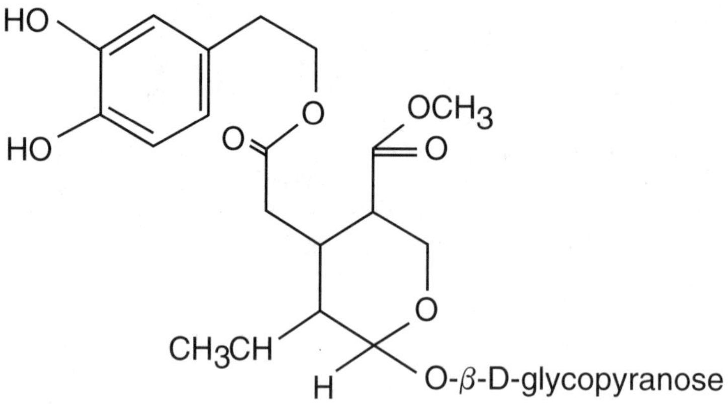

Figure 7–6. Structure of oleuropein from olives.

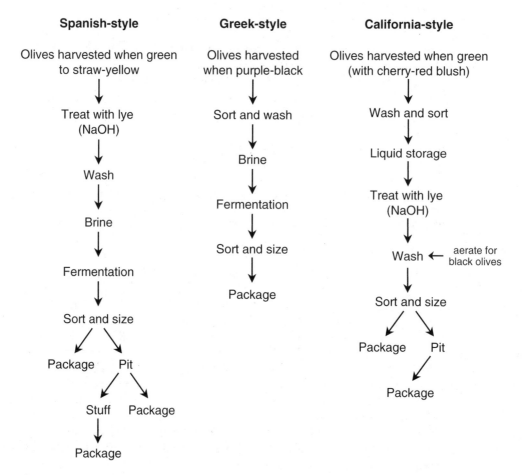

Figure 7–7. Types of olives and their manufacture. Adapted from Harris, 1998 and Romero et. al., 2004.

coliforms, *Pseudomonas, Bacillus, Clostridium* and other Gram negative and Gram positive bacteria during the first two to four days, especially when low salt brines are used, and before a lactic fermentation can begin in earnest. This so-called first stage of the fermentation is then followed by a second stage, initiated primarily by *L. mesenteroides* and *Pediococcus* sp. Once the pH nears 5.0, non-lactics are inhibited, and the brine consists only of lactic acid bacteria. After two to three weeks, *L. plantarum, L. brevis, Lactobacillus fermentum*, and other lactobacilli displace the leuconostocs, and eventually *L. plantarum* dominates the third stage of the fermentation. The appearance of facultative yeasts at the end stage of the fermentation, however, is not uncommon, nor undesirable, since they may produce ethanol, acetaldehyde, and other flavor compounds. In contrast, growth of *Propionibacterium acnes* is to be avoided, since this organism can raise the pH by fermenting lactic acid.

When the fermentation is complete, the final pH after three to four weeks (or longer for barrel-fermented olives) should be within a range of 3.5 to 4.2, with a titratable acidity (as lactic) of 0.8 to 1.0. Following the fermentation, Spanish-style olives can be pitted, pitted and stuffed (e.g., with pimentos), or packed directly into jars. Pasteurization to promote extended shelf-life is optional.

Greek-style

There are several significant differences between Greek-style and other olive types. First, these olives are naturally black when they are harvested, in contrast to California-style black olives that rely on oxidation to generate black pigments. Second, Greek-style olives are not lye-treated, giving them a more bitter flavor. Third, the fermentation is mediated not just by lactic acid bacteria, but also by yeast, non-lactic acid bacteria, and even fungi. Some of these non-lactic organisms (e.g., *Pseudomonas* and Enterobacteriaceae) may actually remain in the brine, albeit at low levels, for several weeks before they begin to decline (Figure 7–8). Thus, the fermentation end-products include not just lactic acid, but also acetic acid, citric acid, malic acid, CO_2, and ethanol. This mixed fermentation may result in a less acidic product with a final pH as high as 4.5 and an acidity less than 0.6%. Lower brine concentrations (5% to 10%) may contribute to the more diverse flora that develop in these olives.

Another type of fermented olive that is made in a very similar manner (i.e., no lye-treatment) is the Sicilian olive. The main difference between these olives and the Greek olives is that the Sicilian olives are green (like Spanish olives). Also, while lactobacilli are the main bacteria involved in the fermentation, the dominant species appears to be *Lactobacillus casei*, an organism not ordinarily associated with fermented olives.

Ripe- or California-style

California-style olives are the most popular olives consumed in the United States. Both the black and green versions are produced from the same starting material: green olives (with a bit of cherry-red blush). For both types, the olives are lye-treated, as described above for Spanish olives, except that several applications are used. In the case of green olives, the lye is removed and the olives are washed in water, and then dilute brine is added. The olives are then canned (after a pitting step, if desired) and thermally processed as a low-acid canned food.

For black olives, the lye-treated olives are heavily aerated between the lye applications to promote darkening reactions. Aeration can be accomplished either manually, by stirring, or mechanically, by direct injection of air into the tanks. The latter method is preferred because it is faster and reduces opportunities for spoilage. Air, or more specifically, oxygen, promotes a chemical oxidation reaction in which phenolic compounds are polymerized to form first brown, then black pigments. During air exposure, the concentration of two phenols, hydroxytyrosol and caffeic acid in particular,

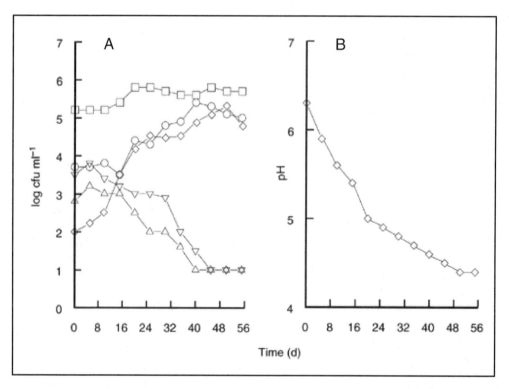

Figure 7–8. Fermentation of black olives. Panel A shows growth of total viable bacteria (□), lactic acid bacteria (□ ◇), *Pseudomonas* sp. (△), Enterobacteriaceae (▽), and yeasts (○) during the fermentation of Greek-style black table olives. The decrease in pH is shown in Panel B. From Nychas et al., 2002.

decreases, as the black color increases. Ionic iron can form complexes with the oxidized phenols; thus, iron salts are usually added to these olives to fix the color and prevent fading.

Defects and spoilage

Olives are, generally, more susceptible to microbial spoilage than other fermented vegetables. Initially there is a diverse microflora present in raw olives and that flora is well maintained during the early stages of the fermentation. Reflective of this heterogenous mix, spoilage organisms include aerobic bacteria and fungi, facultative bacteria and yeasts, strict anaerobes, as well as sporeforming bacteria. If the lactic fermentation is delayed due to residual lye or limiting glucose, or if the pH remains high (i.e., above 4.5), organisms capable

of causing spoilage may be given ample opportunity for growth and production of spoilage products.

Several of the spoilage defects that occur in olives are not unlike those that occur in pickles and other fermented vegetables (Table 7–4). Excessive gas production and tissue softening are major problems in olives, just as they are for pickles. In olive production, for example, gas produced by coliforms early in the fermentation can accumulate as gas pockets inside the olive surface, causing an unsightly blister-like appearance called "fish-eyes." Likewise, pectin hydrolysis by pectinolytic *Penicillium*, *Fusarium*, *Aspergillus*, and other fungi or, less frequently, by coliforms, results in a soft tissue defect.

Another defect that is mostly specific for olives is termed zapatera spoilage. This occurs late in the fermentation and is characterized by foul, fecal, cheesy-like odors. It is caused by *Clostridium sporogenes, Clostridium butyricum*, and other putrefactive clostridia whose presence in fermented foods is never a good thing. These bacteria, along with other anaerobes, produce butyric acid, sulfur dioxide, and other sulfur-containing compounds. Putrescine, cadaverine, and other putrefactive and malodorous end products may also be produced. *Propionibacterium* sp. also are associated with zapatera spoilage, although they may be indirectly involved. Growth of *Propionibacterium zeae* and other related species can occur at low salt concentrations ($<$5%), which leads to an increase in the brine pH (as lactic acid is consumed). If the pH is high enough, clostridia are no longer inhibited.

Control of spoilage organisms can be accomplished by making sure the salt concentration is sufficiently high ($>$7.5%) and the pH is sufficiently low ($<$4.0). Olives can be given a quick heat treatment or partially acidified with lactic acid (prior to fermentation) to reduce and control the background flora. Although seldom practiced, the addition of a lactic starter culture can provide additional assurance that a prompt fermentation will occur.

Fermented Vegetables and Biogenic Amines

Nearly all fermented foods can potentially contain biogenic amines, and fermented vegetables are no exception. This is because the series of events that culminate in biogenic amine formation are essentially the same for fermented vegetables as they are for other fermented foods. Specifically, accumulation of amines occurs whenever there is an available pool of free amino acids and a source of amino acid decarboxylase enzymes. The amino acids are formed from protein in the food by the action of microbial proteinases and peptidases. The decarboxylases are also produced by microorganisms, including lactobacilli that may be present in fermented vegetables. Thus, the amount of biogenic amines that are actually produced depends on both the concentration of the amino acid substrates and the expression and activity of the relevant decarboxylases. The biogenic amines that have been found in fermented vegetables (mainly sauerkraut, kimchi, and fermented olives) include histamine (from histidine), tyramine (from tyrosine), and putrescine and cadaverine (from lysine, arginine, and glutamine). In most cases, however, the concentrations that have been reported were less than the level ordinarily thought to cause food disease symptoms (1 g/Kg).

Bibliography

Ayres, J.C., J.O. Mundt, and W.E. Sandine. 1980. Microbiology of Foods. Chapter 9, Lactic fermentations, p. 198–230. W.H. Freeman and Company, San Francisco.

Fernández, A.G., M.J.F. Díez, and M.R. Adams. 1997. *Table Olives: Production and Processing*. Chapman and Hall, London.

Harris, L.J. 1998. The microbiology of vegetable fermentations, p. 45–72. *In* B.J.B. Wood (ed.) *Microbiology of Fermented Foods, Volume 1*, Blackie Academic and Professional (Chapman and Hall).

Nout, M.R.J., and F.M. Rombouts. 1992. Fermentative preservation of plant foods. J. Appl. Bacteriol. Symp. Supple. 73:136S–147S.

Nychas, G.-J.E., E.Z. Panagou, M.L. Parker, K.W. Waldron, and C.C. Tassou. 2002. Microbial colonization of naturally black olives during fermentation and associated biochemical activities in the cover brine. Lett. Appl. Microbiol. 34:173–177.

Pederson, C.S. 1960. Sauerkraut. *In* C.O. Chichester, E.M. Mrak, and G.F. Stewart (ed.), Advances in Food Research. 10:233–291.

Randazzo, C.L., C. Restuccia, A. D. Romano, and C. Caggia. 2004. *Lactobacillus casei*, dominant species in naturally fermented Sicilian green olives. Int. J. Food Microbiol. 90:9–14.

Romero, C., M. Brenes, K. Yousfi, P. García, A. García, and A. Garrido. 2004. Effect of cultivar and processing method on the contents of polyphenols in table olives. J. Agri. Food Chem. 52:479–484.

8

Bread Fermentation

"... cut for yourself, if you will, a slice of bread that you have seen mysteriously rise and redouble and fall and fold under your hands. It will smell better, and taste better, than you remembered anything could possibly taste or smell, and it will make you feel, for a time at least, newborn into a better world than this one often seems."
From How to Rise Up Like New Bread *by M.F.K. Fisher 1942*

Introduction

The fermentation that occurs during bread manufacturing is different from most other food fermentations in that the purpose is not to extend the shelf-life of the raw materials, per se, but rather is a means of converting the grain or wheat into a more functional and consumable form. In fact, in contrast to dairy, meat, vegetable, or wine fermentations, where the starting material is much more perishable than the finished product, the raw material for bread-making, i.e., cereal grains, are better preserved than the bread that is ultimately produced. It is also interesting to note that in the bread fermentation, again in contrast to other lactic acid or ethanolic fermentations, essentially none of the primary fermentation end products actually remain in the food product.

In the United States, bread manufacturing is a $16 billion industry. Despite its commercial and dietary importance, however, consumption of bread was substantially higher a century ago than it is currently. At the end of the nineteenth century, for example, Americans consumed more than 100 Kg of wheat flour (used mostly for bread) per person per year. Consumption of wheat, however, then began to decline steadily for nearly a hundred years, reaching an all-time low of just 50 Kg per person per year in the

1960s. During the next thirty years, per capita grain consumption slowly increased, to nearly 70 Kg (Figure 8–1), only to fall back slightly in 2000 to just under 65 Kg, presumably as a result of the popularity of low-carbohydrate diets (more than one-third of Americans think that bread is "fattening").

Given the recommendations made in the USDA's revised Food Pyramid and 2005 Dietary Guidelines, one might expect that consumption of bread (and whole grain breads, in particular) will begin to increase. Still, some countries already have much higher per capita consumption of bread, compared to the United States (Figure 8–2), with annual consumption of more than 140 Kg per person (more than 0.8 pounds per day). Compared to European and other high bread-consuming countries, the United States ranks near the bottom for bread consumption, at only 25 Kg per person per year (less than 2.5 ounces per day).

History

It is impossible to know exactly when humans first made and ate bread. Ancient artifacts and writings discovered in the Middle East suggest that bread-making had its origins in the eightieth century B.C.E., but it is possible that bread may have been produced even

Figure 8–1. Wheat flour consumption (per person per year) in the United States. Adapted from the USDA Economic Research Service.

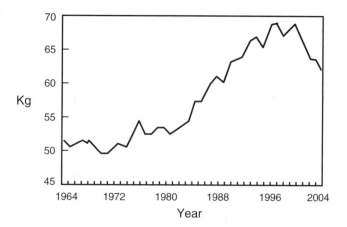

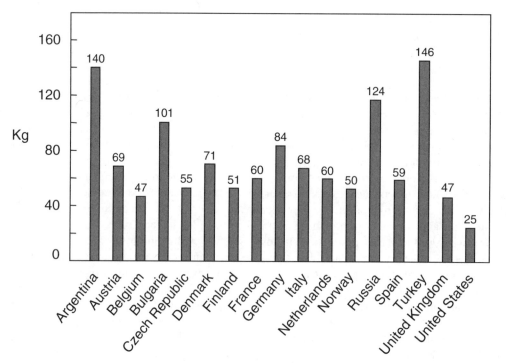

Figure 8–2. Worldwide consumption of bread (per person per year). Adapted from Association Internationale de la Boulangeri Industrielle (www.aibi-online.org) and Federation of Bakers (www.bakersfederation.org), 2001-2002 statistics.

several thousands of years earlier. The first breads were probably flat breads, with little or no leavening (i.e., fermentation), similar to those consumed even today in many parts of the world. It seems likely that humans began making leavened bread at very near the same time they began to make other fermented products. Leavening must certainly have been an accidental discovery made when a flour-water mixture became contaminated with wild airborne yeasts, leading to a spongy dough that was transformed by baking into a light and airy, aromatic, and flavorful product. Perhaps more so than any other fermented food, bread has played a major role in the history and culture of human civilization (Box 8-1).

Despite obvious increases in the size and scope of the bread-baking industry, the historical record has revealed that ancient bread-making is not all that different from modern bread manufacturing practices. Given the simplicity of the bread manufacturing process, perhaps this observation is not surprising. After all, bread-making requires only a few ingredients, a few simple mixing and incubation steps, and an oven for baking. In this chapter, the chemical, physical, and biological properties of these ingredients and the processes used for converting ingredients into doughs and breads will be described.

Box 8–1.   Civilization and Bread

It is no coincidence that the history of bread making parallels the history of human civilization. Bread, the oft-quoted "staff of life" is still one of the principle staple foods throughout the world, and has sustained human beings for thousands of years.

Bread may have been one of the earliest of all "processed" foods and was certainly one of the first foods to be produced on large scale. Archaeological evidence (recounted in Roberts, 1995 and Wood, 1996) indicates that a rather sophisticated bread manufacturing "industry" had developed in Egypt as long ago as 3,000 B.C.E., in large part to provide nourishment for the substantial labor force necessary to build pyramids and other structures.

The very first breads were probably flat breads that had been made using barley, which does not work well for leavened breads. Therefore, the production of leavened breads likely coincided with the development of suitable ancestral wheat varieties, such as emmer and kamut. Given the open environments in which these factories operated, the dough fermentation was undoubtedly due to the combined effects of yeasts and bacteria, suggesting that these early breads were probably sourdough breads (discussed later). Rather than baking the breads on pans in ovens, the doughs were apparently developed and baked in clay pots arrayed around charcoal-fired pits. However, the use of enclosed ovens eventually became more common as this early industry evolved.

The importance of grain and bread during the course of human history is evident in the Bible, in literature and mythology, and throughout the historical record. Osiris, the Egyptian god of the underworld, was also responsible for vegetation and agriculture; Neper was the Egyptian god of grain. During the ancient Greek spring festival of Thargelia, bread is offered to the mythological gods Artemis and Apollo, symbolizing the first fruits of the year and thanking them for providing fertile soil.

According to the Old Testament story, when Joseph, son of the Hebrew patriarch Jacob, predicted that a famine was soon to occur in Egypt, the Pharaoh ordered that grain be stored, ensuring a sufficient supply of wheat when the famine eventually occurred. Later, when Hebrew slaves fled from Egypt more than 4,000 years ago, their hasty exodus left no time for their dough to rise. Instead, they had to eat unleavened bread, a form of which is eaten even today as a symbol of the Jewish holiday of Passover. The Eucharist, one of the sacraments of Catholicism, requires consecration and consumption of bread during Holy Communion. Communion in the Protestant tradition also includes bread, typically unleavened. A blessing over bread is commonly recited as part of many religious rites.

During the Roman era, bakers had an elevated status and were widely respected. Bread was so important to the citizenry that it was either provided free or was heavily subsidized. Roman soldiers were even equipped with portable bread-making equipment. Milling technologies were also developed that could separate wheat bran from endosperm (discussed in more detail later in this chapter), which made it possible to produce white flour bread.

However, genuinely effective technologies for producing refined white flour were not developed until the 1800s, when the roller mill was invented. For many centuries, this more expensive

(Continued)

Box 8–1. Civilization and Bread *(Continued)*

white flour was prized by society's upper class, while dark, whole grain breads were left to the peasantry. In the Middle Ages, baker's guilds, the forerunners of trade unions, were formed in Britain and Europe to set quality standards and fair prices and to protect their overall interests. At the same time, laws were enacted that also established prices and weights for bread.

These laws were apparently necessary because some bread manufacturers resorted to questionable, if not dangerous, practices, such as adding potentially toxic whitening agents to flour. These unscrupulous tactics led some bakers to garner a rather poor reputation. In Geoffrey Chaucer's the Miller's Tale, one of the great early pieces of literature (and one that all college students no doubt remember), the miller is hardly portrayed as a pillar of the community.

Because bread is such an important food, wheat has become one of the most important agricultural crops produced on the planet. For thousands of years, and even today, when wheat harvests are poor, due to weather or plant diseases, famines and starvation conditions often result. Political unrest, trade embargos, and other economic factors have also contributed to shortages of flour and bread. History books are replete with instances in which rebellions and uprisings, and even revolutions, occurred when a steady supply of affordable bread became unavailable.

Queen Marie-Antoinette's infamous remark about the lack of bread for the hungry children of France ("let them eat cake") proved to be her undoing, figuratively as well as literally. In the modern era, it was only as recently as the 1980s when citizens of the former Soviet Union had to wait in long lines just to buy a loaf of bread. This situation no doubt led to significant public discontent and a loss of confidence in the government, both of which may have even contributed to the fall of the USSR.

Reliance on cereal grains for bread making is so crucial that during times of famine and war, even mold-infested grains have been used to make bread. Such was the case in the Soviet Union during the 1930s and 1940s and through World War II, when farmers were unable to promptly harvest cereal grains, which then over-wintered and became infected with mycotoxin-producing field fungi. Consumption of bread made from infected flour led to thousands of deaths, and many more were sickened. In more modern times, wheat has even been used as a "weapon," as grain embargoes against several U.S. adversaries are still in place even today. Ironically, European and other countries have threatened to boycott U.S. wheat, if genetically-engineered varieties are produced.

References

Roberts, D. 1995. After 4,500 years rediscovering Egypt's bread-baking technology. National Geographic 187:32–35.

Wood, E. 1996. *World Sourdoughs from Antiquity*. Ten Speed Press. Berkeley, California, U.S.A.

Wheat Chemistry and Milling

The most common starting material for most breads is wheat flour. Breads are also commonly made from a wide variety of other cereal grains, including rye, barley, oats, corn, sorghum, and millet. However, as will be discussed in more detail below, gluten, a protein complex that gives bread its structure and elasticity and is necessary for the leavening process, is poorly formed or absent in most non-wheat flours.

Thus, although accommodations can sometimes be made to account for the absence of gluten in wheat-free doughs, most commercial breads contain at least some wheat.

Wheat is one of several cereal grasses in the family Poaceae (or Gramineae). The main variety used for bread is *Triticum aestivum*. The wheat plant contains leaves, stems and flowers. It is within the flowers (referred to as spikelets) that the wheat kernels are formed. The kernels are

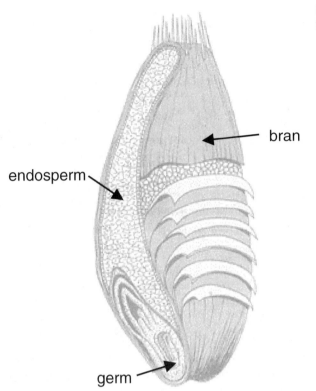

Figure 8–3. Structure of a wheat kernel. Adapted from the Montana Wheat & Barley Committee.

essentially seeds, and are surrounded by a hard covering material designed to protect the would-be plant from the external elements. The original or wild wheat seed that grew many thousands of years ago had a tough, hard-to-break husk which made it difficult to use. However, the development of crude hybrids (the mixing of different wheat varieties) led to the "domestication" of wheat with more manageable properties. Moreover, the wheat kernel develops in the husk, but the husk is removed during harvest, unlike barley or oats, where the husk remains attached to the kernel.

The modern day wheat kernel consists of three main constituents: the germ, bran, and endosperm (Figure 8-3). The germ portion contains the embryonic plant, along with oils, vitamins, and other nutrients. It represents 2% to 3% of the total kernel weight. The bran portion, or coat, represents 12% to 13% of the kernel, and is comprised of multiple distinct layers. The bran contains mostly cellulose and other fibrous carbohydrates, along with proteins, minerals, and vitamins. The remaining portion, about 85%, is called the endosperm and is the main constituent of wheat flour. It consists of mostly protein and starch, along with some water (12% to 14%) and a small amount of lipid (1%). Kernels that are crushed or milled without regard to separation of the individual fractions yield a "whole" wheat flour containing the germ, bran, and endosperm. Usually, however, these three constituents are separated, either partially or nearly completely, in the milling step, yielding refined flours of endosperm, with little or no germ or bran (Box 8-2).

Flour Composition

The specific composition of flour is critically important because it has a major influence on the fermentation as well as the physical structure of the dough and finished bread. Wheat flour, after all, is the primary ingredient in most

Box 8–2.   Principles of Milling Wheat

The object of wheat milling is to extract the flour from the wheat kernel. This is such an integral step in the manufacture of bread that an entire milling industry has developed whose purpose is to do just that—convert wheat into flour. Originally, some thousands of years ago, milling was done in a single step, as the kernels were ground or pounded manually by a crude mortar and pestle or between stones.

Even today, single-pass milling is performed to make whole wheat flour, although the milling technologies have obviously improved and are now mechanized and automated. Still, stone-ground flours are widely available and are prized for their quality. Functionally, however, stone-grinding has no advantages over other size reduction techniques, such as a hammer mill.

The modern milling operation begins with various wheat handling steps, including sampling, inspection, and testing; screening; and conditioning (Catterall, 1998). Following these steps, which ensure that the wheat has the appropriate composition and quality and is free of debris and impurities, the wheat is ready for milling. Although the whole wheat flours produced by a single-pass grinding step have wide appeal, most breads are made using flours that contain reduced levels (or even none) of the germ and bran. It is the job of the wheat miller to separate these wheat constituents from the main component, the endosperm.

In principle, this can be done simply by successive grinding and separation steps, such that the resulting fractions can either be removed or further processed. In actual practice, the wheat kernels are passed through a series of paired-rollers or stones whose gap distances between each pair is successively more narrow. Modern grinding devices consist of cast iron break rollers, constructed with flutes or ridges to enhance the grinding or shearing action (Figure 1). The rolls run at different speeds to give a shearing, rather than a crushing, action. In between each so-called break steps the ground wheat is passed through sieves that retains the coarse material (containing bran still stuck to the endosperm), while permitting the fine flour to pass. Along the way, an air purifying device is also used to further separate the different fraction. Eventually, the germ and bran-free white flour that is the most common type used for bread manufacture is produced.

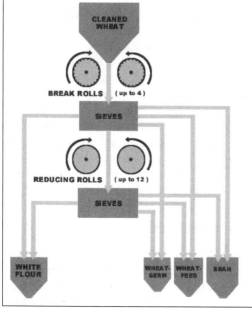

Figure 1.   A simplified diagram of wheat milling. Adapted from www.fabflour.co.uk.

Box 8–2. Principles of Milling Wheat *(Continued)*

The milling operation yields dozens of different product streams (Calvel, 2001). Flour removed from the early and late separations is called clear or common flour. It is generally less refined, higher in protein (>14%), and somewhat darker in color. Patent flour is obtained from the intermediate separation steps. It has the least amount of germ and bran, and, therefore has the whitest appearance. Patent flour is still high in protein (13%) and is considered to have the best bread-making quality (at least in the United States and Canada).

In many parts of the world, it is a standard practice to re-combine the different flour fractions to give flours of varying composition, ranging from a straight flour containing all of the separated fractions, to different patent flours containing 60% to 80% of the original wheat. This allows baker to choose a flour based on desired strength (i.e., protein content) or functional properties (i.e., dough elasticity, bread flavor, and oven spring).

It is important that a high yield be achieved during the milling process. The yield is given as the extraction rate, and is expressed as the amount of flour that is obtained from the wheat. In general, extraction rates of about 72% are achieved, meaning that 72 Kg of straight flour can be obtained from 100 Kg of wheat. The balance (about 28%) includes mainly bran and germ. However, since 85% of the wheat is endosperm, some of the remaining material consists of endosperm that adheres to the bran layers (Pyler, 1988).

References

Calvel, R. 2001. *The Taste of Bread.* (R.L. Wirtz, translator and J.J. MacGuire, technical editor). Aspen Publishers, Inc. Gaithersburg, Maryland.

Catterall, P., 1998. Flour milling, p. 296–329. *In* S.P. Cauvin and L.S. Young, (ed.), *Technology of breadmaking.* Blackie Academic and Professional (Chapman and Hall). London, UK.

Pyler, E.J., 1988. Wheat flour, p. 300–377. In *Baking Science and Technology. Volume 1, 3rd edition*, Sosland Publishing Co. Merriam, Kansas.

breads. As noted above, however, most of the flour used for bread manufacture in the United States is refined white flour, containing only the endosperm portion. It consists of two main components—protein and starch—with a small amount of hemicellulose and lipid.

Table 8.1. Major types of wheat and their applications.

Type	% Protein	Application
Hard winter	10 – 13	Bread
Hard spring	13 – 16	Bread
Soft winter	8 – 10	Cakes, cookies, pastries
Durum	12 – 14	Pasta

Protein

About 8% to 15% of wheat flour is protein. This wide range reflects the various types or classes of wheat used as a source of the flour (Table 8-1). Protein content, as well as protein quality of a given flour, dictates the use of that flour. High protein flours, derived from hard wheat, generally contain more than 11% protein and are best used for bread. In contrast, low-protein flours, derived from soft wheat, contain 9% or less protein, and are most often used for cakes, cookies, and pastries. It is the protein than provides the support matrix necessary to retain the carbon dioxide made during fermentation. Therefore, in bread making, protein content has a major impact on dough expansion and loaf volume.

The protein fraction actually contains several different proteins. The most important of these are gliadin and glutenin, which account for 85% of the total protein. When gliadin and glutenin are hydrated and mixed, they form a complex called gluten, which is a key component of bread dough. The remaining proteins (15% of the total) consist of other globulins and albumins. Several enzymes are also present, in particular, α-amylase and β-amylase, that both play an important role (discussed below).

Carbohydrates

Carbohydrates represent the main fraction of flour, accounting for up to 75% of the total weight. This fraction is largely comprised of starch, although other carbohydrates are also present, including a small amount (about 1%) of simple sugars, cellulose, and fiber. The main carbohydrate component, however, is starch, which consists of amylose, an α-1,4 glucose linear polymer (about 4,000 glucose monomers per molecule) and amylopectin, an α-1,4 and α-1,6 glucose branched polymer (about 100,000 glucose monomers per molecule). About 20% to 25% of the starch fraction is amylose and 70% to 75% is amylopectin. Properties of these two fractions are described in Table 8-2.

In its native state (i.e., before milling), wheat starch exists in the form of starch granules. The amylose and amylopectin are contained within these spherical granules in a rigid, semi-crystalline network. Native starch granules are insoluble and resist water penetration. However, some of the starch granules (3% to 5%) are damaged during milling, which enhances absorption of water and exposes the amylose and amylopectin to hydrolytic enzymes, such as α-amylase.

Yeast Cultures

The yeast used for bread manufacture is *Saccharomyces cerevisiae*, often referred to as simply bakers' yeast. Although ale (Chapter 9), wine (Chapter 10), and various distilled alcoholic beverages are also made using *S. cerevisiae*, the cultures for each of these products are not interchangeable. The bakers' yeast strains of *S. cerevisiae* clearly have properties and performance characteristics especially suited for bread manufacture (Table 8-3). For example, bread strains are selected, in part, on their ability to produce CO_2, or their gassing rate. It is the CO_2, evolved during fermentation, that is responsible for leavening. Yeast strains should also produce good bread flavor. In addition, stability and viability during storage is also important in selecting suitable yeast strains.

Modern industrial production of bakers' yeast starts with a pure stock culture, which is then scaled up through a series of fermentors until a large cell mass is produced (Figure 8-4). The culture is propagated initially in small (1 L to 5 L) seed flasks, and then in progressively larger fermentors of about 250 L to 1,000 L. Eventually the culture is inoculated into large (250,000 L) production fermentors. The growth medium usually consists of molasses or another inexpensive source of sugar and various ammonium salts (e.g., ammonium hydroxide, as a cheap source of nitrogen). Other yeast nutrients include ammonium phosphate, magnesium sulfate, calcium sul-

Table 8.2. Physical and chemical properties of amylase and amylopectin.

Property	Amylose	Amylopectin
Monomers per molecule	300 — 1,000	> 5,000
Linkage	α-1,4	α-1,4 and α-1,6
Conformation	Linear	Branched
% in wheat starch	25	75
Retrogradation rate	Fast	Slow
Products of:		
α-amylase	Dextrins	Limit dextrins
β-amylase	Maltose	Maltose, limit dextrins

Table 8.3. Desirable properties of bakers' yeasts.

Gassing power
Flavor development
Stable to drying
Stable during storage
Easy to dispense
Ethanol tolerant
Cryotolerant

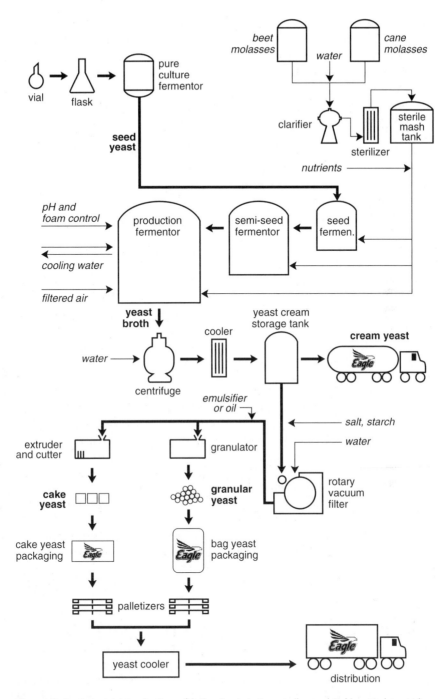

Figure 8–4. Industrial production of bakers' yeast. From Lallemand Baking Update, Volume 1/Number 9 (with permission).

fate, and lesser amounts of the trace minerals zinc and iron. Since the goal is to produce cell mass as the end product (rather than metabolic products), growth occurs under highly aerobic conditions by providing continuous agitation and oxygen (or air). In addition, the yeast propagation step is performed at optimum temperature and under pH control (usually around 30°C and at a pH of 4.0 to 5.0), with nutrients provided on a continuous basis. At the end of the growth fermentation, the yeast is collected by centrifugation, resulting in a yeast "cream" that contains about 20% total yeast solids and a yeast concentration of about 10^{10} cells/g.

Bakers' yeast preparations are available in several forms. The yeast cream can be used directly, although this form is highly perishable. Most commercial bakers use compressed yeast cultures. These are produced by pumping the yeast cream through a filtration press or vacuum filter to remove more of the water. The yeast is collected in the form of moist cakes, separated by wax paper. Compressed yeast cakes (about 30 cm x 30 cm x 2 cm) still have a high moisture content (70% to 75%), require refrigeration, and last only a few weeks. However, because the cells are metabolically active, once they are introduced into the dough, fermentation can occur very quickly.

Compressed yeast can be further dried to about 90% solids to provide dry active yeast. This is the form that is familiar to consumers who make homemade bread, but small manufacturing operations or those located where compressed yeast is not available also use dry yeast preparations. Dry active yeast preparations last six months or longer, even at room temperature. They do require a hydration step, and in general, are not as active as compressed yeast, although improved drying technologies have greatly enhanced the activity of dried yeast. In addition, dry active yeasts can be "instantized" such that they rehydrate quickly.

Bread Manufacturing Principles

In general, bread manufacture involves just a few steps. First the ingredients are assembled, weighed, and mixed to make a dough and the "bulk" dough is allowed to ferment. The fermented dough is then portioned and shaped, given a second opportunity to ferment, and then baked, cooled, sliced, and packaged. Of course, the actual procedures used by modern bakeries are often more involved, as will be described later. Described below is what is essentially the "straight dough" system.

Ingredients

Most ingredient statements on a loaf of commercial bread are a paragraph long and list thirty or more ingredients. However, only four ingredients—flour, water, salt, and bakers's yeast—are actually required to make a perfectly acceptable, if not exceptional, bread. Although the other ingredients do indeed provide important functional properties, not only in the finished product, but also during the dough development steps, they are strictly optional. Of course, one could argue that consumer preferences for breads with a soft crumb texture; long, mold-free shelf-life; and resistance to staling have led to such widespread use of some of these ingredients that they can hardly be considered "optional." Moreover, large scale manufacturing operations that demand high throughput could not function nearly as well without these extra ingredients and process aids.

Wheat flour is the main ingredient of bread, representing about 60% to 70% of a typical formulation. The flour contains the proteins that are essential for dough formation and the starch that absorbs water and serves as an energy source for the yeasts. Water, added at about 30% to 40%, acts as the solvent necessary to hydrate the flour and other ingredients. Salt, at 1% to 2%, toughens the gluten, controls the fermentation, and gives a desirable flavor. Finally, the yeast, also added at about 1% to 2%, provides the means by which leavening and flavor formation occur.

Many optional ingredients are often included in bread formulations, depending on the needs and wants of the manufacturer.

- **Sugars.** In the United States, many large-scale bakeries add either sucrose or glucose (about 2% to 3%) as an additional source of readily fermentable sugars. They also supply flavor and, when the dough is baked, color. Wheat flour contains only modest amounts of maltose and glucose, and although yeasts do express amylases that can release fermentable sugars from starch, sugar availability may still be growth-limiting.

- **Enzymes.** Another way to increase the amount of free sugars in the dough is to add α- and β-amylases, enzymes that specifically hydrolyze the α-1,4 glucosidic bonds of amylose and amylopectin (Figure 8–5). Flour naturally contains both of these enzymes, but α-amylase, in particular, is present at very low levels. These enzymes are also available in the form of microbial preparations or in the form of malt. Not only do these enzymes collectively increase the free sugar concentration by hydrolyzing amylose and amylopectin to maltose, but they also can enhance bread quality. Controlled hydrolysis of starch—especially the amylopectin portion—by α-amylase tends to increase loaf volume and, by virtue of softening the crumb texture, staling is delayed. Of course, if too much enzyme is added, the dough will be sticky and unmanageable and the excess free sugars that are formed could contribute to over-browning during baking.

  One other important factor that must be considered when enzymes are used as process aids is their temperature inactivation profile. Heat-resistant amylases, such as those produced by bacteria, may survive baking and continue to hydrolyze starch, which could result in a soft and sticky bread. Fungal enzymes are generally heat labile and are inactivated during baking.

  Increasing the amylase activity in dough, and thereby increasing the free sugar concentration, can also be accomplished by adding malt powder. Malt, an essential ingredient in beer manufacture (see Chapter 9), is prepared from germinated grains, usually barley and wheat. Malt contains both α- and β-amylases that are both somewhat more heat stable than fungal enzymes.

  Finally, the addition of proteolytic enzymes has been advocated as a means of achieving partial hydrolysis of the gluten. In theory, this would reduce mixing time and would make a softer dough.

- **Fat.** The addition of 0.1% to 0.2% fat, as either a shortening or oil, is now commonplace in most commercial breads. Whereas a generation or two ago, animal fats with excellent "shortening" properties were used in breads, the use of such fats has about disappeared in the United States. Now most breads are made with partially hydrogenated vegetable oils. The composition of these oils is not that different from animal fats, in that they contain mostly long chain, saturated fatty acids, with a high melting point. Not surprisingly, they impart similar properties as the animal fats, giving a soft, cake-like texture. The precise mechanism, however, for how these functional effects occur is not clearly known. Fats with high-melting points (i.e., containing long, saturated fatty acids, such as palmitic and stearic acids) clearly provide discernable improvements, especially for short-proof breads. There is currently a movement to use non-hydrogenated oils to eliminate trans fatty acids. This does change the dough properties compared to the use of hydrogenated vegetable shortenings.

- **Yeast nutrients.** Various nutrients can be added to the dough mixture to enhance growth of the yeast, including ammonium sulfate, ammonium chloride, and ammonium phosphate, all added as sources of nitrogen. Phosphate and carbonate salts may also be added to adjust the acidity or alkalinity, and calcium salts can be added to increase mineral content and water hardness.

- **Vitamins.** There has been a flour and bread enrichment program in the United States for more than sixty years. Flour currently is fortified with four B vitamins—thiamine, riboflavin, niacin, and folic acid—and one mineral, iron. However, non-enriched flour is also available and bread manufacturers can enrich

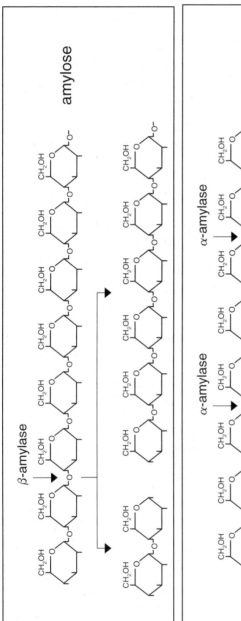
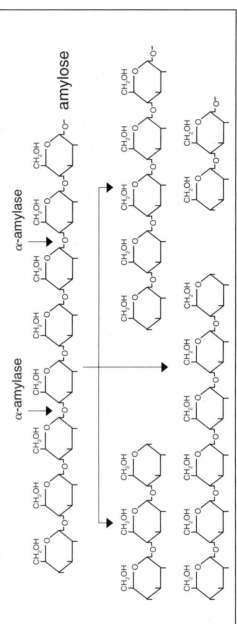

Figure 8–5. Hydrolysis of amylose and amylopectin by α- and β-amylases. The top and middle panels (A and B) show reaction products of β-amylase and α-amylase on amylose. The bottom panel (C) shows reaction products of β-amylase and α-amylase on amylopectin.

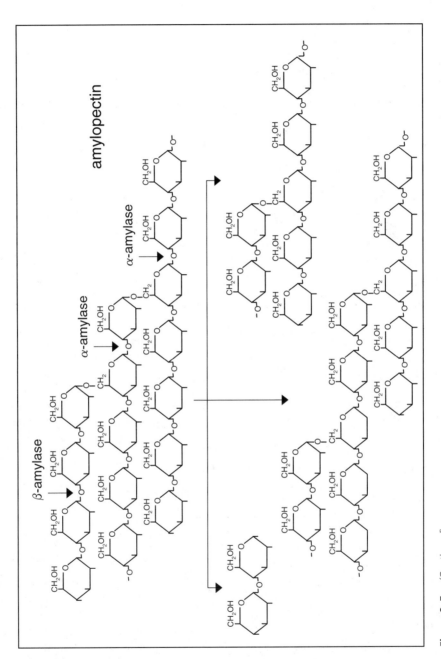

Figure 8–5. *(Continued)*

bread by adding these nutrients directly to the dough.

- **Dough improvers.** Mixing the ingredients and making an elastic dough can often be a slow process, due to the sticky gluten protein that forms when the flour is hydrated. Reducing agents, such as cysteine, are used to speed up dough mixing, by decreasing the number of disulfide cross-links that make gluten so highly elastic (as described in the next section). In other words, these agents weaken the dough structure and decrease mixing times. Alternatively, other bread dough improvers include oxidizing agents, such as ascorbic acid and potassium bromate, that can improve dough structure by increasing the number of disulfide bonds (and cross-linkages) in the gluten network. Their effect is to increase elasticity and gas retention. Bromates have long been considered as the most effective oxidizing agent; however, recent questions regarding their safety has led some government authorities to discourage or prohibit the use of this additive in bread manufacture.

- **Biological Preservatives.** Biological spoilage of bread (in contrast to staling or other chemical-physical causes) is invariably caused by fungi. Thus, mold inhibitors are routinely added to most breads. The use of these inhibitors is somewhat less common today than twenty years ago, however, due to the increased consumer demand for natural foods that are free of chemical preservatives. By far, the most common anti-fungal agents are the salts of weak organic acids; they include potassium acetate, sodium diacetate, sodium propionate, and calcium propionate. The latter is the more effective and is the most widely used bread preservative in the United States. As illustrated in Box 8-3, the mode of action of these agents against target cells is similar and depends on the form of the acid and the pH of the food. They are typically added at 0.15% to 0.2% of the flour weight; the maximum amount allowed for acetates and propionates in the United States is 0.32% and 0.40%, respectively (less in Europe). It is important to note that, despite their effective-

ness against fungi, these agents are only inhibitory and are not cidal. They also are not inhibitory to bakers' yeast, and, therefore, have no effect on the fermentation.

- **Emulsifiers.** Although bread dough does not ordinarily require emulsification, emulsifying agents may improve the functional properties of the dough by increasing water absorption and gas retention, decreasing proofing times, and reducing the staling rate. As such, bread emulsifiers are perhaps best considered as dough conditioners. By far, the most commonly added emulsifiers are mono- and diglycerides, derived from hydrogenated vegetable oils. In the United States, they are added at a concentration of 0.5% (flour weight).

- **Gluten.** Gluten, as discussed below, is formed naturally during dough hydration and mixing steps. However, it is also a common practice to add dried gluten, in the form of vital wheat gluten, as an ingredient to the dough mixture. Not only does the added gluten increase the protein content of the bread, but it also increases loaf volume and extends shelf-life. In addition, high gluten breads may have improved texture properties by virtue of strengthening the dough and bread structure. In general, vital gluten is most commonly added to flour in crop years when protein quality or quantity are low; it is also added to whole grain and specialty breads to increase loaf volume.

Hydration and Mixing

Depending on the nature of the ingredients (solid or liquid), they are either weighed or metered into mixing vats. Once all of the ingredients are combined, they are vigorously mixed. This is a very important step because the flour particles and starch granules are hard and dense and water penetrates slowly. Mixing, therefore, is the primary driving force for the water molecules to diffuse into the wheat particles. Mixing also acts to distribute the yeast cells, yeast nutrients, salt, air, and other ingredients throughout the dough. Mixing, in other words, is necessary to develop the dough.

Box 8–3. How Propionates Preserve Bread

Calcium propionate (Figure 1) is one of the most widely used antimicrobial preservatives in the fermented foods industry, due to its widespread application as an anti-fungal agent in bread. It also inhibits rope-producing species of *Bacillus*. Although it is the propionic acid moiety that is the active agent, the calcium salt of this weak organic acid is used commercially since it is readily soluble and is easier to handle. Calcium propionate is a GRAS substance (generally recognized as safe), and occurs naturally in various cheeses (e.g., Swiss cheese).

The inhibitory activity of calcium propionate against fungi is due to its ability to penetrate the cytoplasmic membrane and gain entry inside the fungal cell. However, propionate can only traverse the membrane when it is in the acid form. This is because the lipid-rich, non-polar, hydrophobic cytoplasmic membranes of cells are largely impermeable to charged, highly polar molecules, such as the anion or salt form of organic acids. Rather, only the uncharged, undissociated, lipophilic acid form can diffuse across the membrane.

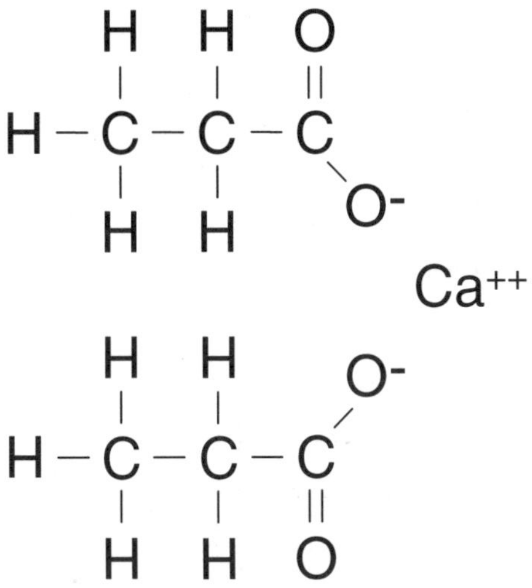

Figure 1. Structure of calcium propionate.

How then, does calcium propionate exerts its anti-fungal activity? Recall that weak organic acids can exist in either the acid (undissociated) or salt (dissociated) form (Figure 2, upper panel). When the pH is equal to the pKa (the dissociation constant) of the acid, the concentration of undissociated and dissociated forms (at equilibrium) will be the same. When the pH is above the pKa, the equilibrium shifts to the salt or dissociated form. Conversely, when the pH is below the pKa, the acid or undissociated form predominates. It is this uncharged, undissociated, lipophilic form of the acid that can diffuse across the cytoplasmic membrane. Of course, once the acid reaches the cytoplasmic environment, where the pH is usually near neutral, it dissociates, in accord with its pKa, forming the anion and releasing a proton. Not only does the organism have to spend valuable energy (in the form of ATP) to expel these protons, but their accumulation in the cytoplasm eventually will result in a decrease in the intracellular pH to a point that is inhibitory to the cell.

(Continued)

Box 8–3. How Propionates Preserve Bread *(Continued)*

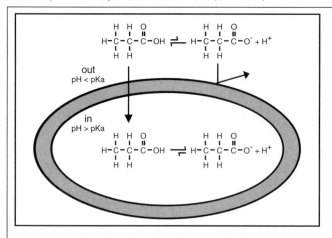

Figure 2. Mode of action of propionate. The cell membrane is permeable only to the acid form (top panel). The ratio of propionate (A^-) to propionic acid (HA) at a typical bread pH of 5.2 can be calculated from the Henderson-Hasselbalch equation (middle panel). If 0.2% propionate is added to bread, the concentrations (%) of propionate and propionic acid at other pH values can be similarly determined (lower panel).

1. $pH = pKa + \log \dfrac{[A^-]}{[HA]}$

2. $5.20 = 4.87 + \log \dfrac{[A^-]}{[HA]}$

3. $0.33 = \log \dfrac{[A^-]}{[HA]}$

4. $2.14 = \dfrac{[A^-]}{[HA]}$

pH	$[A^-]/[HA]$	$[A^-]$	$[HA]$
4.40	0.34	0.05	1.15
4.80	0.85	0.09	0.11
5.20	2.14	0.13	0.07
5.60	5.37	0.17	0.03
6.00	13.49	0.19	0.01

The pKa of propionic acid is 4.87, and, according to the Henderson-Hasselbalch equation, calcium propionate would be in the acid form below this pH and in the salt form above this pH. It should be noted that the ordinary pH range of bread is between 5.0 and 6.0 (i.e., above the pKa of propionic acid). However, enough of the acid form will still be present to diffuse into the cell and cause inhibitory effects (Figure 2, middle and lower panels).

In the home, a good strong pair of hands can do the mixing and kneading, but commercial bread manufacturers rely on large mixers that knead and tumble the dough. There are actually many different types of mixers, ranging from low- or slow-speed mixers to high-speed devices that require as little as five minutes to complete the job. The latter are used for processes, such as no-time bread manufacturing methods (discussed later) that rely on mechanical development of the dough and shorter fermentations. This is in contrast to traditional low-speed mixing methods in which dough development (defined loosely as the cumulative changes that give the dough its necessary physical and chemical properties) depends on an extended period of yeast activity and fermentation.

Several events occur during mixing. Upon hydration, gliadin and glutenin form a visco-elastic gluten complex. The sulfur-containing amino acid cysteine is present in both gliadin and glutenin, where it forms disulfide bonds or crosslinks between poypeptide chains. However, in gliadin the disulfide crosslinks occur within a single protein molecule, whereas it is generally believed that the disulfide crosslinks in glutenin form between different protein strands (according to most cereal chemists). The more crosslinking, the more elastic the dough. Dough mixing and elasticity are affected by oxidizing and reducing agents, as described above. Eventually, as the dough is mixed and the gluten complex is formed, a continuous network or film will surround the starch granules. Naturally occurring or exogenous amylases then begin to hydrolyze the starch, generating maltose, maltodextrins, and other sugars.

Fermentation

Yeast growth is initiated as soon as the flour, water, yeasts, and other ingredients are combined and the dough is adequately mixed. A lag phase usually occurs, the duration of which depends on the form of the yeast and the availability of fermentable sugars. Bakers' yeast (*S. cerevisiae*) has a facultative metabolism, meaning that it can use glucose by either aerobic (i.e., via the tricarboxylic acid or TCA cycle) or anaerobic pathways (Figure 8–6, upper panel). The former pathway yields much more cell mass and more ATP per glucose than the anaerobic pathway. However, despite the incorporation of oxygen into the dough during the mixing step, oxidative metabolism of carbohydrates occurs only briefly, if at all. Instead, carbohydrates are metabolized by the glycolytic fermentative pathway. This is due, in part, to the presence of glucose, which inhibits synthesis of TCA cycle enzymes via catabolite repression, but also because the dough quickly becomes anaerobic due to the evolved CO_2. Also, as shown in Figure 8–6 (lower panel), the ethanol-forming reduction reaction generates oxidized NAD, which is necessary to maintain glycolysis.

Sugar metabolism by bakers' yeast

Because several sugars are ordinarily present in the dough (mainly glucose and maltose), and others may be added (sucrose, high fructose corn syrup), the yeast has a variety of metabolic substrates from which to choose. In addition, more maltose may be formed in the dough during the fermentation step via the successive action of endogenous α- and β-amylases present or added to the flour and that act on damaged starch granules. The order in which these different carbohydrates are fermented by *S. cerevisiae* is not random, but rather is based on a specific hierarchy, with glucose being the preferred sugar.

For the most part, regulation is mediated by catabolite repression, acting at early steps in various catabolic pathways. Thus, in most strains of *S. cerevisiae,* glucose represses genes responsible for maltose transport and hydrolysis, as well as the invertase that hydrolyzes sucrose to glucose and fructose (Figure 8–7). Consequently, in a dough containing glucose, sucrose, and maltose, the disaccharides will be fermented only when the glucose is consumed. Moreover, maltose represses invertase expression, so sucrose

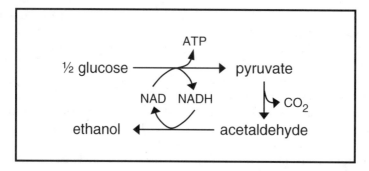

Figure 8–6. Biochemical reactions in *Saccharomyces cerevisiae*. Aerobic metabolism (upper panel, reaction 1) occurs via the tricarboxylic acid cycle, resulting in CO_2 and cell mass, and an ATP yield of 38 moles per mole of glucose oxidized. Under anaerobic conditions (reaction 2), the Embden-Meyerhoff-Parnas glycolytic pathway yields ethanol and CO_2, an ATP yield of 2 moles of ATP per mole of glucose fermented, and little cell mass. The regeneration of oxidized NAD by acetaldehyde dehydrogenase is essential (lower panel) to maintain glycolytic flux.

would be the last sugar fermented in this mixture. Depending on how the cells had previously been grown, however, it is possible that invertase had already been induced, resulting in rapid formation of sucrose hydrolysis products. Finally, there exist strains of *S. cerevisiae* whose expression of catabolic genes is constitutive, meaning that they are not subject to catabolite repression. Such strains may be particularly useful, since constitutive expression of genes coding for maltose permease and maltase means that these strains will ferment maltose even in the presence of glucose. In most doughs, however,

glucose is metabolized rather quickly, and the maltose utilization genes would soon be de-repressed and induced, even in non-constitutive strains.

Sugar transport

As shown in Figure 8-7, glucose and fructose are both transported by common hexose transporters. Genetic evidence has revealed that there may actually be as many as twenty such transporters in *S. cerevisiae*. Although it is not known how different these transporters are to one another, they all transport their substrates

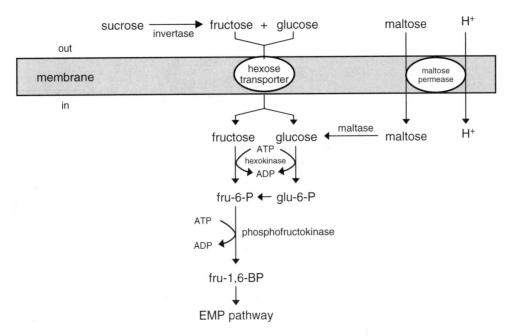

Figure 8–7. Transport and metabolism of glucose, fructose, maltose, and sucrose by *Saccharomyces cerevisiae*. Fructose and glucose are transported by one of several different hexose transporters via facilitated diffusion. Sucrose is hydrolyzed by a secreted invertase. Maltose transport occurs via a proton symport-mediated maltose permease. Once inside the cell, maltose is hydrolyzed by maltase to give free glucose. The accumulated monosaccharides are phosphorylated to glucose-6-phosphate (glu-6-P) and fructose-6-phosphate (fru-6-P) by hexokinases, then to fructose-1,6-bisphosphate (fru-1,6-BP) which feeds directly into the Embden-Meyerhoff-Parnas (EMP) pathway.

via facilitated diffusion. No energy is required and instead the driving force is simply the osmotic gradient. It is also clear that they vary with respect to their substrate affinities. For example, there are hexose transporters that have high affinity for glucose (with affinity constants or K_m values <5 μM) or very low affinity ($Km > 1$ M). The presence of low-affinity and high-affinity systems provides the cell with the versatility necessary to transport sugars under a wide range of available concentrations. Thus, yeasts are especially suited to grow in environments containing glucose and fructose, since their uptake costs are cheap (in terms of energy expenditure), and they have carriers that function even when substrates are not plentiful (for example, in lean doughs where no exogenous source of sugar is added).

Glycolysis

Following transport, the accumulated monosaccharides are rapidly phosphorylated by hexokinases or glucokinases. As for the hexose transport systems, hexokinase enzymes (there are at least two, hexokinase PI and PII) have broad substrate specificity, being able to phosphorylate glucose and fructose, as well as mannose. The sugar phosphates then feed into the glycolytic or Embden-Meyerhoff (EM) pathway (Figure 8–8), with the eventual formation of pyruvate. Reducing equivalents, in the form of NADH, are formed from oxidized NAD by the glycolytic enzyme, glyceraldehyde-3-phosphate dehydrogenase. To maintain glycolytic metabolism during active growth, therefore, it is necessary to restore the NAD. As described in Chapter 2, lactic acid bacteria reduce pyruvate

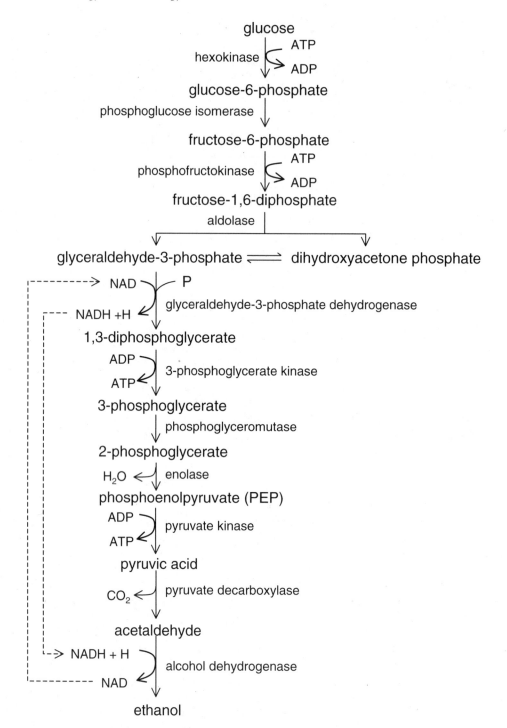

Figure 8–8. The Embden-Meyerhoff-Parnas glycolytic pathway used by *Saccharomyces cerevisiae*

and re-oxidize NAD via the lactate dehydrogenase reaction. In contrast, in *S. cerevisiae*, this is done via a two-step series of reactions, in which pyruvate is first decarboxylated by pyruvate decarboxylase, generating acetaldehyde and CO_2.

Next, the acetaldehyde is reduced by one of several NADH-dependent alcohol dehydrogenases to form ethanol and oxidized NAD. The net effect, then, of glycolytic metabolism by bakers' yeast is the formation of two moles of CO_2 and two moles of ethanol per mole of glucose fermented (Figure 8–6). The oxidized and reduced forms of NAD are in balance, and the cell obtains a net gain of two moles of ATP. Yeast growth itself is quite modest during glycolysis, because most of the glucose carbon is used for generating energy rather than for producing cell mass. Therefore, the initial population will increase only a few generations during the entire fermentation process.

End products

The most obvious and most important end product of the bread fermentation is CO_2. It is, after all, the CO_2 that is mainly responsible for transforming the stiff, heavy, dough into a spongy, elastic, and airy material. Some of the CO_2 dissolves in the water phase (even though CO_2 solubility is generally low), causing a slight decrease in pH. Eventually, the CO_2 saturates the aqueous environment surrounding the yeast microcolonies. At that point, the CO_2 evolves into the dough. The CO_2 causes the gluten proteins to stretch, and some escape, but most of the gas is retained and is trapped within the matrix. This process is known as leavening.

The retained CO_2 ordinarily collects in the dough to form large irregular gas cells that are dispersed in a non-uniform heterogenous manner. This is one reason why doughs are "punched down" during the course of the fermentation. This step re-disperses the gas, forming smaller, more regular, and evenly dispersed gas cells. About 45% of the gas is lost. However, re-mixing causes the yeast cells and sugars to be re-distributed in the dough such that more substrate is made available to the yeast and additional CO_2 can be formed.

Other metabolic products are also produced during the yeast fermentation, including various acids, that give the dough a slightly acidic, but pleasant flavor and aroma. The yeast itself can produce acids, as well as other organic compounds. Lactic acid bacteria, which are inevitably present, either in the yeast cake or in the flour, also begin to grow, ferment sugars, and produce acids. Ultimately, the pH of the dough will drop from about 6.0 to 5.0. This decrease in pH, however, has little effect on yeast growth and metabolism; in fact, pH 5 is very near the optimum for *S. cerevisiae*. Only when there is enough acid produced in the dough to lower the pH to 4.0 or less (as is the case for sourdough breads, discussed below) will inhibition of bakers' yeast strains occur. Despite their modest accumulation in dough, the acids produced by yeasts and lactic acid bacteria make important contributions to the flavor, as well as rheology of the dough. For example, low pH improves the water-binding capacity and swelling of gluten, making it more elastic and pliable.

Factors affecting growth

Temperature and relative humidity have an important effect on yeast growth and fermentative capacity. Most strains of *S. cerevisiae* used for baking have an optimum temperature of about 36°C to 39°C, although seldom are such high temperatures used during the dough fermentation. Rather, doughs are ordinarily held at temperatures of about 25°C to 28°C (and sponges slightly lower). Although higher temperatures can accelerate fermentation and gassing rates (defined as the amount of CO_2 produced per unit time), elevated temperatures also can enhance growth of microbial contaminants, including wild yeasts and mold. Also, the temperature at which yeast activity declines is only a few degrees higher than its optima. Relative humidity must also be controlled, usually at about 70% to 80%. Lower levels of moisture in the air may cause the dough to dry at the surface, leading to formation of a crust-like material, as well as inhibiting the fermentation within the areas near the surface.

Dividing, Rounding, and Panning

The fermentation described above for a straight dough or bulk fermentation (see below) usually occurs in troughs. Once the fermentation is considered complete, the dough is then ready to be divided, on a volumetric basis, into loaf-sized portions. Dividing is usually a simple process that involves cutting extruded dough at set time intervals, such that each piece has very near the same weight. Other devices, driven by suction, are also common. Despite the differences in design, the main requirement of the dividing step is that it be done quickly, since the fermentation is still ongoing and the weight-to-volume ratio (i.e., density) of the dough pieces may change.

The divided dough is then conveyed to a rounding station, where ball-shaped pieces are formed. At this point the dough is given a short (less than twenty minutes) opportunity to recover from the physical strains and stresses caused by being cut, compressed, and bounced about. A portion of the gas is also lost during the dividing and rounding steps. Thus, not only does the dough have a chance to rest (structurally, that is), but the fermentation also continues, adding a bit more gas into the dough. Following this so-called intermediate proofing step, the dough pieces are delivered into a molding system that first sheets the dough between rollers, then rolls the dough into a cylinder shape via a curling device, and finally, shapes the dough into the desired final form. The shaped loafs are then transferred to pans for proofing and baking.

Proofing

Most of the CO_2 is expelled during the sheeting step, and it is during the final proofing step where the dough is re-gassed and the fermen-

tation is completed. For some bread production systems (e.g., no-time processes described below), the entire fermentation takes place during the proofing step. Proofing is usually done in cabinets or rooms between 35°C and 42°C, giving dough temperatures near the optimum for *S. cerevisiae* (35°C to 38°C). Proofing rooms are also maintained at high relative humidity (>85%). Total proofing times vary, depending on temperature, but are usually less than one hour. The increase in dough volume, measured as the height of the loaf, is often used to determine when the dough is sufficiently proofed.

Baking

A remarkable confluence of physical, chemical, and biological events occurs when the dough loaves are placed into a hot oven. Into the oven goes a glutenous, sticky, spongy mass, with a pronounced yeasty aroma and inedible character, and out comes an airy, open-textured material, with a unique aroma and complex flavor. It is a transformation like none other in food science.

Most modern commercial bakers use continuous conveyer-type ovens, rather than batch-type, constant temperature ovens. Most of these tunnel-like ovens are configured as single or double-lap types, in which the dough loaves are loaded and unloaded (after baking) at the same end. Alternatively, in other ovens, the dough is fed at one end and the baked breads are discharged from the other end. It usually takes about twenty-five to twenty-eight minutes for a loaf to traverse through the oven.

The temperature with these ovens is not constant, but rather increases in several stages along the route, starting at about 200°C for six to eight minutes, and then increasing to about 240°C for the next twelve to fourteen minutes. Finally, the temperature is reduced slightly to about 220°C to 235°C for the remaining four to eight minutes. It should be evident that the temperature of the dough itself will always be less than the oven temperature. The actual temperature of the loaf during baking depends on those general factors affecting heat transfer kinetics, including dough composition, moisture content, loaf size and shape, and the type or form of heating applied. Although ovens may transfer heat by convection currents (i.e., via the air and water vapor), heat transfer within the dough is via conduction. Thus, there is a decided temperature gradient within the loaf, with the interior being much cooler than the surface. Moreover, many of the readily observed physical changes that occur in the bread during heating are localized, the formation of crust color being the most obvious example.

Several events rapidly occur when the dough is placed into the first stage or zone of the oven. First, as the interior dough temperature increases (about 5°C per minute), there is an immediate increase in loaf volume ("oven spring") due to expansion of the heated CO_2. When the dough temperature reaches 60°C, all of the dissolved CO_2 is evolved, causing further expansion of the dough. Ethanol also is volatilized and lost to the atmosphere. At very near the same time, there is a transitory increase in yeast and amylase activity, at least up to dough temperatures of 55°C to 60°C, at which point the yeasts are killed. Starch swelling and gelatinization increases, and the gluten begins to dehydrate and denature, causing the bread to become more rigid and more structured. All of these events occur within about six to seven minutes.

In the meantime, the surface temperature will have reached close to 100°C, and the early stages of crust formation will have begun. During the next heating stage, evaporation of water, gelatinization of starch, and denaturation and coagulation of gluten continue. Remaining enzymes and microorganisms are inactivated as the internal temperature approaches 80°C. Finally, when the temperature reaches 95°C, the dough takes on a crumb-like texture and the crust become firmer, due to dehydration at the surface. A brown crust color is formed as a result of caramelization and Maillard non-enzymatic browning reactions. The latter reaction also generates volatile flavor and aroma compounds.

Cooling and Packaging

Once out of the oven, the baked bread is susceptible to microbial spoilage. Therefore, cooling must be performed under conditions in which exposure to airborne microorganisms, particularly mold spores, is minimized. The bread also must be cool enough so that condensate will not form inside the package, a situation that could also lead to microbial problems. Various cooling systems are used, including tunnel-type conveyers in which slightly cool air passes counter-current to the direction of the bread, as well as forced air, rack-type coolers. For sandwich breads, the cooled loaves are sliced by continuous slicing machines. Ultimately, commercial breads are packaged in moisture impermeable polyethylene bags.

Modern Bread Technology

Although hundreds of variations exist, most leavened breads are manufactured by one of four general processes (Figure 8-9). These processes vary in several respects, including how the ingredients are mixed, where and for how long the fermentation occurs, and the overall time involved in the manufacturing process. Of course, how the bread is made also has a profound effect on the quality characteristics of the finished product.

Straight dough process

The homemade, one-batch-at-a-time method described above is generally referred to as the straight method or straight dough process. Basically, the overall procedure (outlined in Figure 8-9) involves mixing all of the ingredients and then allowing the dough to ferment for several hours (with intermittent punching down). The developed dough is then divided, formed into round balls, given a brief (intermediate) proof, shaped into loaves, and placed in baking pans. Finally, after a final fermentation or proof, the bread is baked. The straight

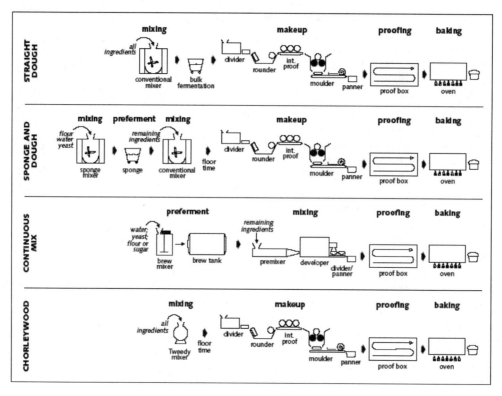

Figure 8–9. Major bread manufacturing processes. From Lillemand Bakery Update, Volume 1/Number 10 (with permission).

dough method is little used by the baking industry, the exceptions being mostly small or specialty bakeries. The process results in a bread that is chewy, with a coarse cell structure and a moderate flavor. Although the quality of these breads (Box 2) is quite good, the limitation of this process is that it lacks flexibility and is sensitive to time. In other words, when the fermentation is sufficiently complete, the dough must soon be baked. Otherwise, if fermentation is prolonged, the bread will be yeasty, excess air cells will be formed, and the structure will be weak.

Sponge and dough process

The most commonly used method in the bread industry is the sponge and dough process. It is used by small, medium, and most large bakeries in the United States. The basic

principle of this process relies on the use of a "sponge," a partially concentrated portion of a flour-water dough that is allowed to ferment and then is mixed with the remaining dough ingredients. The main advantage of this method is it that it is tolerant to time. In other words, once the sponge is developed, it does not have to be used immediately, but rather can be used over a period of time. Also, bread made by this method has a fine cell structure and well developed flavor. As outlined in Figure 8-9, the process involves making a sponge that consists of part of the flour (60% to 70%), part of the water (40%), and all of the yeast and yeast nutrients. The sponge is mixed and allowed to ferment for three to six hours at 16°C to 18°C (80°F). The remaining ingredients are then added and mixed, and the dough is allowed to develop. After a brief floor time period, the dough is then divided, molded, panned, given a final proof, and baked.

Liquid sponge process

The traditional methods of bread making, as described above for the straight dough and sponge and dough methods, require sufficient time for the initial or bulk fermentation to occur and for the dough to develop. In the past several decades, an emphasis on speed and economy of scale has led to newer methods of bread manufacture. These methods rely less on bulk fermentation and natural dough development, and more on mechanical dough development and a relatively short fermentation period. The straight dough method, for example, can be shortened considerably by adding more yeast and giving the dough just a few minutes of floor time, or perhaps even eliminating the bulk fermentation step altogether. The sponge and dough method can be similarly modified by reducing the flour-to-water ratio in the sponge (i.e., adding more water), thereby shortening the pre-fermentation time and producing a pumpable sponge.

One of the best examples of the quick type of process is the continuous bread-making procedure, sometimes called the liquid sponge process (Figure 8-9). Developed in the 1960s, it was adopted by large-scale bakeries in the United States to enhance throughput and to mass produce loaves in as short a time as possible. The method saved time and labor and was very economical. As much as 35% to 50% of the industry used this method, but its popularity eventually waned, and now more bread is probably made by sponge-and-dough methods. In principle, the process is simply an extension of the sponge-and-dough rationale.

Instead of making a thick sponge (i.e., mostly flour), a thin, liquid sponge (mostly water) is made. This mixture of water, yeast, and a very small portion of the flour is allowed to ferment in a tank to form a liquid "pre-ferment." This pre-fermented material is then metered into high speed mixers, where it is rapidly mixed with the remaining dough ingredients. The dough is then divided, panned, given a final proof, and baked.

Although the process is fast—three hours versus up to six hours for sponge and dough methods—and produces bread with a uniform structure and texture, many experts have argued that bread quality is not nearly as good as that made by other methods. By growing mainly in the liquid material and not in the actual dough, the yeasts have less time to generate flavor compounds or to react with flour components. Moreover, yeast has an oxidizing effect on gluten proteins during fermentation in the straight dough or sponge and dough methods, leading to increased elasticity and stronger dough structure. This is lost in the continuous mix system. Thus, these breads are typically softer and spongy, less elastic, and have a weak, cake-like texture. The flavor is rather bland and not well-developed. One noted bread researcher suggested that these breads are simply wrappers for meats, cheeses, and jams. Today, continuous mix systems are used almost exclusively for hot dog and hamburger buns.

Chorleywood process

In the United Kingdom, the Chorleywood process is also a rapid, high throughput method and is probably the main commercial system for bread manufacture (Figure 8-9). It is little used, however, in the United States. This is a no-time process in which rapid mixing is a critical feature. Whatever fermentation there is, prior to the final proof, occurs during the brief floor time (at temperatures near 40°C). Thus, the start-to-finish process takes only two hours.

Sourdough Fermentation

Most yeast breads have a slightly acidic pH, due, in part, to acids produced by the yeast and, to a lesser extent, to dissolved CO_2 in the dough. In addition, acids also are produced by lactic acid bacteria that ordinarily are present in the flour, in the yeast preparations, or that simply naturally contaminate the dough. However, the presence of endogenous lactic acid bacteria has, under most circumstances, only a modest effect on dough and bread acidity. The perception of sourness for most consumers does not occur unless the pH is near 4.0, which is far lower than the normal pH range of 5.0 to 6.0 for most yeast breads. However, if lactic acid bacteria are deliberately added and allowed to grow, enough acids can be produced to lower the pH to below 4.0 and to cause a

distinct sour but appealing flavor. Not only do these breads have a unique flavor, but they are also better preserved due to their high acidity and low pH. In fact, there are several functional advantages to the sourdough fermentation, particularly for breads made from rye and other flours (Box 8-4).

Despite the fact that sourdough breads have only recently been studied, they actually have

Box 8-4. Beyond Sourness: Functional Advantages of the Sourdough Fermentation

Although a desirable sour flavor is likely the most noticeable property of sourdough breads, the use of sourdough cultures in bread manufacture offers other important advantages (Thiele et al., 2004; Table 1). Arguably, the most important benefit of the sourdough fermentation is due to the organic acids that are produced by the sourdough bacteria and the ensuing low pH that results in enhanced preservation and increased shelf-life of sour dough breads. However, the sourdough fermentation also has a major impact on dough functionality, in particular, when rye or other whole grain flours are used as an ingredient.

Table 1. Functional advantages of sourdough cultures.

Property	References
Preservation	Messens and De Vuyst, 2002.
Production of anti-fungal agents	Gänzle and Vogel, 2002.
Flavor	Thiele et al., 2002.
Reduced staling rate	Armero and Collar, 1998.
Increased loaf volume	Collar Esteve et al., 1994.
Lower glycemic index	Liljeberg and Björck, 1996.
Increased gluten tolerance	Di Cagno et al., 2004.
Increased mineral availability	Lopez et al., 2003.
Improved texture	Korakli et al., 2001.

Next to wheat, rye is the second most common cereal grain used to make bread. Rye, however, has properties that pose particular challenges when used in bread making. First, rye contains a high concentration of pentosans, a heterogenous mixture of pentose-containing polysaccharides consisting mostly of xylose and arabinose. They constitute as much as 10% of rye flour, which is four to five times more than that found in wheat.

In bread making, the pentosans may have both positive and negative effects. For example, pentosans have high water-binding capacity and may, therefore, decrease retrogradation and delay staling. On the other hand, pentosans may interfere with gluten formation, giving an inelastic dough that retains gas poorly. Perhaps the major problem with rye is that the rye proteins do not form a viscoelastic dough. As a result, breads made with rye as the main grain typically have a small loaf volume and a dense crumb texture. In addition, rye flour contains more α-amylase than is present in wheat, and this amylase is particularly active at the temperature at which starch gelatinizes. This results in excessive starch hydrolysis in the dough and bread, giving a poor texture and further reducing loaf volume.

The addition of sourdough cultures to rye doughs can compensate, to a large extent, for these complications. First, as the pH decreases due to the lactic fermentation, the pentosans become more soluble. They also begin to swell and form a gluten-like network that enhances dough elas-

Box 8–4. Beyond Sourness: Functional Advantages of Sourdough Fermentation *(Continued)*

ticity and gas retention. In other words, at low pH, the pentosans assume the role normally performed by gluten. In addition, the sourdough organisms are stimulated in rye flours by the availability of free sugars liberated from starch via the α-amylase. Moreover, this enzyme begins to lose activity at the low pHs achieved during the sourdough fermentation, so excessive hydrolysis is avoided. Some sourdough bacteria also have the ability to ferment pentoses released from pentosans, producing heterofermentative end products, including acetic acid (Gobbetti et al., 2000). Finally, the subsequent acidic conditions enhance the water-binding capacity of the starch granules, which further decreases the rate at which staling occurs.

Another potential problem in breads made from rye and other whole grain flours is due to the presence of phytic acid. Phytic acid (also called myoinositol hexaphosphate) is a highly phosphorylated, negatively charged molecule that is capable of binding zinc, iron, calcium, and other divalent metal cations, preventing their absorption. Thus, phytic acid, by reducing the bioavailability of essential minerals, has "anti-nutritional" properties. Cereal grains can contain as much as 4% phytic acid, and although levels in whole grain bread are usually less than 0.2%, this is still enough to be a nutritional concern.

Degradation of phytic acid in bread ordinarily occurs via the enzyme phytase, which is present in flour and is also produced by yeasts. Recently, it has been shown that *Lactobacillus sanfranciscensis, Lactobacillus plantarum,* and other sourdough bacteria also produce this enzyme (De Angelis et al., 2003; Lopez et al., 2000). Moreover, phytase activity, and in particular, the phytase produced by *L. sanfranciscensis,* has an optima at low pH (pH 4 in the case of *L. sanfranciscensis*). Thus, the sourdough fermentation also enhances the nutritional quality of rye and other whole grain breads.

References

Armero, E., and C. Collar. 1998. Crumb firming kinetics of wheat bread with anti-staling additives. J. Cereal Sci. 28:165-174.

Collar Esteve, C., C. Benetido de Barber, and M. Martinez-Anaya. 1994. Microbial sour doughs influence acidification properties and breadmaking potential of wheat dough. J. Food Sci. 59:629–633, 674.

De Angelis, M., G. Gallo, M.R. Corbo, P.L.H. McSweeney, M. Faccia, M. Giovine, and M. Gobbetti. 2003. Phytase activity in sourdough lactic acid bacteria: purification and characterization of a phytase from *Lactobacillus sanfranciscensis* CB1. Int. J. Food Microbiol. 87:259-270.

Di Cagno, R., M. De Angelis, S. Auricchio, L. Greco, C. Clarke, M. De Vincenzi, C. Giovannini, M. D'Archivio, F. Landolfo, G. Parrilli, F. Minervini, E. Arendt, and M. Gobbetti. 2004. Sourdough bread made from wheat and nontoxic flours and started with selected lactobacilli is tolerated in celiac sprue patients. Appl. Environ. Microbiol. 70:1088-1096.

Gänzle, M.G., and R.F. Vogel. 2003. Contribution of reutericylin production to the stable persistence of *Lactobacillus reuteri* in an industrial sourdough fermentation. Int. J. Food Microbiol. 80:31–45.

Gobbetti, M., P. Lavermicocca, F. Minervini, M. De Angelis, and A. Corsetti. 2000. Arabinose fermentation by *Lactobacillus plantarum* in sourdough with added pentosans and α-L-arabinofuranosidase: a tool to increase the production of acetic acid. J. Appl. Microbiol. 88:317-324.

Korakli, M., A. Rossmann, M.G. Gänzle, and R.F. Vogel. 2001. Sucrose metabolism and exopolysaccharide production in wheat and rye sourdoughs by *Lactobacillus sanfranciscensis*. J. Agric. Food Chem. 49:5194–5200.

Liljeberg, H.G.M., and I.M.E. Björck. 1996. Delayed gastric emptying rate as a potential mechanism for lowered glycemia after eating sourdough bread: studies in humans and rats using test products with added organic acids or an organic salt. Am. J. Clin. Nutr. 64:886-893.

Lopez, H.W., V. Duclos, C. Coudray, V. Krespine, C. Feillet-Coudray, A. Messager, C. Demigne, and C. Remesy. 2003. Making bread with sourdough improves mineral availability from reconstituted whole wheat flour in rats. Nutrition 19:524-530.

(Continued)

Box 8–4. Beyond Sourness: Functional Advantages of Sourdough Fermentation *(Continued)*

Lopez, H.W., A. Ouvry, E. Bervas, C. Guy, A. Messager, C. Demigne, and C. Remsey. 2000. Strains of lactic acid bacteria isolated from sour doughs degrade phytic acid and improve calcium and magnesium solubility from whole wheat flour. J. Agric. Food Chem. 48:2281–2285.

Messens W., and L. De Vuyst. 2002. Inhibitory substances produced by Lactobacilli from sour doughs—a review. Int. J. Food Microbiol. 72:31–43.

Thiele, C., M. G. Gänzle, and R.F. Vogel. 2002. Contribution of sourdough lactobacilli, yeast, and cereal enzymes to the generation of amino acids in dough relevant for bread flavor. Cereal Chem. 79:45–51.

Thiele, C., S. Grassl, and M. Gänzle. 2004. Gluten hydrolysis and depolymerization during sourdough fermentation. J. Agric. Food Chem. 52:1307–1314.

been made for many years. In fact, it seems likely that the very first breads made thousands of years ago were sourdough breads. Until the advent and availability of bakers' yeast cultures just a century ago, it also is probable that sourdough breads were the main type of bread consumed throughout Europe and North America. This is because doughs were naturally fermented by wild cultures, which inevitably contained lactic acid bacteria, including heterofermentative strains that could produce sufficient CO_2 to cause the dough to rise. In addition, most wild sourdough cultures also contained wild yeasts that would further contribute to the leavening process (see below). These mixed bacteria-yeast wild type cultures could easily be maintained and used for many years, if not decades.

Only in the past thirty years, however, have the lactic acid bacteria and yeasts that participate in the sour dough fermentation been identified. In the early 1970s, researchers at the USDA Western Regional Research Laboratory in Albany, California (located about ten miles from San Francisco) isolated a strain of *Lactobacillus* from a commercially-manufactured, locally-produced sourdough sponge. The isolate appeared to be a unique species, based on its physiological properties and DNA-DNA hybridization, leading the discoverers to propose the name, *Lactobacillus sanfrancisco* (now called *Lactobacilllus sanfranciscensis*). This organism, which has become known as the sourdough bacterium, was also present in sourdough sponges from other area bakeries. Interestingly, this strain appears to be identical to *Lactobacillus brevis* var. *lindneri*, a *Lacto-*

bacillus culture used in European sourdough breads.

In addition to having isolated *L. sanfranciscensis* from sourdoughs, the same USDA researchers discovered that the sourdough sponge also contained a unique yeast strain. This strain was identified as *Torulopsis holmii* (the imperfect form of a *Saccharomyces* species that was once classified as *Saccharomyces exiguus*), and was later reclassified as *Candida milleri*. This yeast was found in other *L. sanfranciscensis*-containing sourdough sponges from the local area, and always in a yeast:bacteria ratio of about 1:100. The same yeast-bacteria duo has subsequently been found in sourdoughs from around the world. That these organisms are found consistently in sourdough breads from such distant geographical locations reflects their remarkable adaptation to the dough environment (Box 8–5). This also accounts for the durability and stability of sourdough sponges, some of which have reportedly been carried for more than 100 years.

Many traditional sourdough bread manufacturers continue to maintain their own particular sponge, which contains a mixture of undefined, yet distinctive strains. In fact, some artisanal bakeries consider their sourdough starters to be uniquely responsible for the quality of their bread and guard their cultures as valued assets. Analyses of these cultures has shown that they contain many different species of *Lactobacillus* and other lactic acid bacteria, as well as several species of yeast (Table 8–4). This diversity of sourdough starter strains likely accounts for the variety of sour-

Box 8–5. Happy Together: Sourdough Bacteria and Yeast

Almost all wild sourdough cultures, especially those that have been maintained for a long time, contain both lactic acid bacteria and yeasts. Although many species of lactic acid bacteria and yeast have been isolated from sourdoughs, the most common appear to be *Lactobacillus sanfranciscensis* (formerly *Lactobacillus sanfrancisco*) and *Candida milleri* (Kline and Sugihara, 1971 and Sugihara et al., 1971). It is remarkable that the interaction between these organisms is not only stable, but that both organisms derive benefit from this unique ecological association. Moreover, despite their saccharolytic and fermentative metabolism, these organisms appear to observe a non-compete clause, and instead share the available sugars present in the dough.

Several physiological and biochemical properties are now known to be responsible for the stable, symbiotic relationship between *L. sanfranciscensis* and *C. milleri* and their ability to grow together in bread dough. First, *L. sanfranciscensis* is heterofermentative, producing acetic and lactic acids, ethanol, and CO_2. The pH of the dough can drop to as low as 3.5, a pH that is inhibitory to competing organisms, including bakers' yeast strains of *Saccharomyces cerevisiae*. In contrast, *C. milleri* (and related species) is acid-tolerant and benefits by the lack of competition from other acid-sensitive yeasts. The yeast, in turn, releases free amino acids and other nutrients needed by the *Lactobacillus*.

There is also a fascinating metabolic component to the *L. sanfranciscensis-C. milleri* symbiotic partnership. Most of the fermentable carbohydrate found in sourdough is in the form of maltose. Whereas *S. cerevisiae*, the ordinary bakers' yeast, readily ferments maltose, *C. milleri* is unable to ferment this sugar, leaving this organism without an obvious substrate for growth. In contrast, *L. sanfranciscensis* not only prefers maltose as a carbohydrate source, but ferments this sugar in a particularly relevant manner.

Ordinarily, one might expect that metabolism of the disaccharide maltose would result in both glucose moieties being fermented. However, when *L. sanfranciscensis* grows on maltose it uses the cytoplasmic enzyme maltose phosphorylase to hydrolyze accumulated maltose, yielding free glucose and glucose-1-phosphate. The latter is isomerized to glucose-6-phosphate, which then feeds directly into the heterofermentative phosphoketolase pathway (Chapter 2). The glucose generated by the reaction is not further metabolized, but rather is released into the extracellular medium, or in this case, the dough. Glucose excretion occurs, in part, because maltose is transported and hydrolyzed so rapidly that more glucose is formed than the cell can handle (Neubauer et al., 1994).

Weak expression of hexokinase also contributes to slow glucose use. The net effect of this unique metabolic situation is that maltose is available for the sourdough bacteria, glucose is available for the sourdough yeast, and neither organism competes against each other for fermentation substrates.

References

Gobbetti, M., 1998. The sourdough microflora: interactions of lactic acid bacteria and yeasts. Trends Food Sci. Technol. 9:267–274.

Kline, L., and T.F. Sugihara. 1971. Microorganisms of the San Francisco sour dough bread process. II. Isolation and characterization of undescribed bacterial species responsible for the souring activity. Appl. Microbiol. 21:459–465.

Neubauer, H., E. Glaasker, W.P. Hammes, B. Poolman, and W.N. Konings. 1994. Mechanisms of maltose uptake and glucose excretion in *Lactobacillus sanfrancisco*. J. Bacteriol. 176:3007–3012.

Sugihara, T.F., L. Kline, and M.W. Miller. 1971. Microorganisms of the San Francisco sour dough bread process. I. Yeasts responsible for the leavening action. Appl. Microbiol. 21:456–458.

dough breads produced by different bakeries. However, sourdough cultures, comprised of one or more defined strains, are now commercially available. In addition to *L. sanfranciscensis*, available strains include *L. brevis*, *L. delbrueckii*, and *L. plantarum*. Some of these

Table 8.4. Microorganisms isolated from sourdoughs[1].

Bacteria	Yeast
Lactobacillus alimentarius	Saccharomyces exiguus[2]
Lactobacillus brevis	Saccharomyces cerevisiae
Lactobacillus casei	Candida milleri
Lactobacillus curvatus	Candida humilis
Lactobacillus delbrueckii	Issatchenkia orientalis
Lactobacillus fermentum	
Lactobacillus hilgardii	
Lactobacillus pentosus	
Lactobacillus plantarum	
Lactobacillus pontis	
Lactobacillus panis	
Lactobacillus reuteri	
Lactobacillus sanfranciscensis	
Pediococcus acidilactici	

[1]Data from Okada et al., 1992; Pulvirenti et al., 2004; and Vogel et al., 1994

[2]This species is now classified as *Saccharomyces cerevisiae*

strains are obligate homofermentors (e.g., *L. delbrueckii*), and although the sourdough breads made from these cultures are indeed sour, they often lack the characteristic acetic acid flavor and have a somewhat less complex flavor profile.

In actual practice, authentic sourdough breads are almost always made via a sponge and dough process. The initial sponge is maintained by regular (daily or even more frequent) additions of flour-water mixtures. Portions are removed and are successively built up by adding water-flour mixtures to the sponge followed by an incubation period, until the appropriate size is achieved. Incubations can be long (twenty-four hours) or short (three hours), depending on the temperature, culture activity, and the presence of salt. The sponge can be held for a period of time, and is eventually mixed with the remaining dough ingredients at a rate of about 10% to 20% of the total dough mixture.

It should be noted that a type of sourdough bread can be made simply by adding vinegar or a combination of acetic and lactic acids directly to the dough. This eliminates the need for a sourdough fermentation. Although such breads are quite common in the retail market, they certainly lack the complex flavor and functional characteristics associated with real sourdough bread. Alternatively, some bakers add bakers' yeast to the dough to accelerate production of CO_2. While this practice may shorten the time necessary for dough development and leavening, it likely results in bread with a weak sourdough flavor.

Bread Spoilage and Preservation

Despite the fact that bread has a moderately low water activity and pH and contains few microorganisms when it leaves the oven, it is still, relative to other fermented foods, a highly perishable product. This is because the shelf-life of bread depends not only on microbial activities, but also on physical-chemical changes in the bread. Specifically, it is the phenomenon called staling that most frequently causes consumers to reject bread products. For some fresh baked products stored under ambient conditions, shelf-life can be as short as just two or three days, due to the onset of staling. However, the longer the bread is held, the more likely it is that microorganisms, and fungi, in particular, will grow and produce visible and highly objectionable appearance defects. Thus, maintaining bread in a fresh condition remains a major challenge for the bread industry.

Staling

It would be an understatement to say that staling is a complicated phenomenon. Entire books and extensive review articles have been written on this topic alone. Recently, however, thanks to the application of modern molecular methods, including infrared, near infrared, and nuclear magnetic resonance spectroscopy, x-ray crystallography, and various microscopic techniques, a clearer understanding of the details involved in staling has emerged. Simply stated, staling refers to the increase in crumb firmness that makes the bread undesirable to consumers. In addition, staling is associated with an increase in crust softness and a decrease in fresh bread flavor.

Staling is basically a starch structure and moisture migration problem. The reactions that eventually lead to staling actually start when the bread is baked, as starch granules in the dough begin to adsorb water, gelatinize, and swell. The amylose and amylopectin chains separate from one another and become more soluble and less ordered. Then, when the bread is cooled, these starch molecules, and the amylopectin, in particular, slowly begin to re-associate and re-crystallize (see below). This process, called retrogradation, results in an increase in firmness due to the rigid structures that form. Amylose retrogrades rapidly upon cooling, while amylopectin retogrades slowly. It is the slow retrogradation of amylopectin that is now thought to be primarily associated with staling of the crumb. Furthermore, moisture migration from starch to gluten and from crumb to crust make the crumb dryer and more firm. Although staling is an inevitable process, a number of strategies have been adopted to delay these reactions and extend the shelf-life of bread (Box 8-6).

Biological spoilage

Microbiological spoilage of bread is most often associated with fungi, and occurs when fungal mycelia are visible to the consumer. Some strains of *Bacillus subtilis, Bacillus mesentericus,* and *Bacillus licheniformis* can spoil high-moisture breads via production of an extracellular capsule material that gives the infected bread a mucoid or ropy texture. There are also wild yeasts capable of causing flavor defects in bread after baking; however, bacterial and yeast spoilage of bread is relatively rare, and it is fungi that are, by far, the most common microbial cause of bread spoilage (Table 8-5). In large part, this is because the water activity of bread is usually less than 0.96, which is below the minimum for most spoilage bacteria. In addition, baking ordinarily kills potential spoilage bacteria, the exception being spore-forming bacilli mentioned above.

The baking process kills fungi and their spores. Thus, when molds are present in bread, it is invariably a result of post-processing contamination. Fungal spores are particularly widespread in bakeries due to their presence in flour and their ability to spread throughout the production environment via air movement. When the baked breads leave the oven, their transit through the cooling, slicing, and packaging operations leave plenty of opportunity for infection, either indirectly by airborne spores or directly by contact with contaminated equipment. Fungi are aerobic and ordinarily grow only on the surface of loaf bread. However, slicing exposes the internal surfaces to mold spores, enabling growth within the loaf. Once packaged, moisture loss in the bread is negligible; thus, the water activity remains well within the range necessary for growth of fungi. Bread that is packaged while still warm is very susceptible to mold growth, due to localized areas of condensate that form within the package.

Although there are many different species of fungi associated with bread spoilage (Table 8-5), the most common are species of *Penicillium, Aspergillus, Mucor,* and *Rhizopus.* Bread serves as an excellent growth substrate for these fungi; visible mold growth may appear within just a few days. What one actually sees is a combination of vegetative cell growth (the mycelia), along with sporulating bodies (Figure 8-10). The latter are responsible for the characteristic blue-green or black color normally associated with mold growth. The ability of fungi to grow on bread and which species predominate depend on several factors, including bread pH and water activity, and storage temperature and atmosphere. Finally, it is important to recognize that some mold strains not only can grow on bread, causing spoilage and economic loss, but, under certain conditions, specific strains can also produce mycotoxins. Fortunately, visible mold growth ordinarily precedes mycotoxin formation, so in the very unlikely event that a mycotoxin were present, most consumers would reject the product before ingesting it.

Box 8–6. Fresh Ideas for Controlling Staling of Bread

The precise mechanisms responsible for bread staling are complex. According to current models (Figure 1), staling occurs when unstructured and gelatinized starch (mainly amylopectin) that is formed during baking reorganizes or retrogrades, first after cooling and continuing during aging, into rigid, crystalline structures (Gray and Bemiller, 2003). As starch retrogrades, intramolecular hydrogen bonding between amylopectin branches occurs, causing additional structural rigidity. In addition, the outer chains of adjacent amylopectin molecules may form double helices. Efforts to reduce staling, therefore, necessarily involve reducing the rate at which these crystallization and hydrogen-binding reactions occur.

Because staling is influenced by many factors (Table 1), no single approach aimed at delaying or preventing staling is likely to be completely effective. Rather, bread manufacturers have adopted a more multifaceted strategy that considers the entire bread-making process, from ingredient selection and product formulation to processing and packaging steps (Zobel and Kolp, 1996). In the discussion below, the role of ingredients, enzymes, and processing conditions on controlling staling are described.

Table 1. Factors affecting staling.

Factor	Effect
Ingredients	
Flour	High protein flours maintain crumb softness
Fats	Delay staling by increasing loaf volume
Surfactants	Increase crumb softness
Enzymes	Amylases (mainly α) hydrolyze starch and reduce retrogradation
Moisture	Low moisture increases staling
Packaging	Prevents moisture loss and crumb firming
Manufacturing	Long fermentation times increase loaf volume and softness
Baking temperature	High temperature, short time baking increases staling rates
Storage temperature	Refrigeration increases staling, freezing decreases staling rates

Ingredients

Several of the ingredients that are added to bread dough either retard staling rates and/or soften bread crumb and crust. Included are surfactants and other emulsifying agents, dextrins, and selected oligosaccharides. Mono- and diglycerides are probably the most common surfactant-type agents used by the bread industry and the most effective as anti-staling agents. They appear to function by binding to amylose and amylopectin and forming complexes with lipids, thereby reducing starch retrogradation rates (Knightly, 1996).

Enzymes

Enzymes are widely used by the bread industry. Amylases, in particular, are often used to hydrolyze starch and to increase the concentration of fermentable sugars in the dough. These sugars can also contribute to desirable flavor and color changes via the Maillard reaction. However, specific α-amylases are now also used for their anti-staling properties.

How amylases actually delay staling is not clearly known, although several explanations have been proposed (Bowles, 1996). It could be that the simple sugars released by α-amylases could act directly by interfering with crystal formation; however, this seems unlikely since sugar addition to breads has no effect on retrogradation. Instead, it appears that the anti-staling function of α-amylases is due to the hydrolysis of amylopectin, especially near branch-points. This would reduce the number of cross-links that contribute to crumb and crust firmness.

Box 8–6. Fresh Ideas for Controlling Staling of Bread *(Continued)*

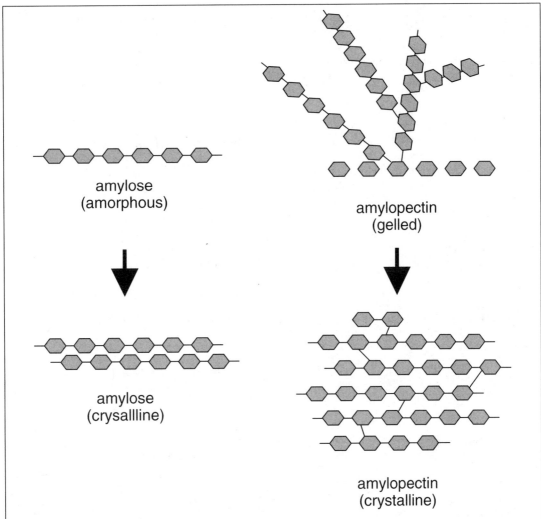

Figure 1. Model of bread staling. In fresh bread, amylose and amylopectin exist in amorphous or gelled forms. During storage, moisture is lost, the amylose and amylopectin retrograde, and crystalline forms appear, leading to firmness and staling. Adapted from Zobel and Kulp, 1996 and Lallemand Baking Update, Volume 1/Number 6.

The commercially-available α-amylases used as anti-staling agents are derived from either bacterial or fungal sources. Most are produced by species of *Bacillus* and *Aspergillus*, and consist of either pure enzymes or mixtures of enzymes. These enzymes have optimum activity within a moderately high temperature range (50°C to 70°C), and are most active during the early stages of baking. However, bacterial α-amylases are generally thermal-resistant and are inactivated only at high baking temperatures (>80°C).

If too much hydrolysis occurs during dough formation and baking, the dough will be gummy and sticky, and the baked bread may also have a similar texture. Depending on the specific application, residual activity may remain in the bread even after baking, resulting in an increase in

(Continued)

Box 8–6. Fresh Ideas for Controlling Staling of Bread *(Continued)*

starch hydrolysis and free sugars during storage. Thus, there is interest in identifying α-amylases that are less thermal-resistant. Although the fungal-produced α-amylases are more heat-labile (or have intermediate heat tolerance) than bacterial enzymes, they typically have less activity at the baking temperatures normally used.

Processing

The manner in which the dough is mixed and fermented has considerable impact on bread staling. In general, bread freshness may be poorly retained during storage if the dough is over-mixed or under-mixed. Likewise, if the fermentation time is too short or too long, bread freshness is reduced. Thus, all other factors being equal, bread made from no-time doughs or long-time (i.e., straight) doughs usually will be less soft than bread made using continuous or sponge and dough processes.

References

Bowles, L.K. 1996. Amylolytic enzymes, p. 105—129, *In* Hebeda, R.E., and H. F. Zobel. (ed.). *Baked goods freshness*. Marcel Dekker, Inc. New York.

Gray, J.A., and J.N. Bemiller. 2003. Bread staling: molecular basis and control. Comp. Rev. Food Sci. Food Safety 2:1–21.

León, A.E., E. Durán and C.B. de Barber. 2002. Utilization of enzyme mixtures to retard bread crumb firming. J. Agri. Food Chem. 50:1416-1419.

Knightly, W.H. 1996. Surfactants. p. 65—103, *In* Hebeda, R.E., and H. F. Zobel. (ed.). *Baked goods freshness*. Marcel Dekker, Inc. New York.

Zobel, H.F., and K. Kulp. 1996. The staling mechanism, p. 1—64, *In* Hebeda, R.E., and H. F. Zobel. (ed.). *Baked goods freshness*. Marcel Dekker, Inc. New York.

Table 8.5. Spoilage organisms of bread.

Organism	Appearance or defect
Fungi	
Aspergillus niger	Black
Aspergillus glaucus	Green, grey-green
Aspergillus flavus	Olive green
Penicillium sp.	Blue-green
Rhizopus nigricans	Grey-black
Mucor sp.	Grey
Neurospora sitophila	Pink
Yeasts	
Saccharomyces cerevisiae	Alcoholic-estery off-odors
Pichia burtonii	White, chalky appearance
Bacteria	
Bacillus subtilus	Ropy
Bacillus licheniformis	Ropy
Bacillus mesentericus	Ropy

Adapted from Pateras, 1998

Preservation

Given that biological spoilage of bread is caused primarily by molds, it is not surprising that preservation strategies have focused on controlling fungi, both in the production environment and in the finished product. As noted previously, mold and mold spores are present in flour and other raw materials and may be widespread in bakeries. Therefore, rigorous attention to plant design and sanitation is essential. The post-production environment (i.e., baked products) should be separated from pre-production environments. Air handling systems should be designed (including the use of filters or ultraviolet lamps and positive air pressure) such that airborne mold spores cannot gain entry to the product side.

Several different approaches have been adopted to control mold growth in bread. In

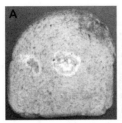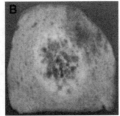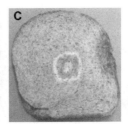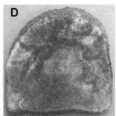

Figure 8–10. Spoilage of bread by fungi. Shown are *Penicillium commune* (A), *Aspergillus niger* (B), *Penicillium roqueforti* (C), and *Rhizopus stolonifer* (D). Photos courtesy of A. Bianchini and L. Bullerman, University of Nebraska.

the United States, the most common means of bread preservation is to use propionate salts (Box 8-3). Calcium propionate is especially effective against most of the molds associated with bread, and is widely used in commercially-produced products. Sorbate and acetate salts are also used as anti-mycotic preservatives in bread. However, these acids also have inhibitory activity against baking yeasts, whereas propionates are much less inhibitory. Moreover, calcium propionate has no flavor or toxicity, is active even against rope-forming bacteria, and is effective in most varieties of bread. There are, however, some fungal strains (e.g., *Penicillium roquefortii*) that are associated with rye breads and that are insensitive to propionates.

Due to interest in chemical additive-free breads, alternative approaches for bread preservation have been considered. Bread can be exposed to ultraviolet, infrared, or microwave radiation to inactivate mold and mold spores or packaged in modified or vacuum atmospheres to inhibit their growth. These methods, however, are not widely used. In contrast, one effective method used for bread preservation that is very popular and does not involve addition of chemical preservatives instead relies on growth of lactic acid bacteria in the dough. As discussed above, these bacteria produce lactic and acetic acids and lower the pH to levels inhibitory to most fungi.

Another indirect way to extend the shelf-life of bread is via freezing. Many bread manufactur-

ers freeze the baked and packaged breads as a means of preserving the bread prior to delivery. Another freezing method that has been adopted is to freeze un-baked breads (Box 8-7). This process is gaining popularity, since it allows retailers to do the baking and then sell "fresh"-baked breads (aroma and all) at the retail level.

Bread Quality

Assessing the quality of bread is a very subjective process. Cultural, ethnic, and personal attitudes certainly influence the sort of bread an individual prefers. For example, the hard, crusty baguette eaten in France bears little resemblance to the soft, doughy, plastic-wrapped French bread preferred by many U.S. consumers. Whole grain peasant breads—so-named because centuries ago expensive refined flours went to the upper classes, while the poor were left with unrefined, whole flours—are now popular in specialty bakeries serving an affluent, upscale clientele.

The criteria commonly used to assess bread quality are based primarily on appearance, texture, and flavor attributes. Appearance refers both to external (i.e., crust), as well as internal appearance (i.e., crumb). Depending on the type of bread, the intact loaf should have a particular volume, or in some cases, height and length. The color of the crust also is an important property, and is judged mainly on the basis of the expected color established by the

Box 8–7. Frozen Doughs and Yeast Cryotolerance

One of the hottest trends in the baking industry is the use of frozen doughs. These doughs are especially appealing to small retail bakery operations (like those in grocery stores), because they eliminate the need for dough production equipment and labor, while at the same time making it possible to offer fresh-baked bread products to customers.

Frozen doughs are particularly convenient for the end-user. The frozen dough is simply removed from the freezer, thawed overnight in the refrigerator, given a final proof, and baked. Despite these advantages, however, the quality of breads made from frozen doughs can be quite variable, due to the loss of yeast viability during storage. Doughs made using cold-sensitive yeasts not only require longer proof times, but the bakers' yeast strains ordinarily used for bread manufacture are so cryosensitive that they may fail to provide any leavening at all after the dough-thawing step. Therefore, for frozen dough manufacture, it is essential that the yeast strain be cryotolerant.

What makes some yeast strains resistant to the effects of freezing? And what can be done to improve cryotolerance? These questions are now receiving much attention, and the answers are, at least in part, inter-related.

First, it appears that cryotolerance in *Saccharomyces cerevisiae* is due the ability of the organism either to transport extracellular cryoprotectant solutes from the environment or to synthesize them within the cytoplasm. These solutes, much like osmoprotectants, are then accumulated to high intracellular concentrations, without causing detrimental effects on the enzymatic or reproductive machinery within the cell.

At a molecular level, these agents prevent dehydration, presumably by re-structuring the bound water in the cytoplasm. Microbial cryoprotectants are typically small, polar molecules and include several amines (e.g., betaine, carnitine, proline, and arginine), as well as various sugars. In *S. cerevisiae*, the disaccharide trehalose (which consists of two glucose moieties linked α-1,1; Figure 1, upper panel) is probably the most important cryoprotectant (although some amino acids also have cryoprotective activity). When the temperature decreases (or when other environmental stresses are applied), the cell responds by increasing *de novo* synthesis of trehalose. Extracellular trehalose, if available, may also be transported via a high affinity trehalose (α-glucoside) transport system. Ultimately, intracellular levels can reach up to 20% of the total weight of the cell (on a dry basis).

The trehalose concentration in yeasts is not always maintained at such a high level. Rather, there is a balance between trehalose synthesis and metabolism, such that under non-stress conditions, the enzymes used to synthesize trehalose (the trehalose synthetase complex) are turned off and the hydrolytic enzymes (trehalases) are turned on (Figure 1, lower panel). Trehalose, it is important to note, can also serve as a storage carbohydrate in *S. cerevisiae*, providing a readily metabolizable source of energy for when exogenous sugars are in short supply.

The coordination of the catabolic and anabolic enzymes, combined with the trehalose transporter, therefore, allows the cell to modulate the intracellular trehalose concentration, depending on the particular circumstances in which it finds itself. This system, however, can also be manipulated genetically for very practical purposes. If, for example, the trehalose degradation pathway is blocked, then higher levels of trehalose can be maintained, making the yeast more cryotolerant. One recently described strategy involved inactivation of genes encoding for an acid trehalase (Ath1p) and a neutral trehalase (Nth1p). When yeasts containing these defective genes were used to make frozen doughs, cryotolerance was enhanced, as observed by the increase in gassing power (CO_2 produced per minute) in the thawed doughs.

Despite its importance in cryotolerance, trehalose is not the only cryoprotectant molecule accumulated by *S. cerevisiae*. As noted above, arginine can also serve this role. Thus, an approach similar to that described for trehalose has recently been used to enhance cryotolerance in a bakers' yeast strain of *S. cerevisiae*. In this case, the *CAR1* gene, encoding for the arginine-hydrolyzing enzyme arginase was inactivated. This led to increased intracellular concentrations

Box 8–7. Frozen Doughs and Yeast Cryotolerance *(Continued)*

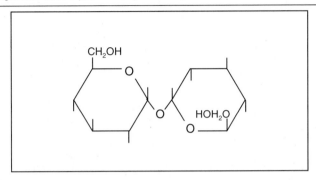

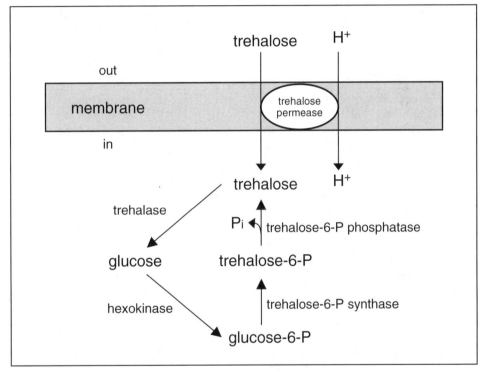

Figure 1. Trehalose metabolism by *Saccharomyces cerevisiae*. Structure of trehalose (α-D-glucosyl-1→ 1-α-D-glucose) is shown in the upper panel. Transport is mediated by a symport system, but synthesis from glucose-6-phosphate (P) may be the major route for trehalose accumulation (lower panel). Trehalose not only alleviates stress, but it can also serve as a carbon source following its hydrolysis to glucose by the enzyme trehalase.

of arginine and an increase in freeze tolerance and gassing power compared to the parent strain.

Despite the promising experimental results with genetically modified strains, such strains are not used commercially for frozen bread manufacture. Instead, the industry relies more on manipulating manufacturing conditions to enhance yeast performance. It is essential, for example,

(Continued)

Box 8–7. Frozen Doughs and Yeast Cryotolerance *(Continued)*

that naturally cryotolerant strains or mutant strains derived from wild type strains be used (Codón et al., 2003). Of course, increasing the amount of yeast is also important to compensate for any loss of activity. Finally, minimizing the extent of fermentation prior to freezing appears to improve yeast viability after thawing.

References

Codón, A.C., A.M Rincón, M.A. Moreno-Mateos, J. Delgado-Jarana, M. Rey, C. Limón, I.V. Rosado, B. Cubero, X. Peñate, F.Castrejon, and T. Benítez. 2003. New *Saccharomyces cerevisiae* baker's yeast displaying enhanced resistance to freezing. J.Agri. Food Chem. 51:483–491.

Shima, J., A. Hino, C. Yamada-Iyo, Y. Suzuki, R. Nakajima, H. Watanabe, K. Mori, and H. Takano. 1999. Stress tolerance in doughs of *Saccharomyces cerevisiae* trehalose mutants derived from commercial baker's yeast. Appl. Environ. Microbiol. 65:2841–2846.

Shima, J., Y. Sakata-Tsuda, Y. Suzuki, R. Nakajima, H. Watanabe, S. Kawamoto, and H. Takano. 2003. Disruption of the *CAR1* gene encoding arginase enhances freeze tolerance of the commercial baker's yeast *Saccharomyces cerevisiae*. Appl. Environ. Microbiol. 69:715–718.

manufacturer. Internal appearance is also based, in part, on color, specifically, the absence of streaks or uneven color. The size, frequency, and distribution of air cells are important determinants of bread quality, especially if there are large, irregular-shaped holes that are often considered as defects.

In contrast to the appearance properties, which are difficult to assess objectively, methods do exist to measure texture properties. Instrumental devices, such as the Instron Universal Testing Machine, can be used to measure crumb softness, resistance, and compressibility. Other compression-type instruments also exist, such as the Baker Compressimeter and the Voland Stevens Texture Analyzer, which can be used to measure firming changes and staling rates. Various spectrophotometric methods, including near infrared and infrared spectroscopy can detect changes in starch crystallization and conformation, and are especially useful for measuring retrogradation and staling rates. However, X-ray diffraction is now generally considered the most definitive technique for detecting changes in crystallization.

Bread flavor is no less easy to describe or quantify. It can also be difficult to separate flavor from texture and other sensory attributes. In general, bread should have a fresh, some-what yeasty flavor and aroma, with acidic and wheaty notes. Salt also provides an important flavor in bread, as do the alcohols, aldehydes, and other organic end products produced during yeast metabolism.

Many of the important flavor compounds of bread are formed during baking and are derived from caramelization and Maillard reactions; the Maillard reaction alone results in more than 100 volatile aroma compounds. Importantly, these reactions also generate pigments and are responsible for crust color formation. Although both types of reactions require heat, are non-enzymatic, and use sugars as the primary reactant, they differ in that caramelization reactions occur at higher temperatures than Maillard reactions and only reducing sugars react in the Maillard reaction. The latter also requires primary amino acids. The Maillard reaction is probably predominant during baking, since the temperatures achieved generally are not high enough to induce extensive caramelization.

Bibliography

Calvel, R. 2001. *The Taste of Bread*. (R.L. Wirtz, translator and J.J. MacGuire, technical editor). Aspen Publishers, Inc. Gaithersburg, Maryland.

Cauvain, S.P. 1998. Breadmaking Processes, p. 18—44. *In* S.P. Cauvain and L.S. Young (ed.), *Technol-*

ogy of Breadmaking. Blackie Academic and Professional (Chapman and Hall). London, UK.

Gobbetti, M., and A. Corsetti. 1997. *Lactobacillus sanfrancisco* a key sourdough lactic acid bacterium: a review. Food Microbiol. 14:175-187.

Gray, J.A., and J.N. Bemiller. 2003. Bread staling: molecular basis and control. Comp. Rev. Food Sci. Food Safety. 2:1-21.

Kline, L., and T.F. Sugihara. 1971. Microorganisms of the San Francisco sour dough bread process. II. Isolation and characterization of undescribed bacterial species responsible for the souring activity. Appl. Microbiol. 21:459-465.

Lallemand Baking Update, Lallemand Inc., Montreal, QC, Canada.

Okada, S., M. Ishikawa, I. Yoshida, T. Uchimura, N. Ohara, and M. Kozaki. 1992. Identification and characteristics of lactic acid bacteria isolated from sour dough sponges. Biosci. Biotech. Biochem. 56:572-575.

Pateras, I.M.C. 1998. Bread spoilage and staling, p. 240—261. In S.P. Cauvain and L.S. Young (ed.), *Technology of Breadmaking*. Blackie Academic and Professional (Chapman and Hall).

Pulvirenti, A., L. Solieri, M. Gullo, L. De Vero, and P. Giudici. 2004. Occurrence and dominance of yeast species in sourdough. Lett. Appl. Microbiol. 38:113-117.

Pyler, E.J., 1988. *Baking Science and Technology. Volumes 1 and 2, 3rd edition*. Sosland Publishing Co. Merriam, Kansas.

Röcken, W., and P.A. Voysey. 1995. Sour-dough fermentation in bread making. J. Appl. Bacteriol Symp. Supple. 79:38S-48S.

Stear, C.A. 1990. *Handbook of Breadmaking Technology*. Elsevier Applied Science, New York.

Stolz, P., G. Böcker, R.F. Vogel, and W.P. Hammes. 1993. Utilization of maltose and glucose by lactobacilli isolated from sourdough. FEMS Microbiol. Lett. 109:237-242.

Sugihara, T.F., L. Kline, and M.W. Miller. 1971. Microorganisms of the San Francisco sour dough bread process. I. Yeasts responsible for the leavening action. Appl. Microbiol. 21:456-458.

Vogel, R.F., G. Böcker, P. Stolz, M. Ehrmann, D. Fanta, W. Ludwig, B. Pot, K. Kersters, K.H. Schleifer, and W.P. Hammes. 1994. Identification of lactobacilli from sourdough and description of *Lactobacillus pontis* sp. nov. Int. J. System. Bacteriol. 44:223-229.

9

Beer Fermentation

"I would give all my fame for a pot of ale."
King Henry V, Act 3, Scene 2, William Shakespeare
"I would kill everyone in this room for a drop of sweet beer."
Homer Simpson

Introduction

In 1992, a research team from the Applied Science Center for Archaeology at the University of Pennsylvania made a discovery that was publicized within archaeology circles (and college campuses) around the world. These researchers had analyzed a small amount of organic residue from inside an ancient pottery vessel that had been retrieved from the Zagros Mountains of Western Iran. When the residue from this clay pot (which itself was dated circa 3500-3100 B.C.E.) was analyzed, the results revealed the presence of oxalate ion. Since oxalate, and calcium oxalate in particular, collect in only a few places, this finding could mean only one thing—this pot had been used to brew beer. The 5,000-year-old residue, the archaeologists concluded, was the earliest chemical evidence for the origins of beer in the world.

Although this discovery attracted its share of headlines, it was but one of many events in the remarkable historical record of beer making (Figure 9-1). Beers were widely prepared and consumed in Egypt and other Middle Eastern counties from ancient days and then spread west into Europe and the British Isles. Despite the introduction of Islam in the eighth century, and its prohibition against alcohol consumption, beer making and consumption continued to grow throughout Europe during the Middle Ages.

During that time, beer making was considered an art, performed by skilled craftsmen. Many of the early breweries were located within various European monasteries, which created a tradition of brewing expertise and innovation. The use of hops in beer as a flavoring agent, for example, was first practiced by monks, as was the use of bottom-fermenting yeasts (discussed below). Although beer manufacture was practiced throughout Europe, it was of particular cultural and economic importance in Great Britain and Germany, which became the epicenters for brewing technology.

By the late eighteenth century, and especially by the mid-1800s, beer making was one of the first food processes to become industrialized. Although monastery-based breweries were still common, many small breweries began to form, serving their product directly on the premises (much like the brew pubs that are popular even today). Eventually, some of the larger breweries contained not only production facilities, but also the beginning of what we might now consider to be quality control laboratories. And although these breweries were located mainly in Europe and England, beer making had also spread to North America and the American colonies.

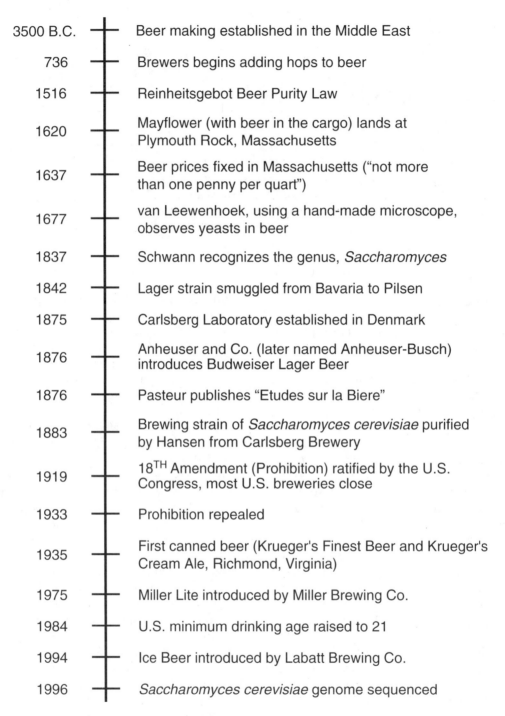

Figure 9–1. Milestones in the history of beer.

In fact, English ale-style beer had been among the provisions carried by Pilgrims on the Mayflower, and there are reports, written in the actual travel logs, of the ship running low on beer and the worries that caused the passengers. ("We could not now take time for further search . . . our victuals being much spent, especially our beer. . .") During the latter half of the nineteenth century, the number of breweries in the United States increased three-fold, from 400 to 1,300. Most of these breweries produced German lager-style beer (see below), reflecting the huge immigrant population from Germany during that era (including the brewers themselves, e.g., Adolphus Busch, Eberhard Anheuser, Adolph Coors, Frederick Miller, Joseph Schlitz).

Beer Spoilage and the Origins of Modern Science

Most fermentation microbiology students are aware that cheese, sausage, and other fermented foods evolved, in part, because these products had unique and desirable sensory characteristics. Likewise, they might also appreciate the many pleasant attributes of malted and hopped beverages. However, it is important to recognize that, while our ancestors undoubtedly enjoyed beer for many of the same reasons as today's consumers, they also understood that beer, like other fermented foods, was somehow better preserved than the raw materials from which they were made. For hundreds of years, for example, early trans-Atlantic voyagers (like the Pilgrims mentioned above) relied on beer because it was a well-preserved form of nourishment. In fact, until relatively recently, beer was often safer to drink than water, was less likely to cause water-borne disease, and was less susceptible to spoilage. Of course, we now know that the microbiological stability and "safety" of beer is due, in part, to its ethanol content, as well as other anti-microbial constituents and properties (discussed, in more detail, later in this chapter).

Despite the stability and general acceptance of beer as a well preserved product, it can indeed become spoiled. As beer manufacturing in Europe grew from small, craft-oriented production into a large brewing industry in the middle of the nineteenth century, the prevention of beer spoilage became an important goal of industrial brewers, who were often troubled by inconsistencies in product quality. These difficulties attracted the attention of chemists and other scientists, who were enlisted to solve some of these technical problems.

Beer manufacture was, in fact, one of the first industrial fermentations to be studied and characterized, and was the subject of scientific inquiry by early microbiologists and biochemists. The very development of those scientific disciplines coincides with the study of beer and other fermented foods (Box 9–1). In particular, the science of beer making was revolutionized in 1876 by Pasteur, who not only showed that yeasts were the organisms responsible for the fermentation, but also that the presence of specific organisms were associated with specific types of spoilage. Pasteur also developed processes to reduce contamination and preserve the finished product. It is worth noting that even today, preventing beer spoilage by microorganisms is still an important challenge faced by the brewing industry.

Scientific interest in brewing extended throughout Europe, leading to establishment of research laboratories in Copenhagen (the Carlsberg Laboratory) and Bavaria (the Faculty of Brewing), as well as laboratories located within several breweries. Perhaps the most noteworthy discoveries of the late nineteenth and early twentieth centuries were made by Emil Christian Hansen at the Carlsberg Laboratory. He developed pure culture techniques for yeasts, which eventually led to methods (still in use today) for propagation and production of yeast starter cultures free of contaminating bacteria and wild yeasts.

The Modern Beer Industry

Today, beer has one of the largest dollar values of all fermented food products, with U.S. retail sales in 2002 of more than $65 billion dollars.

Box 9–1. Pasteur, the Origins of Microbiology, and Beer

It is hard to imagine, given the current age of scientific specialization, that one person could have been as accomplished in so many fields as was Louis Pasteur in the latter half of the nineteenth century. He was trained as a chemist and, at the age of only 26, made important discoveries in stereochemistry and crystallography (specifically, while examining tartaric acid crystals in wine). Pasteur then became interested in the not yet named field of microbiology, and devoted nearly two decades of his life disproving the theory of spontaneous generation and establishing microorganisms as the causative agent of fermentation and putrefaction.

His work on vaccines against animal anthrax and human rabies, as well as the establishment of the germ theory of disease, brought Pasteur worldwide recognition and fame (Schwartz, 2001). During his career, Pasteur addressed practical agricultural, food, and medical problems, but did so by relying on basic science ("There is no such thing as applied science, only applications of science," as quoted in Baxter, 2001). He was a meticulous researcher, who rigorously defended the scientific method, saying, "In the field of observation, chance favors only the prepared mind."

There were certainly other well-known scientists of his era, including Schwann, Koch, and Lister, whose works also helped establish the field of microbiology. But the collective contributions of Pasteur led to his recognition as one of the "most distinguished microbiologists of all time" (Barnett, 2000), or as stated by Krasner (1995), the "high priest of microbiology."

Pasteur was already an accomplished scientist, having completed his work on the stereochemistry of tartaric acid, when he turned his attention to solving spoilage problems plaguing the French wine industry. In reality, he was "commissioned" by Emperor Napoleon III in 1863 to serve as a "consultant" and to study these so-called wine diseases. Pasteur correctly diagnosed the spoilage conditions, which he called tourne (mousy), pousse (gassy), and amertume (bitter), as being caused by bacteria. He then showed that the responsible organisms could be killed, and the wine stabilized, by simply heating the wine to 50°C to 60°C. These findings were published in 1866, as "Études sur le Vin."

Nearly ten years later, Pasteur began studying the fermentation of beer, another industrially-important product that was also suffering from spoilage defects. This work led to a second treatise on alcoholic fermentations, "Études sur la bière," published in 1876 (the English-translated version is available and well worth reading). His motivation this time had less to do with his interest in beer (he apparently was not a beer drinker), than it did for his animosity for all things German (Baxter, 2001).

In the early 1870s, France and Germany were at war (the Franco-Prussian War), whose outcome resulted in France having to cede the hop-producing region of Alsace-Lorraine to Germany. Already Germany produced the best beer and was the dominant producer in Europe. German beers were better from most English and other beers for several reasons, but mainly because these beers were fermented at a low temperature (6°C to 8°C). The resulting "low beers" (so-named because of the low fermentation temperature, but also because the yeast would sink to the bottom of the tank) were lighter-colored and less heavy than "high beers," but most importantly, they were also better preserved.

Low beers (which we now refer to as lagers) could be produced in the cool winter months, and, provided they were stored cold (in cellars or by ice), be consumed during the summer. High beers (or ales) had to be consumed shortly after manufacture and did not travel well. This situation, in which French beers were at a commercial disadvantage, was intolerable to Pasteur and he was determined to improve French-made beers or as he himself stated, to make the "Beer of the National Revenge."

Pasteur started out by building a brewing "pilot plant" in the basement of his laboratory and by visiting commercial breweries. Then, using microscopic techniques, he showed that specific yeasts were necessary for a successful beer fermentation. He also observed that, based on microscopic morphology, certain contaminating microorganisms were associated with specific

Box 9–1. Pasteur, the Origins of Microbiology, and Beer *(Continued)*

types of spoilage conditions (or "diseases"). For example, long rod-shaped or spherical-shaped organisms in chains (probably lactic acid bacteria) were responsible for a sour defect. In Pasteur's own words, "Every unhealthy change in the quality of beer coincides with a development of microscopic germs which are alien to the pure ferment of beer" (Pasteur, 1879). The corollary was also true: "The absence of change in wort and beer coincides with the absence of foreign organisms." These foreign organisms had gained entrance into the beer either by virtue of their presence in the production environment or via the yeast (which were used repeatedly and were subject to contamination).

To solve these problems, Pasteur offered several recommendations. He showed that a modest heating step (55°C to 60°C) would destroy spoilage bacteria and render a beer palatable even after nine months. He also emphasized that elimination of contaminants, both environmental and within yeast preparations, would also be effective. Oxygen, Pasteur realized, enhanced growth of spoilage organisms, thus, he devised a brewing protocol that precluded exposure of the wort to air (described in a U.S. patent; Figure 1).

Pasteur demonstrated through actual industrial scale experiments that these strategies would work, and indeed these practices were readily adopted. And although the German beer industry was hardly affected by his personal "revenge," Pasteur's influence on brewing science and microbiology is considerable even today, more than 125 years later.

References

Barnett, J.A., 2000. A history of research on yeasts 2: Louis Pasteur and his contemporaries, 1850–1880. Yeast. 16:755–771.

Baxter, A.G. 2001. Louis Pasteur's beer of revenge. Nature Rev. 1:229–232.

Krasner, R.I. 1995. Pasteur: high priest of microbiology. ASM News. 61:575–579.

Pasteur, L. 1879. *Studies on fermentation. The diseases of beer, their causes, and the means of preventing them.* A translation, made with the author's sanction, of "Études sur la bière" with notes, index, and original illustration by Fran Faulkner and D. Constable Robb. Macmilan and Co. 1879, London and Kraus Reprint Co., New York, 1969.

Schwartz, M. 2001. The life and works of Louis Pasteur. J. Appl. Microbiol. 91:597–601.

(Continued)

In 2002, U.S. consumers drank nearly 190 million barrels (1 barrel = 117 L = 31 gallons), with per capita consumption of 81 L or 21.5 gallons per person per year (or nearly forty six-packs per man, woman, and child!). That's also more, on a volume basis, than wine (8 L), juice (34 L), milk (72 L), and bottled water (42 L). Although one might assume that this is a lot of beer, beer consumption in the United States is half as much as that of several other European countries (Figure 9-2). For example, per capita consumption (for 2002) in Germany and Ireland was 121 L and 125 L per person per year, respectively, and Austria (109 L) and the U.K. (100 L) are not far behind. The Czech Republic, however, leads all other countries, with per capita consumption of 162 L (2003).

The importance of beer to the world economy is also considerable. In the United States alone, the overall economic impact is estimated to be more than $144 billion, and the industry pays more than $5 billion directly in excise taxes to federal and state government (according to the U.S. Beer Institute). The industry also spends about $1 billion per year for advertisements and promotions.

The beer industry, like many other segments of the food industry, both in the United States as well as internationally, has become highly consolidated. In the past fifty years, many medium

Box 9–1. Pasteur, the Origins of Microbiology, and Beer *(Continued)*

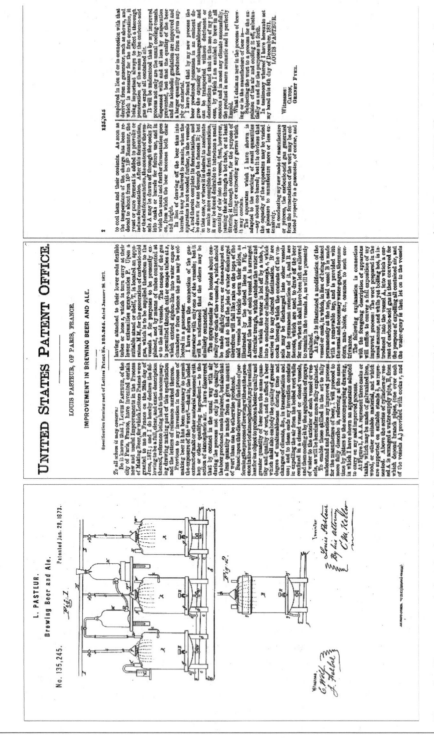

Figure 1. "Improvement in brewing beer and ale," a U.S. patent (No. 135,245) issued to Louis Pasteur of Paris, France in 1873. The patent is not for "pasteurization," but rather describes a method for excluding air from the boiled wort prior to inoculation with yeast. The resulting beer has improved stability, or, according to the patent, "possesses in an eminent degree the capacity for unchangeableness."

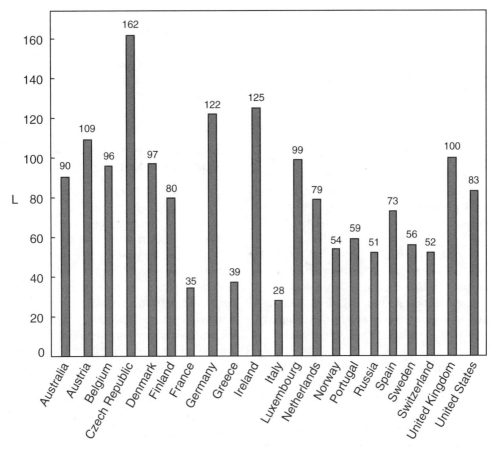

Figure 9–2. Worldwide per capita beer consumption (L per person per year). Adapted from Brewers of Europe 2002 Statistics and Beer Facts.

sized breweries were either acquired by larger breweries or were unable to compete and went out of business. Where there were once several hundred small, privately owned breweries operating in the United States as recently as 1950, now more than 80% of U.S. beer is produced by just three companies. Despite this centralization of the beer industry, during the past twenty years there has been a remarkable increase (from zero to more than 1,000) in the number of small breweries (small enough to earn the designation "microbreweries") that are now operating in all regions of the United States. These breweries are important to mention here, not only because they have captured a small but noticeable share of the market, but also because

they have adopted many of the traditional brewing practices that will be described later in this chapter.

Beer Manufacturing Principles

Only four ingredients are necessary to make beer: water, malt, hops, and yeast. Despite its ancient origins and long history, and this seemingly short list of ingredients, the manufacture of a quality beer remains a rather challenging task. In part, this is because beer making consists of several different and distinct processes that are not always easy to control. In addition, some steps taken to improve one aspect of the process—for example, filtering the finished beer to enhance clarity—may at the same time remove desirable flavor and body constituents.

The actual brewing process involves not only the well-studied yeast fermentation, but also entails other biological, as well as chemical and physical reactions. It is, therefore, convenient to consider the beer manufacturing process as consisting of several distinct phases or steps.

The primary purpose of the first phase, called *mashing*, is to transform non-fermentable starch into sugars that the yeast can ferment. The process, which is enzymatic in nature, involves a series of biochemical events that starts with the conversion of cereal grain, usually barley, into malt. The malt is then used to make a mash, ultimately resulting in the formation of a nutrient-rich growth medium, called wort. Although other grains, such as sorghum, maize, and wheat can also be malted, barley is by far the most frequently malted cereal grain.

In the second, or fermentation, phase, which is microbiological in nature, sugars, amino acids, and other nutrients present in the wort are used to support growth of yeasts. Yeast growth, under the anaerobic conditions that are soon established, is accompanied by fermentation of sugars and formation of the end products ethanol and CO_2. Technically speaking, the fermentation step results in beer that could then be consumed. However, at this point, the beer contains yeast cells, insoluble protein-complexes, and other materials that cause a cloudy or hazy appearance. In addition, carbon dioxide is lost during the fermentation. Beer at this stage is also microbiologically unstable and susceptible to spoilage. Therefore, additional measures are almost always taken to remove yeasts and other microorganisms and any other substances that would otherwise affect product quality and shelf-life and to provide carbonation in the final product.

In this chapter, therefore, a third phase, consisting of important post-fermentation activities, also will be discussed. These latter steps of the beer-making process, some might argue, are among the most important, since they have a profound effect on the appearance, flavor, and stability of the finished product. The general beer manufacturing process is outlined in Figure 9-3. As the reader will soon be aware, many beer-specific terms, such as mashing, are used to describe the brewing process. Some of these terms are defined in Box 9-2.

Enzymatic Reactions: Malting and Mashing

The first part of beer manufacture, the enzymatic steps, actually begins far from the brewing facility, in the malting houses that convert barley into malt. It is the malt that serves as the source of the amylases, proteinases, and other enzymes necessary for hydrolysis of large macromolecules, such as starch and protein. For most beers, the malt also serves as the substrates for those enzymes (i.e., malt contains the starch and protein hydrolyzed by malt enzymes). In addition, malt is the primary determinant of color and body characteristics in beer, and influences flavor development. Considering the critical functions that malt contributes to the beer-making process, it is evident that malt quality has a profound influence on beer quality.

The goal of the maltster is to convert barley to malt such that enzyme synthesis is maximized and enzyme activity is stable and well-preserved. The process starts with selection of barley. Several different barley cultivars are used to produce malt. In North America, six-row barley is generally preferred, whereas in Europe and the U.K., two-row barley is used (however, some six-row is also used in Europe and the U.K. and some two-row is used in the United States). Six-row barley contains more protein and less starch than two-row barley. The former also contains a higher level of starch-degrading enzymes, making it more suitable for beers containing adjuncts (discussed below). The harvested barley is then cleaned, graded, and sized. Unless the barley is to be used right away (which is usually not the case), it is dried from about 25% moisture to 10% to 12% so that it can be safely stored.

| Malting | Mashing | Fermentation | Post-Fermentation |

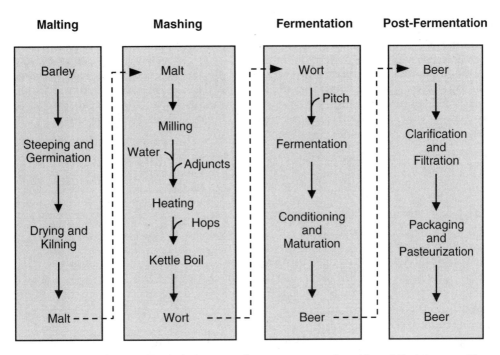

Figure 9–3. Manufacture of beer. The beer manufacture process consists of four distinct stages: malting, in which barley is converted to malt; mashing, in which enzyme and substrate extraction and reactions occur and a suitable growth medium is prepared; fermentation, in which wort sugars are fermented to beer; and post-fermentation, in which the beer is made suitable for consumption.

In the next step, the barley is allowed to germinate or to develop the beginning of a root system. Barley is essentially a seed—if it is held under warm, moist conditions, it will begin to sprout or root, just as it would if it were growing in nature. For malting purposes, germination is done by steeping the barley in cool water at 10°C to 20°C for two to three days, or long enough to increase the moisture from 10% to 12% to about 45%. The steep water is usually changed every twelve hours, in large part to minimize the impact of potential spoilage organisms. During this step, the material is well aerated to promote the germination process. The moist barley is then removed from the steep water and incubated in trays or drums under cool, humid, and well-aerated conditions for two to eight days. The barley initially swells, then germinates, such that

"rootlets" or sprouts appear. Germination is accompanied by the synthesis of myriad enzymes that the barley grain theoretically needs for subsequent growth into new barley plants. The barley has reached its maximum enzyme activity when the sprout reaches a length of one-third the size of the grain.

Next, it is necessary to arrest further germination and to stabilize and preserve the enzymatic activity. This is done by a step-wise drying process, in which the germinated "green" barley is dried incrementally within a temperature range of 45°C to 60°C. The purpose of this slow drying process is to remove water without inactivating enzymes. The dried malt, therefore, becomes a stable source of enzymes. Since moist heat is more detrimental to enzymes than dry heat, it is important to dry slowly at the start (when the grains are still

Box 9–2. Beer Terminology 101

The manufacture of beer, like that for wine making, bread making, and other food processing technologies, has evolved its own peculiar vocabulary to describe many of the manufacturing steps. Thus, the brewing jargon contains words and phrases unique to beer making, but not so familiar to the casual reader. Given that the origins of modern beer manufacture arose in Germany, it should not be surprising that many of the terms are derived from German. In fact, much of the overall beer vocabulary used by brewers even today contain German expressions. Below are some of the more common terms used in brewing.

Coppers—Vessels used for wort-boiling, called coppers because of the construction material used in their manufacture; even though these kettles are now made from stainless steel, they are still referred to as coppers.

Diastase, diastatic power—Refers to the overall starch hydrolyzing activity present in the malt.

Fass—The process of pumping beer into kegs.

Kiln—The oven used for drying and cooking malt.

Krausen—The carbon dioxide layer that forms at the top of the fermentation tank; "high krausen" refers to the period at which the krausen reaches its maximum.

Krausening—The addition of krausen (containing very active, log phase cells) to beer to promote a secondary fermentation.

Lagering—The act of storing or conditioning the beer at low temperature to promote maturation.

Lauter tun—A tank, containing a false bottom, used to promote clarification and separation of the spent grains from the wort.

Mash—The heated malt-water mixture, consisting of malt-derived enzymes and substrates.

Mash off—At the end of mashing, when the temperature is raised to about 170°C to inactivate enzymes.

Mash tun—Where the mashing step takes place; may contain a false bottom and be used for wort separation.

Oast house—The facility used for drying whole hops.

Pitch—The step when wort is inoculated with yeasts.

Trub—The precipitated material obtained from the wort after boiling; it is rich in protein and hop solids.

wet), and only later, when the moisture is reduced, to dry at higher temperatures.

During the drying step, the moisture drops from 45% at the start to about 15% to 18% at the end. The dried malt is then further dried—or "cured" or "kilned"—at temperatures as high as 80°C, and the moisture decreases to less than 5%. At this point the water activity of the malt is usually reduced to 0.3 or lower, so enzymes are well preserved and no microbial growth is expected to occur. At this point, the

malt contains mostly carbohydrate and protein (Table 9-1). Finally, the malt is placed in large trucks or train cars for large breweries or packaged in bags and shipped to smaller brewers.

Although the main purpose of the drying and kilning steps is to arrest germination and preserve enzyme activity, another important series of reactions also takes place. It is during kilning that non-enzymatic browning and associated heat-generated flavor reactions occur, which is readily apparent in the malt and espe-

Table 9.1. Approximate composition of malt.

Component	% dry weight
Starch	58
Sucrose	4
Reducing Sugars	4
Hemicellulose	6
Cellulose	5
Lipid	2
Protein	12
Amino acids/peptides	1
Minerals	2
Other	6

cially later in the finished beers. Dark beers (e.g., stouts, porters) use darker, more flavorful malt, whereas paler beers use lightly-colored, less roasted malt (Figure 9-4).

The flavors contributed by dark malts are often described as coffee- or chocolate-like, with toasty nutty notes. It would be a mistake, however, to conclude that darker beers, because of their stronger, more pronounced flavor and color, contain more ethanol than pale-colored beers. In fact, the darker the malt, the less amylolytic (or diastatic) activity is present. Ultimately, less sugar substrates are generated from starch, and less ethanol can theoretically be produced from the mashes (see below) made using dark malts. For this reason, stouts and other dark beers are inevitably made with a combination of dark, highly-roasted malt, to provide color and flavor, and lighter base malts, to supply enzymatic activity.

Although the malted and kilned barley contains starches, proteins, and the enzymes necessary for their degradation, these materials are not readily extracted from the intact malt kernel. Rather, to facilitate extraction, the malt must first be milled to break the kernel and expose the interior portion to the aqueous medium. Most breweries purchase whole or unground malt and do the milling themselves, according to their particular specifications. The kilned malt is most commonly milled in roller mills or hammer mills that can be set to deliver

Light   Golden   Amber   Dark

Figure 9–4. Types of malt used in brewing. Shown (from left to right) are four types of malt, ranging from light to dark. These malts are used to make beers with similar colors. Figure courtesy of Rich Chapin (Empyrean Ales, Lincoln, Nebraska) and John Rupnow (University of Nebraska).

products ranging from a very fine to very coarse grind.

Prior to milling, the malt is surrounded by a tough husk, which separates from the grain during milling, but generally is only partially degraded. The fineness of the malt and the extent of husk degradation are extremely important and depend on the preference of the brewer. The finer the malt, the more easily enzymes and other materials are subsequently extracted during the next step, the mashing phase. However, the insoluble residue remaining after mashing is difficult to separate, due to its small particle size. In contrast, the spent malt material is more easily separated from the mixture if coarse malt, with minimum husk damage, is used. In other words, extraction of enzymes, starches, protein, nutrients, and flavors is greater with finely milled malts, but the filtration rates are slow. Conversely, good filtration rates but poor extraction occurs with coarsely milled malts.

Finally, the ground malt or "grist" is ready to be used in the step called mashing. The purpose of mashing is two-fold. First, the enzymes, starches, proteins, and other substances are extracted and solubilized in a hot aqueous solution. Second, the extracted enzymes, starches, and proteins react to form products that ultimately serve as substrates to support growth of yeasts during the fermentation phase. The precise steps involved in mashing—in particular, the composition of the water, the temperature, and the heating rate—are critical factors, subject to the preferences of the individual brewer.

Mashing begins when the malt is mixed with brewing water (usually at ambient temperature) in a specialized tank called a "mash tun." About 20% to 30% malt is usually added. Water quality has a profound influence on the quality of the beer (it is, after all, the main ingredient), thus geographical considerations can be very important.

The reputations of many beers are based, in part, on the source, location, and properties of the water used to make the beer. For example, the famous Pilsener beers were made from the very soft water in the Pilsen region of the Czech Republic, whereas the great English-style pale ales developed in Burton-on-Trent in England, where the relatively hard water is high in minerals and bicarbonate. And although many beer manufacturers continue even today to make quality claims (and television commercials) based on the water from which the beer is made, others argue that by making adjustments for mineral content, pH, and ionicity, water variability is much less a factor in determining beer quality.

In general, brewing water should have medium hardness, containing about 100 ppm each of calcium and magnesium salts and 50 ppm or less of bicarbonate/carbonate. The pH should be between 6 and 7. The specific functions of these components are described in Table 9-2.

The malt-water mixture, or mash, is gradually heated from 20°C to 60°C to 65°C by infusion, decoction, or a combination of the two. Infusion heating is a simple step-wise process in which the mash is heated to an incremental temperature, held for a period of time, then the temperature is raised again, held, and so on until the final temperature is reached.

In decoction heating, a portion of the mash is separated and boiled in a separate kettle, then added back to the main mash, thereby raising the combined mash temperature. The main advantage of decoction mashing is an increase in starch extraction. Decoction mashing is used mainly for lager beers; in the United States a combination of infusion and decoction heating is most common.

One might ask why the heating process during mashing is done in a particular order, rather than simply heating the mash to some given temperature and then holding it there for a pre-determined time. Remember that the purpose of mashing is to extract and react. In reality, there are many enzymes in the malt, not just a single amylase or a single proteinase. Likewise, there are many sugar-containing polymers and starches with varying degree of branching, as well as heterogenous proteins, that will eventually serve as substrates. Pigments, flavors, and other substances are also

Table 9.2. Components of brewing water and their function in beer[1].

Component	Concentration (ppm)	Function
Calcium	10—250	When wort is heated, insoluble calcium phosphate complexes are formed, resulting in a decrease in pH during mashing and an increase in the activity of enzymes active at low pH (e.g., β-amylase). Calcium may also increase thermal stability of α-amylase.
Magnesium	5—100	Required for activity and stability of enzymes produced by the yeasts. At high concentrations, magnesium may compete with calcium for phosphate, reducing calcium phosphate levels.
Bicarbonate	10—250	High bicarbonate concentrations increase the pH of the mash or wort. High pH then causes a decrease in the activity and stability of amylases and proteinases.
Sulfate	5—500	Required for biosynthesis of sulfur-containing amino acids, which are essential for yeasts. At high concentrations, sulfates may be reduced to sulfur dioxide and hydrogen sulfide, which confer undesirable aroma and flavor characteristics in beer.

[1]Adapted from www.rpi.edu/dept/chem-eng/Biotech-Environ/beer/water2.htm

extracted during mashing. The important point is that each of the enzymatic reactions have optimum temperatures that are likely different from one another. By extracting the reactants in step-wise fashion, it is possible to coordinate the optimum or near-optimum temperature of a particular enzyme with the appearance of its substrate. For example, a particular α-amylase may hydrolyze its substrate at 45°C, but a different amylase (perhaps a β-amylase) may have a temperature optimum of 60°C. If the mash was heated directly to 60°C, then the α-amylase would not have had enough time at 45°C to perform its job.

Malt enzymology

The carbohydrate fraction of malt is mostly in the form of starch. Approximately one-fourth of malt starch is amylose, a linear polymer consisting of glucose, linked α-1, 4. The remaining starch, three-fourths of the total, is amylopectin, which contains not only linear glucose, but also glucose in branched α-1, 6 linkages. The main starch-degrading enzymes synthesized during malting are α-amylase and β-amylase. The former is an endoenzyme, acting primarily at in-

tramolecular α-1, 4 glucosidic bonds. Products formed by α-amylase are dextrins (short chain glucose-containing α-1, 4 linear oligosaccharides) and limit dextrins (short chain glucose-containing α-1, 4 and α-1, 6 branched oligosaccharides). In contrast, β-amylase acts at the end or near the end of amylose and amylopectin chains. The main products are maltose and small branched dextrins.

Adjuncts

The mash, as described so far, contains only the malt and brewing water. The malt, as stated above, supplies the enzymes and the enzyme substrates. Many of the European beers, as well as the micro-brewed beers produced in the United States, are made using 100% malt mashes. However, the most widely-consumed beers in the United States, those made by the large breweries, contain an additional source of starch-containing material in the form of corn, rice flakes, or grits. These materials are called adjuncts, and can account for as much as 60% of the total mash solids. Adjunct syrups, which contain sucrose, glucose, or hydrolyzed starch, are also commonly used. Adjuncts are allowed in most of the world, but their use is actually forbidden in some countries, including Germany, where centuries-old beer purity laws are still enforced (Box 9-3).

Adjuncts have several functions. First, they dilute the strongly flavored, dark-colored, "heavy" characteristics of the malt. Although these properties are preferred by some beer drinkers, most U.S. consumers favor the paler color and milder flavor associated with adjunct-containing beers. Second, adjuncts increase the carbohydrate content and provides the amylase enzymes with an additional source of substrate, and, ultimately, more fermentable sugar for the yeast. Third, they reduce the carbohydrate-to-protein ratio, such that less haze-forming proteins are present in the finished beer. Finally, adjuncts are less expensive than malt as a source of carbohydrate and, therefore, reduce the ingredient costs. In the United States, 70% of the adjuncts are corn or rice grits. The two brands that dominate the U.S. market, Budweiser and Miller, both incorporate adjuncts in their formulations.

It is important to recognize that adjuncts are not essential, since malt contains as much as 70% soluble carbohydrate, which is more than enough to satisfy the energy and carbon requirement of the yeast. As noted above, adjuncts lower the protein:carbohydrate ratio, so if too much adjunct is added, relative to malt (e.g., >1:1 ratio), there may not be enough protein for the yeast. When adjuncts are used, they are usually added to brewing water in a separate tank (cereal cooker). The mixture is brought to a boil to pre-gelatinize the starch and is then added to the mash tun.

Wort

Eventually, the temperature of the mash is raised to 75°C, which effectively inactivates nearly all enzymatic activity. The insoluble material in the mash liquid is then separated by one of several means (discussed below). This is an important step because the mash still contains grain solids and insoluble proteins, carbohydrates, and other materials. In addition,

within the mash solids is a reasonable amount of soluble materials, including fermentable sugars, that would otherwise be discarded with the spent grains.

For some breweries, the mash tun can also provide filtration. These tanks contain a false bottom that allows most of the mash liquid to flow out of the tank and into a collection vessel, while at the same time retaining a portion of the mash containing the spent grains. The spent grains form a filter bed that enhances flow rate (depending on the fineness of the malt). This material is then stirred and sparged with hot liquid to extract as much of the soluble material as possible. The liquid is added to that already removed from the mash tun. An alternative process involves pumping the mash into a separate tank called a lauter tun, which also contains a false bottom and sparging system and operates much like the mash tun, except it provides greater surface area and faster and more efficient filtration. Because the initial liquid material obtained from the lauter tun (or the mash tun) often still contains solids, it may be recycled back until a better filter bed is established and the expected clarity is achieved.

Finally, the mash separation step can be performed using plate and frame type filtration systems. These systems consist of a series of connected vertical plates, housing filters of various composition and porosities. The mash liquid is pumped horizontally through the system. As for the mash and lauter tuns, the filtrate can be recycled and the entrapped material can be sparged to enhance extraction.

The liquid material or filtrate that is collected at the end of the mash separation step is called "wort." Since it is the wort that will be the growth medium for the yeast, and which will ultimately become beer, its composition is very important (Table 9-3). The main component of wort (other than water) is the carbohydrate fraction (90%). Most (75%) of the carbohydrates are in the form of small, fermentable sugars, including maltose, glucose, fructose, sucrose, and maltotriose. The rest, about 25% of the total carbohydrate fraction, are longer, nonfermentable oligosaccharides that include dextrins (α-1-4

Box 9–3. The Reinheitsgebot

Among the most ancient of all laws related to foods is the German Purity Law known as Reinheitsgebot. These brewing laws were established in 1516 by Bavarian dukes Wilhelm IV and Ludwig X in response to the frequent occurrences of what we would now call adulteration. During this time, brewers had begun to add questionable, if not dangerous, substances to beer in an effort to disguise defects and deceive consumers as to the quality. Brewers had been known to add tree bark, various grains, herbs, and spices to make the beer more palatable. These unscrupulous beer traders were giving the legitimate brewers of Bavaria a rather bad name. Thus, these laws were written to protect the brewing industry and perhaps to protect consumers. It is interesting to note that the ingredients clause (in bold) is but a part of the Reinheitsgebot—most of the law deals with the price that brewers can charge for beer.

Below is an English translation of the Reinheitsgebot, taken from the article "History of German Brewing" by Karl J. Eden, published in Zymurgy, Vol. 16, No. 4 Special 1993.

"We hereby proclaim and decree, by Authority of our Province, that henceforth in the Duchy of Bavaria, in the country as well as in the cities and marketplaces, the following rules apply to the sale of beer: From Michaelmas to Georgi, the price for one Mass[1] or one Kopf[2], is not to exceed one Pfennig Munich value, and From Georgi to Michaelmas, the Mass shall not be sold for more than two Pfennig[3] of the same value, the Kopf not more than three Heller[4]. If this not be adhered to, the punishment stated below shall be administered.

Should any person brew, or otherwise have, other beer than March beer, it is not to be sold any higher than one Pfennig per Mass. Furthermore, we wish to emphasize that in future in all cities, markets and in the country, the only ingredients used for the brewing of beer must be Barley, Hops and Water. Whosoever knowingly disregards or transgresses upon this ordinance, shall be punished by the Court authorities' confiscating such barrels of beer, without fail. Should, however, an innkeeper in the country, city or markets buy two or three pails of beer (containing 60 Mass) and sell it again to the common peasantry, he alone shall be permitted to charge one Heller more for the Mass of the Kopf, than mentioned above. Furthermore, should there arise a scarcity and subsequent price increase of the barley (also considering that the times of harvest differ, due to location), We, the Bavarian Duchy, shall have the right to order curtailments for the good of all concerned."

Note that, in contrast to the often-quoted statement that only four ingredients are permitted, the Reinheitsgebot actually restricts beer making to just three ingredients: barley, hops, and water. This is because the fourth ingredient, the yeast, had not yet been "discovered." Rather, early brewers either relied on a natural fermentation (i.e., with wild yeast initiating the fermentation) or else used a portion of a previous batch to start the fermentation (i.e., backslopping).

The Reinheitsgebot is still in effect today in Germany, although it has been re-written and is now more explicit. The current laws also have been modernized, such that hop extracts and filter aids are now permitted. In addition, there is now a distinction made between lagers and ales, with exceptions regarding adjuncts and coloring agents allowed for the latter.

Notes

1. One Mass=one mug, or about a liter
2. Kopf is a bowl-shaped container for fluids, not quite one Mass
3. Two Pfennig=two pennies
4. One Heller=one-half Pfennig

linked glucose molecules) and limit dextrins (α-1,4 and α-1,6 glucose molecules). A small amount of β-glucans (β-1,4 and β-1,6 linked glucose molecules), derived from the cell walls of the barley, may also be present.

In addition to the carbohydrate fraction, wort also contains 3% to 6% nitrogenous matter, including proteins, peptides, and free amino acids. Proportionally less protein, relative to the total solids, will be present if adjuncts are used. About half of the amino acids will eventually be used to support yeast growth, and the other half remaining in the wort contributes to flavor and browning reactions. Various ions (e.g., ionic calcium, magnesium, and carbonates) are also present. The final pH of the wort is around 5.2.

The average molecular weight of the protein fraction is very important, because it determines the "palate fullness" or mouth feel, and also affects the physical stability (cloudiness), color, and the foamability. In general, the greater the protein hydrolysis, the less cloudy the beer. However, foam stability also is reduced. In contrast, the less hydrolyzed the protein, the more likely there will be cloudiness problems.

Table 9.3. Approximate composition of mash, wort, and beer (g/100 ml)[1,2].

Component	Mash[3]	Wort[4]	Beer
Water	75	88	92
Carbohydrates	15	10	2.3
Starch	10	<0.1	<0.1
Maltose	1	5	<0.1
Maltotriose	0.5	1	0.2
Dextrins	0.2	2	2.0
Sucrose	0.5	0.3	<0.1
Glucose	1	1	<0.1
Fructose	0.5	0.2	<0.1
Others	3	0.2	<0.1
Proteins + peptides	3	0.3	0.3
Amino Acids	0.2	0.2	<0.1
Lipid	<0.1	<0.1	<0.1
Ethanol	<0.1	<0.1	3.5
pH	5.6	5.4	4.2
Specific Gravity (°Plato)	<0.1	12	3

[1]Assumes an all-malt mash (no adjuncts)
[2]From multiple sources
[3]At start of mashing
[4]After kettle boiling

Hops

In the final step prior to fermentation, the wort is pumped into a special heating tank called the *brew kettle*. It is here that the wort is boiled and other important reactions occur. Before the wort is heated, however, one more essential beer ingredient, hops, is added to the wort. Hops are derived from the plant *Homulus lupulus* (in the family, Cannabinaceae), and although they were not part of the "original" beer formula, they have been added to beer since the Middle Ages. Why hops came to be used in beer making is not known, but it seems likely that they were added as flavoring agents, then later additional benefits were realized.

Hops provide two main sensory characteristics to beer—flavor and aroma. A large number of different hop cultivars or varieties are commercially available, and they are distinguished mainly on the basis of their aroma and flavor properties. By far, the predominant flavor is bitterness, which is due primarily to the presence of α-acids, such as humulone and cohumulone, that are contained within the resin fraction. Although another group of related compounds, the β-acids, are also found in hops and also contribute some bitterness, the α-acids are the more important group. In fact, brewers usually select hops based on their α-acid content (as a function of the whole hop weight), which can range from 5% (low bitterness) to as high as 14% (very bitter). While the intact α-acids do impart some bitterness, they actually serve as precursors for isomerization reactions (see below) that result in formation of the main iso-α-acids, isohumulone and isocohumulone, which are considered the real bittering compounds.

The amount of bitterness ultimately contributed by the hops is expressed in terms of International Bitterness Units (IBU). One IBU is approximately equal to 1 mg of iso-α-acids per liter of beer. Most U.S. lager beers have bitterness intensities of less than 15 IBU, compared to nearly 50 for some of the ales produced in the United Kingdom (hence, the name "bitters" for a typical British ale).

The other major component of hops is the essential oil fraction. This fraction, which is comprised of various terpenoids, esters, ketones, and other volatiles, is responsible for the aroma or bouquet properties of hops. Most of these substances are volatile and, if added at the beginning of the kettle boil step, would mostly be lost by the end of the step. Therefore, when high-oil hops are used (when the brewer desires a beer with a strong hoppy aroma), the hops are usu-ally added near the end of the kettle boil step. In addition to providing flavor and aroma characteristics, hops also enhance preservation and increase shelf-life of beer. This is because the iso-α-acids formed during isomerization have considerable antimicrobial activity and inhibit lactic acid and other bacteria capable of causing beer spoilage (Box 9–4).

Box 9–4. Antimicrobial Activity of Hop α-acids and Mechanisms of Resistance

It has long been known that beer is less prone to microbial spoilage than many other aqueous beverages. Thus, in some situations, beer has historically often been considered better preserved than even water (e.g., the Pilgrims on the Mayflower). It is now known that the preservation properties of beer are due to the presence of several components or constituents that are inhibitory to a wide variety of microorganisms. Ethanol concentrations above 3% and pH levels below 4.6 are particularly inhibitory to spoilage organisms. Carbon dioxide, low oxygen levels, and limiting nutrient concentrations also restrict growth of microorganisms. The hops are, perhaps, one of the more potent antimicrobial constituents in beer, and are, in large part, responsible for its long shelf-life. In particular, hops inhibit many of the lactic acid bacteria that spoil beer due to their production of acids, diacetyl, and other products that confer flavor and aroma defects.

Although the antibacterial properties of hops, and the α-acid fraction, specifically, were recognized more than seventy-five years ago, the physiological means by which hop acids inhibit spoilage bacteria in beer has only been determined recently (Sakamoto et al., 2002, Simpson, 1993). As shown in the model below (Figure 1A), the mechanism is actually similar to how other weak organic acid preservatives, such as benzoic and propionic acids, inhibit microorganisms in foods.

Hop acids, like other weak acid preservatives, have ionophoric activity and are inhibitory due to their ability to decrease intracellular pH and to disrupt proton gradients. Given the low pH of beer (<5) and their low pKa (3.1), isomerized iso-α-acids are in their undissociated or acid form (abbreviated as HA), and hence, are able to diffuse across the lipophilic cell membranes of spoilage lactic acid bacteria. Since the cell cytoplasm of these bacteria is near-neutral, the iso-α-acids dissociate, releasing the anion (A^-) and a free proton (H^+). The latter then lowers the cytoplasmic pH. The cell may respond by activating ATP-dependent pumps that efflux the accumulated protons, but the continued proton cycling is expensive (energy-wise) and also may lead to dissipation of the pH gradient portion of the proton motive force (the PMF or proton gradient) that the cell uses to drive various transport systems. Thus, energy depletion, cytoplasmic acidification, and reduced nutrient uptake all occur as a result of hop-derived α-acids and account for their inhibitory activity.

However, it is possible that a second mechanism also may be involved, because it appears that the dissociated form (A^-) of the α-acids binds to cytoplasmic divalent cations, such as Mg^{2+}, and the entire complex is effluxed from the cell. This results in the loss of a necessary nutrient that must then be re-transported, with the concurrent cost of energy.

Despite the effectiveness of hop α-acids as antimicrobial agents, some bacteria have been isolated from spoiled beer that appear insensitive or resistant to hops. At least three resistance mechanisms have been identified (shown in Figure 1B as dashed lines). In *Lactobacillus brevis*, a common spoilage organism, resistance is mediated, in part, by two multiple drug resistance (MDR) systems. Both consist of membrane-integrated proteins. The HorA system is classified as an ATP-binding cassette (ABC) transport system, using ATP hydrolysis to drive efflux of iso-α-acids. The MDR system also pumps out iso-α-acids, but uses the PMF as the driving force. A third system relies on increased activity of the proton translocating ATPase (H^+-ATPase) to pump out accumulated protons.

(Continued)

Box 9–4. Antimicrobial Activity of Hop α-acids and Mechanisms of Resistance *(Continued)*

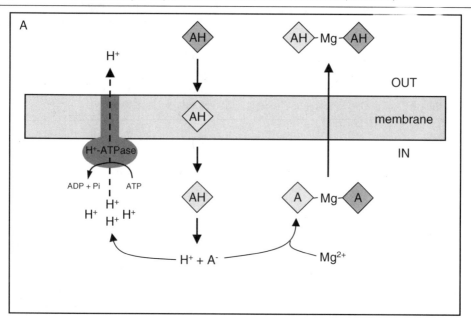

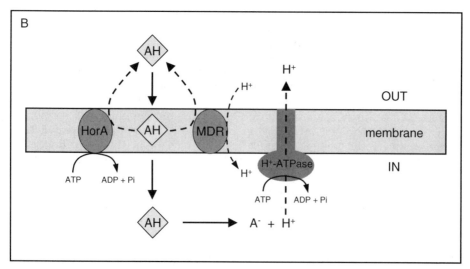

Figure 1. Inhibition of lactic acid bacteria by hops. In A, the means by which hop iso-α-acids inhibit sensitive organisms is shown. A model for hop resistance is shown in B. Adapted from Sakamoto et al., 2002 and Sakamoto and Konings, 2003.

References

Sakamoto, K., H.W. van Veen, H. Saito, H. Kobayashi, and W. N. Konings. 2002. Membrane-bound ATPase contributes to hop resistance of *Lactobacillus brevis*. Appl. Environ. Microbiol. 68:5374–5378.

Sakamoto, K., and W.N. Konings. 2003. Beer spoilage and hop resistance. Int. J. Food Microbiol. 89:105–124.

Simpson, W.J., 1993. Ionophoric action of *trans*-isohumulone on *Lactobacillus brevis*. J. Gen. Microbiol. 139:1041–1045.

Hops are usually obtained from specialty suppliers in one of several forms, including liquid hop extracts, hop pellets, or as dried whole flowers or cones. Although the latter are still used, hop extracts and pellets have gained popularity due to their consistency and ease of use. Hop products can also be obtained as isomerized extracts or pellets, containing iso-α-acids. This allows the brewer to add the hops to the wort during kettle boil, as well as later stages, so that the desired hops flavor can more easily be achieved.

Kettle Boil

After the hops are added to the wort, the mixture is boiled for one to one-and-a-half hours. Boiling accomplishes at least seven functions. First, it kills nearly all of the microorganisms remaining after mashing, making the wort, for all practical purposes, sterile. Second, boiling inactivates most of the enzymes still active after mashing or reduces their activity to barely detectable levels. Third, the boiling step enhances extraction of oils and resins from the hops and accelerates isomerization of hop acids. Fourth, proteins, tannins, and other materials that would ordinarily cause clarity and cloudiness problems precipitate during the boiling step. Removing this precipitate (known as "hot break" or "hot trub") helps to prevent such defects. Fifth, wort boiling enhances color development by catalyzing formation of Maillard reaction products. Sixth, undesirable volatile components, such as sulfur-containing aroma compounds, are removed. Finally, prolonged boiling causes evaporation of water and concentration of the wort. Because the mash or wort is diluted by sparge water, evaporation returns the density or specific gravity back to the desired level.

As noted above, a trub or precipitate forms during the kettle boil step. In addition, the wort may also contain insoluble hop debris, especially if whole hops were used. The latter are removed via strainers or screen-type devices that may also allow for sparging and re-circulation. For hot trub removal, several separation systems can be used. The simplest method is sedimentation, but this is not a very efficient process. A much more efficient method is to centrifuge the hot wort using continuous centrifuges. Such units, however, are expensive and require frequent maintenance. The most common means of separating out the hot trub is the whirlpool separator, which is typically a low-height, conical-shaped (cone pointing downward) tank whose geometry promotes whirlpool-type movement of the wort. The trub or precipitate collects in the center of the tank.

After the trub is removed, the wort is cooled in plate-type heat exchangers. During the cooling stage, when the wort reaches about 50°C, more precipitated material forms (known as the "cold break"), which is separated by centrifugation or filtration. When the wort temperature is reduced to 10°C to 15°C, it is pumped into fermentation vats, where it is, at last, ready to be made into beer.

The Beer Fermentation

All of the steps described above are performed to provide a suitable growth medium for the ethanol-producing yeast. Before yeast is added, however, the wort is first aerated or sparged with sterile air. Even though the beer fermentation eventually becomes an anaerobic process, creating an aerobic environment at the outset jump-starts the yeast and promotes a more rapid entry into a logarithmic growth phase. Importantly, oxygen also is necessary for biosynthesis of cell membrane lipids that are essential for growth of yeast in wort.

Fermentors

The fermentation of the wort occurs in fermentor vessels of varying composition, size and configuration, and in either batch or continuous modes. Although most modern fermentors are now constructed of stainless steel, the traditional materials were wood, concrete, or copper. Size and shape depend on a number of considerations, but especially on whether the fermentation is top fermenting, as for ales, or bottom fermenting, as for lagers (see below).

For example, lager fermentation vessels are typically cylindric, with cone-shaped bottoms (cylindroconical), so that when the yeasts flocculate and settle, the cells collect within the conical region. Traditional fermentors, at least for ales, were also open or uncovered, so that the CO_2 foam that developed during fermentation provided the only protection against the elements (e.g., oxygen in the air, airborne yeasts and bacteria, and other contaminants). Thus, although the yeast can be skimmed from the top and re-used, the CO_2 is lost.

Currently, enclosed, pressurized, cylindroconical fermentors are used for lagers, as well as ales, with capacities of nearly 10^6 L. Enclosed fermentors obviously have the advantage of reducing exposure to air and airborne contaminants; however, it is not possible in these fermentors to recover the evolved CO_2. Whereas open fermentors, whether round or rectangular, are necessarily more horizontal (i.e., width $>$ height), enclosed fermentors are usually constructed in a vertical orientation, so less floor space is required. Regardless of shape, however, most modern fermentors are jacketed to provide efficient cooling.

Inoculation

After the wort is cooled and aerated, the yeast culture is, at last, added to the wort, in a step called *pitching*. Brewing yeast strains are dramatically different from the *Saccharomyces cerevisiae* strains used in the laboratory in many important respects (Box 9-5). In addition, there are two types or strains of brewing yeasts that are used, depending on the type of beer being produced, and these strains also differ physiologically, biochemically, and genetically. Many of these differences are highly relevant to the beer fermentation (Table 9-4).

Specifically, ale beers are made using selected strains of *Saccharomyces cerevisiae*, otherwise known as the "ale" or "top-fermenting" yeast. Lager beers are fermented by *Saccharomyces pastorianus* (formerly called *Saccharomyces carlsbergensis*), also known as the "lager" or "bottom-fermenting" yeast. Despite this termi-

nology, growth of these yeasts in wort is not necessarily confined to the top or bottom regions of the fermentor. Rather, top-fermenting yeasts, as they grow in the wort, tend to form low density clumps or flocs that trap CO_2 and rise to the surface. In contrast, when lager yeasts flocculate, the flocs sediment or settle to the bottom. The ability of brewing yeast to flocculate and the point during the fermentation at which flocculation occurs are very important factors, as will be discussed later. It should also be noted that the distinction between top- and bottom-fermenting yeast is beginning to be less relevant, as the use of enclosed cylindroconical fermentation vessels results in even ale yeasts dropping to the bottom of the fermentor (discussed below).

Another major difference between ale and lager beer concerns the temperature at which the fermentation occurs. Ales are normally fermented at fairly high temperatures, from 18°C to as high as 27°C, well within the range at which *S. cerevisiae* grows. It is no coincidence, therefore, that ales were more commonly produced in warmer climates that are typical in the British Isles; hence, many of the classic English-style beers are ales. In contrast, lager style beers are fermented by yeast capable of growing at temperatures below 15°C. Not surprisingly, these beers evolved from Germany and other northern European areas, where the ambient climates were cooler. Interestingly, when German immigrants moved to the United States in the late 1800s and early 1900s, they brought with them German beer-making technology. The American beer industry subsequently became dominated by large breweries making lager beers (e.g., Schlitz, Anheuser-Busch, Coors, Miller, Stroh, Heileman, Pabst, and others).

Although beer yeast cultures are commercially available, most breweries use their own house cultures. Strains can be bred, much like plants or seeds, to provide hybrids with specific traits. These cultures are maintained and propagated in the laboratory and, when needed, grown in volumes necessary for inoculation into the fermentation tanks. Some breweries may also use the yeast slurry left over from the

Box 9–5. Classification of Brewing Yeasts

Yeast classification can be confusing for microbiologists who are often more familiar with classification systems used for bacteria. Part of the difficulty in understanding yeast taxonomy is due to the significant biological differences between prokaryotic and eukaryotic organisms. Structurally, yeasts contain a nuclear membrane, whereas this membrane, by definition, is absent in prokaryotic bacteria. Although yeasts mostly exist, like bacteria, as unicellular organisms, they can also develop mycelia and grow as multicellular organisms.

The means by which yeasts replicate provides another major difference. Yeasts can divide, like bacteria, by binary fission, but they can also reproduce by sporulation or by budding. Whereas bacteria contain a single chromosome and are monoploid, yeasts, in their diploid state, contain sixteen pairs of chromosomes. Actually, as discussed in Box 9-10, most brewing yeast strains are polyploid, containing more than two chromosomal copies. Finally, in addition to genes located on the chromosome, yeasts contain mitochondrial DNA, and most strains of *Saccharomyces cerevisiae* also contain multiple copies of a large, extrachromosomal element called 2 μm DNA.

Although *S. cerevisiae* is among the most well-studied of all organisms, the laboratory strains for which genome sequences and other biological information are known are quite different from the strains used to make beer (Table 1). For example, the life cycle of *S. cerevisiae* ordinarily includes a sexual or sporulation phase, with diploid cells undergoing meiosis and leading to formation of asci-containing spores (Figure 1). The haploid spores are of opposite mating types (a and α) and can mate, forming new diploid cells (a/α). Brewing strains of *Saccharomyces*, in contrast, rarely sporulate, and when they do, the spores are usually not viable nor are they able to mate. Rather, strains used for brewing reproduce primarily by multilateral budding, with buds forming across the entire surface of the cell (i.e., in contrast to strictly polar budding).

Table 1. Differences between lab and brewing strains of *Saccharomyces*[1].

Lab strains	Brewing strains
Haploid and diploid	Polyploid and aneuploid
Sporulating	Sporulate poorly
Spores viable	Spores mostly non-viable
Able to mate (a and α mating types)	Mating rare

[1]Adapted from Dufour et al., 2003

Until relatively recently, yeast classification was based on much the same criteria used for fungal taxonomy (yeast, after all, are unicellular fungi and are grouped in the same family, Eumycetes). Thus, the main classification criteria were based on spore formation and the presence/absence of ascospores. However, as noted above, many yeasts (especially brewing and wine strains) do not sporulate, and classifying yeasts on this basis was not very informative.

Instead, yeast classification is now based on morphology, biochemical, physiological, and genetic properties. Brewing yeasts, in contrast, share similar morphological, biochemical, and physiological properties and distinguishing between different strains is not easy (Table 2). This has not been a serious problem for brewers, who routinely classify yeast in a rather practical way as belonging to one of two species. Ale, or top-fermenting yeasts, belong to the *Saccharomyces cerevisae* group. Lager, or bottom-fermenting yeasts, belong to the *Saccharomyces pastorianus* group (formerly classified as *Saccharomyces carlsbergensis*, then as *Saccharomyces uvarum*).

When a more exact identification is necessary (i.e., to the strain level) or when the goal is to differentiate between strains, genetic methods are the most reliable. DNA probes that hybridize specific genes or regions can distinguish between ale and lager strains, but other methods have proven to be easier and more powerful. For example, techniques based on the polymerase

(Continued)

Box 9–5. Classification of Brewing Yeasts *(Continued)*

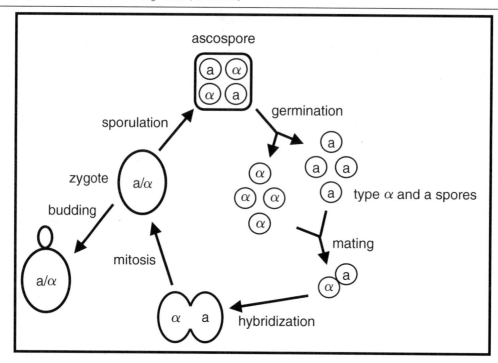

Figure 1. Yeast life cycle. Most lab strains form ascospores containing spores of opposite mating types, a and α. These spores can mate, hybridize, and form diploidal cells capable of budding or sporulating. In contrast, brewing yeasts generally only reproduce by budding.

chain reaction (PCR), restriction fragment length polymorphism (RFLP), amplified fragment length polymorphism (AFLP), pulsed field gel electophoresis (PFGE), and karyotyping all provide unique genetic fingerprints that distinguish not only between ale and lager yeasts, but also between strains of the same group.

Table 2. Characteristics of *Saccharomyces* species used in brewing[1].

Species	Fermentation of:			Assimilation of:					Fructose transport activity	Assimilation of:			Growth at:	
	Suc[2]	Raf	Tre	Suc	Mal	Raf	Man	Eth		Lys	Cad	Ety	30°C	37°C
S. bayanus[3]	+	+	−	+	+	+	v	+	+	−	−	−	+	−
S. cerevisiae	+	+	−	+	+	+	−	+	−	−	−	−	+	v
S. exiguus	+	s	−	+	−	+	−	s	u	−	−	−	+	−
S. paradoxus[4]	+	+	−	+	+	+	+	+	−	−	−	−	+	+
S. pastorianus[5]	v	+	−	+	+	+	−	+	+	−	−	−	+	−

[1]Adapted from Vaughan-Martini and Martini, 1998

[2]Abbreviations: Suc, sucrose; Raf, raffinose; Tre, trehalose; Mal, maltose; Man, mannose; Eth, ethanol; Lys, lysine; Cad, cadaverine; Ety, ethylamine; s, positive, but slow; v, variable; u, undetermined

[3]synonyms: *S. uvarum*

[4]synonyms: *S. cerevisiae*

[5]synonyms: *S. carlsbergensis*

Box 9–5. Classification of Brewing Yeasts *(Continued)*

References

Dufour, J.-P., K. Verstrepen, and G. Derdelinckx. 2003. Brewing yeast, p. 347—388. *In* T. Boekhout and V. Robert (ed.), *Yeasts in food.* Woodhead Publishing Limited, Cambridge, England and CRC Press, Boca Raton, Florida.

Hammond, J.R.M., 2003. Yeast genetics, p. 67—112. *In* F.G. Priest and I. Campbell (ed.), *Brewing Microbiology, 3rd ed.* Kluwer Academic/Plenum Publishers. New York.

Kurtzman, C.P., and J.W. Fell. 1998. Definition, classification and nomenclature of the yeasts. P. 3—5. *In* C.P. Kutrzman and J.W. Fell (ed.), *The Yeast: A Taxonomic Study, 4th Edition.* Elsevier Science B.V. The Netherlands.

Vaughan-Martini, A., and A. Martini. 1998. Saccharomyces Meyen ex Reess, 358–371. *In* C.P. Kurtzman and J.W. Fell (ed.), *The Yeast: A taxonomic study, 4th Edition.* Elsevier Science B.V. The Netherlands.

Table 9.4. Differences between ale and lager yeasts.

Ale *(Saccharomyces cerevisiae)*	Lager *(Saccharomyces pastorianus)*
Flocculated yeast rises to the top	Flocculated yeast settles to the bottom
Optimum growth temperature > 30°C	Optimum growth temperature < 30°C
Minimum growth temperature = 15°C	Minimum growth temperature = 7°C
Maximum growth temperature = 40°C	Maximum growth temperature = 34°C
Cannot metabolize melibiose	Able to metabolize melibiose
Slow assimilation of maltotriose	Efficient assimilation of maltotriose
Sporulating (at low frequency)	Non-sporulating

previous fermentation to pitch the next batch. Only when the fermentation appears slow or sluggish or when quality attributes suffer will a new culture be prepared.

Yeast Metabolism

The usual starting inoculum gives an initial yeast population of abut 5×10^6 cells per ml of wort. Depending on the activity of the yeast inoculum, a lag period of six to eighteen hours may occur. Although no increase in cell number is observed during the lag phase, metabolic activity is well under way. The yeasts are synthesizing sugar and amino acid transport systems, as well as enzymes necessary for their metabolism. As noted earlier, the wort is sparged with oxygen prior to pitching, thus the wort medium is initially highly aerobic. This oxygen-rich environment stimulates synthesis of membrane-associated sterols and unsaturated fatty acids. Sterols, and in particular,

ergosterol, are essential for formation of yeast cell membranes; likewise, the unsaturated fatty acids provide membrane fluidity. Because of these requirements, yeasts may grow poorly in oxygen-deprived worts.

Following the lag phase, cells enter into the logarithmic growth phase, and a primary ethanolic fermentation period begins. Despite the aerobic conditions initially established by sparging, sugar metabolism by yeast occurs not by aerobic respiration (i.e., the tricarboxylic acid or Kreb's cycle), but rather via the Embden-Meyerhoff-Parnas (EMP) glycolytic pathway. Although *Saccharomyces* are facultative anaerobes, enzymes of the Kreb's cycle are not even expressed during growth in wort. The repression of these genes under aerobic conditions occurs only for as long as the sugar concentration is high. This phenomenon, called the glucose effect, is due to catabolite repression by glucose. When the availability of glucose and other fermentable sugars is nearly diminished (i.e., < 1 g/L), however, metabolism in an aerobic environment shifts from fermentation to respiration (Box 9–6).

Catabolite repression by glucose also affects the specific biochemical processes that occur during the fermentation. Although brewing strains of *S. cerevisiae* and *S. pastorium* can use all of the major sugars normally present in wort (glucose, fructose, maltose, and maltotriose), metabolism of these sugars does not occur concurrently. In fact, wort contains more maltose and maltotriose than glucose, yet glucose is still metabolized t. Tl is preferential use of glucose occurs because: (1) at least one of the glucose transporters is constitutively expressed; and (2) the gene coding for the α-glucoside (maltose and maltotriose) transporter (*AGT1*) is repressed by glucose. Only when the glucose is fermented (after about one to two days), therefore, will expression of

Box 9–6. Physiology of Brewing Yeasts

Given the importance of yeast to the brewing industry, it is not surprising that so much attention has been devoted to understanding the physiological properties of *Saccharomyces cerevisiae*. Most brewing strains, as noted elsewhere, are very different from ordinary laboratory strains. This is because the beer and the brewing environment in which these strains have been grown for hundreds of years is quite unlike the docile conditions in the laboratory. Thus, most brewing strains have been screened and selected on the basis of their ability to grow well in wort and on their overall beer-making performance. Strains also have been modified by classical breeding programs such that specific traits are (or are not) expressed. More recently, the application of molecular techniques have made it possible to directly and specifically alter yeast physiology by introducing genes that encode for specific properties.

Although the primary function of the yeast during brewing is to ferment sugars to ethanol and CO_2, it is also responsible for performing many other important duties. The yeast must not only produce adequate amounts of end products, but it must also metabolize all of the available fermentable carbohydrate to fully attenuate the beer. The main end products produced by brewing yeasts are ethanol and CO_2; however, small amounts of higher alcohols, esters, aldehydes, phenolics, and other organic compounds also are produced and may affect beer flavor and quality. Some of these compounds have a negative impact, and their production is discouraged. Of course, the ability of brewing yeast to perform specific metabolic functions depends on how well it tolerates the high osmotic pressure, high ethanol concentrations, and other physiological stresses found in wort and beer. Thus, there are considerable physiological demands on yeasts during the brewing operation.

When the brewer first adds yeast to the wort, the environment is highly aerobic, due to the oxygen-sparging step that occurs just after the wort is boiled and cooled. Thus, metabolism by the facultative yeast might be expected to occur via a respiratory route (i.e., the tricarboxylic acid or TCA cycle). In fact, Pasteur demonstrated more than a century ago that when facultative organisms, such as brewers' yeast, are placed in an aerobic environment, aerobic metabolism is preferred (the so-called Pasteur effect). Aerobic metabolism, after all, is a more efficient process, yielding considerably more ATP than anaerobic metabolism (36 moles versus 2 moles ATP per glucose).

However, in the wort situation just described, yeast growth occurs via fermentative metabolism. Why would glycolysis occur in this environment? In part, because when the concentration of available sugar (i.e., glucose) is high ($>1\%$ or 50 mM), the reverse of the Pasteur effect occurs. That is, even under aerobic conditions, provided there is excess glucose, expediency, rather than efficiency, dictates how metabolism will occur. In fact, the enzymes of the TCA cycle are actually repressed, a situation referred to as the "glucose effect." Another reason, perhaps, for why the fermentation route proceeds under aerobic conditions is that brewing yeast strains are, in general, highly accustomed to fermentative metabolism.

this gene, as well as the α-glucosidase gene coding for maltose and maltotriose hydrolysis, occur.

Sucrose, in contrast, is hydrolyzed by an extracellular invertase, yielding its component monosaccharides, glucose and fructose. The regulation of sugar metabolism becomes especially relevant during the beer fermentation when high glucose adjuncts are used. Under these circumstances, repression of maltose and maltotriose metabolism may exist throughout the entire fermentation, and the beer will be poorly attenuated.

Following transport, intracellular metabolism of wort sugars via the EMP pathway yields mostly pyruvic acid and reduced NADH. Pyruvate is subsequently decarboxylated by pyruvate decarboxylase, forming CO_2 and acetaldehyde. The latter is then reduced by NADH-dependent alcohol dehydrogenase, forming ethanol and re-oxidizing NAD. Because a small amount of the glucose carbon must be used to support cell growth, some of the pyruvate is diverted, via pyruvate dehydrogenase, to the Kreb's cycle, where biosynthetic precursors are formed. Re-oxidation of NAD then requires an alternative source of electron acceptors, leading to the formation of glycerol, higher alcohols, and other minor end-products found in beer.

The logarithmic growth phase period continues for two to three days for ales, and up to six or seven days for lagers, at which point all of the mono- and disaccharides, and most of the maltotriose, will have been fermented. Cell mass usually increases by less than two logs.

Importantly, the fermentation is exothermic, meaning that heat is generated, so the fermentation tanks must be cooled and maintained. Either internal cooling coils or external jackets are used to maintain temperatures of 8°C to 15°C for lagers and 15°C to 22°C for ales.

For ale fermentations, yeast cells, as noted earlier, rise to the surface, along with the evolved CO_2. The yeast can then be skimmed off for re-use later. For lager fermentations, cells remain suspended, but there is enough CO_2 evolved during the early stages of the fermenta-

tion to form cauliflower-shaped clumps. This thick foam layer is called krausen, and as growth and CO_2 formation become more rapid, corresponding to the maximum growth rate of the cells, a period known as "high krausen" is reached. A portion of this material can be collected and re-used to initiate a secondary fermentation, as described below.

Although it is possible to monitor the progress of the fermentation simply by observing CO_2 formation, the preferred means is to measure the specific gravity. As the fermentable sugars are consumed, the specific gravity, expressed as °Plato, decreases. When the specific gravity no longer is decreasing, most of the fermentable sugars will have been depleted, and the fermentation is complete. The beer is then considered to be "fully attenuated." At this point, usually four to seven days after the beginning of the primary fermentation period, the fermentation vessel is quickly cooled to 4°C or less. However, some brewers, prior to cooling, maintain a high temperature to promote use of diacetyl and related vicinal diketones that are themselves produced by yeast and which are responsible for off-flavors in beer (discussed later). This so-called "diacetyl rest" may reduce the necessary duration of the post-fermentation conditioning phase, when diacetyl reduction would ordinarily occur.

Flocculation

Flocculation is the ability of yeast cells to agglomerate or adhere to one another in the form of clumps. When lager yeasts flocculate, the clumps have a density greater than that of the beer and settle to the bottom. Ale yeasts, in contrast, form clumps or flocs that entrap CO_2 bubbles and have a lower density, and, therefore, rise to the surface.

The ability of yeast cells to clump or flocculate, and the time at which flocculation occurs, are very important properties in beer manufacture. In most cases, flocculation should occur at the end of the fermentation, when all of the monosaccharides (glucose and fructose), disaccharides (sucrose and maltose), and trisaccharides (maltotriose) have been fermented. The yeast will have done its job, fermentable sugars are depleted, and the beer is considered to be fully attenuated.

Importantly, the beer will be clear or "bright," even in the absence of additional clarification steps. If, however, flocculation occurs prematurely, before the end of the fermentation, fermentable sugars will remain in the beer (a situation brewers refer to as a "hanging fermentation"). These residual sugars can affect maturation and flavor. In particular, these unattenuated beers have a lower than normal ethanol concentration and are relatively sweet which, depending on the intent of the brewer, may or may not be desirable.

Of course, the presence of sugars in the beer also provides growth substrates for other organisms. In contrast, yeast cells that fail to flocculate, and instead remain in the beer, are difficult to remove, causing cloudiness problems. These yeasts may later autolyze in the beer, releasing enzymes that contribute to yeasty flavor defects. The onset of flocculation appears to be triggered by several factors, including entry into stationery phase, low pH, low temperature, low sugar concentration, and high ethanol concentrations. Hops and calcium (see below) also promote flocculation activity.

Despite the importance of flocculation in the brewing process, the physical-chemical basis of this property is not well understood. Ale yeasts, in general, have been reported to be more hydrophobic than lager yeasts, due to differences in cell surface charges, but it is not clear that this property affects flocculation. Rather, it now appears that flocculation occurs as a result of lectin-like domains, contained within cell surface proteins, that bind in the presence of ionic calcium to mannans (mannose-containing chains) located on the surface of adjacent cells. Flocculation is also a heritable property, meaning it has a genetic basis. Several genes have been identified in *S. cerevisiae* that code for proteins involved in flocculation, and efforts are now under way to manipulate floc gene expression during beer making (Box 9-7).

Post-fermentation Steps

As noted earlier, beer manufacture, at first glance, would appear to be a rather simple exercise: barley is converted to malt, which is converted to wort, which is then fermented by yeast. The problem is that at this point the beer still contains yeasts, other microorganisms, and other insoluble and non-dissolved materials that give it an undesirable cloudy and hazy appearance. In addition, this so-called "green" or immature beer may contain chemical constituents that impart off-flavors. Finally, little or none of the carbon dioxide produced during the primary fermentation is retained in the beer, leaving the product without one of its key components. Thus, the challenge for the brewer is to promote acceptable flavor development and maturation, to remove the undesirable cloud- and haze-forming materials, and to introduce carbonation into the beer. Indeed, it can be argued that the post-fermentation steps are as important as any of the preceding activities.

Flocculation should occur as the fermentation ends, or more precisely, when the fermentable sugars are depleted and the beer is fully attenuated. Once the yeasts have flocculated, they are usually collected promptly and removed from the beer. It is important to recognize that, even though most of the yeast cells are removed by the sedimentation or skimming step, the beer still contains yeast cells. In addition, this green beer typically contains undesirable flavor compounds, including sulfur dioxide and diacetyl, as well as proteins and tannins that can potentially form haze complexes. Therefore, it is essential that the beer be "conditioned" to enhance sedimentation of the remaining yeasts and haze-forming proteins, promote dissipation of off-flavors, and produce beer with a mature or finished flavor. Since yeast growth also occurs during the conditioning period, a secondary fermentation may take place, resulting in formation of CO_2. Depending on the specific conditions, enough CO_2 may be produced to naturally carbonate the beer.

Box 9–7. Flocculation—A Case of Beer Yeasts Sticking Together

The ability to flocculate, consistently and at the right time during the fermentation, is one of the most important traits of a good brewing yeast. If flocculation occurs too early or too late, beer flavor, clarity, and overall quality are bound to suffer. Yeast floccuation is now thought to be mediated via a lectin-like mechanism in which the N-terminal region of a cell wall-associated protein recognizes and binds, in a calcium-dependent manner, to mannose or trimannoside oligosaccharides located on the surface of other cells (Figure 1; Kobayashi et al., 1998; Javadekar et al., 2000; Verstrepen et al., 2003). Although both flocculating and non-flocculating cells contain this mannan receptor, only flocculating cells produce the binding protein. Importantly, flocculation is influenced by several physiological and environmental factors, including growth phase, ethanol and sugar concentrations, and calcium ion availability (Table 1). These observations suggest that flocculation depends not only on physical interactions, but is also subject to genetic regulation.

Five genes (*FLO1, FLO5, FLO9, FLO10,* and *FLO11*) currently have been identified in *Saccharomyces cerevisiae* that encode for proteins involved in adhesion (Halme et al., 2004). At least two of these genes, *FLO5* and *FLO9*, have high sequence homology to *FLO1*. However, several

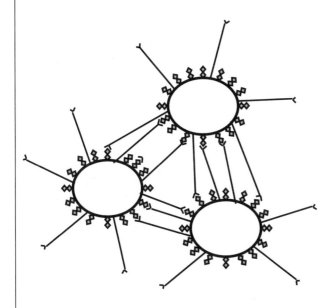

Figure 1. Model for yeast flocculation. Flocculating cells produce cell wall-attached lectin-like proteins that recognize and bind to mannan regions on the surface of receptor cells. The mannan receptors are present on most cells; thus, it is the expression of the lectin-like proteins that is responsible for the flocculation phenotype. Adapted from Verstrepen et al., 2003.

Table 1. Factors affecting flocculation.

Factor	Effect
Fermentable sugars	Inhibitory
Nitrogen, other nutrients	Little effect
Temperature	Strain-dependent, with broad range
pH	Optimal flocculation at pH 3.5—5.8
Oxygen	No direct effect
Ethanol	Strain-dependent, affects cell surface
Cell age	Old cells flocculate more than young cells
Inoculum handling	Ambient temperature increases flocculation

(Continued)

Box 9–7. Flocculation—A Case of Beer Yeasts Sticking Together *(Continued)*

different adhesion phenotpyes exist, including cell-to-cell adhesion (i.e. flocculation), as well as adherence to other surfaces.

The expression of these dominant genes is complex and unstable, and appears to be controlled by both genetic and epigenetic mechanisms (Halme et al., 2004). Whereas *FLO11* is ordinarily expressed (in a lab strain), the other four genes are silent. Expression of *FLO11* gives a non-flocculating (but adhesion positive) phenotype, whereas expression of the silent *FLO* genes (via a heterologous promoter) results in flocculating adherence. Epigenetic silencing of *FLO11* also causes a morphological change in the appearance of the cells, from yeast cells in a filamentous form to yeasts that grow as discreet cells.

Depending on growth and environmental conditions, four different flocculation phenotypes have been described for brewing strains of *S. cerevisiae* (Soares et al., 2004). In cells with a *FLO1* phenotype, flocculation is inhibited by mannose. The NewFlo phenotype is inhibited by mannose and glucose. The MI phenotype is mannose insensitive. Finally, flocculation can be ethanol-dependent. While flocculation by most *FLO1* yeasts is constitutive, flocculation by NewFlo cells occurs during stationary phase. Thus, most lab strains have the *FLO1* phenotype and most brewing strains have the NewFlo phenotype (Javadekar et al., 2000).

Brewing yeasts ordinarily do not flocculate in the presence of glucose and other fermentable sugars. Furthermore, when fermentable sugars are added to flocculant cells, the flocs will disperse and subsequent flocculation is lost (Soares et al., 2004). It also appears that, under experimental conditions, the rate of fermentation-dependent loss of flocculation correlates with growth rate and sugar use. That is, the faster the cells grow, the faster the flocs come apart. Despite these findings, however, it is not clear at a molecular level how the cells sense sugar availability and then respond by inducing or repressing flocculation. Recently, it was suggested that the onset of flocculation occurs as a result of the combined effects of nutrient shortage (of either fermentable sugars or nitrogen) and the presence of ethanol (Sampermans et al., 2005).

Efforts to modify or improve flocculation properties of brewing yeast are being pursued actively as part of current strain improvement programs. Originally, plasmid vectors were used to deliver flocculation genes (e.g., *FLO1*), but because expression was constitutive, these transformed strains had limited application for brewing. In one recent study, however, researchers sought to manipulate the onset of flocculation by using an inducible promoter (Verstrepen et al., 2001). A non-flocculating wild-type strain, harboring a genomic copy of the *FLO1* gene (encoding

Conditioning

Conditioning can occur by one of several methods. In traditional ale manufacture, the beer is pumped from the fermentor into wooden casks ranging in size from 20 L to more than 200 L. Stainless steel casks have replaced many of the wood casks, although the latter are still available. Sugar is added to induce the secondary fermentation, and additional hops and fining agents (see below) may also be added. The casks are held at 12°C to 18°C for up to seven days. Following the maturation period, this cask-conditioned or drought

beer is ready for consumption, without any additional clarification steps (thus, the yeast sediment is still present).

For lager beers, conditioning or lagering occurs in tanks held at lower temperatures for a longer time. Typically, temperatures as low as 0°C, for as long as three months, can be used. A special variation of the lager method involves adding a portion of wort obtained at high krausen (i.e., cells in their most active growth phase) to the green beer. Since the krausen also contains wort sugars, this step, called krausening, essentially serves as a source of both fast-growing cells, as well as fermentable

Box 9–7.   Flocculation—A Case of Beer Yeasts Sticking Together *(Continued)*

for the lectin-like protein, *FLO1p*), was used as the host organism. The expression of *FLO1* was made inducible by replacing the *FLO1* promoter with an inducible promoter derived from the *HSP30* gene.

The latter is a heat shock gene that responds not only to heat, but also to nutrient depletion and high ethanol concentrations. Importantly, the *HSP30* promoter (*HSP30p*) is also induced by stationary phase (i.e., a late-fermentation). Thus, expression of *FLO1* behind *HSP30p* would be expected to occur in the transformant only at the end of the fermentation. Indeed, this strategy was successful, as flocculation was evident in the transformants during stationary phase (at about 95% of the wild-type level). As expected, heat shock or ethanol additional also induced flocculation. The phenotype was stable for 250 generations, and when inoculated into wort, about 90% of the transformed cells sedimented at the end of the fermentation, indicating that flocculation had occurred.

These results indicate that it is now possible to improve the flocculation properties of yeast strains. It is noteworthy that the actual strategy used to modify gene expression in this organism involved a simple self-cloning or benign approach. Thus, the transformants contained no foreign DNA that might warrant legal or public scrutiny.

References

Halme, A., S. Bumgarner, C. Styles, and G.R. Fink. Genetic and epigenetic regulation of the *FLO* genefamily generates cell-surface variation in yeasts. Cell 116:405-415.

Javadekar, V.S., H. Sivaraman, S.R. Sainkar, and M.I. Khan. 2000. A mannose-binding protein from the cell surface of flocculent *Saccharomyces cerevisiae* (NCIM 3528): its role in flocculation. Yeast. 16:99-110.

Kobayashi, O., N. Hayashi, R. Kuroki, and H. Sone. 1998. Region of FlO1 proteins responsible for sugar recognition. J. Bacteriol. 180:6503-6510.

Sampermans, S., J. Mortier, and E.V. Soares. 2005. Flocculation onset in *Saccharomyces cerevisiae*: the role of nutrients. J. Appl. Microbiol. 98:525-531.

Verstrepen, K.J., G. Derdelinckx, F.R. Delvaux, J. Winderickx, J.M. Thevelein, F.F. Bauer, and I.S. Pretorius. 2001. Late fermentation expression of *FLO1* in *Saccharomyces cerevisiae*. J. Am. Soc. Brew. Chem. 59:69-76.

Verstrepen, K.J., G. Derdelinckx, H. Verachtert, and F.R. Delvaux. 2003. Yeast flocculation: what brewers should know. App. Microbiol. Biotechnol. 61:197-205.

substrates. Since the vats are closed to the atmosphere, the CO_2 that evolves is trapped, and the beer becomes naturally carbonated.

Traditional conditioning of lagers and ales is time-consuming and expensive, and therefore, used mainly for premium beers that can bear higher production costs. Most modern breweries have adopted faster methods, referred to as accelerated lagering or brewery-conditioning. In these systems, the fully attenuated beer is pumped into storage tanks at 0°C to 2°C and held for one week or less. Small wood chips (less than 3 cm $\times$ 30 cm) may be layered on the bottom of the tank to promote maturation (e.g., "beechwood aging"), as well as serve a clarification function (see below). During the fermentation and later during conditioning, many volatile flavor compounds are formed, which have a major impact on the overall quality of the beer (Table 9–5).

Clarification

Whether the beer has been conditioned via casking or lagering, suspended yeast cells will still be present. Because consumers have come to expect beers to be haze- or cloud-free, with excellent clarity, clarification and filtration

Table 9.5.   Important volatile flavor compounds in beer. ⭕

Esters	Alcohols	Acids	Carbonyls	Sulfur compounds
isoamyl acetate	propanol	acetic acid	diacetyl	hydrogen sulfide
2-phenylethyl acetate	2-methylpropanol	caproic acid	2,3-pentenedione	sulfite
ethyl acetate	2-methylbutanol	caprylc acid	acetaldehyde	methanethiol
ethyl caproate	3-methylbutanol	capryc acid		dimethyl sulfide
ethyl caprylate	isoamyl alcohol	lactic acid		
	amyl alcohol			
	isobutanol			
	ethanol			
	2-phenylethylethanol			

Adapted from Dufour et al., 2003 and Hough et al., 1971

steps are almost always performed. Clarification can occur via physical separation methods, such as centrifugation, or by the addition of fining agents.

Fining agents act by promoting aggregation or flocculation of yeast cells, and include wood chips, gelatin, and isinglass. The latter, which is derived from fish bladders (really), is more commonly used for ales, especially in the U.K., whereas gelatin is frequently used in the United States for lager-style beers. The suspended cells and other insoluble debris then settle to the bottom of the tank or cask. The clarified, or bright, beer that remains is not cell-free. Therefore, the beer is filtered to remove residual yeast or bacterial cells that still remain.

Several filtration configurations exist, including plate and frame systems, leaf filters, and cartridge filters. The filters or filter sheets do not necessarily exclude cells on their own, that is, the pores within these filters are usually not so small that they directly restrict passage of cells. Yeast and bacterial cells have approximate diameters of about 10 μm and 2 μm, respectively, and typical cellulose filters have much larger pore sizes. To remove cells by filtration, then, filtration aids are used.

Filtration aids are inert, insoluble materials, such as cellulose or diatomaceous earth, that act to form a filter bed that traps cells without causing the filter to become plugged or fouled with a layer of cells. Cell counts of less than 10 per ml can be achieved in the filtered beer. Other filter aids, such as silica gel and polyvinylpolypyrrolidone (PVPP), can also be used, but mainly to remove haze-forming proteins and polyphenolic materials, rather than cells (see below).

Despite the effectiveness of these bulk filtration systems, however, there are now many breweries that further filter the beer through membrane filters with pore sizes less than 0.45 μm. Since bacteria and yeasts cannot pass through these filters, the beer becomes sterile. Many of the large breweries now produce filter-sterilized beer. Because filtration is done at low temperature, this process is also referred to as cold-pasteurized beer (which also has the marketing advantage of distinguishing this process from heat-pasteurized beer).

If a sterile (or polish) filtration step is to be performed, the beer must first be pre-filtered (as described above), to remove cells and other material that would otherwise foul the membranes. Although most brewers recognize that filtration is necessary to produce beer with the expected clarity and stability properties, they also realize that filtration may remove flavor, color, body characteristics, and other desirable components. Thus, filtration systems must be carefully designed to achieve the desired appearance without sacrificing flavor and body.

Process aids

Various approved additives are frequently used during the post-fermentation steps. These include agents to improve flavor, color, appearance, and stability. Hop extracts are sometimes added late in the process to provide additional hop flavor. Another important additive is proteolytic enzymes, which are used as "chill-proofing" agents. These enzymes hydrolyze proteins that would otherwise precipitate and form complexes with tannins and other polyphenolic compounds at low temperatures and give a cloudy or hazy appearance when the beer is chilled. Hydrolysis of these proteins prevents this "haze" or cloudiness.

Several commercially available types of enzymes are used. Papain, obtained from papaya, is commonly used, as are fungal- and bacterial-derived proteinases. The main requirements are that the enzyme be rather specific for haze-forming proteins (and not the foam-forming proteins), and that they are inactivated by a subsequent heating step, so residual activity in the finished beer is absent. Although widely used in the United States, chill-proofing enzymes are not allowed in many European countries.

Non-enzymatic, chill-proofing agents are also used, including tannic acid, bentonite, silica gels, and PVPP. Again, the use of these agents is permitted in some countries, and prohibited in others. In the United States, regulations regarding permitted additives and process aids are described in the U.S. Code of Federal Regulations, Title 21, Part 173. Other general regulations on beer manufacturing are contained in Title 27, Part 25.

Carbonation

Of all the post-fermentation steps, perhaps none is as important as carbonation. Carbonation provides sensual appeal by enhancing mouth feel, flavor, body, and foam (or head). The CO_2 preserves the beer by reducing pH and the oxidation-reduction potential (Eh), such that various aerobic, acid-sensitive spoilage organisms are inhibited. Carbonation of beer can occur naturally, via a secondary fermentation, or mechanically, by directly adding CO_2 after the conditioning and filtration steps.

As described earlier, beer can be conditioned via several procedures. In the case of traditional ales, conditioning is commonly done in casks, whereas traditional lagers are conditioned, or lagered, in enclosed tanks. Some specialty beers (and those made by home-brewers) are conditioned directly in bottles. In all of these processes, the beers become naturally carbonated due to a secondary fermentation that yields CO_2 (as well as additional ethanol).

Several requirements are necessary to induce a successful secondary fermentation. First, there must be sufficient fermentable sugar present in the beer to serve as substrate for the yeast. The cask, tank, or bottle must be able to withstand the pressure that accumulates as a result of CO_2 gas formation. Finally, either naturally present yeasts must remain in the beer, even after flocculation, or more yeast must be added. For lagers, in particular, the krausening technique can be used to initiate a secondary fermentation. Recall that the krausen is wort obtained during the most active phase of the fermentation and that it contains sugars as well as log phase yeasts. The CO_2 bubbles that form by krausening are believed to be "finer" compared to those produced by other procedures. However, the krausening technique is now only infrequently used.

The amount of CO_2 that forms during a secondary fermentation and is retained in the beer depends on several factors. The amount of available sugar present or added to the beer directly influences CO_2 production, but rarely is the sugar a limiting factor. Rather, the correct amount of carbonation is more a function of the counter-pressure and temperature in the tank. Obviously, the more back pressure applied, the greater the dissolved CO_2 in the beer. Likewise, as the temperature decreases, solubility of CO_2 increases. Proper adjustment of temperature and back pressure, therefore, is critical to achieving the desired level of carbonation in the finished beer.

Typically, ales have less carbonation that lagers. In the United States, for example, lager-style beers contain about 2.6 volumes of CO_2 per volume of beer (that is, if the dissolved CO_2 was completely dissipated and collected, it would have a volume 2.6 times that of the "flat" beer). Traditional ales, in contrast, contain about half as much CO_2.

It should be noted that beers that have undergone a secondary fermentation still need to be clarified or filtered (or both) to remove cells and haze-forming material, but without losing any of the accumulated CO_2. Traditional cask-conditioned beer can be treated with fining agents to enhance sedimentation, leaving the sediment in the bottom of the cask (where it remains even during dispensing). In contrast, lagers or tank-conditioned beer must be processed under pressure to retain the CO_2.

In the absence of a secondary fermentation, mechanical carbonation provides a convenient and more controllable means for introducing CO_2 into the beer. Mechanical carbonation is now widely used in the beer industry, not just by large breweries, but by small and even micro-breweries. There are essentially two sources of CO_2. As noted earlier, the CO_2 that evolves during the primary fermentation can be collected, cleaned and purified, and then added back to the beer. This process is not inexpensive, however. Therefore, the more common means of mechanical carbonation is simply to pump CO_2 from pressurized tanks directly into the beer. As for when a secondary fermentation occurs, the CO_2 back pressure and temperature dictate the necessary volume of CO_2 pumped into the beer. In some cases, CO_2 is even added to beer that has undergone a secondary fermentation to achieve the desired level of carbonation.

Packaging and pasteurization

Following carbonation, the beer is ready to be packaged. Perhaps the simplest form of packaging is to fill the beer directly into kegs. Kegs are constructed of aluminum or stainless steel and vary in size between 50 L and 100 L. Kegs not only mimic the traditional cask-style beers (in terms of perceived quality), but provide a convenient means of delivering non-bottled or canned product to the consumer. Thus, kegs are widely used by bars and restaurants for serving draught (otherwise known as draft) beer.

Beer destined for kegging is processed like other beers, in that it is clarified and filtered. Most kegged beer is filter sterilized, although some is heat pasteurized (see below). Therefore, its shelf-life is usually considerably longer than traditional casked beer (about three months versus one month). Kegged beer that is neither filter- or heat-pasteurized has a shelf-life similar to that of casked beer.

Importantly, since kegs are re-usable containers, breweries must pay particular attention to the keg cleaning and sanitizing operation before the kegs are re-filled. The interior of the keg, including the valve housing assembly where microorganisms may collect, is typically steam sterilized. And although it is outside the control of the brewer, the dispensing system used to pump beer from the keg to the glass is also a source of potential spoilage organisms.

The most common package for beer in the United States is the bottle or can (sometimes referred to as "small packs"). Bottles are usually constructed of glass, although plastic (e.g., polyethylene terephthalate or PET) bottles have become popular. Most cans are made from aluminum. Beer that is to be bottled or canned is pasteurized, either before or following packaging.

If the beer is pasteurized prior to filling, two options exist. The beer can be filter-sterilized, as described above, or it can be flash pasteurized via a plate-type, regenerative heat exchanger (similar to that used for pasteurization of milk). In fact, the time-temperature conditions for heat pasteurization of beer, 71°C to 75°C for fifteen to thirty seconds, are very near that used for milk. In beer, these conditions achieve a very high level of microbial killing (>5 logs), such that the product is stable at room temperature for at least six months.

Whether the beer is filter-sterilized or flash pasteurized, the packaging steps must be performed under rather stringent aseptic conditions to prevent post-processing contamination. Bottles (whether glass or plastic) and filling and capping (i.e., crowning) equipment must be sterile, and surroundings protected such that microorganisms are excluded. Sterile N_2 or CO_2 can be used to flush the environment, and sterile water used to ensure that the relevant equipment, especially the areas around the fillers, remain free of microbial contaminants. The beer filling and packaging operation is essentially no different from that used for milk, juice, or other fluid products.

Beer can also be heat pasteurized after it has been filled into cans or bottles. In fact, most beer consumed in the United States is processed via tunnel pasteurization systems in which filled and sealed bottles or cans are heated by hot water. Tunnel pasteurizers operate in a continuous mode, with the temperature being gradually raised to as high as 62°C and held for up to twenty minutes. The collective effect of longer heating at lower temperature has the same kinetic effect on the destruction of microorganisms in beer as flash-pasteurization.

Pasteurizing the beer after it is already in the sealed package prevents post-pasteurization contamination; however, the longer exposure to heat may promote undesirable flavor changes in the beer, including formation of cooked and oxidized flavors. Of course, some beer manufacturers eschew any type of heat treatment and instead rely on filtration systems for preservation of beer. However, even filtration is not benign, as flavor and other beer constituents may be removed during the filtering process. Thus, the challenge for modern brewers is to produce not only a beer with desirable flavor, body, and appearance characteristics, but also one that meets the preservation requirements necessary to operate successfully in a highly competitive market.

Beer Defects

Despite the low pH, high ethanol content, and hop antimicrobials ordinarily present in beer, microorganisms are responsible for many (but certainly not all) of the defects that occur in beer. Chemical and physical defects are also common and can cause significant problems for brewers. However, preventing or minimizing entry and growth of microbial contaminants throughout the beer-making process is absolutely essential for consistent manufacture of high quality beer. This is no simple matter, because fungi, wild yeasts, and bacteria are naturally present as part of the normal microflora of the raw ingredients, the brewery environment, and the brewing equipment. Moreover, even if a heat or filtration step is included at the end of the beer-making process, the damage may have already been done.

Although the water activity of stored barley is too low to support the growth of fungi, yeasts, and bacteria, those organisms begin to grow as soon as the barley is steeped, with increases of about 1 to 3 logs. Kilning eventually inactivates some microorganisms; however, the finished malt may still contain as many as 10^6 bacteria, 10^3 fungi, and 10^4 yeast per gram. Various species of lactic acid bacteria that account for the majority of bacteria found in malt and that may cause several serious defects in the finished beer are particularly important (discussed below). However, even fungi that are present at relatively low levels can cause problems. For example, some strains of *Fusarium* can infect barley in the field and produce the toxin deoxynivalenol. Aside from its possible toxicity to humans, deoxynivalenol is correlated with the "gushing" defect in packaged beer. Gushing is characterized by excessive gassing or over-foaming when packaged beer is opened. Although the exact cause of gushing is not clear, it appears that deoxynivalenol or another mold-produced product (usually from *Fusarium*, but other fungi may also be involved) serves as a seed or nucleation site, at which carbon dioxide microbubbles are formed.

Even though the beer fermentation requires yeast, it is also true that wild yeasts can cause several different types of product defects. Wild yeasts, including species of *Zygosaccharomyces*, *Kluyveromyces*, and other *Saccharomyces*, may compete with brewing yeast during the fermentation, and, in some cases, reach high levels. Since these yeasts typically do not flocculate as do brewing yeasts, they remain in the beer, creating cloudiness and filtration problems. Some of these yeasts also produce killer toxins that can potentially inhibit the brewing yeast culture and dominate the fermentation. Other yeasts, referred to as diastatic yeasts (of which *Saccharomyces cerevisiae* var. *diastaticus* is the most

well-known), can hydrolyze and ferment dextrins, and are particular post-fermentation problem contaminants for two main reasons. First,
they often produce phenolic off-flavors in the
beer, often characterized as medicinal. Second,
by fermenting the body-forming dextrins, they
make the beer "thinner." Finally, aerobic yeasts,
which include members of the genera *Pichia*,
Debaryomyces, and *Brettanomyces*, can produce acetic acid, volatile esters and alcohols, and
other oxidized compounds, and cause turbidity
problems.

Although high mashing temperatures (along
with low pH) restrict growth of most bacteria,
some heat-tolerant bacteria can survive and
grow during the mashing step. However, far
more serious are those bacteria that contaminate the wort after the kettle boil step and
grow in the beer during the fermentation and
post-fermentation steps. Lactic acid bacteria
represent the most bothersome organisms in
the brewing operation and are the cause of the
most common defects. Of particular concern
are species of *Lactobacillus*, including *Lactobacillus brevis*, *Lactobacillus casei*, *Lactobacillus plantarum*, and *Lactobacillus delbrueckii*.

During growth in wort, these bacteria ferment wort sugars and produce several undesirable end-products, especially lactic acid. Of
these bacteria, *L. brevis* is heterofermentative, meaning that it produces acetic acid,
CO_2, and ethanol, in addition to lactic acid.
Some lactobacilli produce 3-hydroxypropi-
onaldehyde, which serves as a precursor for
synthesis of acrolein, a compound that imparts a bitter flavor. Importantly, species and
strains of *Lactobacillus*, as well as other lactic acid bacteria—most notably *Pediococcus
acidilactici*, *Pediococcus inopinatus*, and *Pediococcus damnosus*—are responsible for
perhaps the most objectionable defect in
beer, namely, the production of diacetyl.

Diacetyl is a 4-carbon molecule that imparts
a buttery flavor and aroma that, while pleasant
in cultured dairy products, is highly offensive in
beer. Pediococci (and *P. damnosus*, in particular) produce appreciably more diacetyl than

other lactic acid bacteria and are perhaps the
most serious microbial contaminants that affect
beer quality. They are capable of growth over a
wide range of temperatures, including those
used during both the ale and lager fermentations. These bacteria are also tolerant of low pH,
ethanol, and, depending on strain, hop α-acids.

The diacetyl defect caused by these bacteria
(referred to as "sarcina sickness" due to the
original classification of pediococci as belonging to the genus, *Sarcina*) is detectable at concentrations as low as 0.2 ppm (or 0.00002%).
In addition to flavor defects, some pediococci
and lactobacilli are also capable of producing
extracellular polysaccharides that cause a ropiness defect. This ropy or "slime" material increases beer viscosity and gives the beer an undesirable mouth feel.

Although the lactic acid bacteria are arguably
the most serious spoilage organisms in beer,
Gram negative bacteria also can contaminate
beer and cause spoilage problems. In fact, the
spoilage organisms identified by Pasteur were
probably species of the genus *Acetobacter*,
Gram negative bacteria that oxidize ethanol to
acetic acid. Strains of *Acetobacter* are used in
the manufacture of vinegar, thus, they give beer
a highly objectionable vinegary flavor. It is no
wonder that the French beer industry had enlisted the technical advice of Pasteur to solve
this problem. *Acetobacter* and the related
genus, *Gluconobacter*, consist of aerobic rods
that grow well in the presence of high ethanol
concentrations and are resistant to hop α-acids.
Because these bacteria grow poorly, if at all, under the anaerobic conditions established during
the beer fermentation, they are more likely to
be a problem only when the beer is exposed to
oxygen, such as during storage of beer in casks.

Other Gram negative bacteria may also occasionally contaminate and spoil beer. Included are *Zymomonas*, ethanol-tolerant, facultative, or obligate anaerobes associated with
ale spoilage, and *Citrobacter, Klebsiella,* and
other bacteria belonging to the family *Enterobacteriaceae.* The former are prolific ethanol
producers that also make lesser amounts of
lactic and acetic acid, acetaldehyde, and glyc-

erol. They may also produce sulfur-containing compounds, such as dimethyl sulfide and dimethyl disulfide. As spoilage organisms, they cause a fruity flavor defect that occurs most frequently in casked beer.

Among the Enterobacteriaceae that are found in beer, the most common is *Obesumbacterium proteus*. However, its role in beer spoilage is not clear. It reportedly interferes with normal growth of the yeasts, leading to poor attenuation. The anaerobic Gram negative rods include species from the genera *Pectinatus*, *Selenomonas*, and *Zymophilus*. They are capable of producing organic acids, H_2S, and turbidity in packaged beer. They are easily killed by heat, thus they probably gain entrance into beer during post-pasteurization packaging.

Beer spoilage does not occur only by microorganisms; a number of chemical and physical defects are also common in beer. One of the more frequent defects (although less so in recent years) is referred to as "skunky" beer flavor or light-struck flavor. It is caused by sunlight-induced photooxidation. This defect, which can range from slight to nauseous, is most noticeable in beer bottled in clear glass. The most likely mechanism by which this defect occurs is via ultraviolet or sunlight-induced cleavage of side chains on hop-derived iso-α-acids. These reactive side chains then react with sulfur-containing compounds to form products with a skunky aroma. One thiol in particular, 3-methyl-2-butene-1-thiol or MBT, has a low flavor threshold (just a few ppb) and is characteristic of this defect. Given the potential for beer to become skunky, how is it that there are still many popular beers packaged in clear glass? In general, these beers are "skunk-proofed" to prevent the light-induced reactions described above. Specifically, hop extracts (obtained via either liquid or supercritical CO_2 extraction) are used that are processed such that the isomerized iso-α-acids are converted to a reduced form. Depending on the extent of the reduction (i.e., how many hydrogen atoms are incorporated), dihydro-, tetrahydro-, or hexahydro-iso-α-acids are formed. Importantly,

these modified iso-α–acids are very light-stable and non-reactive.

Another flavor defect is referred to as stale or oxidized. Stale beer is variously described as having cardboard-like, rotten apple, cooked, or toffee-like flavors. It is invariably caused by autooxidation reactions, not unlike the staling reactions that occur in other foods. Staling is generally time-dependent and increases during storage. Therefore, it is often the main factor that determines shelf-life of beer. The best way to control staling is simply to keep oxygen out of the finished beer.

Among the physical defects in beer, perhaps the most common is haze, which, as discussed above, occurs during cooling, and is caused primarily by proteins that form complexes with polyphenols and carbohydrates. Treatment with proteinases can reduce this problem. In addition, β-glucan-containing polysaccharides can also cause haze, gel, and filtration problems, as well as increasing viscosity. Bacterial β-glucanases can be added to the mash or wort to degrade these materials. Foam instability problems are also common and result in loss of foaming or head. Adding proteins, hop components, and gums increases viscosity and surface tension, and can therefore stabilize foaming.

Waste Management in the Brewing Industry

Although waste management affects all segments of the fermented foods industry, the issue is particularly important in the brewing industry. This is because beer manufacturing uses nearly a half-billion tons of grains each year and, although some components of that grain are fermented, much of the grain is left behind. These spent grains represent a considerable disposal problem. Currently, this material is used in one of several ways. It can be further processed and used as a specialty food ingredient (e.g., as a fiber supplement), however this market is way too small to make much of a dent in the supply of spent grains. It is more common for breweries (small and large) to contract with local farmers, who use the spent grains as livestock feed. Spent grains can also be used as fertilizer. The grains can be dried to extend shelf-life and reduce shipping cost, but drying requires energy and is expensive. Still, both wet and dry grains are used. Another waste product that poses problems for the industry is the spent filter material used as filter aids. Alternative filtration systems that do not require filter aids or that use materials that can be recycled are now available.

Finally, brewing uses large volumes of water. As much as 5 L to 10 L of water are used for every liter of beer. Much of this water contains biological material, and its disposal into municipal water supplies may be restricted and expensive. Therefore, water handling sys-tems, such as anaerobic digesters that treating and/or recycle this water are critical.

Recent Developments in the Beer Industry

The beer industry is one of the most competi-tive segments of the food and beverage indus-try. This competition has led to new technolo-gies, new innovations, and new products. At the same time, there has been remarkable growth in the microbrewing industry, and a return to traditional or craft brewing prac-tices. It is now possible to find nearly every type of beer at the local pub or retail outlet (Box 9-8). Thus, the beer industry, from the smallest to the largest brewer, continues to

---

Box 9–8. Beer from A to Z

There are two general types of beer, ales and lagers. However, while it is certainly true that ales and lagers represent the major beer categories, this simple distinction is not really adequate to describe the many varieties that are produced throughout the world. In reality, there are so many different versions of these two styles, that to refer to a particular beer simply as an ale or a lager provides little useful information about the actual nature of that beer. In addition, beers made in a similar manner, but from different geographical regions, often have quite different characteristics, and are classified accordingly. These differences in beer characteristics are due, in part, to the manner in which the beer is produced, but perhaps more importantly, to the spe-cific ingredients used in the particular recipe. For example, one of the key ingredients that dis-tinguishes one beer from another is the water used in the brewing process.

Of course, hops and malt have a major impact on the flavor, aroma, and color properties of the beer. Thus, beers are often classified based on their malt and hop type and content. For example, the Saaz hops used in Belgian pale ales and Czech Pilsner impart a flowery and fruity flavor to the beer. Listed below are some of the major categories or styles of beer and brief descriptions for each.

Ale—Top-fermented beer, produced throughout the world, but most commonly in the U.K. Ales generally have more hop aroma and bitter flavor than lagers, and are usually less carbonated.
Altbier—A German style ale, historically made in Düsseldorf. It is copper-colored, with a mod-est bitter flavor, and only slight bitterness. Altbiers are lagered (and served) at lower temper-atures than typical ales.
Bitter—A typical British-style ale, with a strong bitter flavor.
Bock—A German lager with strong malt flavor and sweetness, but little hop bitterness and little or no hop aroma. The color ranges from light to dark brown, and these beers generally are full-bodied.
Brown Ale—An ale originated from Newcastle upon Tyne (in northeastern England), that is characterized by its full body, dark color, malty and generally sweet flavor, and low to moder-ate hop aroma and bitterness.

Box 9–8.   Beer from A to Z *(Continued)*

Cream Ale—An ale made in the United States; it has a mild sweet flavor.

Dunkel—The classic Munich lager; characterized by its dark brown color, sweet and nutty malt flavor, and full body. Bitterness and hop aroma are usually mild.

Lager—Bottom-fermented beer that originated in Bavaria, probably in the fifteenth century. Lagers generally are less hoppy and are more carbonated than ales, although exceptions exist. In the United States, lagers are amber or golden in color, with medium body and flavor.

Lambic—A specialty beer that originated in Belgium that is noted for several unique characteristics. First, it contains unmalted wheat as a major ingredient (as much as 30% or more). Second, aged hops, having decreased iso-α–acids, are used, minimizing the antimicrobial activity and bitter flavor. Third, microbial growth (including lactic acid bacteria, Enterobacteriaceae, and various yeast), and subsequent acid development is encouraged. Finally, the fermentation is entirely natural or spontaneous—no yeast is added, and it may take place over a period of several months. Lambic beer variously described (even by its followers) as "sour, thin, leathery, fruity, cheesy."

Malt Liquor—An American-style, pale colored lager, with low hop bitterness or aroma, thin body, and little flavor, but with a higher-than-normal alcohol content (>4.5%)

Pale Ale—Typical British ale, with a medium body and malty flavor and light bronze color. Pale ales are moderately dry and bitter, and have a hoppy aroma. Carbonation is generally light and ethanol content high. Adjuncts are often added. India Pale Ale is similar in most respects to conventional pale ale, except it contains more hops, which promotes shelf-life (hence, this style was developed for export from England to India).

Pilsner—A classic lager that originated in Pilzen, a city in the Bohemia region of what is now the Czech Republic. Pilsners are among the most popular lagers in Germany (Bohemia was once part of the Austrian Empire and has a heavy German influence), as well as other regions in Europe. Pilsner (or Pils) has a strong malt flavor and hop aroma, medium body and sweetness, and a golden color. German Pilsners generally are lighter in color and body, and are less sweet compared to Czech Pilsner.

Porter—A British ale that has characteristics of both stouts and pale ales. Porters have light-to-medium body and malty flavor, are well-bittered, have dark color, and usually contain adjuncts.

Steam beer—Introduced (and trademarked) by the Anchor Brewery Co. in San Francisco in the late 1800s. Steam beers have both ale and lager properties, in that they are bottom fermented, but at a high ale-like temperature. They are fermented in wide, shallow tanks.

Stout—A type of ale that is noted for its very dark color (due to heavily roasted malt), rich malt and hop flavor, and moderate to high bitterness. Stouts have a creamy head, and vary from sweet to dry. The Irish stout, Guinness, is the most well-known example.

Zwickelbier—A German beer that is unfiltered.

develop new and innovative products and processes, many of which involve biotechnology and bioprocessing.

Low-calorie beer

One of the most successful and influential products developed by the beer industry was low-calorie beer. Introduced in the 1970s, this "new" type of beer had a dramatic impact, not only in the beer industry, but throughout the food industry, as "light" became one of the most widely used descriptors for reduced calorie foods. In fact, low-calorie beers had been around for at least ten to twenty years prior to the introduction of Miller Lite in 1975, but these products were not very popular. Undoubtedly, highly successful advertising and marketing campaigns, along with more calorie-conscience consumers, led to what is now a major part of the beer industry. Currently, low-calorie beers have about 45% of the total U.S. beer market, and three of the top five selling brands of all beer are in this category (Budweiser Light, Miller Lite, and Coors Light). When first introduced for beer, "light" had no official meaning. The widespread use of this term, however, eventually led the FDA to define light foods as those that are significantly reduced in either fat, calories, or sodium. As applied to beer, "light" means that there must be one-third fewer calories.

The most obvious way to reduce the caloric content of beer would be to simply dilute regular beer with water. A typical American-style lager beer ordinarily contains about 150 calories in a 340 ml (12 ounce) serving, so adding 25% water would result in a beer with only 112 calories. Doing so, however, would obviously not only reduce calories, but would also reduce flavor, color, body, and the ethanol content. Instead, to make a "light" beer with fewer calories, but with a minimal loss of flavor and body characteristics, manufacturers had to consider beer composition and the components that contributed calories.

In general, a typical U.S. beer contains (on a weight basis) about 3.7% ethanol, 1% protein, and 0.5% fat. Thus, in 340 ml of beer, the caloric contribution of those components is about 90 (from ethanol) + 15 (from protein) + 15 (from fat) = 120 calories. What is the source of the additional 30 calories (i.e., 150 to 120)?

As discussed earlier in this chapter, the beer fermentation is considered complete or fully attenuated when the fermentable carbohydrates have been depleted. However, that is not to say that there are not carbohydrates in beer. In fact, wort contains a mixture of fermentable sugars (i.e., glucose and maltose), as well as more complex, non-fermentable sugars, in particular, dextrins and limit dextrins. The latter are a result of the incomplete hydrolysis of starch during the mashing step. Whereas the simple sugars are readily fermented, the complex carbohydrate fraction is not. Importantly, however, these carbohydrates are caloric, meaning they can be hydrolyzed to glucose during digestion, adsorbed into the blood stream, and either used as energy or stored as fat. Calories can be reduced by somehow reducing or removing the nonfermentable but caloric carbohydrates from beer.

The strategy that was adopted, and which is still used today, adds enzymes that hydrolyze a portion of the nonfermentable carbohydrates to form free sugars that are then fermented during the primary fermentation step. The enzymes are commercially available and inexpensive glucoamylases (derived from *Aspergillus niger* and other fungi). Added to the wort during mashing, the exogenous enzymes work in concert with the endogenous enzymes from the malt. Ideally, the fungal enzymes should be temperature labile so there is no residual activity after kettle boil.

Because more fermentable carbohydrate in the wort necessarily results in more ethanol in the beer (and at 7 calories per gram, would defeat the purpose, as well as add to the alcohol excise tax), the extra ethanol is ordinarily removed. The carbohydrate content of these beers is reduced from 9 g per serving to less than 3 g, resulting in a calorie reduction of 25 or more. Thus, a typical light beer contains 100 to 120 calories. Recently (in 2003), even lower calorie beers, containing fewer carbohydrate than conventional light beers, were introduced. These so-called "low-carb beers" are targeted to dieters on carbohydrate-restrictive diets. One popular brand contains only 2.6 g of carbohydrate and 95 calories per 340 ml serving.

An alternate strategy to the use of enzymes replaces even more of the malt or starch adjuncts with adjuncts consisting of simple sugars. For example, if sucrose, fructose, or glucose syrups were used as adjuncts, at the expense of malt, the dextrin fractions would also be reduced. Both of these approaches, however, provide thin-bodied beers, since it is the dextrins that contribute mouth feel and body properties to beer.

Several other approaches for reducing the dextrin (and caloric) content of beer have been considered, and are now being studied. Malt preparations with greater amylolytic activity can be used so hydrolysis of starch is more complete. Alternatively, brewers' yeast strains with greater amylolytic activity can be developed, either via conventional yeast breeding programs or by molecular techniques. The goal is to increase activity of amyloglucosidase, which breaks down α-1,4 and α-1,6 glucosidic linkages in dextrin.

Finally, there is a perception that low-calorie or low-carbohydrate beers also have a much lower alcohol concentration than normal beer. In fact, the alcohol content for most of these beers is somewhat lower, generally by 0.5% to 1.0% (or about 20% less). For example, three of the most popular brands of light beer in the United States each contain 4.2% alcohol (on volume basis), compared to 4.7% to 5% for their conventional counterparts. For other brands, the differences are less than 0.5%

Ice and dry beers

Two other types of beer that gained a following in the 1990s, but whose popularity has since faded somewhat, are ice beers and dry beers. Although made by different processes, both ice beer and dry beer contain more ethanol than conventionally-processed beers and are thought to impart a smoother flavored beer with less aftertaste. Ice beers are manufactured according to the freeze concentration principle—namely, by cooling the beer to temperatures as low as $-4°C$, ice crystals will form, which can then be removed. The actual technology for the manufacture of these beers is not new—a German version called Eisbock has been made for many years. The resulting beer contains less water and more ethanol, and is generally sweeter, with good body and color. Dry beers, in contrast, are less sweet (hence the term "dry"), and are made by using less malt and more readily fermentable adjuncts. These beers are rather similar to low-calorie beers, in that they have less flavor, color, and body compared to conventionally-made

products. They are also produced in a similar manner, in that enzymes are used to hydrolyze dextrins. This results in a more complete fermentation, less residual sugars, and more ethanol.

Non-alcoholic beer

Low- or non-alcoholic beers were first produced in the United States more than eighty years ago (during Prohibition), and have been available ever since. However, the relatively low demand for these products did not drive the industry to devote very much research effort into new technologies. Due to a marked increase in the consumer demand for low- or non-alcohol products, the technology for making these beers has improved dramatically in the last decade. The quality of these beers, not surprisingly, has also improved (although there are still many detractors who would argue otherwise), as has the availability of many different domestic and imported products.

Non-alcoholic beers must contain less than 0.5% ethanol. Historically, several different processes have been used to produce non-alcoholic beer. The earliest methods generally involved removing the ethanol from normal beer by evaporation or distillation. Although these methods are still used, various filtration configurations, in particular, dialysis and reverse osmosis, are now more widely used. Beer quality is arguably much better by the latter processes because they operate at low temperatures ($<10°C$) and the heat-generated reactions that affect flavor do not occur.

In reverse osmosis systems, the beer is pumped or circulated at high pressure (more than 5,000 Pascals) through small pore membranes with average molecular weight cutoffs of less than 0.01 μm. The ethanol molecules pass through the membrane as part of the permeate, along with a portion of the water. The retentate contains nearly all of the beer solids and most of the water, but little or none of the ethanol. Water can either be added back to its normal level, or the beer can be diluted beforehand to account for the water lost from the retentate. Dialysis operates in a similar manner, except that the driving force is simply the concentration difference across the membrane. Despite the less intrusive nature of these processes, however, flavor compounds may still be lost.

Another entirely different way to make these beers relies not on ethanol removal, but rather on modifying the fermentation so that ethanol is not made as an end product. The fermentation can simply be abbreviated, such that yeast growth is stopped or curtailed before much ethanol has been produced. Alternatively, the fermentation can be conducted at a low temperature ($<5°C$) that restricts yeast growth and ethanol formation. However, these methods may result in a beer that is too sweet and microbiologically unstable, due to high levels of residual sugars. Modifying the wort composition by removing fermentable sugars prior to fermentation is another way to limit ethanol production. Finally, metabolic engineering of the yeast, such that the ethanol pathway is blocked, may be an ideal way to reduce the ethanol content in beer, but without leaving behind fermentable sugars (Box 9-9).

Wheat beer

Wheat beers have long been popular in Europe, but have only recently become known to U.S. consumers. They are made using wheat malt combined with barley malt in ratios varying from 1:3 to as high as 3:1. These beers are generally thinner and more sour than barley malt beers, and are usually high in phenolic compounds. They also often have a somewhat cloudy appearance. However, their unique flavor, due in part, to vinyl guaiacol generated from ferulic acid, is particularly appreciated by some consumers.

High-gravity fermentations

Another technology that has captured interest among brewers, especially large, high-volume producers, involves developing a wort with a specific gravity considerably higher than that used ordinarily. The concentrated wort, once made, can be diluted before or, more commonly, after fermentation. The net effect is an increase in brewing capacity and product throughput without having to incur addition capital expenses for equipment (e.g., mash tanks, lauter tanks, and fermentors). Moreover, the production of high-gravity worts reduces energy, labor, and effluent costs. Although high-gravity fermentations place extra demands on the yeast (e.g., the osmotic pressure and ethanol concentrations are higher), the quality of the beer is considered to be comparable to conventionally brewed beers.

Biotechnology and the Brewing Industry

For the first 5,000 years that humans made and consumed beer, little was known about the actual scientific principles involved in its manufacture. Beer making was an art, practiced by craftsmen. Only in the last 150 years have biochemists and microbiologists identified the relevant organisms and metabolic pathways involved in the beer fermentation. In the past ten years alone, the entire genome of *Saccharomyces cerevisiae* (albeit, a lab strain, not an actual brewing strain) has been sequenced, with nearly half of the genes now having an assigned function.

This sequence information is now being used to understand yeast physiology, especially as it relates to brewing (Box 9-10). However, despite this knowledge, and the thousands of years of "practice," the brewing process is still far from perfect, and producing consistent, high-quality beer is still a challenge, even for large, highly sophisticated brewers. In addition, economic pressures, quality concerns, and perhaps most importantly, new market demands, have led the industry to consider new ways to improve the beer-making process. Advances in molecular genetics and biotechnology have made it possible to address these challenges via development of new brewing strains with novel traits and tailored to perform in a specific manner.

Box 9–9. Metabolic Engineering Approaches in the Manufacture of Non-alcoholic Beer

Two general processing approaches have been used to make non-alcoholic beer. One involves removing or separating ethanol from beer by physical means (e.g., distillation, evaporation, and reverse osmosis). The other approach has been to curtail the fermentation such that little or no ethanol is produced. Both of these approaches have their pitfalls, as already described.

The availability of molecular tools has now made it possible to consider biological strategies to produce non-alcoholic beer. Although the selection of brewers' yeasts has been based, for hundreds of years, on their ability to ferment wort sugars to ethanol, recent studies have shown that it is possible to redirect metabolism away from ethanol production and toward synthesis of other end products.

During the beer fermentation, several end products, other than ethanol, are ordinarily produced. Glycerol, in particular, can reach concentrations of more than 2 g per liter. Lesser but still relevant amounts of acetaldehyde, 2,3-butanediol, and acetoin are also formed. These products are derived from the glycolytic intermediates, dihdyroxyacetone phosphate (DHAP) and glyceraldehyde-3-phosphate (G-3-P), at the expense of ethanol (Figure 1).

Glycerol, for example, is formed from G-3-P via glyceraldehyde-3-phosphate dehydrogenase (GPD) and glycerol-3-phosphatase (GPP). By diverting even more of the glucose (or maltose) carbon to these alternative pathways, less ethanol would theoretically be produced. In the approach adopted by Nevoigt et al., 2002, the *GDP1* gene encoding for GPD was cloned and over-expressed in an industrial lager yeast strain. The transformants produced more than four times the amount of GPD, compared to the parent strain, and more than five times more glycerol. Importantly, the ethanol concentration in a typical brewing wort during a simulated beer fermentation was reduced by 18%, from 37 g/L to 30 g/L. This reduction, however, was less than half that achieved previously when GPD was over-expressed in a laboratory strain of *Saccharomyces cerevisiae* (Nevoigt and Stahl, 1996).

There were also large increases in the concentrations of acetaldehyde and diacetyl during the primary fermentation, and although these levels decreased during a subsequent secondary fermentation, they were still high enough to affect the flavor in a negative way. Thus, more metabolic fine-tuning of the competing pathways will be necessary to engineer a yeast capable of producing nonalcohol beers.

A completely different strategy for making non-alcoholic beer also was described by Navrátil et al. (2002). Their approach was based on the knowledge that: (1) non-alcoholic beer is sensitive to microbial spoilage; (2) low wort pH is inhibitory to contaminating microorganisms; and (3) addition of lactic acid or lactic acid bacteria to the wort stabilizes the beer. Because acidification of wort with lactic acid or lactic acid bacteria is either not allowed or is difficult to control, another way to promote acidification was needed.

Therefore, strains of *S. cerevisiae* defective in enzymes of the tricarboxylic acid (TCA) pathway and known to produce elevated concentrations of organic acids were used under simulated batch or continuous fermentation conditions (and using free or immobilized cells). In all cases, the beer pH was 3.25 or less when the mutant cells were used (compared to pH 4.1 to 4.2 for the control strain). Although the mutations were not located within ethanol production genes, the test strains produced very low amounts of ethanol (<0.31% for free cells and <0.24 for immobilized cells). The latter result presumably was due to the inhibition of ethanol formation, specifically, pyruvate decarboxylase and alcohol dehydrogenase, at low pH. Although other end products, including diacetyl, were also produced, informal sensory analysis suggested that the beer compared favorably to conventionally-produced non-alcoholic beer.

(Continued)

Box 9–9.   Metabolic Engineering Approaches in the Manufacture of Non-alcoholic Beer *(Continued)*

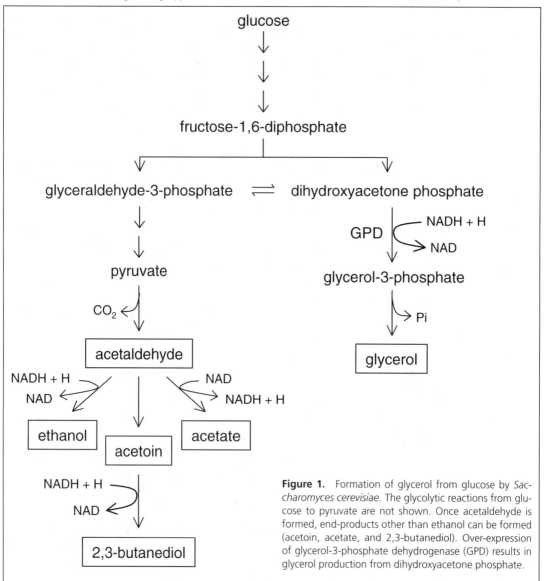

Figure 1.   Formation of glycerol from glucose by *Saccharomyces cerevisiae*. The glycolytic reactions from glucose to pyruvate are not shown. Once acetaldehyde is formed, end-products other than ethanol can be formed (acetoin, acetate, and 2,3-butanediol). Over-expression of glycerol-3-phosphate dehydrogenase (GPD) results in glycerol production from dihydroxyacetone phosphate.

References:

Navrátil, M., Z. Dömény, E. šturdík, D. šmogrovićová, and P. Gemeiner. 2002. Production of non-alcoholic beer using free and immobilized cells of *Saccharomyces cerevisiae* deficient in the tricarboxylic acid cycle. Biotechnol. Appl. Biochem. 35:133–140.

Nevoigt, E., R. Pilger, E. Mast-Gerlach, U. Schmidt, S. Freihammer, M. Eschenbrenner, L. Garbe, and U. Stahl. 2002. Genetic engineering of brewing yeast to reduce the content of ethanol in beer. FEMS Yeast Res. 2:225–232.

Nevoigt, E., and U. Stahl. 1996. Reduced pyruvate decarboxylase and increased glyceraldehyde-3-phosphate dehydrogenase [NAD$^+$] levels enhance glycerol production in *Saccharomyces cerevisiae*. Yeast 12:1331–1337.

Strain improvement strategies

Since the beginning of beer making, all the way to the present, brewing strains have been used continuously, being passed down from batch to batch. These strains are highly adapted to wort and beer and are not very amenable to classical strain improvement strategies. That is, trying to select strains, either spontaneously or following mutagenesis, with improved fermentative, flavor-producing, or other relevant properties, is not an easy proposition. In addition, whereas laboratory strains are usually diploid and capable of sporulating and forming haploid spores or asci, brewing strains are polyploid, containing multiple alleles for many genes. Brewing strains also sporulate poorly. Thus, even mating and hybridization techniques, the standard means for modifying yeast in the laboratory, are often unsuccessful in brewing strains. Moreover, changes in one trait often have pleitropic effects and lead to undesirable changes in other performance characteristics.

Although strain improvement programs still rely on these classical approaches, the ability to target specific genes or traits is now possible using recombinant DNA and genetic engineering techniques. However, such approaches

Box 9–10. Beer-omics

Of all the microorganisms used in food and beverage fermentations, *Saccharomyces cerevisiae* was the first whose genome was sequenced (Goffeau et al., 1996). The sequenced strain (S288C) is considered a lab strain, as opposed to the industrial strains used for ale or lager manufacture. Although there is only limited sequence data for the latter strains, genome structure analysis has revealed significant differences between the laboratory and brewing strains.

Most importantly, whereas most lab strain genomes exist as diploids (i.e., containing two copies of each chromosome), beer yeasts are poly- or alloploidal, containing multiple chromosomal copies (i.e., more than one "genome" per cell). Moreover, it appears that the genome of the lager yeast, *Saccharomyces pastorianus* (or *Saccharomyces carlsbergensis*), is a hybrid, consisting of one genome that evolved from an S288C-like strain of *S. cerevisiae* and another whose ancestral progenitor is still unknown (Casaregola et al., 2001). Despite these differences, however, an incredible amount of practical information has been generated from the original sequence data, as well as that subsequently made available via the web-based *Saccharomyces* Genome Database (SGD, located at www.yeastgenome.org).

The *S. cerevisiae* S288C genome consists of sixteen chromosomes and a total of 12,068 kilobases of DNA. Based on the original computer predictions, there were 5,885 genes encoding for proteins and another 455 that encode for ribosomal, small nuclear, and transfer RNA molecules. However, a more recent analysis of the *S. cerevisiae* genome indicates that there are probably about 500 fewer genes for a revised total of 5,538 (Kellis et al., 2003). Also, the genomes of three other species of *Saccharomyces* (*Saccharomyces paradoxus*, *Saccharomyces mikatae*, and *Saccharomyces bayanus*), as well as the related yeast, *Schizosaccharomyces pombe*, have been sequenced, providing even more information on the structure and function of yeast genomes (Kellis et al., 2003; Wood et al., 2002). Numerous computational tools for comparing these sequences with those obtained from industrial strains have been described and are now available via the SGD (Christie et al., 2004).

While genome information can predict the genes that are present in the chromosome(s) of an organism, it is often more informative to identify genes that are actually transcribed. The so-called transcriptome can be determined using DNA macro- or microarray technology. It also may be interesting to determine how temporal gene expression is influenced by environmental conditions or growth phase. For example, in the case of beer strains of *S. cerevisiae*, one might wish to identify genes expressed during growth on maltose or that are induced when ethanol is present. In fact, it is possible to monitor gene expression throughout the entire beer fermentation process,

(Continued)

Box 9–10. Beer-omics *(Continued)*

generating an expression profile of all the proteins produced at any given time during the course of the fermentation (the proteome, commonly determined by two-dimensional gel electrophoresis).

Finally, the quantitative, global collection of small metabolic products (the metabolome) that appears in the medium during beer manufacture (as measured by NMR, mass spectroscopy, or other analytical methods) provides yet another molecular picture of the beer making process. Several recent reports provide excellent examples of how "omic" approaches can be used to gain a better understanding of the genetic, cellular, and metabolic events that occur during the beer fermentation.

Beer transcriptomics

As reported in two independent investigations, one using a macroarray representing 6,084 ORFs (Olesen et al., 2002) and the other using a microarray representing 6,300 ORFs (James et al., 2003), the gene expression patterns of brewing yeasts (lager strains) change during growth in wort. However, the number of differentially-expressed genes was relatively modest (about 20% of the total ORFs). Furthermore, functions for the majority of the induced or repressed genes could not be assigned.

It should also be noted that the mRNA used in these studies was from beer strains and that the arrays had been derived from a lab strain of *S. cerevisiae*. Thus, there may have been induced or repressed genes that would not hybridize with any of the orfomers on the arrays. Still, in both studies, there were several interesting results. For example, there was a significant increase in the expression of genes that encode for putative enzymes involved in sterol and lipid biosynthesis. This is consistent with the well-known observation that yeast cells synthesize sterols and membrane unsaturated fatty acids following aeration of the wort just prior to the pitching step. Similarly, there was a marked increase in the expression of mitochondrial- and peroxisome-related genes, which may reflect the necessity of anabolic pathways to generate products required for anaerobic growth.

Not surprisingly, genes encoding for glycolytic enzymes were also induced after inoculation of the yeast into the wort. However, the most unexpected result was the finding by James et al. (2003) that glycolytic genes were subsequently repressed after about forty-eight to seventy-two hours and respiration genes were activated. In addition, James et al., (2003) showed that protein synthesis and most stress response genes were repressed after the first day of fermentation, despite the increase in the ethanol concentration. Based on these collective reports, the lager yeast transcriptome clearly changes during the beer fermentation, reflecting the catabolic and anabolic adjustments the cell evidently makes in response to chemical and physical changes in the growth environment.

Beer proteomics

The yeast transcriptome, as described above, provides a means of identifying the mRNA transcripts that are produced at a given time during growth. In contrast, the proteome represents the complete set of actual functional gene products (i.e., the proteins) that are synthesized during growth. Proteome maps of industrial lager and other yeast strains during growth in synthetic medium revealed that about 1,200 polypeptides are produced, although many appear to be duplications (Joubert et al., 2000). There is also a high degree of similarity between the proteins produced by lager strains and lab strains, supporting the notion that lager strains are hybrids and contain a genome derived from *S. cerevisiae*. However, as many as thirty-two other proteins are made by lager yeasts that do not appear in the *S. cerevisiae* proteome (Joubert et al., 2001). Analysis of these non-*S. cerevisiae* proteins by peptide mass fingerprinting and mass spectroscopy techniques revealed that many are involved in maltose metabolism, glycolytic pathways, and production of ethanol.

Box 9–10. Beer-omics *(Continued)*

In another recent report, induction of protein synthesis by lager yeasts was studied during the lag and early log phase of growth (Brejning et al., 2005). Interestingly, the induced protein expression pattern differed from the gene expression patterns (at least for the genes investigated), indicating that post-transcriptional regulation was involved. Among the early expressed proteins were those involved in amino acid and protein synthesis, glycerol metabolism, glycolysis, and ergosterol biosynthesis. Moreover, the expression profile for cells grown in minimal medium was consistent with those grown under brewing conditions.

Beer metabolomics

It would be understatement to suggest that wide variations exist in the chemical composition of different beers. After all, differences in composition account, to a large extent, for the many different types of beer that are produced around the world. Still, determining on a quantitative basis what the actual chemical composition is for a given beer can provide considerable information regarding the brewing process, yeast metabolism, and overall beer quality. This chemical profile, or the metabolome, which can be determined by principal component analysis, has been shown to distinguish between ales and lagers, as well as between different manufactured beers (Duarte et al., 2002; Duarte et al., 2004). For example, the presence or absence of particular beer constituents containing aliphatic and aromatic regions (as determined by NMR spectra), were associated with ales or lagers.

References

Brejning, J., N. Arneborg, and L. Jespersen. 2005. Identification of genes and proteins induced during the lag and early exponential phase of growth of lager brewing yeasts. J. Appl. Microbiol. 98:261–271.

Casaregola, S., H.-V. Nguyen, G. Lapathitis, A. Kotyk, and C. Gaillardin. 2001. Analysis of the constitution of the beer yeast genome by PCR, sequencing and subtelomeric sequence hybridization. Int. J. Syst. Evol. Microbiol. 51:1607-1618.

Christie, K.R., S. Weng, R. Balakrishnan, M.C. Costanzo, and 19 other authors. 2004. *Saccharomyces* Genome Database (SGD) provides tools to identify and analyze sequences from *Saccharomyces cerevisiae* and related sequences from other organisms. Nucl. Acids Res. 32 (Database issue):D311–D314.

Duarte, I., A. Barros, P.S. Belton, R. Righelato, M. Spraul, E. Humpfer, and A.M. Gil. 2002. High-resolution nuclear magnetic resonance spectroscopy and multivariate analysis for the characterization of beer. J. Agric. Food Chem. 50:2475-2481.

Duarte, I.F., A. Barros, C. Almeida, M. Spraul, and A.M. Gil. 2004. Multivariate analysis of NMR and FTIR data as a potential tool for the quality control of beer. J. Agric. Food Chem. 52:1031–1038.

Goffeau, A., B.G. Barrell, H. Bussey, R.W. Davis, B. Dujon, H. Feldmann, F. Galibert, J.D. Hoheisel, C. Jacq, M. Johnston, E.J. Louis, H.W. Mewes, Y. Murakami, P. Philippsen, H. Tettelin, and S.G. Oliver. 1996. Life with 6000 genes. Science 274:546–567.

James, T.C., S. Campbell, D. Donnelly, and U. Bond. 2003. Transcription profile of brewery yeast under fermentation conditions. J. Appl. Microbiol. 94:432–448.

Joubert, R., P. Brignon, C. Lehmann, C. Monribot, F. Gendre, and H. Boucherie. 2000. Two-dimensional gel analysis of the proteome of lager brewing yeast. Yeast 16:511–522.

Joubert, R., J.-M. Strub, S. Zugmeyer, D. Kobi, N. Carte, A.V. Dorsselaer, H. Boucherie, and L. Jaquet-Gutfreund. 2001. Identification by mass spectrometry of two-dimensional gel electrophoresis-separated proteins extracted from lager brewing yeast. Electrophoresis 22:2969–2982.

Kellis, M., N. Patterson, M. Endrizzi, B. Birren, and E.S. Lander. 2003. Sequencing and comparison of yeast species to identify genes and regulatory elements. Nature 423:241–254.

Olesen, K., T. Felding, C. Gjermansen, and J. Hansen. 2002. The dynamics of the *Saccharomyces carlbergensis* brewing yeast transcriptome during a production-scale lager yeast fermentation. FEMS Yeast Res. 2:563–573.

Wood, V., R. Gwilliam, M.-A. Rajandream, M. Lyne, and 130 other authors. 2002. The genome sequence of *Schizosaccharomyces pombe*. Nature 415:871–880.

must be carefully considered, due to the public perception (often supported by regulatory authorities) that genetically modified organisms pose risks to the consumer or the environment. Rather, it seems that commercial applications can still be realized, but will require the use of more benign techniques (i.e., no foreign DNA used, no antibiotic resistance markers) for strain construction.

Examples of modified brewing strains

In the brewing industry, several traits have attracted the most research attention (Table 9-6). As noted earlier, brewing strains of *S. cerevisiae* generally do not hydrolyze dextrins and limit dextrins during the beer fermentation. These sugars remain in the beer, and although they may have a positive influence on body and mouth feel characteristics, they also contribute calories. Reducing these dextrins provides the basis for making low-calorie or light beers. Most brewers use commercially available enzymes, usually fungal glucoamylases, that are added to the wort during mashing. Another approach is to use yeast strains that express glucoamylase and that degrade these dextrins during the fermentation. Such strains have been isolated (e.g., *S. cerevisiae var. diastaticus*), but, unfortunately, when used for brewing, the beer quality is poor, due, in part, to the production of phenolic off-flavors. Even hybrid strains, obtained by mating brewing yeasts with diastatic yeasts, may still produce off-flavors.

Therefore, a better alternative directly introduces glucoamylase genes (encoding for both glucose-1,4 and glucose-1,6 hydrolysis activities) into brewing strains. Accordingly, *STA2* genes (from *S. cerevisiae var. diastaticus*) and *GA* genes (from *Aspergillus niger*) have been cloned into brewing strains, and the genes were expressed and the enzymes functioned as expected (i.e., dextrins and limit dextrins were used). Instability problems occurred initially, since the cloned genes were first located on plasmids, but this problem has largely been circumvented either by using more stable, yeast-derived plasmids or by using integration vector systems that result in the cloned genes being integrated within the host chromosome.

Another problem that occurs in beer production is due to the β-glucans that are derived from the cell walls of barley malt. Consisting of β-1,4 and β-1,6 linked glucose polymers, these materials form gels and foul filters, and increase beer viscosity and haze formation. They can be digested by commercially available enzymes, but, like the glucoamylase example described above, genes coding for bacterial or fungal β-glucanases can be cloned into brewing strains such that they are expressed during fermentation.

Table 9.6.   Traits targeted for strain improvement.

Trait	Relevant gene[1]	Function
Dextrin utilization	*STA2*	Glucoamylase
Diacetyl utilization	*ALDC*	Acetolactate dehydrogenase
Glucan utilization	*EG1*	β-glucanase
Maltose utilization	*MALT*	Maltose permease
H$_2$S reduction	*MET25*	Sulfhydrylase
Dimethyl sulfide reduction	*mxr1*[2]	Sulfoxide reductase
Enhanced flocculation	*FLO1*	Flocculation
SO$_2$ production	*MET3*	ATP sulphurylase
Ester production	*ATF1*	Alcohol acetyltransferase

[1]Except where indicated, the goal is increased expression
[2]Lowercase indicates the gene is inactivated and not expressed

Bibliography

Bamforth, C.W. 2004. *Beer: Health and Nutrition*. Blackwell Science Ltd., Oxford, United Kingdom.

Campbell, I. 2003. Microbiological aspects of brewing, p. 1-17. *In* F.G. Priest and I. Campbell (ed.), *Brewing Microbiology, 3rd Edition*, Kluwer Academic/Plenum Publishers. New York.

De Keukeleire, D., 2000. Fundamentals of beer and hop chemistry. Química Nova 23:108-112.

Dufour, J.-P., K. Verstrepen, and G. Derdelinckx. 2003. Brewing yeasts, p. 347-388. *In* T. Boekhout and V. Robert (ed.), *Yeasts in Food*. Woodhead Publishing Limited, Cambridge, England and CRC Press, Boca Raton, Florida.

Flannigan, B. 2003. The microbiota of barley and malt, p. 113-180. *In* F.G. Priest, and I. Campbell (ed.), *Brewing Microbiology. 3rd Edition*, Kluwer Academic/Plenum Publishers. New York.

Goldammer, T. 1999. *The Brewers' Handbook*. KVP Publishers, Clifton, Virginia.

Hammond, J.R.M., 2003. Yeast genetics, p. 67-112. *In* F.G. Priest, and I. Campbell (ed.), *Brewing Microbiology. 3rd Edition*, Kluwer Academic/Plenum Publishers. New York.

Hornsey, I.S. 1999. Brewing. Royal Society of Chemistry. Cambridge, U.K.

Hough, J.S., D.E. Briggs, and R. Stevens. 1971. *Malting and Brewing Science*. Chapman and Hall Ltd, London, U.K.

Stewart, G.G. 2004. The chemistry of beer instability. J. Chem. Ed. 81:963-968.

Van Vuuren, H.J.J., and F.G. Priest. 2003. Gram-negative brewery bacteria, p. 219-245. *In* F.G. Priest, and I. Campbell (ed.), *Brewing Microbiology. 3rd Edition*, Kluwer Academic/Plenum Publishers. New York.

10

Wine Fermentation

"I do like to think about the life of wine, how it is a living thing. I like to think about the year the grapes were growing, how the sun was shining that summer or if it rained . . . what the weather was like. I think about all those people who tended and picked the grapes, and if it is an old wine, how many of them must be dead by now. I love how wine continues to evolve, how every time I open a bottle it's going to taste different than if I opened it on any other day. Because a bottle of wine is actually alive—it's constantly evolving and gaining complexity—like your '61—and it begins a steady, inevitable decline."
Maya, from the movie Sideways, *screenplay by Alexander Payne and Jim Taylor*

Introduction

The conversion of raw food materials into finished fermented products is often considered to be one of the best examples of "value-added" processing. If this is the case, then perhaps no other process or product exemplifies this more than the fermentation of grapes into wine. To wit, consider the following: In October, 2004, at a wine auction in Los Angeles, a single bottle of Chateau d'Yquem 1847 (a Sauterne wine from the Bordeaux region in France) sold for $71,675, making it the most expensive bottle of wine ever sold in the world. At the same auction, two bottles of Cabernet from the Inglenook Winery in Napa Valley sold for $24,675 each (which was a record for a California wine). Although these wines are certainly the exception, high quality California wines still average nearly $30 per bottle.

Given these examples, one may assume that wine making must be a highly complicated process, involving sophisticated technologies. In fact, although wine manufacture does indeed rely heavily on modern microbiology and biochemistry, traditional techniques and tried and true manufacturing practices are still important, and in many cases, necessary to produce high quality products. Thus, wine making serves not only as an example of value-added processing, but also as an example of an ancient technology that has adopted twenty-first century science.

History

The history of wine is nearly as old as the history of human civilization. The earliest writings discovered on the walls of ancient caves and in buried artifacts contain images of wine and wine-making instruments. Wine is mentioned more than 100 times in both the Hebrew and Christian bibles and many of the most well-know passages involve wine. The very first vines, for example, were planted by Noah, who presumably was the first wine maker; later Jesus performed the miracle of turning water into wine. Wine also was an important part of Greek and Roman mythology and is described in the writings of Homer and Hippocrates. For thousands of years, even through the present day, wine has had great ritual significance in many of the world's major religions and cultures, and it is an important part of the world economy and commerce.

Grape cultivation (viticulture) and wine making appears to have begun in the Zagros Mountains and Caucasus region of Asia (north of Iran, east of Turkey). Domestication of grapes dates back to 6000 B.C.E., and large-scale production, based on archaeological evidence, appears to have been established by 5400 B.C.E. A fermented wine-like beverage made from honey and fruit appears to have been produced in China around 7000 B.C.E., and rice-based wines, similar to modern day sake, were produced in Asia a few thousand years later (Chapter 12). Wines were imported into France, Italy, and other Mediterranean countries by seafaring traders sometime around 1000 B.C.E., and vines and viticulture techniques were likely introduced into those regions several centuries later.

Wine is also one of the oldest of all fermented products that has been commercialized, mass-produced, and studied. In fact, many of the early microbiologists and chemists were concerned with wine making and wine science. Less than 150 years ago, when the very existence of microorganisms was still being debated, Pasteur showed that not only did microorganisms exist, but that they were responsible for both production and spoilage of wine. Of course, wine preservation has been important since ancient days, when early Egyptian and Roman wine makers began using sulfur dioxide (in the form of burned sulfur fumes) as perhaps the first application of a true antimicrobial agent.

Production and Consumption

Most wines are made in temperate climates, particularly those areas near oceans or seas. About 75% of all wine is made in the Mediterranean areas of Europe (Table 10-1). France, Italy, and Spain are the largest producers, and are responsible for more than half of the nearly 27 billion liters of wine produced from around the world. Not surprisingly, these countries also devote the most acreage to grape production.

However, several relatively new entrants into the global wine market have made a significant impact. In particular, Australia, South Africa, and Chile are now responsible for more than 10% of worldwide wine production. Given that these three countries consume less than half the volume of wine that they produce, they represent a significant part of the export market. The United States is also one of the leading producers of wine (more than 2 billion liters per year), with most production coming from California (which alone accounts for about 90% of all U.S. wine). In fact, according to wine industry statistics (www.wineinstitute.org), the economic impact of the California wine industry is estimated to be $33 billion.

On a volume basis, the top five wine consuming countries are France, Italy, the United States, Germany, and Spain (Table 10-1). On a per capita basis, consumers in Luxembourg, France, and Italy drink the most wine, more than 50 liters per person per year (Figure 10-1). This compares to the world per capita average of about 3.5 liters. In the United States, per capita consumption is about 9 liters. However, in contrast to many of the European countries where wine consumption has either been the same or decreased in recent years, wine consumption in the United States has increased (about 8% since 1997). On average, most of the wine consumed in the United States costs only about $7 per bottle (0.75 L).

Wine Basics

Although wine making, like beer manufacture, involves an alcoholic fermentation, the similarities, for the most part, end there. The wine fermentation requires different yeasts and substrates and yields distinctly different products. And whereas beer is best consumed fresh, most wines improve markedly during an aging period that can last for many years. Finally, although most wines, like most beers, start with relatively inexpensive raw materials, the quality of premium wines depends, more so than any other fermented product, on the quality of the raw material used in their manufacture.

Table 10.1. Global production and consumption of wine[a].

Country	Acreage (acres $\times 10^3$)	Production (L $\times 10^6$)	Consumption (L $\times 10^6$)
Argentina	506	1,584	1,204
Australia	366	1,016	398
Austria	121	253	248
Brazil	156	297	308
Bulgaria	273	226	153
Canada	22	45	280
China	888	1,080	1,095
Chile	440	566	225
France	2,259	5,339	3,391
Germany	257	889	2,004
Greece	321	348	284
Hungary	230	541	320
Italy	2,244	5,009	3,050
Japan	53	110	278
Luxembourg	3	14	26
Portugal	613	779	470
Romania	610	509	471
Russia	173	343	500
South Africa	292	647	397
Spain	3,052	3,050	1,383
Switzerland	37	110	308
Turkey	34	27	34
United Kingdom	2	2	1,010
United States	2,417	2,130	2,417
Uruguay	27	100	98
Total[b]	19,586	26,683	22,785

[a]Adapted from 2001 statistics from the Wine Institute (www.wineinstitute.org)
[b]Total based on 69 countries

Although about 99% of the wine produced throughout the world is made from grapes, juice from other fruits can also be made into wine. Berries, including raspberries, boysenberries, and strawberries, are common substrates. Many wines also are made from tree fruits, including apples and pears. The only requirement is that the fruit and the juice from that fruit must contain enough free sugars to support growth of the yeast and to yield a sufficient ethanol concentration (usually >12% by volume).

Viticulture and Grape Science

The starting material for most wines, as noted above, is grapes. The main wine grape grown in temperate zones throughout the world is *Vitis vinifera*. Another grape, *Vitis labrusca*, grows well in northern regions in the United States and is frequently used for Concord varieties. It is important to note that, despite the existence of only a few major grape species, there are many different grape cultivars grown throughout the world. For example, Cabernet Sauvignon, Chardonnay, Gamay, Mission, Gewurztraminer, Grenache, and Sangiovese all refer to different varieties or cultivars of the *V. vinifera* grape. These grapes not only have different compositions, sugar contents, and pigmentation, they also grow better in different climates and soils and are used for different types of wine. Thus, most Bordeaux wines (those produced in the Bordeaux region of France) are made from the grapes that grow well in that region, namely Cabernet Sauvignon. Those same

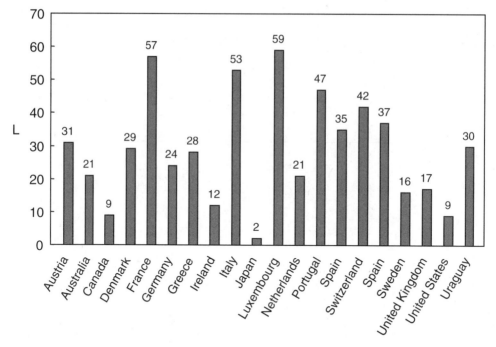

Figure 10–1. World-wide per capita wine consumption (L per person per year). Adapted from the Wine Institute 2001 Statistics (www.wineinstitute.org).

grapes, however, can be grown in California or anywhere else in the world, for that matter, and used to make Bordeaux-style or other types of wine.

Although it is common throughout Europe to name a wine after a region, in other countries, wines are ordinarily named after the grape variety from which the wine was made. How a bottle of wine is actually labeled is itself an important issue. Labels not only must comply with local regulations, but they may also reveal useful information on the precise source of the raw materials, how the wine was made, and the expected quality attributes of the finished product (Box 10–1).

Assuming that it is possible to successfully grow Cabernet Sauvignon grapes, say, in central Minnesota, which is on the same general latitude as central France, could a vintage bottle of Bordeaux be produced? Or to phrase the question another way, what distinguishes wine made from the same grape but from different geographical locations? Several answers are evident. First, different areas have different growing soils and climates. For example, in the famous grape-growing regions in France, different districts within the same region may yield grapes with markedly different composition and properties due to differences in soil composition, temperature, moisture, and sunlight. Even different vineyards within the same district may yield slightly different grapes. In addition, grapes cultivated in the same vineyard but harvested at different times within the same season, or from two different growing seasons, may be quite different and yield distinctly different wines.

Climatic factors have such an important effect on grape quality and maturity that sophisticated computer-generated meteorological models have been developed recently to allow viticulturists to predict how grapes will grow and mature in a given geographical environment.

Box 10–1. Understanding Wine Labels

Few, if any, food labels reveal or imply as much information on the source of the raw materials, when and where the product was made, the manufacturing procedures used, and the overall quality of that food, as do the labels on wine bottles. Wine nomenclature can be so exact that informed and experienced consumers can even predict with some certainty how a particular wine should taste even before the cork is removed from the bottle.

Unfortunately, the lack of international conventions on wine standards and labeling has made it difficult to develop a uniform and consistent set of rules and definitions. For example, French wineries are subject to strict regulatory authority by virtue of Appellation d'Origine Contrôlée (AOC) laws, and, as a result, French wine labels can be extremely informative (at least for those who understand what the labels mean). Moreover, European Union labeling laws prevent use of generic labels (i.e., naming a wine after a specific wine region, when the wine was produced from somewhere else). That is, just as cheese labeled as Parmesan must have been made in the Parma region of Italy, so must a bottle of Champagne have been produced in the Champagne region of France. Also, because authentic Champagne (and Parmesan cheese, for that matter) is made using traditional methods, labeling a bottle of wine as Champagne necessarily tells the consumer how that wine was made.

For the most part, no such rule exists in the United States, making comparisons between wines from different countries, based on labels, a somewhat challenging exercise. What does it mean, for example, for a California-made wine to be labeled as Bordeaux or Champagne? What follows, then, is a general discussion on the terminology and "appellations" used for labeling wines.

In general, there are three types of information on a wine label that provide some indication as to the nature or quality of that wine. Although different countries vary with respect to the specific details (see below), most labels indicate the type of grape (varietal), where the grapes were grown (the appellation), and the owner and/or bottler of the wine (proprietor).

Varietal refers simply to the cultivar (e.g., Chardonnay, Riesling, Cabernet Sauvignon) used to make the wine. For many French wines, the variety is not stated, but can be inferred by virtue of where it was made (Figure 1). In the United States (as well as most other countries), the varietal name can be used only when that variety comprises more than 75% of the grapes used in its manufacture.

Appellations are essentially geographical designations and refer to the location where the grapes were grown, such as Champagne, Chablis, Chianti, Sherry, and Napa Valley. In France and other European countries, even small sub-regions and villages within the larger regions can be designated. In fact, protecting the geographical integrity of wine is one of the main reasons why labeling laws evolved. Although no law prohibits an American wine manufacturer from adopting a European geographical designation, this so-called generic labeling is merely a marketing device and provides no real information. This practice is now much less of a problem than it was previously. To account for the geographical and varietal properties, most countries have adopted a form of the French AOC labeling regulations.

In France, most wines fall into one of three main categories: Appellation d'Origine Contrôlée, Vins de Pays, and Vin de Table. Wines bearing an AOC declaration are those that are (1) produced in one of the designated regions, (2) comprised of specific cultivars, and (3) made according to traditional manufacturing practices. In theory, wines bearing the AOC designation are of the highest quality.

Wines produced from grapes from different regions, outside the established AOC boundaries, are referred to as Vins de Pays (wine of the country). These wines are generally considered to be of somewhat lower quality, but in many cases, they are of similar quality as AOC wines. Finally, wines bearing the Vin de Table (or Vin Ordinaire) designation are of the lowest quality. The label makes no mention of the region or area where the wine was produced or of the grape variety.

(Continued)

Box 10–1.   Understanding Wine Labels *(Continued)*

This wine is made in a village called Pommard. Although not noted on the label (but certainly known to informed consumers), Pommard is located within the Burgundy AOC. This village is part of Cote d'Or, one of five regions in Burgundy (other well-known regions include Chablis and Beaujolais). Note that there is also no mention of the grapes used to make this wine. While varietal labels are common in the U.S., only a few AOC wines (including Rieslings and Gewurztraminers from the Alsace region) contain this information. For other wines, the variety is implicit in the AOC (e.g., Pommard is made from Pinot Noir, Beaujolais is from Gamay).

Wine labels should be informative, but also distinctive and unique. A sketch of the vineyard or an image of the family crest are common. Also, there is no year stated on this label. This particular producer indicates the vintage year on a separate label. However, to include a vintage year, at least 85% (higher in some regions) of the grapes must have been harvested from that year.

Another traditional way to classify AOC wines is via the "cru classé" system. This system is based mostly on geographical, viticultural, and historical considerations, and the actual quality of the wine is somewhat secondary. The highest class in Burgundy is Grand Cru (the terminology varies according to the region). This particular wine, although classified one level lower, as Premier Cru, is still considered to be of very high quality

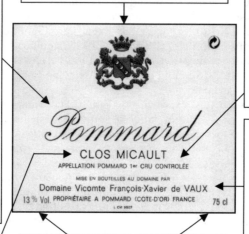

The proprietor's name must be listed on the label. For this bottle, the complete family name is given, a gesture meant to add prestige to the wine and to instill confidence to the customer. Also, it is noted that the wine was made and bottled in this winery from grapes grown in this vineyard (i.e., the grapes were not obtained from another vineyard). Finally, in very small print (bottom center), there is a required code or "serial number" that indicates the lot or barrel from which the wine was obtained.

Not only does the label indicate the village within the AOC region, the actual vineyard in which the grapes were grown is also stated (Vineyard Designates). For this wine, the grapes were grown in a walled or enclosed vineyard called Clos Micault.

The labeling regulations require that the alcohol content and the volume be declared on the label. In France, the alcohol level for AOC wines vary; Pommard wines must contain 13% alcohol (by volume). The bottle volume is 750 ml, and the stamp at the upper right indicates that the producer has paid the required recycling tax. Note that neither sulfite, nor health warning labels are required in France.

Figure 1.   Understanding wine labels. The label in the above panel is from a French Burgundy made in the Cote d'Or region. The label on the right panel is from a bottle of Cabernet Sauvignon, made in the Sonoma County region of California. Labels were provided courtesy of the François-Xavier de Vaux and the Ernest and Julio Gallo wineries.

Other than declaring the wine as "vin blanc," "vin rouge," or "vin rosé," the label simply states the proprietor and the country.

In the United States, regulations regarding wine labeling are established by the Alcohol and Tobacco Tax and Trade Bureau (TTB), a division within the Department of the Treasury (until 2003, this division was called the Bureau of Alcohol, Tobacco, and Firearms). In addition to a health warning on alcoholic beverages and a sulfite warning, there are a minimum of five other TTB requirements that must also be on the label: (1) the brand name or owner name; (2) the

Box 10–1. Understanding Wine Labels *(Continued)*

In contrast to most European wines, where appellation is emphasized, the grape variety is usually a more prominent feature of U.S.-made wines. To use a varietal label, the wine must have been made from at least 75% of that variety. Although varietal labeling is optional, labels must indicate the type or class of wine. This particular wine is a Class 1 wine, as signified by the "Red Wine" declaration. U.S. federal regulations recognize 9 classes; other examples include sparkling wines (Class 2), fruit wines (class 5), and apertif wines (Class 7).

The U.S. version of AOC regions are called American Viticultural Areas. The Barrelli Creek Vineyard, where this wine is made, is located in Alexander Valley, one of 11 AVAs in Sonoma County, California. The notation at the top of the label "single vineyard selection" signifies that the grapes came from one vineyard located within this region, comparable in meaning to the enclosed vineyards of Burgundy.

In addition to stating the wine type or class, other requirements for all U.S. wines include the proprietor name and address; the alcohol content (if 14% or higher), and the metric volume. If the wine was fermented and bottled by the same producer, a phrase such as the one used here "vinted and bottled by" can be used. Otherwise, the producer name would be prefaced by the statement "bottled by."

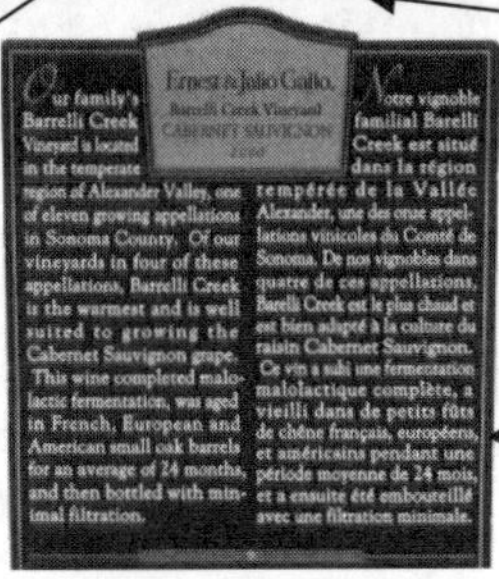

The vintage year (e.g., 2000) can also be stated, provided at least 95% of the grapes were grown and harvested in that calendar year.

Many U.S. wines include optional information regarding the region where the grapes were grown and details on the fermentation and aging conditions.

There are two other label requirements for U.S. wines (although not shown here): the "contains sulfites" declaration and the Surgeon General's warning about alcohol consumption and risks to pregnant women and operators of cars and machinery.

Figure 1. (Continued)

class or type; e.g., Table Wine, Sparkling Wine, Carbonated Wine, Citrus Wine, etc., (3) the alcohol content, by volume; (4) the bottler's name and address; and (5) the fluid volume, in metric measurements.

If the wine is made from grapes harvested and fermented during a particular year and from a particular region, a vintage year may be indicated on the label. Similarly, an appellation or geographical designation may be allowed, provided that more than 85% of the grapes were grown within the appellation boundaries. The boundaries for these appellations, called American Viticulture Areas (AVA), are defined by the TTB. Most of the approved AVA designations are in California (examples include Napa Valley, Sonoma Coast, and Carmel Valley), but twenty-four other states also have AVA regions.

Finally, although some consumers are intimidated by wine labels (especially when they are not written in English), the relevant information can be readily understood, provided one knows a few simple basics (Figure 1).

Vineyard management practices also affect grape maturation. Important factors include vine spacing and density, pruning and thinning, training and trellising, use of canopies, and application of pesticides. Although the combined effects of soil, climate, and season certainly affect grape quality, the overall importance of these factors, the so-called "terroir," on wine quality, is frequently debated (Box 10-2). Certainly, wines made from premium grapes can sometimes be very ordinary, indicating that attention to proper wine-making techniques is also critical. On the other hand, high quality wines can hardly be produced from inferior or immature grapes. The bottom line is that the basic raw material

Box 10–2. The Terroir Theory

No other issue divides European and American wine makers as much as the issue of "terroir." Terroir, translated loosely from the French as "soil" or "earth," refers, at its most simple level, to the ground or location in which wine grapes are grown. In France, the entire appellation control system that forms the basis for how French wines are labeled is based on terroir. Thus, each of the ten AOC regions has its own terroir (e.g., the Bordeaux terroir, the Burgundy terroir, and so on).

In the broader sense, however, terroir theory holds that wine quality, apart from the actual manufacturing steps, is a function of a host of environmental and agronomic factors. Relevant factors include the nutrient composition of the soil in which the grapes were grown, the cultivar of the grapes, the topography of the vineyard, watering conditions, climatic conditions, exposure to sunlight, vine density, and training and pruning practices. Some wine makers might suggest that the specific yeasts that are present on the grapes also contribute to terroir (Box 10-3).

On the surface, one might wonder why controversy regarding terroir exists. Is it not unreasonable to expect that wine quality might be influenced by where the grapes were grown? In fact, it is well established that sugar concentration, phenol content, other grape constituents and overall grape maturity are affected by climate, sunlight, and other growing conditions. Thus, on a general level, there is probably not much argument—environmental conditions have a profound effect on grape composition which ultimately affects wine quality.

Why the issue of terroir raises concern among enologists, it seems, relates to two other issues. First, there is the view held by terroir advocates that the soil itself contributes to wine flavor by somehow translocating minerals and other soil molecules directly into the grapes. Thus, the wine has the "goût de terroir or "taste of the soil." This claim is generally disputed by many viticulturists, who would argue that grape vines and other plants adsorb water, minerals, nutrients, and other inorganic and organic molecules from the soil, but not flavor compounds.

The second issue is more philosophical in nature. Terroir, according to the popular wine writer, Jamie Goode, is considered by many to be an "ethos...a unifying theory encapsulating a certain approach to wine that encompasses the almost metaphysical circle of soil, nature, appellation, and human activity." In other words, it is terroir, and not the skills of the wine-maker, that is responsible for the overall sensory characteristics of a wine. Ultimately, the argument goes, terroir makes a particular wine unique.

By definition, therefore, it would not be possible to replicate a particular wine produced from grapes from one region with a wine made from the same cultivar, but grown even a few meters away. Further, and this is probably what really annoys "new world" wine makers, terroir implies or reinforces the perception that "old world" (and French, in particular) wines are superior to those produced elsewhere. Despite whether or not there actually is a scientific basis for terroir, there are clearly considerable market advantages for French wines bearing AOC designations (which are essentially synonymous with terroir). And even though wines produced in the Napa Valley of California or the Finger Lakes region of New York, or anywhere else for that matter, could also make terroir claims, there is little question that it is difficult to match the history and reputation of Bordeaux, Loire Valley, Champagne, or any of the other terroir regions.

for wine—the grape—is prone to considerable variability, which may have profound effects on the physical, chemical, sensory, and other properties of the finished wine.

Grape Composition

Given that the two major constituents of wine, water and ethanol, have no flavor, color, or aroma, it is not surprising that the other grape components contribute so much to the organoleptic properties of wine. Some of these substances can be problematic, causing a variety of defects. In addition, the composition of grapes changes during growth and maturation on the vine, such that the time of harvest influences the chemical constituents of the grape as well as the wine. For example, the sugar concentration increases as the grape ripens on the vine, due to increased biosynthesis, and to a lesser extent, to water evaporation and subsequent concentration of solutes. In contrast, acid concentrations decrease during maturation. Finally, in discussing the composition of wine, it is often more useful to consider the liquid juice just after the grapes have been crushed as the starting material, rather than the intact grape. As listed in Table 10-2, the juice, or "must," consists of several major constituents and many other minor components that are important, but which are present at relatively low concentrations.

Sugars

Other than water, which is 70% to 85% of the total juice volume, simple sugars represent the largest constituent of grapes or must. Depending on the maturation of the grape at harvest, must usually contains equal concentrations of glucose and fructose, with the latter increasing somewhat in over-ripened grapes. Sucrose is usually present at very low concentrations (less than 1%), except for musts from *V. labrusca* grapes, which can contain as much as 10% sucrose. In general, most grape cultivars contain about 20% sugar (i.e., 10% glucose and 10% fructose), but the actual amount of total

Table 10.2. Constituents of juice and red table wine (g/100 ml)[1].

Compound	Juice	Wine
Water	70—80	80—90
Carbohydrates	15—25	0.1—0.3
Glucose	8—12	0.5—0.1
Fructose	8—12	0.5—0.1
Other[2]	1—3	0.1—0.2
Organic acids[3]	0.5—2.0	0.3—0.6
Inorganic salts[4]	0.3—0.5	0.2—0.4
Nitrogenous	0.1—0.4	0.01—0.1
Phenolics	0.1—0.2	0.2—0.3
Ethanol	0.0	8—15
Other alcohols	0.0	0.01—0.04

[1]Adapted from Amerine et al., 1980 and Boulton et al., 1996
[2]Mainly sucrose and various pentoses
[3]Mainly tartaric, malic, citric, lactic, and acetic
[4]Mainly potassium, magnesium, and calcium

sugar may vary, depending mostly on maturity. Other sugars also may be present, but at very low concentrations, including the sugar alcohol—sorbitol—and the pentoses—arabinose, rhamnose, and xylose. It is common practice among wine makers to refer to the total sugar concentration in units of Brix or °Brix. The °Brix value is actually a measure of density or specific gravity and is easily and quickly determined using a hydrometer. Juice from mature grapes at 20% sugar is ordinarily about 21°Brix to 24°Brix.

As will be discussed later in more detail, glucose and fructose (and to a lesser extent, sucrose) serve as the major growth substrates for the fermenting yeasts. In fact, the main wine yeast, *Saccharomyces cerevisiae*, ferments only these simple sugars, and all of the ethanol that is produced is derived from sugar fermentation. In addition, a small amount of the sugar is converted to esters, aldehydes, higher alcohols, and other volatile organic compounds (formed also from metabolism of fatty acids and amino acids) that contribute important flavor and aroma characteristics. When nearly all of the sugar is fermented, and only residual (0.1% to 0.2%) amounts remain, the wine is considered "dry." In contrast, when the residual sugar concentration at the end of fermentation is 10 g/L or higher, a "sweet" wine is produced. Very sweet wines, such as those consumed as dessert wines, can contain as much as 100 g/L to 200 g/L (10% to 20%). Finally, it should be noted that when the sugar concentration in the must is low, as might occur in grapes grown in cool climates, it is permissible (in some countries, but not in California) to add sugar to the must, a process known as chaptalization.

Organic Acids

Organic acids comprise the second most plentiful non-water constituent in must. Although the organic acids are present at relatively modest concentrations, typically ranging from 1% to 2%, their effect on wine quality is extremely important. These acids are responsible for the low and well buffered pH of the must and the wine (usually between 3.0 and 3.5). That wine is so well preserved is due not only to the ethanol and low pH, but also to the presence of organic acids that have powerful antimicrobial activities. At these pH values, pathogenic and other microorganisms are inhibited, including *Salmonella*, *Escherichia coli*, and *Clostridium* spp. The exceptions are aciduric and acidophilic organisms such as lactic acid bacteria, acetic acid-producing bacteria, and some fungi and yeast. In addition to antimicrobial effects, low pH also stabilizes the anthocyanins that give red wine its color, inhibits oxidation reactions, and contributes desirable flavor.

Wine chemists generally categorize wine acids into two general groups. Volatile acids are those that are volatilized or removed by steam treatment; those that remain are considered fixed acids. There are little, if any, volatile acids in must. During the ethanolic fermentation, small and usually inconsequential amounts (<0.5 g/L) of acetic acid and other short- and medium-chain fatty acids may be produced. However, if higher concentrations are produced, either by oxidation of ethanol by bacteria or as an end product of microbial metabolism, the wine will suffer serious defects and may be unsalable (as discussed later).

The main fixed acids, depending on the grape, condition, and maturity, are tartaric acid and malic acid. The ratio of these two acids is usually about 1:5 (tartaric:malic), but it can be reversed in some grapes. These acids are important in wine for several reasons. Since malic acid contains two carboxylic acid groups, it contributes more protons in solutions, and makes the must more acidic. If the malic acid concentration is too low, as might occur in overly mature grapes grown in warm climates, the wine pH will be too high. The wine will lack the desirable acid flavor and may be more prone to spoilage by bacteria. In contrast, although a minimum acidity is desirable in wine, excess acidity is also a defect and results in an inferior sour-tasting wine. Musts obtained from grapes grown in cool climates may contain high levels of malic acid, and are, therefore, problematic. A natural, biological method for deacidifying wine is commonly used for such musts (see below). Finally, it should be noted that other important fixed acids may be present in wine, including succinic acid and lactic acid, both which are produced during the wine fermentation by yeast or bacteria.

Nitrogenous Compounds

Grapes contain both inorganic and organic sources of nitrogen. Total nitrogen concentrations in grapes (or musts) range from about 0.2 g/L to 0.4 g/L. The ammonium nitrogen is less than 0.1 g/L. Despite their relatively low concentration in juice, the nitrogen content of most musts is generally adequate for rapid growth of yeasts. In fact, the primary role of nitrogen in wine appears to be as a nutrient source for the yeasts, rather than affecting any of the organoleptic or other properties of the wine, per se. Moreover, wine yeasts can assimilate free ammonia into amino acids and can, therefore, use ammonia directly as a source of nitrogen. Nonetheless, nitrogen deficiency can occasionally occur, and is one of the main causes for sluggish or "stuck" fermentations. Some wine makers, therefore, routinely add ammonium salts to the must, in the form of a yeast food (especially for those occasions when the grapes are deficient in nitrogen).

The main organic nitrogen-containing compounds are amino acids, amides, amines, and proteins. The free amino acids are not only synthesized into proteins (following transamination reactions), but several can also be used as an energy source. Among the proteins found in must, the most important are the enzymes. Some grape enzymes serve useful functions, whereas others can present problems. The pectinases and other hydrolases that enhance extraction of juice from the grapes during crushing are especially important. In contrast, phenol oxidases participate in the well-known enzymatic browning reaction. In the presence of oxygen, these enzymes form undesirable brown pigments that seriously discolor the wine. This reaction, however, is effectively inhibited by sulfur dioxide.

In recent years, the presence of potential bioactive nitrogen-containing compounds in wine has attracted concern and interest. When the amino acids histidine and tyrosine are decarboxylated by decarboxylase enzymes produced by wine bacteria, the amines histamine and tyramine are formed. Depending on the dose and individual sensitivity, these biogenic amines can cause headaches, nausea, and allergic-type reactions. Another compound, ethyl carbamate, can be formed via a chemical reaction between urea and ethanol. Ethyl carbamate is suspected of being a carcinogen, and its presence in wine is increased by heating steps, such as pasteurization, and by high urea concentrations.

Polysaccharides

The main polysaccharide in grapes or must is pectin, a structural carbohydrate that provides structural integrity to the plant cell walls. The pectin concentration in the must can be as high as 5 g/L, which could potentially cause the wine to become cloudy. However, most of the pectin is either precipitated out during fermentation or is hydrolyzed to soluble sugars by exogenous

microbial pectinases. The latter are commercially available and are used not only to enhance maceration and pressing, but also to improve wine clarity. Other polysaccharides, such as cellulose and hemicellulose, are not soluble in the juice and are removed along with the other non-soluble solids in the form of pomace.

Sulfur Compounds

Several sulfur-containing substances are found or are formed in grape juice that have a pronounced affect on the wine fermentation and wine quality. Hydrogen sulfide (H_2S) and various organic forms of sulfur, especially the mercaptans which are formed from H_2S, impart highly offensive odors in the wine. They are produced in trace amounts by grape yeasts during fermentation.

The other major group of sulfur compounds found in wine are sulfur dioxide (SO_2) and related aqueous forms that exist as sulfite ions. These substances are produced naturally by yeast, and are invariably present in wine, albeit at concentrations usually less than 50 mg/L. However, sulfur dioxide and bisulfite salts are now commonly added to must due to their strong antimicrobial, antioxidant, and antibrowning properties. It is important to recognize that these activities occur only when the SO_2 is in its free, un-bound form. When bound or fixed with other wine compounds, such as acetaldehyde, reducing sugars, and sugar acids, SO_2 activity is diminished. How SO_2 specifically functions in wine and its important role in wine making will be further discussed later.

Phenols, Tannins, and Pigments

Among the most important naturally occurring substances in grapes and musts are the phenolic and polyphenolic compounds. Some phenols can also be introduced into the wine following aging in wooden casks or via yeast and bacterial metabolism. These chemically diverse compounds contribute color, flavor, aroma, and mouth feel to the wine. They can also react with other grape components and can either improve or diminish wine quality. Finally, many of the phenolic compounds found in wine are thought to be responsible for the putative health benefits associated with wine consumption (Box 10-3).

The phenols important in wine are grouped according to their chemical structure (Figure 10-2). Phenols containing a single phenolic

Box 10–3. L'chaim! Healthful Properties of Wine

However one says it—L'chaim! Salute! a Votre Santé! Cheers!—all are usually said just prior to consuming wine, and all meaning the same thing—to life and good health. But is wine simply a symbol for expression of this sentiment or might there be an actual scientific basis to support the healthful properties of wine? Let's examine some of the epidemiological, as well as mechanistic, evidence collected in recent years.

Epidemiology

If drinking wine promotes health or well-being, one would expect that residents of countries where wine consumption is high would live longer. Rates of heart disease and cancer would also be expected to be lower in those countries compared to countries where wine consumption is less. Indeed, there has long been the suggestion, based on international population comparisons, that moderate wine drinking is good for one's general health (Figure 1).

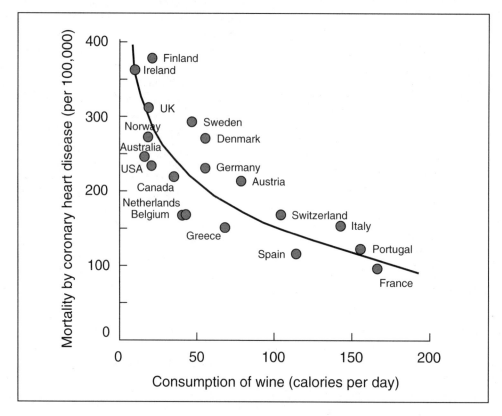

Figure 1. Correlation between wine consumption in eighteen industrialized countries and mortality by coronary heart disease. Adapted from Renaud and de Lorgeril, 1992.

Box 10–3.   L'chaim! Healthful Properties of Wine *(Continued)*

Several recent studies have provided additional epidemiological evidence to support this hypothesis. In the study by Klatsky et al (2003), mortality data of nearly 130,000 California adults who initially had been interviewed between 1978 and 1985 regarding their drinking habits, was obtained through 1998. Compared to groups that either did not drink wine or that drank beer or liquor, wine drinkers had a significantly lower mortality risk. This result was due mostly to the reduced risks of coronary disease and respiratory deaths. Mortality risk reduction did not depend on the type of wine (i.e., red or white) and was consistent for most population subsets (e.g., men versus women, age, and education). Moreover, the risk of coronary disease decreased as wine drinking frequency increased, from less than once per week to daily or almost daily. Finally, it has also been recently suggested (Pinder and Sandler, 2004) that wine consumption might provide a protective effect against dementia and the onset of Alzheimer's disease.

In another epidemiological study, the mortality of 24,000 Danish adults was followed for nearly thirty years (Grønbæk et al., 2000). Although moderate consumption of beer or liquor had little effect on mortality, a significant reduction in mortality risk was associated with wine consumption, even up to twenty-one glasses per week. As for the California study, reduced incidence of heart disease accounted for most of the overall reduction in mortality risks, although reduced cancer risks were also observed. It is important to note that these and other epidemiological studies that show positive associations of wine consumption and reduced mortality do not account for other factors that may influence mortality, such as exercise, diet, other health habits, and genetics. Finally, it should be emphasized that heavy wine drinking may increase mortality risk (due, for example, to cirrhosis or automobile accidents).

Effect of wine consumption on heart disease

High rates of coronary heart disease within a population are associated with those groups that consume high saturated fat diets. Yet, in some countries, such as France and Italy, where saturated fat consumption is high, heart disease rates are significantly lower than in other countries where fat consumption is also high. This so-called French Paradox is thought to be due to the high levels of wine consumed in these countries (Renaud and de Lorgeril, 1992).

Given the strong epidemiological evidence showing an inverse relationship between coronary disease and wine consumption, what biochemical mechanisms would account for this finding? Several possibilities have been suggested (da Luz and Coimbra, 2004). First, alcohol consumption, in general, is known to increase the high-density lipoprotein (HDL) cholesterol concentration in blood, a factor known to reduce heart disease risks. Alcohol may also increase fibrinolytic activity and decrease platelet aggregation, which similarly reduce plaque formation. Wine, however, contains specific components (other than alcohol) that may also account for the reduction in cardiovascular disease. Among the compounds thought to be most important are polyphenols and flavonoids (Stoclet et al., 2004). These compounds have antioxidant activity and can scavenge superoxide and other reactive oxygen species. Thus, formation of oxidized low-density lipoprotein (LDL), which is known to be toxic to endothelial cells and impair vasorelaxation, are inhibited. Similarly, the antioxidant activity of wine flavonoids can prevent the reaction of superoxide and nitric oxide (NO). The latter has antithrombotic activity, prevents atherosclerosis, and inhibits platelet adhesion and aggregation, and, therefore, is necessary for normal endothelial function.

Another explanation for the role of wine in reducing coronary disease and endothelium-dependent constriction, in particular, was recently described (Corder et al., 2001). These investigators reported that polyphenols in red wine extracts suppress aortic endothelial synthesis of endothelin-1 (ET-1), a peptide that has vasoconstriction activity and that is intimately involved in development of atherosclerosis. This inhibition of ET-1 synthesis was specific for

(Continued)

Box 10–3. L'chaim! Healthful Properties of Wine *(Continued)*

red wine, because red grape juice was only 15% as inhibitory and white or rosé wine had less than 5% inhibition. The mechanism by which ET-1 synthesis is suppressed was not due to antioxidant activity, but rather to the specific inhibition of the tyrosine kinase cell signaling machinery involved in the initiation of ET-1 synthesis.

Effect of wine consumption on cancer

Based on epidemiological studies, consumption of alcohol, in the form of beer or spirits, generally increases cancer risk mortality, whereas moderate wine consumption appears to lower risks of some cancers (Bianchini and Vainio, 2003; Grønbæk et al., 2000; Klatsky et al., 2003). Several wine polyphenolic compounds, such as quercetin and catechin, have been shown to have inhibitory effects, *in vitro*, against tumor and cancer cells. Two other compounds found in red wine, resveratrol and acutissimin A, have recently attracted considerable attention as molecules that may have anti-cancer properties.

Resveratrol is a polyphenol that may contribute cardiological benefits (as do other wine phenols), in addition to its protective effects on carcinogenesis. Specifically, resveratrol is thought to interfere with the cell signaling cascades that result in the synthesis of NF-κB and AP-1. The latter are transcription factors that activate expression of genes involved in promotion of cell cycle progression, inflammation, anti-apoptosis, and tumorigenesis. Although the precise mechanisms responsible for the anti-carcinogenic and anti-inflammatory effects of resveratrol are not yet established, it appears that by blocking or modulating NF-κB and AP-1 synthesis, resveratrol down-regulates genes necessary for carcinogenesis.

Another component of red wine that has recently been recognized as having anti-cancer activity is acutissimin A (Quideau et al., 2003). Acutissimin A is a polyphenol with an ellagitannin structure (galloyl units linked to glucose) that was originally isolated from the bark of specific types of oak trees (although not from the oak used to make wine barrels). Its activity as a possible anti-cancer agent is due to its ability to inhibit DNA topoisomerase II, an enzyme involved in growth of cancer cells. In fact, acutissimin A is 250 times more inhibitory to this enzyme (*in vitro*) than etoposide, a drug used clinically to treat some cancers. Finally, it appears that wine, but not grapes, contains acutissimin A. Rather, the presence of acutissimin A in wine is due to its synthesis, during the aging step, from polyphenolic precursors found in grapes. Evidently, the conditions that exist during the aging of wine (i.e., low pH, ethanol, and mildly aerobic atmosphere) are necessary for the synthesis of acutissimin A.

Despite the mostly positive attention that the wine-health connection receives in the popular press, as well as in the scientific literature-gathering headlines, it would be misleading not to note that there are many detractors. Some investigators have argued that dietary diversity, rather than wine, is responsible for the French Paradox, that statistical biases underestimate coronary deaths in France, and that the risk/benefit ratio does not support wine consumption as a means of promoting cardiovascular health (Bleich et al., 2001; Criqui and Ringel, 1994).

References

Bianchini, F., and H. Vainio, 2003. Wine and resveratol: mechanisms of cancer prevention? Eur. J. Cancer Prev. 12:417–425.

Bleich, S., K. Bleich, S. Kropp, H.-J. Bittermann, D. Degner, W. Sperling, E. Rüther, and J. Kornhuber. 2001. Moderate alcohol consumption in social drinkers raises plasma homocysteine levels: a contradiction to the 'French Paradox?' Alcohol Alcohol. 36:189–192.

Corder, R., J.A. Douthwaite, D.M. Lees, N.Q. Khan, A.C. Viseu dos Santos, E.G. Wood, and M.J. Carrier. 2001. Endothelin-1 synthesis reduced by red wine. Nature 414:863–864.

Criqui, M.H., and B.L. Ringel. 1994. Does diet or alcohol explain the French paradox? Lancet 344:1 719-1723.

Box 10–3.   L'chaim! Healthful Properties of Wine *(Continued)*

da Luz, P.L., and S.R. Coimbra. 2004. Wine, alcohol and atherosclerosis: clinical evidences and mechanisms. Braz. J. Med. Biol. Res. 37:1275–1295.

Klatsky, A.L., G.D. Friedman, M.A. Armstrong, and H. Kipp. 2003. Wine, liquor, beer, and mortality. Am. J. Epidemiol. 158:585–595.

Kundu, J.K., and Y.-S. Suhr. 2004. Molecular basis of chemoprevention by resveratrol: NF-κB and AP-1 as potential targets. Mut. Res. 555:65–80.

Grønbæk, M., U. Becker, D. Johansen, A. Gottschau, P. Schnohr, H.O. Hein, G. Jensen, and T.I.A. Sørensen. 2000. Type of alcohol consumed and mortality from all causes, coronary heart disease, and cancer. Ann. Intern. Med. 133:411–419.

Quideau, S., M. Jourdes, C. Saucier, Y. Glories, P. Pardon, and C. Baudry. 2003. DNA topoisomerase inhibitor acutissimin A and other flavano-ellagitannins in red wine. Angew. Chem. Int. Ed. 42:6012–6014.

Pinder, R.M., and M. Sandler. 2004. Alcohol, wine and mental health: focus on dementia and stroke. J. Psychopharmacol. 18:449–456.

Renaud, S., and M. de Lorgeril. 1992. Wine, alcohol, platelets, and the French paradox for coronary heart disease. Lancet 339:1523–1526.

Stoclet, J.-C., T. Chaitaigneau, M. Ndiaye, M.-O. Oak, J.E. Bedoui, M. Chataigneau, and V.B. Schini-Kerth. 2004. Vascular protection by dietary polyphenols. Eur. J. Pharmacol. 500:299–313.

ring and functional groups at carbon 1 or 3 are referred to as nonflavonoids. In contrast, compounds containing two or more phenolic rings connected by pyran ring structures are called flavonoids. Both nonflavonoids and flavonoids can be polymerized to form another important class of phenolic compounds called tannins. Tannins consisting of flavonoid phenols are especially important in wine due to their bitter flavor and color stabilization properties.

Within the flavonoid group are several different types of compounds, including the flavonols, catechins, and anthocyanins. These are extracted from the grape skins and seed just after the grapes have been crushed. In general, the longer the wine is in contact with the seed and skin and the higher the temperature, the more of these substances will be extracted. In the manufacture of red wine, this material is left in contact with the pressed juice for several days, even during fermentation, whereas it is removed almost immediately after crushing for white wine manufacture (see below). Thus, although both red wine and white wine contain phenolic compounds, the concentration in red wines is usually at least four times higher than that of white wine. Importantly, many of these phenolics are responsible for pigmentation of grapes and,

subsequently, the color of the wine. Anthocyanins, for example, have red or blue-red color and make red wines red. Most of the anthocyanins are attached, via ester linkages, to one or two glucose molecules, which enhances their solubility and stability.

Wine Manufacture Principles

Making wine, as far as the actual steps are concerned, looks to be a rather simple and straightforward process (Figure 10-3). Grapes are harvested and crushed, the crushed material or juice is fermented by yeasts and bacteria, the organisms and insoluble materials are removed, and the wine is aged and bottled. In reality, the process is far from easy, and each of these pre-fermentation, fermentation, and post-fermentation steps must be carefully executed if high-quality wine is to be consistently produced.

Harvesting and Preparing Grapes for Wine Making

According to both viticulturists and enologists, the first step in wine making is considered to be one of the most important. Grapes must be harvested at just the right level of maturity. This

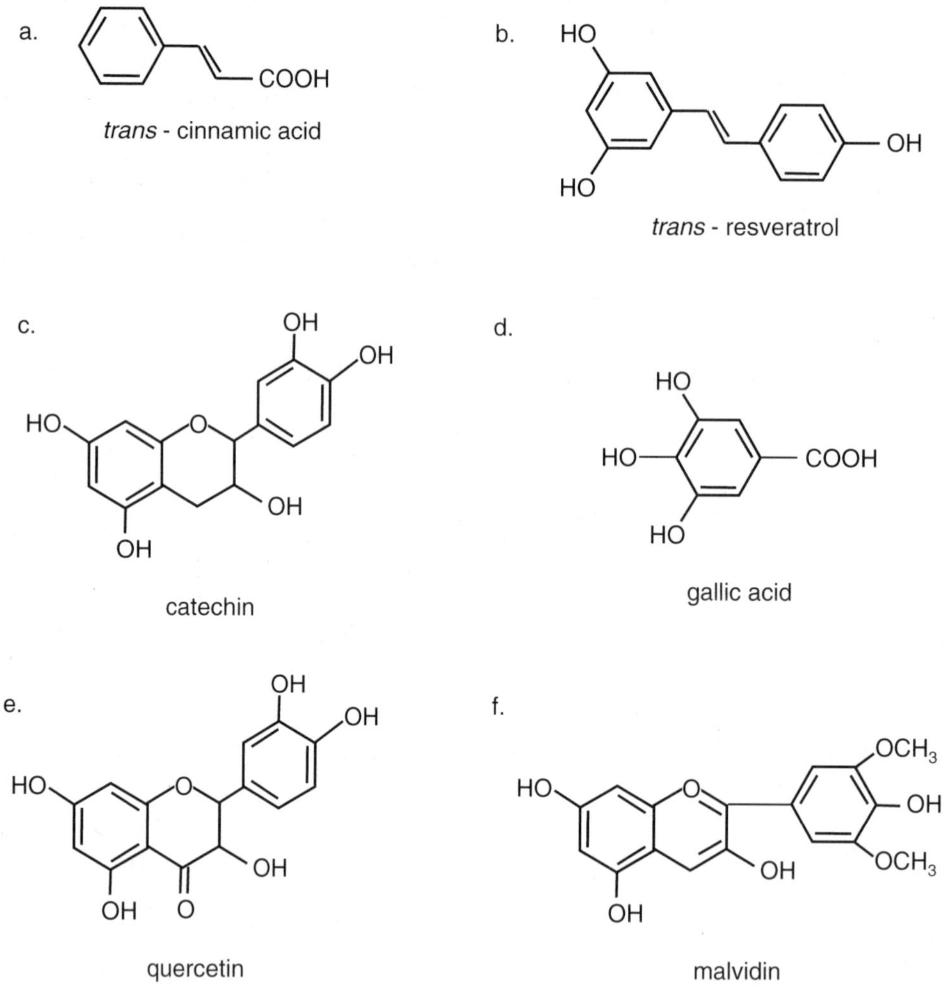

Figure 10–2. Representative phenolic compounds in wine. Examples include: a. cinnamic acid, a phenolic acid; b. resveratrol, a derivative of cinnamic acid; c. catechin, a flavan-3-ol; d. gallic acid, a tannin; e. quercetin, a flavonol; and f. malvidin, an anthocyanin.

means that the concentrations of sugars and acids (and the sugar/acid ratio), pH, the total soluble solids, and even the phenolic constituents must be at just the right level for the particular cultivar and the type of wine being made. In addition, berry size and weight also influence the time at which grapes are harvested. In general, grapes should be sampled sometime before their expected harvest time and their composition assessed (at minimum °Brix and pH should be measured) to make sure that over-ripening does not occur.

Unfortunately, there is no exact or objective set of rules to ensure or predict the optimum time for harvesting grapes. Rather, grapes are frequently harvested based on more subjective criteria. As grapes ripen on the vine, the sugar concentration, as well as flavor and color components, increase, and acids usually decrease, so identifying the correct moment for harvesting can be a real challenge. It is possible, moreover, for grapes to over-ripen, such that the harvested grapes contain too much sugar or too little acid or be too heavily contaminated with

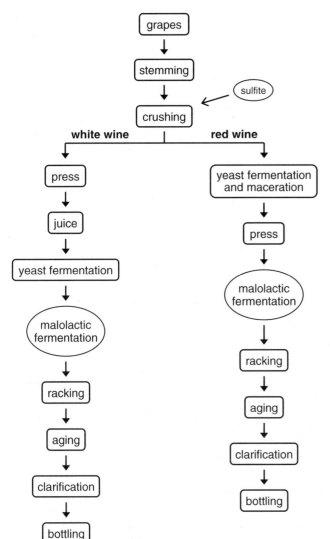

Figure 10–3. Flow chart of wine manufacture. The main difference between the manufacture of red and white wine is that the grapes for red wines are fermented in the presence of skins and seeds (maceration), whereas for white wines, these materials are removed after crushing. Although optional, sulfiting agents are added to most U.S. wines. Whether the malolactic reaction is induced or encouraged depends on the grape composition, the style of wine, and the manufacturer's preferences.

wild yeast and molds. Once the grapes have been deemed properly mature, it is essential that they be picked and harvested quickly, since the composition can continue to change.

Even in this twenty-first century, when so much of modern agriculture has become automated and mechanized, a sizable portion of grapes for wine making is still harvested manually. Only recently has mechanical harvesting begun to displace manual harvesting. In the United States, the majority of grapes are now harvested by mechanical means; however,

manual picking of grapes is still done for premium quality American wines and in much of Europe. Manual harvesting is more gentle on the grapes, and bruising and breaking of the grapes is minimized. For certain wines, such as sweet wines made from noble rot grapes, or wines for which grape harvesting methods are regulated (e.g., French Champagne), manual harvesting is required. Anyone who has seen the odd contours and steep terrains of some of European vineyards will also appreciate the necessity of manual grape picking. On the other hand, mechanical harvesters are faster and cheaper, and, unlike hired laborers, deployable on short notice and available around the clock.

Once the grapes are removed from the vines, they must be transported to the winery. It is important that the grapes not be bruised, crushed, or otherwise damaged either during harvesting or transport, since this encourages growth of microorganisms prior to the actual start of the fermentation. For the same reason, transportation time is also important.

Crushing and Maceration

The purpose of crushing is to extract the juice from the grapes. Before the grapes are crushed, however, leaves, large stems, and stalks are removed. Some wine makers may not remove all of the stems to increase the concentration of tannins and other phenolic compounds that are present in the stems and extracted into the juice. Once the extraneous material is separated and removed, the grapes are crushed by one of several types of devices. Roller crushers consist of a pair of stainless steel cylinder-shaped rollers. Another type of crusher, called the Garolla crusher, not only performs the crushing step, but also removes stems. It consists of a rotating shaft contained within a large horizontal stainless steel cylinder or cage. Arms on the shaft are attached to paddles or blades such that when the shaft turns, the grapes are moved and pressed against the side of the cylinder. Perforations on the walls of the cylinder allow for the juice (along with the skin, seeds, and pulpy material) to pass through into collection vats, whereas the stems gather at the end.

The crushed grape material, as noted above, contains juice, seeds, and skins. Pigments, tannins, and other phenolic compounds are located in the skins and seeds, and their extraction into the juice takes time. Endogenous pectinases and other hydrolytic enzymes within the grapes enhance extraction and must also be given time to work. This extraction step, where the crushed grape material is allowed to sit, is referred to as maceration.

Maceration conditions are not the same for all wines. For red wines, where pigment extraction is especially important, long maceration times at high temperatures are usually employed. In general, maceration is done at around 28°C for up to five days. The shorter the maceration times and the lower the temperature, the less material will be extracted. Thus, lighter red wines, such as Beaujolais, are macerated for just a few days at no higher than 25°C. In contrast, deeper red wines, such as Bordeaux, are macerated for up to twenty-eight days at 30°C. Since fermentation begins shortly after the grapes are crushed, maceration and fermentation essentially occur at the same time. In fact, the ethanol made by fermenting yeasts enhances extraction. This situation only occurs, however, if the musts are not treated with sulfur dioxide (see below).

As noted above, maceration at low temperatures (<15°C) ordinarily results in only moderate pigment extraction and little fermentation. However, if the must is macerated at a low temperature (between 5°C and 15°C), but for longer time, extraction of anthocyanins and aroma and flavor compounds can be enhanced. This technique, called cold maceration, simulates the natural conditions in cooler wine-producing areas, such as the Burgundy region of France. For white wines, the maceration step is done at lower temperature and for much less time. Typically, only a few hours at 15°C is sufficient. For most white wines, the producers remove the seeds and skins immediately after crushing. As for red wine, the maceration conditions used for white wines influence the amount of pigments and tannins that are extracted. Wines made from Sauvignon blanc grapes where little maceration occurs typically have a low phenolic concentration, whereas Riesling and Chardonnay musts, which are often macerated in the cold, may contain appreciable amounts.

Sulfur Dioxide Treatment

This is now a convenient point at which to discuss the use of sulfur dioxide and pure culture technique for wine making. As soon as the integrity of the grapes has been compromised by the crushing step, the sugars in the juice are liberated and made available for whatever microorganisms happen to be present. Ordinarily, the must is populated by epiphytic yeasts (that is, yeasts that reside on the surface of the grapes) and by yeasts that have "contaminated" the crushers, presses, and other wine-making equipment. Although the surface of a single grape may contain only about 10^2 to 10^4 yeast cells, after the grapes have been exposed to the contaminated equipment, the number of cells increases about 100-fold, to about 10^4 to 10^6 cells per ml. Whether this resident microflora actually commences a fermentation, however, depends on the intent of the wine maker.

Two options exist. First, a spontaneous or natural fermentation may be allowed to proceed ("natural" simply means that pure starter cultures are not added). In this case, except for temperature control, essentially no other restrictions are placed on the fermentation, and yeast (and bacterial) growth occurs with just a relatively short lag phase. The other option is to start the fermentation, under controlled circumstances, with a defined yeast starter culture selected by the wine maker. The latter option usually requires that the indigenous microflora be inactivated, so that it does not compete with and possibly interfere with the added culture. This is not, however, always the case, because it is still possible to add a starter culture even in the presence of the background flora. The use of wine starter cultures has now become commonplace, even among many traditional wine manufacturers. Moreover, recent research on yeast ecology suggests that there may be less diversity in wines made by spontaneous fermentations than previously thought and that these natural fermentations involve relatively few strains (Box 10–4).

When starter cultures are used, the naturally occurring or so-called wild yeasts are inactivated in one of two ways. First, the must can be

Box 10–4. Culture Wars and the Role of Diversity in Wine Fermentations

Like all of the fermented products described in this text, wines were made for thousands of years before scientists recognized that microorganisms were responsible for the fermentation. In contrast to dairy, meat, and other food fermentations, where first backslopping, and later pure culture techniques were applied, wine makers have long relied on the indigenous microflora to initiate and complete the wine fermentation. Even today, many of the wines produced by traditional manufacturers, especially those in Europe, are made by a "spontaneous" or natural fermentation carried out by the wild yeasts that are present on the grapes, in the must, or on equipment. Most large, modern wineries, however, have adopted yeast starter culture technology, and many have abandoned the traditional practice of allowing the wild yeasts to commence the fermentation.

The debate

Advocates of wine starter culture technology make several good arguments. First, they contend that using defined yeast strains results in wine having a "pure" or cleaner flavor. In addition, the availability of yeast with particular physiological characteristics such as osmotolerance or the ability to grow at low temperature allows the wine maker to select yeasts particularly suited to the grape characteristics one is using or the wine type one is making (Table 1).

Importantly, the wine maker can expect culture performance to be consistent in terms of fermentation times, flocculation properties, and flavor and end-product formation. In addition, culture yeasts can also be customized to provide traits and performance characteristics that suit the particular needs of the wine producer. Finally, there may be conditions that make pure cultures almost essential, such as when the grapes or musts contain very high sugar or acid concentrations.

(Continued)

Box 10–4.   Culture Wars and the Role of Diversity in Wine Fermentations *(Continued)*

Table 1.   Starter cultures versus natural fermentation for wine-making.

Starter cultures	Natural
Cleaner flavor	More complex flavor
Greater consistency	Unique qualities
Faster	Slower
Low frequency of stuck fermentations	Greater frequency of stuck fermentations
Can customize strains	Cannot customize
Immune to killer yeasts	May be sensitive to killer yeasts

Despite these advantages, it is undoubtedly true that naturally fermented wines have an excellent track record and many are highly regarded by wine experts. Even if one were to concede that the flavor, aroma, and other organoleptic properties are more variable in naturally fermented wines, is that such a bad thing? Certainly, one could argue that it is the variability itself that makes it possible to produce truly exceptional wines. Proponents of natural wine fermentations claim that such wines are more "complex" due to the more complicated metabolic processes that occur within a complex yeast flora. In other words, it is the diverse metabolic flora that leads to a more complex distribution of flavor components in the wine.

To address this issue of metabolic diversity on a rational basis, several questions must first be raised. How many different strains are actually present during a natural wine fermentation? Are some strains more dominant than others? What are the phenotypic properties of these strains, relevant to wine flavor generation? Several recent studies have attempted to answer these questions. In the report by Cappello et al. (2004), between thirty and forty naturally-occurring yeasts were isolated from each of twelve different musts. Although the musts were made from different cultivars (each obtained aseptically), the grapes were grown within the same vineyard. All of the yeast isolates were identified by classical methods and by genetic analyses as *Saccharomyces cerevisiae*, and, based on PCR amplification and mitochondrial restriction analyses, all of the strains within each of the twelve groups appeared to be genetically homogenous. Comparative analysis of representative strains from the twelve groups revealed that there were three genetically distinct groups. However, there were only minor differences in their physiological properties.

This study points out several important features. First, it suggests that the indigenous yeasts present in wine reflect the yeast population within the vineyard, rather than the grape. Second, it would appear that only a few strains, rather than a diverse collection of strains, dominate the wines produced within a vineyard. Finally, these dominant yeasts were well-adapted to the particular local environmental conditions and shared similar physiological properties relevant to wine making. However, because the yeasts were isolated at a single point in time (at the end of the fermentation), it is possible that other strains may have been present at the beginning or at some earlier time during the fermentation.

Another recent study addressed similar questions, but in these experiments the musts were analyzed at the beginning and end of the fermentation (Demuyter et al., 2004). Thus, strain diversity could be determined at two different stages. Over a three-year period, grapes from Alsace (used to produce Gewurztraminer wines) were obtained, crushed, and fermented under three different conditions. For each year, 100 yeast isolates obtained from the beginning and end of the three independent fermentations were identified and classified by polymerase chain reaction and pulsed field gel electrophoresis techniques. Grape and must handling conditions were slightly different for each of the three fermentations, such that the origins of the isolated yeasts could have been from the grapes, the press equipment, or the vat.

The results for the first-year grapes showed that nearly all of the grape-associated strains, both from the beginning and end, belonged to a single homogenous group of *S. cerevisiae*. In con-

Box 10–4. Culture Wars and the Role of Diversity in Wine Fermentations *(Continued)*

trast, the microflora from grapes exposed to crushing equipment or to the vat environment was more heterogenous and contained relatively few *S. cerevisiae* isolates. Instead, these musts contained mostly *Saccharomyces bayanus* subsp. *uvraum.* Results for the second-year grapes showed that *S. bayanus* subsp. *uvraum* was the dominant organism from all three environments at all sampling times. Third-year results for grapes or grapes exposed to crushing equipment indicated that *S. cerevisiae* was dominant at the beginning and end of the fermentation, but grapes exposed to the vat environment contained *S. bayanus* subsp. *uvraum* as the dominant organism (as was the case for the year one results).

Collectively, this study on Alsatian wine shows that a homogeneous group of *S. cerevisiae* strains appears to dominate the grape surface, yet other non-*S. cerevisiae* yeasts still show up in the wine. The latter are associated with both the crushing equipment or the vat environment. Moreover, under the conditions established in this study, those yeasts present in the vat environment may actually be the dominant yeasts in the wine fermentation, since they were present at the beginning and end of all three fermentations conducted over a three-year period.

Although results represent only two limited studies, they nonetheless suggest that the yeast population responsible for the wine fermentation had only modest diversity and that domination by a particular strain or strains may be common. Indeed, these findings are consistent with other ecological studies (Versavaud et al., 1995). Certainly, these reports also confirm previous reports that suggest wine crushing equipment and the wine environment serve as a major source of yeasts involved in the wine fermentation. Thus, it may well be that natural fermentations do not owe their special properties to a diverse population of wild yeasts that reside in the vineyard or on the surface of grapes, but rather to only a few yeasts that contaminate the wine-making equipment and that live in the confines of the winery environment.

References

Cappello, M.S., G. Bleve, F. Grieco, F. Dellaglio, and G. Zacheo. 2004. Characterization of *Saccharomyces cerevisiae* strains isolated from must of grape grown in experimental vineyard. J. Appl. Microbiol. 97:1274–1280.

Demuyter, C., M. Lollier, J.-L. Legras, and C. Le Juence. 2004. Predominance of *Saccharomyces uvarum* during spontaneous alcoholic fermentation, for three consecutive years, in an Alsatian winery. J. Appl. Microbiol. 97:1140–1148.

Versavaud, A., P. Courcoux, C. Roulland, L. Dulau, and J.-N. Hallet. 1995. Genetic diversity and geographical distribution of wild *Saccharomyces cerevisiae* strains from the wine-producing areas of Charentes, France. Appl. Environ. Microbiol. 61:3521–3529.

heat treated, usually via a high-temperature, short-time (or flash) pasteurization process. Although very effective against most organisms found in musts, even moderate heating is often detrimental to the juice and to the wine. Thus, this process is rarely used. The preferred method is to chemically pasteurize the must by adding sulfites. It should be noted that even naturally fermented wines, especially white wines (see below), are often sulfite-treated to control undesirable organisms. The most common sulfiting agents are SO_2 gas and potassium bisulfite salts. Sulfites are cheap, effective, and multi-functional. In addition to their effectiveness against wild yeast, these agents also inhibit growth of acetic acid-producing microorganisms, malolactic bacteria, and various fungi. Sulfites can thus be considered as serving a preservative function in wine. Importantly, they also control several deleterious chemical reactions, particularly oxidation and browning reactions.

The amount of sulfite added and when it is added varies depending on the condition of the grapes, the microbial load, must pH and acidity, and the type of wine (red or white). Musts from mature grapes that often contain high levels of wild yeast require more SO_2, but in general, about 80 mg/L is sufficient. Also, the lower the pH, the less sulfite is necessary for antimicrobial activity. Due to human health concerns, however, there are also regulations that dictate how much sulfite can be present in wine. In France, most red and white wines must contain less than 160 mg/L or 210 mg/L, respectively. In the United States, the limits are 350 mg/L for both red and white wines. Usually, SO_2 or sulfite salts are added to the must just after crushing.

When pure yeast cultures are used, they are usually obtained from commercial sources. So-called house strains, although common for beer, are infrequently used for wine making, due primarily to the expertise and equipment required for strain maintenance and propagation. In contrast, commercial cultures are easy to use and require only modest technical expertise. The main advantage of commercial cultures, however, is that they provide the wine manufacturer with many culture options. This is because a variety of strains is available, and the choice of strain can be selected based on the needs of the customer (Table 10-3). For example, grapes with consistently high sugar concentrations may require an osmotolerant yeast culture. Yeast used to perform the secondary fermentation that occurs in Champagne manufacture must sediment well to facilitate its removal during the disgorgement step (described later in this chapter).

Table 10.3.   Desirable properties of wine cultures.

Able to produce high levels of ethanol
Does not produce off-flavors
Capable of producing unique flavors
Ethanol tolerant
Osmotolerant
Ferment sugars to completeness
Sediments or flocculates well (especially for Champagne yeasts)
SO_2 and sulfite resistant
Cold resistant
Rapid fermentation rates
Able to grow at temperatures below 10°C
Predictable and consistent and genetically stable

Yeasts are usually supplied in the form of active dry yeast, not unlike that used by the bread industry (Chapter 8). Culture suppliers ordinarily recommend inoculum levels sufficient to provide a starting cell density of about 10^6 cells per ml of must. An increase of only about two log-cycles occurs during the fermentation. The manner in which the yeast cultures are produced and preserved is important, however, because how the cells are handled during the culture production steps may influence how they grow later in the juice.

When culture companies produce yeast cultures, cells are grown under highly aerobic conditions to achieve the maximum amount of biomass. These high biomass fermentations also elicit a general stress response that enhances survival and viability during the subsequent drying steps. It appears that resistance to drying is promoted by the synthesis of the disaccharide trehalose by wine yeast (similar to the cryoresistance mechanism by bakers' yeast, described in Chapter 8). Thus, following culture production, yeast cells are induced for aerobic metabolism and anabolic activities, whereas, when these cells are inoculated into the must, fermentative metabolism, with a minimum of biomass, is desired. Therefore, rapid adaptation to the new wine environment is also important.

Other Pre-treatments

It is permissible to add other materials, besides sulfites, to the must to enhance extraction, modify the composition, or promote fermentation. For example, pectic enzymes can added during crushing to facilitate extraction of juice from skins and later during pressing to improve clarification. This practice is quite common for white wines manufactured in the United States. As noted above, it is not uncommon for some grapes, and juices from those grapes, to have either too low or too high of a pH. Thus, either acids, such as tartaric acid, or neutralizing salts, such as calcium carbonate, can be added to adjust must pH. Finally, nutrients that enhance yeast growth and fermentation can be added. These yeast growth factors usually are added to white wine juices, since shorter extraction times result in lower nutrient concentrations. Nutrients added to juices include mainly ammonium salts and various vitamins.

Microbial Ecology and Spontaneous Wine Fermentations

In the absence of SO_2 addition, the indigenous microflora is relied upon to initiate and then carry out a spontaneous or natural fermentation. This is one of the most well studied of all fermentations, and much is now known about the ecology of wine and the yeasts that participate in the wine fermentation. In reality, however, the yeast fermentation is but one of two distinct fermentations that occur in wine making. Yeasts, of course, ferment sugars to ethanol, CO_2, and small amounts of other end products. A second fermentation, called the malolactic fermentation, is carried out by specific lactic acid bacteria that are either naturally present or added for this purpose. The malolactic fermentation, to be discussed later, is now regarded as nearly as important to wine quality as the ethanolic fermentation.

As noted above, the surface of grapes usually contain less than 10^4 yeast cells per grape (or per ml of juice). This number may increase during ripening on the vine, especially if the temperature is warm. Although ten or more yeast genera may be represented, the primary organism most frequently isolated from grape surfaces and the fresh must is *Kloeckera apiculata*. In contrast, *S. cerevisiae,* the yeast most responsible for the wine fermentation, is rarely observed on grapes. Rather, *S. cerevisiae* and other related strains are introduced into the must during grape handling and crushing steps directly from the equipment. The must is inoculated, in other words, by the yeasts originating from the grape surface as well as by those residing on the winery equipment.

Despite the large amount of available carbohydrates and other nutrients, the must is actually quite a selective environment. The pH is typically below that which many organisms can tolerate, and the organic acids present in the must have considerable antimicrobial activity. The high sugar concentration and resulting high osmotic pressure can also inhibit many of the indigenous organisms. Eventually, the CO_2 formed during the early stages of fermentation makes the environment anaerobic, restricting growth of aerobic organisms. Likewise, enough ethanol is produced to provide selection against ethanol-sensitive organisms. Finally, sulfites, if added, have antimicrobial activity against a wide range of yeasts and bacteria. Thus, as the environment changes, yeast species that are numerically dominant may be displaced by other species or strains better suited to the environment at any particular time. Microbial ecologists refer to such arrangements as successions. One reason why the outcome of a natural wine fermentation is so difficult to predict is that successions rarely occur exactly the same way in different musts, since the wild flora is never entirely the same.

Members of the genus *Saccharomyces* represent less than 10% of the initial yeast population. The early yeast population is dominated by *K. apiculata*. This organism produces up to 6% ethanol from glucose and fructose; however, it has a low ethanol tolerance, being inhibited by 3% to 4% ethanol. Thus, even though *K. apiculata* is among the first to grow in the must, its numbers are not usually maintained beyond the first several days. However, it is thought that this organism does produce small amounts of acids, esters, glycerol, and other potential flavor components, some of which may or may not be desirable. As the ethanol concentration increases and the Eh (or redox potential) is reduced, due to CO_2 formation, the environment begins to select for *S. cerevisiae* and other various ethanol-tolerant *Saccharomyces* spp.

According to current yeast taxonomy (Chapter 2), most wine yeasts are now classified simply as *Saccharomyces cerevisiae*. Thus, although different *Saccharomyces* strains may be present in musts during the course of a fermentation or in musts of varying composition (e.g., high sugar, low pH, etc.), these strains are all *S. cerevisiae*. It is interesting to note that

S. cerevisiae is especially well-adapted to the wine environment and appears to have been the predominant wine strain for thousands of years (Box 10-5).

Separation and Pressing

After the maceration step, or in the case of most white wines, almost immediately after crushing, the juice is separated from the seeds, skins, and pulp (collectively referred to as the pomace). For red wines, some fermentation will have already occurred prior to the separation step, whereas for white wine, fermentation follows the separation and clarification steps. The juice that separates from the pomace simply by gravitational forces is called the "free run." Screens are typically used to catch any large particles. The free run juice is pumped into vats or barrels.

Since the free run juice contains less than 75% of the total juice volume and the rest is present within the pomace, the latter is usually pressed to recover the remaining juice. Several types of presses and configurations are used. Hydraulic or pneumatic wine presses squeeze the juice from the pomace. Screw- or auger-type devices force the juice against perforated cylinder walls and have the additional advantage of being continuous. The so-called first press juice can be collected and either added back to the free run juice or kept as a separate portion. The free run fraction is considered to have an appreciably higher quality and is used for premium wines. Juices containing mixtures of free run and pressed fractions are used for lower quality wines. Finally, for white wine, the juice is clarified to remove any remaining solids. Clarification is done via either settling and decantation, filtration, or centrifugation.

Fermentation

The wine fermentation begins as soon as the grapes are crushed. However, when a starter culture is used and SO_2 is added to control the indigenous organisms, limited ethanol fermentation will occur prior to addition of the culture.

In the case of white wine production, the culture is added to the must after pressing and clarification, whereas for red wine, culture addition is done prior to seed and skin removal. Thus, for red wines, fermentation occurs during maceration, just as it would for a natural fermentation. The amount, concentration, and form of the culture depends on the type of wine being produced, the composition of the grapes, and other considerations specific to the wine manufacturer. Of course, the culture's main responsibility is to produce ethanol from sugars (see below), but the criteria for culture selection actually includes many other properties (Table 10-3). For most wines, the culture inoculum, whether in a rehydrated dried or active liquid form, (Chapter 3), should provide about 10^6 cells per ml of must.

Traditionally, fermentations were performed in open barrels or vats with a capacity of 500 L or less. Such barrels still are used today; however, enclosed stainless steel tanks are now more common. The latter have several advantages. They are easy to clean and disinfect and often can be sterilized. Airborne microorganisms are less likely to contaminate the wine. Various control features, including temperature control and mixing and pumping activities (see below), are easily incorporated into the design, and are usually computerized. Finally, modern tanks can be quite large, with capacities of more than 250,000 L.

The temperature of incubation depends on the type of wine being produced. In general, white wines are fermented at lower temperatures than red wines. For example, many wineries control the temperature between 7°C and 20°C for white wine and between 20°C and 30°C for red wines. Some wineries prefer low incubation temperatures for all wines, because less ethanol is lost to evaporation and fewer of the volatile flavors are lost. In addition, low temperature incubations result in overall higher ethanol concentrations and less sugar remaining at the end of the fermentation (assuming time is not a factor). Although *S. cerevisiae* has a lower growth rate as the temperature decreases, the diversity of

Box 10–5. Molecular Archaeology and the Origin of Wine Yeast

How is it that the fermentation of beer, wine and bread all require yeasts with specific physiological and biochemical traits, yet all are made with *Saccharomyces cerevisiae*? Was there a progenitor *S. cerevisiae* strain that somehow evolved over thousands of years into specialized strains that were especially suited for these different fermented foods? If so, where did that progenitor strain arise and what were its properties? These are obviously questions that cannot be answered directly. However, it is possible, using molecular archaeology analyses, to make inferences regarding the evolution of these important fermentative yeasts (Cavalieri et. al., 2003).

As noted previously, wine making appears to have developed in the Caucasus and Zagreb Mountain areas of the Near East (the "Fertile Crescent") more than 7,000 years ago. Ancient clay jars (estimated date about 3150 B.C.E.) from these regions once contained wine, according to chemical and instrumental analyses. Furthermore, DNA residues were found when organic residues (presumably the lees, comprised of dead yeast cells) from these artifacts were analyzed. In fact, the amount of DNA present in the residue was high enough (it could even be seen in a gel) to eliminate the possibility that it had come from a stray microbial contaminant. Given the low water activity, low relative humidity (essentially zero), and overall ideal conditions for storing a biological material, it was also not surprising that the DNA was in such good shape.

The DNA was subsequently extracted and used as a template for PCR. Primers were based on *S. cerevisiae* rDNA sequences, such that the spacer regions between the 18S and 28S rRNA genes would be amplified. Three PCR products (540, 580, and 840 base pairs) were initially sequenced. Based on a BLAST search, all of the sequences had homologies to existing GenBank sequences. The sequence of the 580 bp PCR product was similar to various fungi, but remarkably, there was strong homology (nearly 90% identity) between the 540 bp sequence and a similarly derived sequence obtained from a fungal clone isolated from the clothing of "The Iceman Ötzi." The latter dates back to 3300 B.C.E. The investigators suggest that both of the ancient fungi had been "buried" along with the wine.

Perhaps the most interesting finding from this analysis, however, concerned the 840 bp PCR product. This fragment, as well as smaller pieces obtained using internal PCR primers, had very high sequence similarity with a 748 bp region from chromosome 12 (part of the 5.8S rDNA) from contemporary strains of *S. cerevisiae*, *Saccharomyces bayanus*, and *Saccharomyces paradoxus*. Across this entire region there were only four nucleotide mismatches and no deletions or insertions between *S. cerevisiae* and the ancient wine sequence. This level of similarity should not be so surprising, perhaps, since the DNA regions represented by the PCR products are known to be quite stable. Thus, this stability could account, in part, for the lack of sequence alterations between the ancient and modern strains during the past 5,000 years of yeast evolution. Importantly, these findings suggest the yeast used to carry out the wine fermentation in ancient days was *S. cerevisiae,* just as it is today.

References

Cavalieri, D., P.E. McGovern, D.L. Hartl, R. Mortimer, and M. Polsinelli. 2003. Evidence for *S. cerevisiae* fermentation in ancient wine. J. Mol. Evol. 57:S226–S232.

metabolic end products may actually increase, enhancing flavor development. In addition, lower temperatures may favor growth of *K. apiculata* and other wild yeasts (at least when they are not inactivated by SO$_2$) that produce various volatile compounds, making the wine aroma and flavor appear more complex.

It is critical to recognize that the wine fermentation is exothermic and a considerable amount of heat may be generated. Some of this heat is gradually lost or dissipated into the environment without ill effect. However, in large volume fermentations, in particular, much of this heat is retained, raising the temperature of the wine. For example, if the initial temperature starts at 20°C, the temperature can increase 10°C or more. If the temperature were to rise above 30°C, the yeast may become inhibited or stop growing altogether. The wine will contain less ethanol and more residual sugar.

Such fermentations are said to be "stuck." Although other factors may cause a wine fermentation to become stuck (see below), high temperature is the most common reason. It is, therefore, essential that the appropriate temperature is maintained. For fermentations conducted in modern, stainless steel, jacketed vats, coolant solutions can easily be circulated, externally. Alternatively, internal cooling coils can also achieve the same effect. The cooling requirement can also be met, especially for white wines, by simply locating fermentation barrels in cold rooms or cellars. However, the need for adequate cooling has led even some traditional wine manufacturers to abandon oak barrels and casks in favor of stainless steel vats.

The actual fermentation period is not long. After culture addition, the yeasts enter a short lag phase (from a few hours up to a day or two) that is then followed by a period of active growth (log phase) that lasts for three to five days. If the fermentation is conducted at lower temperatures (10°C to 15°C), the lag and log phases can be extended for several days. Conversely, if the yeast culture is highly active at the outset, by virtue of having been previously propagated under ideal growth conditions, the cells will almost immediately enter log phase. Although one might expect that a natural fermentation would take longer, in fact, growth of the indigenous yeast begins so soon after crushing, that the lag phase is barely noticeable.

During the log phase of growth, when an active fermentation is occurring, a layer of CO_2 forms across the surface. In red wine production, some of the pomace will float to the top and be trapped within this CO_2 layer, forming a dense blanket or cap. Since the pigments and tannins are present in this thick cap layer, a mixing step is required to return these substances back into the fermenting

must. The temperature in the cap can also become elevated, supporting growth of undesirable thermophilic bacteria; thus, mixing serves to maintain a more uniform temperature. Various techniques exist for this mixing step (called pigeage). In the "pumping over" technique, a portion of the must is periodically removed from the vat and pumped onto the cap. Alternatively, the cap can be "punched down," either manually or via mechanical means. In modern wineries, automated pumping and punching systems are used to mix the pomace cap into the fermenting must.

The fermentation of white wine occurs after the must is pressed and clarified. After about the seventh or eighth day of fermentation, cell numbers may begin to decline, representing the end of the primary ethanolic fermentation. Some yeast strains will flocculate or clump together, causing them to settle to the bottom of the tank. This is a desirable property that enhances their removal later during the racking and clarification processes. The fermentation is considered complete when all or most of the sugars are depleted, as determined by a decrease in the Brix value. For red wines this may take as long as five to six weeks. If less than about 0.5% sugar is present, and there is no apparent perception of sweetness, the wine is considered to be dry. Of course, not all wines are intended to be dry. If some sugars are present in the wine after fermentation (or if sugars are added), a sweet wine results. Specialized techniques for the manufacture of sweet wines will be described later.

For red wines, much (if not all) of the fermentation occurs in the presence of the pomace. When the extraction of pigments, tannins, flavor compounds, and other materials is considered sufficient, or when the desired ethanol concentration is reached, the free run juice is separated from the pomace and moved into another tank. The fermentation is then completed (if not already). In the meantime, the pomace is pressed and is either fermented separately from the free run or is mixed with the free run for the final fermentation. Since the pressed wine is rich in pigments and tannins, adding a portion back to the free run wine makes the final product richer in color and flavor.

Yeast Metabolism

It should be evident by now that the main job of the yeasts during wine manufacture is to produce ethanol from the sugars present in the juice. However, if ethanol was the only product formed and if sugars were the only substrates metabolized by the yeast, then wine flavor and aroma would be sorely lacking. In fact, yeast growth and fermentation results in a myriad of metabolic end products that contribute, for better or worse, to the organoleptic properties of the finished wine.

The main wine yeast, *S. cerevisiae*, as well as other species found in natural fermentations, are facultative anaerobes. They have the full complement of enzymes necessary to oxidize sugars via the Kreb's or tricarboxylic acid (TCA) cycle, as well as enzymes of the Embden-Meyerhoff (EM) glycolytic pathway. Metabolism of sugars by wine yeasts is not, however, simply a choice of the TCA versus the EM pathway, but rather depends on the environmental conditions, particularly the redox potential or E_h.

At the start of the wine fermentation, the E_h of the must is positive, and metabolism by either the natural flora or added culture organisms is mostly by the TCA cycle, which is strongly favored under aerobic conditions. Sugar metabolism by the TCA cycle yields CO_2 and H_2O as end products, with a large increase in cell mass. The consumption of oxygen and production of CO_2, however, quickly reduces the E_h such that anaerobic conditions eventually develop. Metabolism then shifts to the glycolytic pathway, yielding equimolar concentrations of ethanol and CO_2, and a relatively modest increase in cell number. Both pathways also generate ATP, but aerobic metabolism yields a net of 36 ATP molecules per glucose compared to only 2 molecules of ATP made per glucose during anaerobic glycolysis. A third route for sugar metabolism, the pentose phosphate pathway, also exists, but it is used mainly in the anabolic direction, for biosynthesis of pentoses.

The biochemical details for sugar metabolism by *S. cervisiae* are now well known. As noted above, environmental conditions dictate which metabolic pathway is chosen. Regulation occurs both at the enzyme level, as well as genetic level. Moreover, the absolute requirement for maintaining redox balance within the cell also has a major influence on specific metabolic reactions and the end products that are formed.

For both the TCA and EM pathways, glucose and fructose metabolism begins with their transport into the cell via specific transport systems. For glucose, at least two such systems exist. One is a low affinity (high Km), facilitated transport system, driven simply by the concentration gradient. It operates when the glucose concentration is high (i.e., at the beginning and during much of the fermentation). A second system has a low Km (high affinity) for glucose, is energy-requiring, and functions only at low substrate concentrations (i.e., at the end of the fermentation, when most of the glucose is depleted).

Next, the accumulated glucose is phosphorylated by one of several kinases to form glucose-6-phosphate, which is then isomerized to fructose-6-phosphate. In the case of fructose, transport is also via a facilitated system (not energy-requiring), and the intracellular fructose is directly phosphorylated to fructose-6-phosphate by a hexokinase. The fructose-6-phosphate is phosphorylated a second time to form fructose-1,6-bisphosphate (FDP), which is split by an aldolase to form triose-phosphates, and eventually pyruvate is formed.

During the glycolytic pathway, a net of two molecules of ATP are synthesized per hexose by the substrate level phosphorylation reactions. Since the glyceraldehyde-3-phosphate dehydrogenase reaction generates NADH from NAD, there must be some means of replenishing the oxidized form of NAD. Otherwise, not enough NAD would be available for this reaction, and metabolism would bottleneck. Under aerobic conditions, the pyruvate is decarboxylated by pyruvate dehydrogenase and the product, acetyl CoA, feeds directly into the TCA cycle. The latter pathway generates reducing equivalents in the form of NADH and FADH, which are subsequently oxidized via the electron transport system. The electron recipient is molecular oxygen, NAD is re-formed, and ATP is subsequently synthesized by oxidative phosphorylation via the ATP synthase reaction. When all is said and done, each molecule of glucose yields 36 of ATP.

In contrast, during anaerobic metabolism, the enzymes of the TCA cycle are not expressed because the absence of oxygen causes repression of the genes that code for those enzymes. However, since glucose and fructose are still metabolized to pyruvate by the glycolytic pathway, generating NADH, an alternative means of regenerating NAD is necessary. Recall that lactic acid bacteria, facing the same situation, use pyruvate itself as the electron acceptor, generating lactic acid via the enzyme lactate dehydrogenase. The route taken by *S. cerevisiae* involves two reactions. In the first, pyruvate is decarboxylated via pyruvate decarboxylase, in a reaction that requires the cofactor thiamine pyrophosphate. Carbon dioxide and acetaldehyde are formed. The acetaldehyde then serves as the electron acceptor, in a reaction catalyzed by NADH-dependent alcohol dehydrogenase, resulting in production of ethanol and NAD. No additional ATP is made by these two reactions, indicating their function is solely to replenish NAD.

Factors Affecting Yeast Metabolism

S. cerevisiae has the genetic capacity to metabolize sugars via either the glycolytic or respiratory (i.e., TCA) pathways. Although oxygen availability affects expression of genes encoding enzymes of these two pathways and is, therefore, an important determinant of which way metabolism will occur, gene expression is also regulated by substrate availability. Although one might expect that in the presence of oxygen, metabolism would always be via the respiratory pathway, this is not the case. If the glucose concentration is sufficiently high, metabolism will be fermentative. This is because transcription of catabolic genes, including genes coding for some of the TCA enzymes, is repressed by glucose, a phenomenon known as the glucose or Crabtree effect. Thus, in the wine fermentation, where sugar concentrations are especially high, metabolism of glucose yields mostly ethanol and CO_2, and oxidative metabolism of sugars is unlikely to occur. Only when the substrate concentration is low (less than 2 g/L), will O_2 repression of glycolysis occur (the so-called Pasteur effect).

Since the wine fermentation, as noted above, is anaerobic, one might reasonably expect that the only end products formed from glucose and fructose would be ethanol and CO_2. If that were the case, then how could one explain the appearance of glycerol, succinic acid, acetaldehyde, acetic acid, and other products, that appear during the wine fermentation? In addition, some of the glucose carbon (albeit only about 1%) is used to form biomass (i.e., cell constituents). Moreover, even after accounting for evaporative effects, the theoretical 50% yield (on a weight basis) of ethanol from glucose is never reached during the fermentation, due to byproduct formation. These byproducts can account for as much as 4% of the total products formed. Synthesis of these byproducts occurs, in part, in response to demands on the cell to maintain Eh balance and to salvage ATP. For example, when the demand for NAD is high, a portion of the dihydroxyacetone phosphate formed from FDP via the aldolase reaction is reduced to glycerol-3-phosphate, and NAD is generated. The glycerol-3-phosphate is subsequently dephosphorylated to glycerol.

Aerobic metabolism, at least at the very start of the fermentation, may also result in products other than ethanol, in particular, TCA intermediates, such as succinic acid. Even when conditions are anaerobic, TCA products may still be formed to provide the cell with carbon skeletons (e.g., α-ketoglutarate) necessary for biosynthesis of amino acids. Another factor influencing sugar metabolism is SO_2 added to control the indigenous yeast population. Sulfur dioxide can bind acetaldehyde, preventing its reduction to ethanol by alcohol dehydrogenase. As a result, NADH accumulates inside the cells and is diverted to other NAD-generating reactions, forming glycerol and other non-ethanol end products. Finally, end product formation is also influenced by the yeast strains naturally present or added to the must.

Sulfur and Nitrogen Metabolism

Although metabolism of carbohydrates is obviously critical to the outcome of the wine fermentation, metabolism of other must components is also important. How wine yeasts metabolize sulfur-containing compounds that are present in the must as normal grape constituents is particularly important. Most of the sulphur in grapes is in the form of elemental sulfur, sulfates, or as sulfur-containing amino acids. Since the range of sulfur-containing metabolic end products includes various sulfides, mercaptans, and other volatile compounds, sulfur metabolism can have a profound influence on wine quality. Yeasts can also produce sulfites, which, as already mentioned, have antimicrobial activity. In fact, even if sulfur dioxide or sulfite salts are not intentionally added, wine invariably contains sulfite due to its production by wine yeast.

Grapes usually contain sufficient ammonia, ammonium salts, and free amino acids to support good growth of most wine yeasts. In addition, wine yeasts can synthesize amino acids and purine and pyrimidine nucleotide from ammonia, and, therefore, have no essential requirement for amino nitrogen or pre-formed nucleotides. During the course of the fermentation, total nitrogen decreases. However, nitrogen-deficient grapes can result in an inadequate amount of nitrogen in the must, and, if not supplemented with ammonium salts, can lead to reduced fermentation rates.

Stuck Fermentations

Despite the apparent simplicity of the wine fermentation, as evidenced by so many successful outcomes, there are occasions when the fermentation fails. Such wines typically contain a significant amount of residual, unfermented sugar and an insufficient concentration of ethanol. The fermentation may actually still be occurring, just more slowly, or it may be at a complete standstill. These slow or halted fermentations are referred to as sluggish or stuck fermentations, respectively. Although they occur infrequently, they represent a significant problem and source of economic loss for the wine producer, since wine quality will inevitably be poor. Stuck wine is especially susceptible to spoilage, since the low ethanol concentration and the availability of fermentable sugar may promote growth of undesirable bacteria and yeasts. Under more severe conditions, the wine may simply have to be discarded. Therefore, despite their rare occurrence, it is important to understand the causes and to know how to prevent sluggish or stuck wine fermentations.

Among the possible causes of a stuck fermentation are those that are due to the must composition, the handling of the must and wine, or the presence of wild yeasts that inhibit desirable wine yeast. The must may contain, for example, an insufficient level of nitrogen or other nutrients necessary to support adequate yeast growth. The sugar concentration in the grapes or must may be too high, resulting in osmotic pressures that inhibit the yeasts. Some yeasts also are inhibited by high ethanol concentration. As noted previously, the ethanol fermentation is exothermic, and if the temperature is not controlled or cooling is inadequate, the resulting high temperature (>30°C) may cause the fermentation to come to an abrupt halt. In contrast, too cool an incubation temperature (e.g., <10°C), as might occur during white wine production, can also result in a stuck fermentation. Most of these situations, however, are easily corrected, either by supplementing the juice with appropriate yeast nutrients, using osmotolerant or ethanol-tolerant yeast strains, or by proper temperature control.

Finally, some wild yeast strains secrete proteins called killer toxins that inhibit or kill other indigenous or starter culture yeasts. There are more that five types of killer toxins, but the most common are K1 and K2. These proteinaceous toxins first attach to cell wall receptors, then integrate and form pores within the cytoplasmic membranes of sensitive cells, thereby disrupting ion gradients and interfering with energy-transducing reactions. Producer strains resist the toxin they produce, but are sensitive to those toxins produced by other strains. Some yeast strains do not produce killer toxins, but are nonetheless resistant. Several yeast genera, including species of *Hansenula*, *Pichia*, and *Saccharomyces,* are able to produce killer toxins. Importantly, so-called killer yeasts are found throughout the wine environment, and even many of the *Kloeckera* and *S. cerevisiae* strains isolated from natural wine fermentations have the "killer" property. In fact, since yeast strains with killer toxin activity can potentially inactivate competing strains, this trait may be desirable for yeast starter cultures (Box 10-6). Certainly, starter culture strains that are immune to these toxins would not be affected by other killer strains and would not be the cause of a stuck fermentation.

Adjustments, Blending, and Clarification

After the fermentation is complete, the wine will contain little or no sugar and about 12% to 14% ethanol. Still, because of differences in

---

Box 10–6. Killer Yeasts

Wine spoilage is often mediated by growth of wild yeasts that contaminate wine while it is aging in wooden barrels. In particular, species of the *Brettanomyces* and *Dekkera* groups are frequently involved in wine spoilage and are difficult to control. This is because these yeasts are naturally present not only on grapes and in musts, but also in the barrels in which the wine is aged (Comitini et al., 2004).

Growth of *Dekkera* and *Brettanomyces* may result in formation of ethylphenolic compounds that have unpleasant, "barny" or "wet-dog" off-odors. They are also a possible cause of the "mousy" off-odor defect (the smell of which, one can imagine). Control measures ordinarily include attention to cleaning and sanitation at the harvest and crushing steps and adequate sulfiting of the grapes. It is more difficult, however, to control *Brettanomyces* and *Dekkera* once the wine reaches the wooden barrels, which cannot be effectively sterilized.

Many yeast strains can produce and secrete exotoxins that can kill other yeast cells. There are numerous differences, depending on the producer, between the many so-called killer toxins that have been characterized (Magliani et al., 1997). They vary with respect to their genetic basis, how they are synthesized and exported, their size and structure, and their mode of action. Genes encoding for killer toxins, for example, may be located on linear plasmids, nuclear gene regions, or double-stranded, virus-like RNAs. In addition, the killer activity in sensitive cells may be due to pore formation and ion leakage, the arrest of G_1 or G_2 phases of cell growth, and the inhibition of cell wall synthesis (Magliani et al., 1997).

Among the yeasts that are known to produce killer toxins are several genera associated with the wine fermentation, including *Saccharomyces*, *Kloeckera*, *Pichia*, and *Kluyveromyces*. Although killer yeast strains are immune to the cognate toxin they produce, they may be sensitive to toxins produced by other strains. If wild killer yeast strains are present during the wine fermentation, and the wine starter yeast (whether part of a pure starter culture or are naturally-occurring) are susceptible to that toxin, a failed or stuck fermentation may result. Thus, it is essential that the killer sensitivity phenotype for wine cultures be known, since, in most natural wine fermentations, killer yeasts can be found (Vagnoli et al., 1993).

Box 10–6. Killer Yeasts *(Continued)*

In contrast to the detrimental effects killer yeasts may have during wine fermentations, they can also be exploited in wine fermentations. In fact, killer yeasts are now routinely used in the wine industry, just as they have been used in other industrial processes to control wild yeasts. Killer toxins have also been shown to be effective in clinical applications to treat fungal and bacterial infections (Buzzini et al., 2004; Conti et al., 2002; Palpacelli et al., 1991; Walker et al., 1995). Yeast cultures with a killer phenotype have the potential to dominate a fermentation. Provided that the killer strains have other desirable wine-making traits, the use of such cultures would be expected to produce a consistent quality wine, due to the inhibition of wild yeast competitors. Killer toxin genes can be introduced into non-killer wine-making strains by mating, protoplast fusion, transformation, or micro-injection techniques.

A wine preservation strategy based on the application of killer toxins also was recently reported (Comitini et al., 2004b). As noted above, spoilage of wine by *Brettanomyces* and *Dekkera* during barrel aging is a serious and difficult-to-control problem. In the study by Comitini et al. (2004b), however, it was shown that these yeasts can be controlled by specific killer toxins. Two toxins in particular, Pikt and Kwkt (produced by *Pichia anomala* and *Kluyveromyces wickerhamii*, respectively), inhibited species of *Brettanomyces* and *Dekkera*. Their activities were stable within a typical wine pH range of 3.5 to 4.5. When the producer strains were grown in co-culture with *Dekkera bruxellensis* (at a 1:10 ratio), growth of the latter was delayed; however, when inoculated at higher rate (1:1), complete inhibition occurred. Finally, these researchers showed that direct addition of the killer toxins to wine effectively inhibited growth of *D. bruxellensis* for at least ten days.

Because not all killer toxin-producing yeasts grow well in wine (e.g., *P. anomala* and *K. wickerhamii*), and it is unlikely that killer toxin preparations could be added directly to wine, other means of applying killer toxin technology must be considered. As noted above, killer toxin genes could be genetically introduced into other wine-adapted yeast strains, but it would be necessary for these strains be present during the aging step. It is also conceivable that the toxins could be used to treat the interior surfaces of wine barrels and thereby inactivate the *Brettanomyces* and *Dekkera* that reside there.

References

Bizzini, P., L. Corazzi, B. Turchetti, M. Bratta, and A. Martini. 2004. Characterization of the *in vitro* antimycotic activity of a novel killer protein from *Williopsis saturnus* DBVPG 4561 against emerging pathogenic yeasts. FEMS Microbiol. Lett. 238:359–365.

Comitini, F., N. Di Pietro, L. Zacchi, I. Mannazzu, and M. Ciani. 2004a. *Kluyveromyces phaffii* killer toxin active against wine spoilage yeasts: purification and characterization. Microbiol. 150:2535–2541.

Comitini, F., J.I. De, L. Pepe, I. Mannazzu, and M. Ciani. 2004b. *Pichia anomala* and *Kluyveromyces wickerhamii* killer toxins as new tools against *Dekkera/Brettanomyces* spoilage yeasts. FEMS Microbiol. Lett. 238:235–240.

Conti, S., W. Magliani, S. Arseni, R. Frazzi, A. Salati, L. Ravanetti, and L. Polonelli. 2002. Inhibition by yeast killer toxin-like antibodies of oral streptococci adhesion to tooth surfaces in an *ex vivo* model. Mol. Med. 8:313–317.

Magliani, W., S. Conti, M. Gerloni, D. Bertolotti, and L. Polonelli. 1997. Yeast killer systems. Clin. Microbiol. Rev. 10:369–400.

Palpacelli, V., M. Ciani, and G. Rosini. 1991. Activity of different 'killer' yeasts on strains of yeast species undesirable in the food industry. FEMS Microbiol. Lett. 84:75–78.

Vagnoli, P., R.A. Musmanno, S. Cresti, T. di Maggio, and G. Coratza. 1993. Occurrence of killer yeasts in spontaneous wine fermentation from the Tuscany region of Italy. Appl. Environ. Microbiol. 59:4037–4043.

Walker, G.M., A.H. McLeod, and V.J. Hodgson. 1995. Interaction between killer yeasts and pathogenic fungi. FEMS Microbiol. Lett. 127:213–222.

grape composition, microflora, and wine manufacturing practices, variations in wine composition and sensory quality are to be expected. Therefore, adjusting the wine after fermentation (and sometimes before) is a normal step. The pH and acidity, in particular, can vary markedly, as can the color and flavor. Therefore, some wineries adjust the acidity of wine by acidification or deacidification steps (e.g., by adding acids or neutralizing agents).

When acidity of wine or must is due to malic acid, deacidification is managed via the malolactic fermentation, which is discussed in detail later in this chapter. Adjustment to wine color and flavor can also be done, if legally permitted, by filtration and enzyme treatments, respectively. Filtration techniques, for example, are most often used to "decolorize" wine by removing undesirable pigments.

Except for very small wineries, which may have only a few vats of wine, most modern wineries have many individual vats of wine. Each one is unique, in that a particular vat may contain wine made from grapes harvested at a time or place different from the grapes in a neighboring vat. Wines within a single winery may be made from different grape varieties. Therefore, another common procedure, especially for premium wines, is to blend different wines to optimize or enhance the organoleptic properties. Blending also produces wines with consistent flavor, aroma, and color from year to year. Perhaps more so than any other wine-making step, however, blending is a tricky business, and is a highly subjective process. Success relies on the imagination, creativity, and skill of the wine blending specialist.

At the end of the fermentation, the wine contains non-soluble proteins and protein-tannin complexes, as well as living and dead microorganisms. These materials give the wine a cloudy, hazy, undesirable appearance. The clarification step removes these substances from the wine without removing desirable flavor and aroma components. It is particularly important that the cells are removed. If left in the wine, these cells can lyse, releasing enzymes that may catalyze formation of off-flavors and odors (although an exception to this rule exists, as described below). Inducing precipitation of tartrate salts and tannin-protein complexes is also commonly done to facilitate their removal before they precipitate later during aging. It is important to recognize that, in some cases, cell lysis may be a good thing. Intracellular constituents released during cell lysis include amino acids and nucleotides, providing nutrients that are later used by bacteria in secondary fermentations or that contribute to the sensory properties.

According to traditional practices, wine is clarified by simply allowing the sediment, containing the yeasts and bacterial cells, as well as precipitated material, to settle naturally in barrels or vats. The wine could then be removed from the sediment (or "lees") by decantation. This process, called "racking", is usually done for the first time after three to six weeks following the end of the fermentation. Racking can be repeated several times over a period of weeks or months until the wine is nearly crystal clear. During the racking step, the wine is also aged (see below). Racking can now be done in enclosed tanks using automated transferring systems.

Filtration is another method used to clarify wine. This can be especially effective if fining agents, such as bentonite, albumin, or gelatin, are used as filtration aids. If micropore filtration membranes are used, it is even possible to sterilize wines. Clarification may occur after racking or after aging.

In contrast to removing the sediment shortly after the fermentation has ended, some wines are intentionally left in contact with the lees for an extended time before the first racking occurs. This traditional maturing practice, known as "sur lies," enhances the flavor, character, mouth feel, and complexity of the wine. It is more common for white wines than red.

Aging

Aging actually begins just after fermentation. Thus, aging occurs when the wine is racked, as well as beyond. Aging conditions vary considerably. Some wines are aged for several years, whereas others are "aged" for only a few weeks. Some wines are aged in expensive oak barrels, others in stainless steel, and yet others depend on bottle-aging, or a combination of all of the above. Whether a wine is aged for a long time in oak barrels or is quickly bottled and sent to market depends, in part, on marketing considerations, but also on the original composition of the grapes and how they are made into wine. Thus, long, careful aging should be reserved for only premium wines made from high quality grapes. By analogy, Cheddar cheese manufactured for the process cheese market cannot be expected to develop into a flavorful, two-year Cheddar, no matter how carefully it may have been aged.

Of course, some excellent quality wines, like excellent cheeses, are meant to be consumed in a "fresh" or un-aged state, so whether or not a wine is aged does not distinguish wine quality, per se. For example, Beaujolais nouveau, a popular wine from the Burgundy area in France, is meant to be drunk after only a few weeks after the grapes are harvested. These wines are fruity and "gulpable"; no amount of aging will lead to their improvement.

To say that the actual events that occur during aging are complicated would be quite the understatement. Hundreds of enzymatic, microbiological, and chemical reactions occur, and as many as 400 to 600 volatiles, including esters, aldehydes, higher alcohols, ketones, fatty acids, lactones, thiols, and other compounds are formed (Table 10-4). The wine interacts with the wood and wood constituents

in the barrel, oxygen in the air, and even the cork. It is important to recognize that not all of these reactions are beneficial in terms of wine quality, and some wines may actually deteriorate during aging. In fact, long aging is not good for most wines.

Ordinarily, in large wineries, fermentations occur in large tanks (exceeding 250,000 liters), and then the wine is moved into wooden 200 liter barrels for aging. However wine can also be aged, at least initially, directly in tanks, and then later moved into oak barrels for final aging.

In many European and other traditional wineries, in contrast, the entire aging period is conducted in oak barrels. The oak barrel or "cooperage" is so important to wine quality that entire industries devoted to oak tree production and cooperage construction have developed. This is because the oak barrels are not inert containers used simply to store wine, but rather they are a source of important flavor and aroma compounds. In fact, one of the major steps in barrel construction involves heating or "toasting" the barrels to promote pyrolysis. This generates a number of flavor and aroma volatiles. In the presence of wine, these compounds, along with tannins, phenolics, lignins, and lactones, are extracted from the wood and solubilized in the wine. Some of these compounds impart unique flavor notes, including vanilla and coconut. Aging wine in oak cooperage is not, however, without a downside. Oak barrels are expensive, and, even if carefully maintained, do not last forever. Loss of wine volume (and hence profit) due to evaporation can also occur. Thus, alternative

Table 10.4. Effects of aging on wine.

Reaction or step	Effects
Tannin precipitation	Color darkens; astringency increases initially, then decreases
Wood cooperage	Phenolic and other flavors extracted
Ester hydrolysis	Fruitiness decreases
Oxidation	Browning and flavor reactions induced
Evaporation	Concentration of nonvolatile solutes; color and flavor intensifies, but aroma volatiles decrease

materials, in particular, stainless steel, have displaced oak cooperage in many wineries. While it certainly does not contribute flavor and aroma compounds, stainless steel is less expensive, easier to clean and maintain, and can be fabricated to accommodate size and shape preferences. It is still possible for wine aged in stainless steel cooperage to obtain desirable oak-derived flavors by adding oak shavings or chips to the aging wine.

Malolactic Fermentation

A certain amount of acidity is expected and desirable in wine. Red wines typically have a pH of 3.3 to 3.6; white wines are usually slightly more acidic. Some grapes, and the musts made from those grapes, however, may contain high levels of organic acids, such that the pH is too low (i.e., <3.5). Wines made from those grapes will suffer from excess acidity, a serious and readily noticeable flavor defect.

Of the organic acids ordinarily present in grapes, malic acid is particularly important because of its ability to influence pH. This is because malic acid is a four carbon dicarboxylic acid, meaning it contains two carboxylic acid groups and can release or donate two protons. Thus, musts containing high concentrations (0.8% to 1.0%) of malic acid are acidic and have a low pH. High malic acid concentrations are especially common in grapes grown in cooler, more northern climates, such as those in Oregon, Washington, northern California, and New York. Although many of the vineyards in Europe are located in warmer regions and produce grapes with less malic acid (a situation that may lead to the opposite problem—too little acidity), grapes from Germany, Switzerland, and even some regions in France can still contain significantly high malic acid levels. Also, some grape cultivars ordinarily contain more malic acid than others.

One way to reduce the malic acid levels and to "deacidify" the wine is to promote the biological decomposition of malic acid. This deacidification process occurs via the malolactic fermentation pathway that is performed by

specific species and strains of lactic acid bacteria. These bacteria may be naturally present in wine and may, therefore, initiate the fermentation on their own. It has now become common to add selected malolactic strains, in the form of a pure culture, directly to the must.

The malolactic fermentation has been the subject of extensive research in the last two decades. The pathway was initially thought to involve two separate reactions (Figure 10-4A), in which malate is first decarboxylated by an NAD-dependent decarboxylase yielding pyruvate and reduced NADH. In the second reaction, pyruvate serves as the electron acceptor yielding lactic acid. Is it now known that the reaction is catalyzed instead by a single malolactic enzyme that decarboxylates malic acid directly to lactic acid (Figure 10-4B). Although this enzyme requires NAD (and manganese), no intermediate is formed. The net effect of the malolactic reaction is that malic acid, a dicarboxylic acid, is converted to lactic acid, a monocarboxylic acid, thereby reducing the acidity of the wine.

Several lactic acid bacteria have the enzymatic capacity to perform this fermentation. The most well-studied are species of *Oenococcus*, particularly *Oenococcus oeni* (formerly *Leuconostoc oenos*). Several species of *Lactobacillus* also have malolactic activity. Although some of these bacteria are found naturally in musts, commercial cultures are now available and are commonly used.

The malolactic fermentation was one that interested microbiologists, not only because of its industrial importance, but also because there seemed to be no obvious reason for bacteria to perform this conversion. In other words, how do malolactic bacteria benefit, or more to the point, how do they gain energy, from the conversion of malic acid to lactic acid? The pathway, after all, contains no substrate level phosphorylation step that would lead to ATP formation, nor is there a change in the redox potential. As it turns out, there is a means of generating ATP via the malolactic pathway, but it is indirect.

As shown in Figure 10-5, malic acid is transported into the cell via one of two ways. In

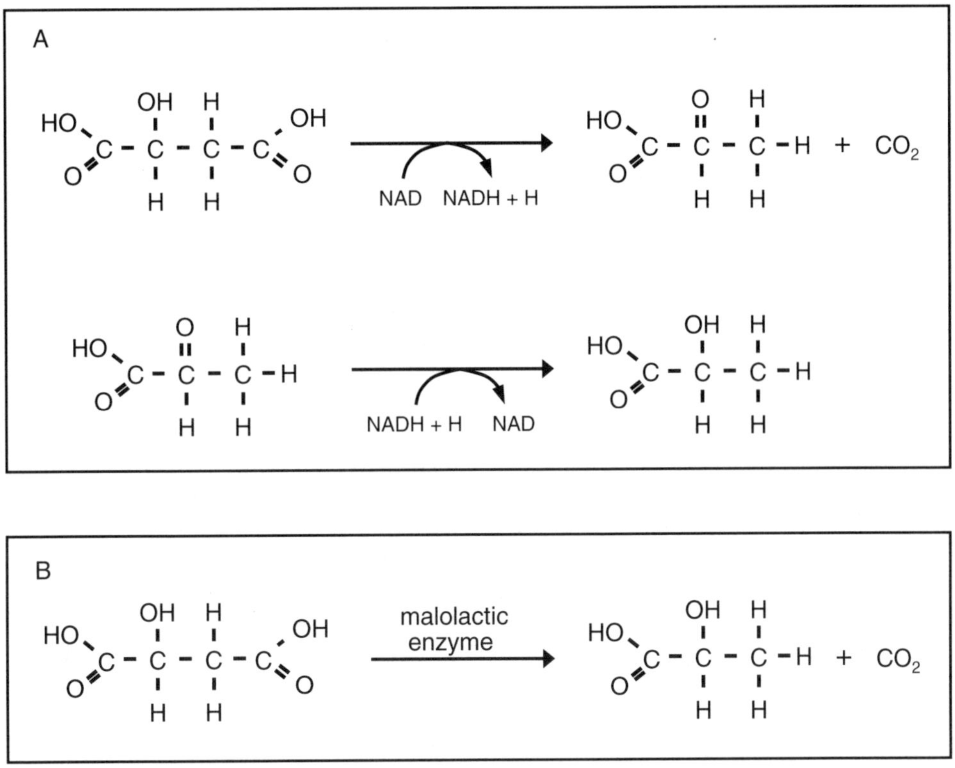

Figure 10–4. The Malolactic Reaction. The conversion of malic acid to lactic acid was originally thought to be catalyzed by two separate reactions, with pyruvic acid formed as an intermediate, and with NAD required as a co-factor (A). It is now well established that the malolactic reaction occurs as a single step, with lactic acid formed directly from malic acid via a decarboxylation reaction catalyzed by malolactic enzyme (B). No intermediate is formed.

Oenococcus, the malate permease is a uniporter, whereas in lactococci and lactobacilli, uptake of malate is mediated by an antiporter that exchanges an incoming malic acid for an outgoing lactic acid. No energy is spent for malate transport in either system. The exchange reaction, however, is not electroneutral. This is because one extracellular molecule of malic acid, carrying a net electric charge of -2, is exchanged for one of lactic acid that carries a net charge of only -1. This charge difference arises as a result of the decarboxylation reaction, which consumes an intracellular proton. The bottom line is that the cell is able to extrude a proton (or its equivalent) without having to spend energy to do so. That is, the cell conserves energy, in the form of ATP, that it

would ordinarily spend to pump protons from the cytoplasm to the extracellular medium. The proton gradient that forms, or the proton motive force, can then either perform other work (e.g., nutrient transport) for the cell or be used to drive ATP synthesis by the proton-translocating $F_0 F_1$ ATPase.

It is important to note that in low-acid grapes, the malolactic fermentation is undesirable, since some acidity is desired in wine. Thus, under some circumstances, the presence of naturally occurring malolactic bacteria is undesirable and the source of potential defects (see below). In contrast, the malolactic fermentation not only is performed for deacidification, but also to promote flavor stability and balance. Moreover, malolactic bacteria often

A

B

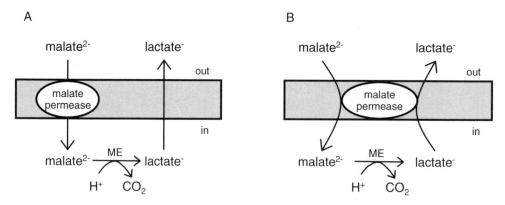

Figure 10–5. Malate transport in wine lactic acid bacteria. Transport of malate by *Oenococcus oeni* is mediated via a uniport malate permease (A). The accumulated malate is then decarboxylated directly to lactate by malolactate enzyme (ME); lactate efflux occurs via passive diffusion. The malolactic reaction also consumes a proton, such that the cytoplasm becomes more alkaline and more negatively charged. Thus, the electochemical gradient across the membrane (the proton motive force) is increased, and ATP is conserved. In lactobacilli and lactococci, a similar process occurs (B), except that malate permease acts as an antiporter, such that lactate efflux drives malate uptake. Adapted from Konings, W.N. 2002.

produce diacetyl from citrate, which may, at the appropriate concentration (generally between 1 mg/L and 4 mg/L), be desirable in some wines.

Types of Wine

Aside from the rather broad distinction of classifying wines based on their color—red, white, or rose—there are obviously more descriptive means for grouping different types of wine. Similarly, the procedures used for manufacture of wine have, so far, been described in a mostly generic manner. In this section, the manufacture of several well-known categories or styles of wine will be discussed, including sweet wines, fortified wines, sparkling wines, and distilled wines.

Sweet wines

Sweet wines are simply those that contain unfermented sugar (either fructose, glucose, or sucrose). There are several ways to produce a sweet wine. The easiest (and least expensive) way is to simply add sugar (up to 2% to 4%) or sucrose syrup to a dry wine. This results in a

wine that is definitely sweet, but not necessarily one that would be acceptable to many consumers. A similar and more common approach is to add unfermented juice, preferably from the same grapes used to make the wine. Alternatively, sweet wines can be made by stopping the fermentation before all of the glucose and fructose have been fermented. In some cases, sugar may be added to the juice prior to fermentation, such that when the fermentation is complete, residual sugar (and sweetness) remain. This is one of the more common practices in the United States, and many of the sweet wines from New York state are produced this way.

In either case, arresting the fermentation is the key step. Generally, this is done by rapidly cooling the wine, and then filtering out the yeast. One recent innovation is to encapsulate the yeasts within alginate beads, which are placed inside permeable bags. At the desired time, the bag (containing the yeast) is simply removed (see Chapter 3 for encapsulated cultures). Culture activity can also be stopped by adding alcohol spirits, as is the case for fortified wines (see below). Finally, sweet wines can be manufactured via one of several traditional techniques. These are based on concentrating the sugar in the juice or grapes. The juice, for example, can be partially concentrated by heating. In contrast, the grapes can be dried via atmospheric drying, or frozen so that the ice can be removed. However, the most well-known traditional technique for concentrating sugars in grapes is to dehydrate the grapes while they are still on the vine. These so-called botrytized wines are made throughout Europe, with the Sauternes, white wines from the Loire Valley of France, being the most widely prized.

The traditional process for making sweet botrytized wines starts in the vineyard. Grapes left on the vine past their optimum harvesting will invariably be infected by fungi, including the ubiquitous fungal organism *Botrytis cinerea*. Depending on climatic conditions, there can be two outcomes to this infection. If the climate is sunny and moderately dry during the day but humid during the evening, *B. cinerea* will grow to just the right extent on the surface of the grape, resulting in what is referred to as "noble" rot (Figure 10–6). If conditions are not "just right," overgrowth of *B. cinerea* or infection by other fungi can occur, resulting in simply rotten grapes that have no value. Thus, the manufacture of these wines carries considerable risk.

Ultimately, noble rot growth of *B. cinerea* results in dehydration of the grapes and concentration of the sugars. Dehydration occurs due to secretion of pectinolytic enzymes by *B. cinerea* that degrade the pectin-containing cell walls of the grapes. Water evaporation then occurs, provided the atmosphere is sufficiently dry. The decrease in moisture is important not only because it concentrates the solids, but also because it controls growth of undesirable fungi. Eventually, the mold-covered grapes begin to take on the appearance of raisins (moldy ones at that). The moisture level of the grapes will be reduced by about 50% and, even though the mold metabolizes some of the sugars in the grapes, the sugar concentration in the juice is still increased by 20% to 40%. After the grapes are harvested and gently crushed, portions of the free run and the pressed juice are combined. The fermentation is then generally the same as for other white wines, with a temperature range of 18°C to 20°C. At the end of the fermentation, the ethanol concentration will range from 9% to 13%, and the wine will contain as much as 10% total unfermented sugar.

Fortified wines

Fortified wines are those to which distilled spirits (containing as much as 95% ethanol) are added to raise the total ethanol concentration to 15%. Not only do these wines contain higher concentrations of ethanol, the source of the ethanol (e.g., brandy) is also important since they may contribute unique flavor compounds to the finished product. Aside from this common feature, however, a variety of quite different fortified wines exist. Included are whites and reds, dry and sweet. Fortification usually occurs during or just after the fermentation. In some wines, the added ethanol may inhibit the yeast and prevent the complete fermentation of sugars, resulting in sweet dessert wines. The most well-known of the fortified wines are sherry and port.

Figure 10–6. Grape clusters infected with *Botrytis cinerea*, the so-called noble rot. Photo courtesy of David Mills, University of California, Davis.

Sherry originated in Spain around 200 years ago, and is now produced in the United States and throughout the world. One of the main features used in the traditional manufacture of Spanish sherry (but rarely outside of Spain) is a blending technique known as the solera system. According to this technique, wines of different ages are progressively and uniformly blended to achieve wines of consistent quality. Another unique feature of most types of sherry is the presence of film yeasts that grow on the surface of the wine while it is aged in special casks called "butts." The *S. cerevisiae* strains that comprise this film, called a flor, are thought to be the same yeasts involved in the primary alcoholic fermentation, although it has also been suggested that a separate group of yeasts actually comprise the flor yeast population. In either case, the secondary yeast growth occurs under aerobic conditions, resulting in oxidation of some of the ethanol and subsequent formation of unique flavor and aroma components (e.g., esters, aldehydes, higher alcohols, and lactones).

In contrast to the traditional (and costly) solera system for manufacture of sherry, other techniques are now widely used in the United States and other countries. In the submerged flor procedure, the wine is aerated and mixed in fermentors to enhance growth of the flor yeast. Baked sherries are made by heating fortified but non-aged wine to 50°C to 60°C for ten to twenty weeks. A number of flavor and aroma compounds, including acetaldehyde and furfural, are generated by baking due to caramelization and non-enzymatic browning reactions.

Port, first produced in Portugal, is another fortified wine. It is made by adding distilled wine (i.e., brandy) to the red base wine, thereby raising the ethanol concentration to nearly 20%. Since fortification occurs well before all of the sugars have been fermented, and this much ethanol inhibits yeast, this step effectively ends the fermentation. Thus, Port contains as much as 10% sugars. Port and other fortified sweet wines made in this manner (e.g., Madeira and Marsala) are often referred to as dessert wines. Because fortification not only arrests the fermentation, it

also reduces the time available for extraction of anthocyanin pigments and phenolic flavor compounds. Mixing pressed wine (pomace) with the free run wine, therefore, is necessary to provide adequate color and flavor in the finished wine. In the United States, heat (e.g., steam) is sometimes used to extract pigment from the grapes. Finally, as for sherry, blending and aging are also important steps for port production. Some ports are aged for as many as fifty years.

Sparkling wines

Sparkling wines are those which contain carbon dioxide, providing bubbles and effervescence. For some sparkling wines, CO_2 pressures as high as 600 kPa atmospheres can be reached (by comparison, the pressure inside a can of soda pop is less than 200 kPa). Although sparkling wines are made throughout the world, there are several manufacturing methods that are used to produce the CO_2, and these methods define, to a certain extent, the type of sparkling wine being produced.

Clearly, the most well known sparkling wine is Champagne, which is traditionally made, not surprisingly, via the Champagne method. For other sparkling wines, CO_2 is introduced by other methods, as described below. It is worth mentioning that Champagne, like Parmesan and Roquefort cheeses, enjoys a protected status in the European Union. Only wine produced in the Champagne region of France— just to the southeast of Paris—and made according to the centuries-old process and aged in the caves in that region can be called Champagne. All other carbonated wines, regardless of how CO_2 is introduced, must be referred to as sparkling wine. In contrast, wine makers in the United States, which are under no obligation to follow EU rules, can label any sparkling wine as a champagne.

The manufacture of sparkling wines starts out no differently than other wines. The main difference is that at some point carbon dioxide is introduced into the "still" (i.e., not carbonated) wine. There are four possible ways to perform this step. The first is to simply add CO_2

directly to the wine, followed by bottling. This is not unlike soda pop production and is definitely the most economical means. At about $5 a bottle, this "champagne," although bearing little resemblance to the real thing, is certainly the favorite of budget-minded celebrants.

The other processes all involve a second ethanolic-CO_2 fermentation performed by yeasts (the exception, however, are those wines that contain CO_2 as a result of the malolactic fermentation). However, in contrast to the primary fermentation, which is done in ambient atmosphere (i.e., open), the second fermentation occurs in an enclosed environment such that the CO_2 is trapped and becomes supersaturated in the wine. One seemingly simple way to initiate such a fermentation is to stop the primary fermentation before all of the sugar has been fermented, and then to bottle the wine and allow the remaining sugar to be fermented by the residual yeast. The difficulty with this method is that stopping and starting the fermentation is not always so easy. Still, this method has long been practiced in France, Germany, and Italy.

Rather than rely on the endogenous sugar and yeast from the primary fermentation, an alternative and more common means of initiating the second fermentation involves the direct addition of an additional source or "dose" of sugar, as well as an inoculum of yeast, to the still wine. The yeast need only ferment a modest amount of substrate (usually sucrose) to generate enough CO_2 to make the wine bubbly. The secondary fermentation can be performed either in enclosed vats or directly in bottles. For either method, however, there remains the problem of the residual yeast, which, if not removed, will make the wine cloudy or turbid. Sparkling wines, especially champagnes, are prized for their absolute clarity and perfect appearance. Thus, removing the yeast following the second fermentation is absolutely necessary. Several methods have been devised to perform this step.

In the bulk or Charmat method, the second fermentation occurs in a pressurized vat, then the wine is filtered to remove residual yeasts and filled into bottles under pressure. In the transfer process the wine is placed into thick-walled bottles, sugar and yeasts are added, and the bottle is corked. After the second fermentation in the bottle is complete, the wine is transferred into tanks, filtered to remove yeast and then re-bottled. In the Champagne method, the process is similar to the transfer system, except that no transfer occurs. Rather, the second fermentation occurs in the bottle, where the wine remains. How then, if the wine does not leave the bottle, can it be clarified and made free of residual yeasts? The answer to this question and other unique features of Champagne manufacture are described below.

Champagne

Champagne is made from Chardonnay, Pinot Noir, and Pinot Meunier grapes grown in the Champagne district. This is a northern grape-growing region and the still wines made from individual cultivars (the Pinots make red wines and the Chardonnay is used for white) are not particularly remarkable (some might call them insipid). However, when the base wines are appropriately blended (a skill first perfected by the monk Dom Pérignon), the wine assumes the best qualities of each individual cultivar. Manufacture starts, as for other white wines, with a rather fast pressing of the grapes, such that pigment extraction is minimized. The first press juice, called the cuvée, is then inoculated with selected yeasts (these are almost always proprietary strains), and the juice is fermented at about 18°C in oak barrels (stainless steel vats are also used). Because the base wines can be rather acidic (<pH 3.0), a malolactic fermentation may be desirable and the appropriate cultures can be added (or growth of indigenous malolactic strains encouraged). However, promotion of the malolactic fermentation in Champagne is not a universally accepted practice, because some bacteria-generated flavor products may affect the delicate flavor and aroma balance of the finished wine.

After fermentation to about 12% ethanol, the critical blending step is performed (using high quality wines, free of defects). Base wines from a single year are blended to create vintage Champagne, whereas wines from different years are more often used to create non-vintage Champagne. Next, the blended wine is bottled, the "tirage" or sugar solution (about 20 g to 25 g of sucrose per liter) and a yeast mixture is added, and a cork closure is inserted. A fastener or clamp is placed over the cork to prevent its displacement. The bottles are then held at about 10°C to 12°C. The yeast inoculum for the secondary fermentation usually contains strains of *S. cerevisiae* (formerly *Saccharomyces bayanus*) that are selected based on their ability to grow at high ethanol concentrations and low pH and temperature. The bottles are also constructed differently from ordinary wine bottles (thicker, with a tapered neck), since they must be able to withstand the high CO_2 pressure.

When the secondary fermentation is complete (after about seven weeks), the wine may be allowed to age in the bottles for up to three years. Aging or maturation occurs while the lees are still present in the wine, a critical step that distinguishes Champagne from other sparkling wines fermented by the bulk method. Next, the time to remove the yeast cells (most of which are now dead) and sediment material has arrived. This step, called riddling (or remuage), involves gradually shifting the position of the bottle, starting from near horizontal and ending up near vertical. At the end of this daily or every-other-day jolting, turning, and twisting process (about one to three months), the sediment will have collected within the neck of the bottle, specifically settling on the inside of the cork. Although the riddling process is still done manually in several of Champagne houses, mechanical riddling is now common.

Once the sediment has settled on the cork, it is removed by a process called disgorgement. First, the now vertical bottles (neck down) are cooled to about 7°C, then conveyed through a freezing solution (at −20°C) such that the material in the neck, not more than about 2 cm above the cork, is quickly frozen. The bottle is inverted to a 45° angle (neck up), and the frozen plug, containing the cork, sediment, a small amount of slushy wine, is removed. About 10 to 50 ml of a sugar-wine solution (called the dosage) is then quickly added to replace the wine lost in the disgorgement process, and a fresh cork is applied and fastened with a wire "crown." The amount of sugar in the dosage, from less than 1.5% to more than 5%, determines whether the champagne is dry (brut), medium-dry (sec) or sweet (doux). The bottle is gently shaken to mix the wine and dosage, and the bottles are then stacked horizontally for up to three months. Champagne does not improve much beyond this aging period, and is essentially ready for consumption.

Brandy

Brandy is produced by distilling wine. The wine can be made from other fruits, but when made from grapes, white wines are used as the base. The most well known brandy is Cognac, made from the Cognac district of France. In the United States, brandy must conform to a standard of identity that describes the starting fruit or juice, the ethanol concentration, the duration of aging, and other compositional and manufacturing details. Most American beverage-type brandies contain less than 50% ethanol (100° proof). In contrast, brandy used for fortification purposes usually contains 70% to 95% ethanol (140° proof to 190° proof). Due to evaporation, long aging can substantially reduce the ethanol concentration.

In general, the base wine production follows that for other white wines. Skins and seeds are removed immediately after crushing and pressing to minimize pigment extraction. Distillation can occur as soon as the wine is fermented. Two types of distillers or stills are used, pot stills and continuous stills. The pot still, used for Cognac, consist of a single pot or boiler that is directly heated. A portion of the vapor is condensed in a reflux condenser and returned to the boiler, and the remaining vapor condenses and is collected. After distillation, the brandy is aged in oak barrels, which results in extraction of oak flavor, aroma, and color compounds. Most American brandies are aged for two to three years; however, some Cognacs and other premium brandies are aged for more than twenty years.

Wine Spoilage and Defects

Wine making is fraught with risks. It can reasonably be said that as many things that can go right during wine manufacture, just as many can go wrong. The risk is exacerbated by the considerable investment that must be made to produce the wine (thus the old joke: How does one make a million dollars in the wine business? Start with $2 million). Grapes must be cultivated over several seasons before a reasonable crop can be harvested. Disease, climate, insects, and other factors can cause serious problems even before the first grape has been harvested and crushed. Once wine making begins, growth of desirable yeast and bacteria and inhibition of others is not always easy to manage. Finally, aging of wine may result in a product ranging from truly spectacular to totally undrinkable.

Spoilage of wine, like other fermented (and even non-fermented) foods, is often due to chemical and physical activities. Oxidation reactions, whether induced by oxygen, sunlight, or metals, can be especially damaging to wine, leading not only to development of off-flavors and aromas, but also to discoloration and pigmentation defects. High temperatures are also detrimental to wine quality, catalyzing chemical and biochemical reactions that result in caramelization, browning, and oxidation reactions. Physical phenomena, such as precipitation of proteins and tannins, are responsible for cloudiness and haze formation. A particular type of haze, called casse, occurs in white wines and is caused by precipitation of metallic salts. Despite these various chemical-physical defects, however, wine spoilage is most commonly caused by microorganisms. In fact, fungi, yeasts, and bacteria can all be responsible for spoilage at some point during the wine manufacturing process.

Spoilage by fungi

Fungal growth and spoilage rarely occurs during the wine fermentation, since most fungi are aerobic and sensitive to ethanol. Rather, molds are most important before and after the wine is made. Fungal growth on grapes is one of the most serious problems encountered in grape viticulture, causing considerable loss of crop. If not controlled, rots, mildews, and other fungal diseases can wipe out an entire vineyard. As noted earlier for sweet botryized wines, a fine line sometimes separates spoiled, rotten grapes from desirable, noble rot growth of *B. cinerea*. Some fungi, such as *Penicillium*, *Aspergillus*, *Mucor*, and *Rhizopus*, can grow on freshly harvested grapes during transport to the winery. Pesticides, sulfiting agents, and other antimycotics can be applied to help control this problem, but care must be exercised to minimize their impact on the wine and during fermentation.

Post-fermentation problems with mold are usually due to contaminated cork closures. Cork, the bark from the cork tree, can contribute so-called cork taints. This is one of the most serious defects in bottled wine and one which has attracted wide attention among cork producers, wineries, and consumers (Box 10-7). The defect is now thought to occur as a result of growth of various fungi (including *Penicillium, Aspergillus,* and *Trichoderma*). Visible mold growth is rarely evident (those corks would not be used), so cork taints occur earlier in the cork making process, when exposure to fungi is common. These fungi then produce musty- or mushroom-smelling compounds that diffuse into the wine after the corks are inserted into the bottles. Wine corks can now be treated to remove the offending taint. However, there has been a strong trend toward the use of plastic and other alternative wine enclosures, the use of which may significantly reduce the incidence of cork-related defects.

Box 10–7.   Cork Taints in Wine: Are Corks the Cause?

Imagine you've just purchased an expensive bottle of wine and are anticipating the fine bouquet about to be released from the soon-to-be opened bottle. Upon removing the cork, however, you discover that the wine has an obnoxious musty and moldy aroma that renders the wine undrinkable. As unfortunate as this scenario might be, it is by no means rare. In fact, cork taint, the most likely cause of this particular defect, is considered to be the most common cause of wine spoilage, affecting as much as 12% of all wine (this spoilage rate has been disputed, however, as discussed below).

How often consumers actually return or even reject cork-tainted wine is not known, but it seems likely they will remember their experience with that particular brand when they purchase their next bottle of wine. Not surprisingly, given the frequency with which this problem occurs, cork taint has become one of the most important (and perhaps most controversial) issues facing the wine industry. Annual economic losses of more than $10 billion have been estimated (Peña-Neira et al., 2000). Therefore, understanding what causes cork taint and developing strategies to prevent it are the focus of considerable academic and industry-sponsored research.

Cork technology and microbiology

Corks have been used as closures for wine bottles at least since the 1600s. Not only are cork stoppers effective at preventing leaking, they also permit a very small amount of gas exchange that is believed to promote flavor and aroma development in aged wine. Cork is obtained from the bark of the *Quercus suber* oak tree, which grows in regions around the western Mediterranean Sea. Portugal is the main producer of cork (and has the biggest stake in the cork taint problem).

The manufacture of cork follows a traditional process that starts with its harvest (Jackson, 2000). The first, or virgin bark, stripped from the tree when it is about twenty to thirty years old, is not used; neither is the second growth obtained after another seven to ten years. Bark is not cork-worthy for another nine years, when it is called reproduction bark (subsequent strippings can occur every nine years). The harvested cork slabs are then boiled, cut into strips, and the cylindric cork units are punched out and washed in water. Next, the corks can be treated either with a hypochlorite solution or with peracetic acid. Both of these treatments inactivate microorganisms, although they are not without important side reactions (see below). The corks are dried, sorted, and packaged. The name of the winery or vintage can also be etched onto the corks.

Among the microorganisms associated with cork, fungi, and *Penicillium* species, in particular, appear to be the most common (Lee and Simpson, 1993). Other fungi include *Trichoderma*, *Aspergillus*, *Mucor*, and *Monillia*. Yeast may also be present. The boiling and bleaching steps effectively reduce the microbial load. However, fungi growth is possible prior to these steps or after, if re-contamination occurs. In fact, mold growth on cork has long been thought to be necessary for cork maturation and is, therefore, encouraged as part of traditional cork processing (Silva Pereira et al., 2000). Still, fungi have long been suspected of being involved in wine taint formation, although direct evidence establishing this link has only recently emerged.

Cork Taints

The actual chemical agent responsible for cork taint has been identified as 2,4,6-trichloroanisole or TCA (Figure 1). Not only does TCA have a musty, disagreeable odor, but its threshold for detection is extremely low. Sensitive tasters can detect TCA at levels as low as a few parts per trillion, although for most individuals, higher concentrations must be present before TCA is noticed (Peña-Neira et al., 2000). Although other chloroanisoles, as well as other phenol- and pyrazine-containing compounds, may also contribute to cork taint (Peña-Neira et al., 2000; Simpson et al., 2004), TCA is invariably present in tainted wine (Coque et al., 2003).

Box 10–7. Cork Taints in Wine: Are Corks the Cause? *(Continued)*

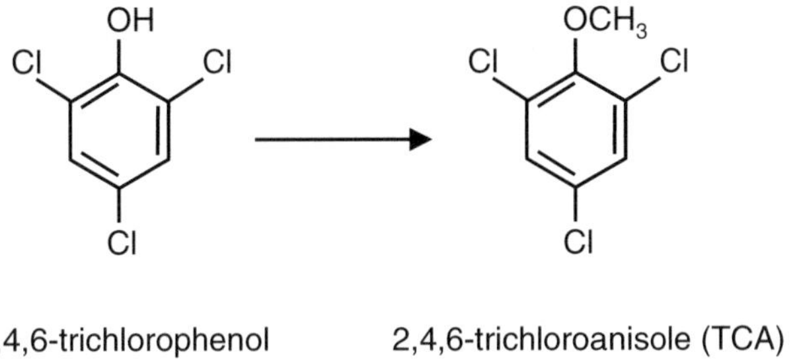

2,4,6-trichlorophenol   2,4,6-trichloroanisole (TCA)

Figure 1. Conversion of 2,4,6-trichlorophenol to 2,4,6-trichloroanisole (TCA).

Biochemical evidence suggests that TCA is synthesized from chlorophenol precursors by a methylation reaction (Álvarez-Rodríguez et al., 2002). Chlorophenols are highly toxic, and this methylation reaction has been thought to be involved in detoxification (TCA, although odorous, is non-toxic). It has recently been reported that cork-associated fungi can perform this reaction and produce TCA in cork, provided 2,4,6-trichlorophenol was present as a substrate (Ílvarez-Rodríguez et al., 2002). In this particular study, a strain of *Trichoderma longibrachiatum* produced the highest level of TCA (nearly 400 ng/g of cork), and several other fungi produced more than 100 ng/g of TCA. It is important to emphasize that TCA synthesis required 2,4,6-trichlorophenol as a substrate and that no TCA was formed in these experiments when other chlorinated phenols were present. Thus, these investigators concluded that fungi are responsible for TCA and cork taint formation. When TCA is actually made during cork processing is not known, nor is it clear from where the trichlorophenol originates. Chlorinated polyphenols were once used as agricultural herbicides and fungicides and were likely applied in cork-producing areas. It has been suggested that chlorine treatment of harvested cork may further increase the trichlorophenol content, and subsequently, the TCA level.

Despite these recent findings, there is still considerable debate regarding the true incidence of cork-taint in wine, as well as the source of the tainted aroma. Cork producers and some cork researchers have argued that cork taint occurs far less often than the reported frequencies (Silva Pereira et al., 2000). They claim that TCA can arise from other sources—not just cork—and that other non-cork taints continue to be referred to as cork taint. Poor storage and handling of corks can indeed result in other off-flavors; however, these defects are controllable. Cork stoppers, their proponents contend, are still the best way to seal bottles and preserve the quality of wine (Silva Pereira et al., 2000). Finally, some wine writers have suggested that the cork taint problem has been exaggerated, in part by the industry itself and also by other wine writers (Casey, 1999).

Remedies and alternatives

Regardless of whether the frequency of cork taint is 1% or 12% or somewhere in between (2% to 5% is the figure most often quoted), this can hardly be considered an acceptable defect rate. No industry could long tolerate a situation where even one unit out of 100 fails to meet quality standards. The wine industry is no different and that is why non-cork alternatives have become so popular in the past ten years. Efforts by the cork industry to improve cork and prevent cork taint are also under way. For example, supercritical CO_2 treatment can reportedly extract TCA from cork to undetectable levels (by sensory analysis). Analytical methods have also been developed to screen corks for TCA so that tainted corks are not used.

(Continued)

Box 10–7. Cork Taints in Wine: Are Corks the Cause? *(Continued)*

The main alternatives to corks for bottled wine include cork agglomerates and hybrids, plastic enclosures, and metal screw caps (bag-in-the-box packages might also qualify, but will not be further discussed). Agglomerates consists of cork granules (also called "dust") stuck together with glue and extruded into cork shapes. They have long been used by sparkling wine manufacturers and have many of the desirable features of cork, but there are reports they may also contain TCA. A new generation of agglomerates are now available that reportedly are TCA-free (via supercritical extraction). Hybrid corks are similar to the agglomerates, but contain plastic particles mixed with natural, ground cork. Plastic or synthetic corks are comprised of ethylene vinyl acetate. They have resilience similar to cork, but contain no cork material (and no TCA). Detractors have suggested that plastic corks impart plastic flavors to the wine (i.e., plastic taint).

Metal screw caps for wine were introduced in the 1990s. They are similar to those used for carbonated beverages. Screw caps are easy to open and re-close, and are TCA-free. They have all of the attributes of an ideal enclosure with no real technical downside. Wines from screw-capped bottles always score well in taste tests (although rubber-like flavors, from the enclosure seal, have been reported to migrate into the wine).

Ultimately, the type of enclosure used by a particular winery depends only in part on the performance characteristics of the particular material. Rather, perhaps the more important determining factor is the attitude or perception consumers have toward specific types of wine enclosures. Surveys indicate that consumers prefer cork, that they expect to hear a "pop" when the bottle is opened, and that tradition is a key feature of the wine-drinking experience. Many wine drinkers may not even recognize off-flavors in wines, including cork and other taints. Moreover, plastic corks, and especially screw cap enclosures, are often associated with economy wines, regardless of what is actually in the bottle.

Despite these attitudes, non-cork enclosures continue to gain popularity, and now are used for as much as 50% or more of the wine produced in some countries. For example, Australia and New Zealand have championed the use of screw cap enclosures. Many California wineries use cork only for their premium wines, and synthetic composite or screw caps for their other wines. Some experts have predicted that the trend away from cork will continue.

References

Álvarez-Rodríguez, M.L., L. López-Ocaña, J.M. Lípez-Coronado, E. Rodríguez, M.J. Martínez, G. Larriba, and J.J.R. Coque. 2002. Cork taints of wines: role of the filamentous fungi isolated from cork in the formation of 2,4,6-trichloroanisole by O-methylation of 2,4,6-trichlorophenol. Appl. Environ. Microbiol. 68:5860–5869.

Casey, J. 1999. T'aint necessarily so. Wine Industry Journal. 14:(6)49–56.

Coque, J.-J.R., M.L. Álvarez-Rodríguez, and G. Larriba. 2003. Characterization of an inducible chlorophenol O-methyltransferase from *Trichoderma longibrachiatum* involved in the formation of chloroanisoles and determination of its role in cork taint of wines. Appl. Environ. Microbiol. 69:5089–5095.

Lee, T.H., and R.F. Simpson. 1993. Microbiology and chemistry of cork taints in wine, p. 353–372. *In* Fleet, G.H. (ed.), *Wine Microbiology and Biotechnology*. Harwood Academic Publishers. Chur, Switzerland.

Jackson, R.S. 2000. *Wine Science: Principles, Practice, Perception, 2nd Ed.*, Academic Press. San Diego, California.

Ribéreau-Gayon, P., D. Dubourdieu, B. Donèche, and A. Lonvaud. 2000. *Handbook of Enology, Volume 1: The Microbiology of Wine and Vinifications*. John Wiley and Sons, Ltd. West Essex, England.

Peña-Neira, A., B.F. de Simón, M.C. García-Vallejo, T. Hernández, E. Cadahía, and J.S. Suarez. 2000. Presence of cork-taint responsible compounds in wines and their cork stoppers. Eur. Food Res. Technol. 211:2 57–262.

Silva Pereira, C.S., J.J.F. Marques, and M.V.S. Romão. 2000. Cork taint in wine: scientific knowledge and public perception—a critical review. Crit. Rev. Microbiol. 26:147–162.

Simpson, R.F., D.L. Capone, and M.A. Sefton. 2004. Isolation and identification of 2-methoxy-3, 5-dimethylpyrazine, a potent musty compound from wine corks. J. Agric. Food Chem. 52:5425–5430.

Spoilage by yeasts

Yeasts represent a major cause of wine defects and spoilage. Moreover, since yeasts are an expected part of the natural flora of grapes and must, their growth before, during, and after the wine fermentation is difficult to control. For example, *Kloeckera apiculata*, one of the yeasts involved in the early stages of a natural fermentation, can produce high enough levels of various esters (mainly ethyl acetate and methylbutyl acetate) to cause an ester taint, which has a vinegar-like aroma. Once vigorous growth of *S. cerevisiae* begins, other yeasts are generally unable to compete and grow. However, if *S. cerevisiae* does not become well-established (i.e., during natural fermentation), other yeasts, including *Zygosaccharomyces bailii*, can grow and produce acetic and succinic acids. Growth of this organism is especially a problem in sweet wines, due to its ability to tolerate high osmotic pressure and high ethanol concentrations.

Growth of yeasts during aging of wine, either in barrels or bottles, is a particularly serious spoilage problem. The main culprits are species of *Brettanomyces/Dekkera,* and *Brettanomyces bruxellensis* in particular (note that *Brettanomyces/Dekkera* exists in one of two forms, where *Brettanomyces* is the asexual, nonsporulating form and *Dekkera* is the sexual, sporulating form of the same yeast). These yeasts are common contaminants of wineries and the oak barrels used for aging. Growth of these organisms may lead to volatile phenol-containing compounds that give the wine a disagreeable "mousy" aftertaste (although mousy taints may also be caused by bacteria).

Brettanomyces may produce a variety of other taints, variously described as "barnyard," "horse sweat," "band-aid," and "wet dog." A common marker or signature chemical for "Brett" spoilage is 4-ethyl phenol, whose presence in suspect wines can be routinely monitored. There are also various film yeasts, including species of *Pichia, Candida,* and *Hansenula,* that grow on the surface of wine during barrel aging, much like the flor yeast whose growth in sherry making is desirable. Under ordinary circumstances, however, growth of film yeast can lead to oxidation of ethanol and formation of acids, esters, acetaldehyde, and other undesirable end products. Most of these spoilage yeast can be controlled or managed by SO_2 addition, maintenance of proper anaerobic conditions (topping off of barrels), barrel management, and good sanitation practices, or in extreme cases, by sterile filtration.

Spoilage by bacteria

Probably the most common and most disastrous types of microbial spoilage of wine are those caused by bacteria. Two distinct groups are of importance: the acetic acid bacteria and the lactic acid bacteria, both of which contain species able to tolerate the low pH, high ethanol conditions found in wine. These bacteria are responsible acidic and other end products that seriously affect wine quality.

The acetic acid bacteria that are most important in wine spoilage belong to one of three genera: *Acetobacter, Gluconoacetobacter*, and *Gluconobacter*. The main species involved in wine spoilage are *Acetobacter aceti*, *Acetobacter pasteurianus*, and *Gluconobacter oxydans.* They are Gram negative, catalase-positive rods capable of oxidizing alcohols to acids. These bacteria also are considered as obligate aerobes; however, it now appears that limited growth and metabolism can occur even under the mostly anaerobic conditions that prevail during wine making. Although acetic acid bacteria are generally found at relatively low levels in vineyards and in must (<100 cells per g), moldy or bruised grapes can contain appreciably higher levels. If the ethanol fermentation occurs soon after harvesting and crushing, then growth of these organisms, especially *G. oxydans,* is inhibited and numbers may actually decline. When the fermentation is complete and the wine is drawn off and subsequently transferred and racked, aeration inevitably occurs, activating growth of these bacteria.

Aging in oak barrels may also promote growth of acetic acid bacteria, in part because barrels may actually be contaminated with these bacteria, but also because oxygen diffusion into small oak barrels can be significant. Under these conditions, then, acetic acid bacteria can oxidize ethanol in the wine, producing enough acetic acid (>0.7 g/L) to give the wine a pronounced vinegar flavor and aroma. Although lesser amounts can sometimes be tolerated, other end products resulting from growth of acetic acid bacteria, including ethyl acetate, acetaldehyde, and dihydroxyacetone, may also contribute to spoilage defects. Low sulfur dioxide levels will further enhance growth of acetic acid bacteria.

The other group of bacteria associated with microbial spoilage of wine are the lactic acid bacteria. This cluster of Gram positive, facultative rods and cocci consists of twelve genera, however, only four, *Lactobacillus, Oenococcus, Leuconostoc,* and *Pediococcus*, are involved in wine spoilage. Due to their saccharolytic metabolism, it is not surprising that lactic acid bacteria are found on intact grapes, albeit at low populations (10^2/g to 10^3/g) that eventually increase ten- to 100-fold during harvesting and crushing. Although most lactic bacteria do not grow during the ethanolic fermentation (the exception are the malolactic bacteria), some strains are tolerant of high ethanol concentrations and are able to grow later during post-fermentation steps. High temperature and pH and low SO_2 concentrations favor growth of lactic acid bacteria. Growth of these bacteria in wine can result in several spoilage conditions, including acidification and deacidification, as well as production of various metabolites that cause off-flavors, aromas, and other defects.

As described in Chapter 2, lactic acid bacteria have either a homofermentative or heterofermentative metabolism, or in some strains, the metabolic capacity for both. When homofermentative lactic acid bacteria, such as *Pediococcus* and *Lactobacillus plantarum*, grow in wine and ferment glucose, they produce lactic acid. This acid causes the wine to become excessively sour. In contrast, metabolism of pentose sugars, present in the grapes as well as extracted from oak barrels, occurs via the pentose phosphate pathway (also

known as the phosphoketolase pathway), yielding acetic acid, ethanol, and CO_2. These same end-products are also produced by several heterofermentative lactic acid bacteria found in wine, including *Leuconostoc mesenteroides, Oenococcus oenus*, and *Lactobacillus brevis*. Although these bacteria produce considerably less acetic acid than *Acetobacter* or *Gluconobacter*, enough may be produced to impart a detectable vinegar-like flavor to the wine.

Several lactic acid bacteria have the capacity to convert malic acid to lactic acid via the malolactic pathway. In musts containing high malic acid concentrations and having low pH, the malolactic bacteria perform a desirable, even essential function, by deacidification of the wine. However, if the must acidity is already low and the pH high, the malolactic fermentation may result in an increase in wine pH such that too little acidity remains. The wine also will be more susceptible to spoilage since a major barrier to microbial growth, low pH, is diminished.

Spoilage by lactic acid bacteria can also be caused by other metabolic end products. Metabolism of fructose by heterofermentative *L. brevis*, for example, can cause mannitol to form, and generate acetic acid as a side reaction. In a fermentation analogous to the malolactic pathway, in terms of effect, *L. brevis* can metabolize tartaric acid, causing an increase in wine pH and formation of volatile acids. Mousy taints, similar to those produced by the spoilage yeast *Brettanomyces*, can also be produced by *Lactobacillus* sp., *Leuconostoc* sp., and other heterofermentative lactic acid bacteria. A flowery off-odor referred to as geranium taint also may be produced by lactobacilli.

Glycerol oxidation by *Leuconostoc mesenteroides, Lactobacillus brevis,* and other lactobacilli generates acrolein, which reacts with tannins and anthocyanin phenolics to form bitter compounds. This defect, referred to as amertume, is more common in red wines, due to the higher tannin and phenolic levels in red wines. Its microbiological cause was first noted by Pasteur. Several lactic acid bacteria, including strains of *Pediococcus* and *Lactobacillus,* produce diacetyl, which imparts a buttery aroma in wine that, at high concentrations (generally above 5 mg/L), is considered undesirable. Finally, the formation of glucose-containing polysaccharides, such as dextrins and glucans, can give an oily, viscous and objectionable mouth feel. Ropiness usually occurs only in sweet wines and is caused by *Pediococcus, Oenococcus,* and *Leuconostoc spp.*

Bibliography

Amerine, M.A., H.W. Berg, R.E. Kunkee, C.S. Ough, V.L. Singleton, A.D. Webb. 1980. *The Technology of Wine Making, 4th Ed.*, Avi Publishing Company, Inc., Westport, Connecticut.

Bartowsky, E.J., and P.A. Henschke. 2004. The 'buttery' attribute of wine-diacetyl-desirability, spoilage and beyond. Int. J. Food Microbiol. 96: 235–252.

Boulton, R.B., V.L. Singleton, L.F. Bisson, and R.E. Kunkee. 1996. *Principles and Practices of Winemaking.* Chapman and Hall, New York, New York.

Coates, C. 2000. *An Encyclopedia of the Wines and Domaines of France.* University of California Press. Berkeley, California.

Fleet, G.H. 2003. Yeast interactions and wine flavour. Int. J. Food Microbiol. 86:11–22.

Fugelsang, K. C. 1997. *Wine Microbiology.* Chapman and Hall, New York, New York.

Giudici, P., L. Solieri, A.M. Pulvirenti, and S. Cassanelli. 2005. Strategies and perspectives for genetic improvement of wine yeasts. Appl. Microbiol. Biotechnol. 66:607–613.

Jackson, R.S. 2000. *Wine Science: Principles, Practice, Perception, 2nd Ed.*, Academic Press, San Diego, California.

Loureiro, V., and M. Malfeito-Ferreira. 2003. Spoilage yeasts in the wine industry. Int. J. Food Microbiol. 86:23–50.

Konings, W.N. 2002. The cell membrane and the struggle for life of lactic acid bacteria. Antonie van Leeuwenhoek 82:3-27.

Pretorius, I.S., and F.F. Bauer. 2002. Meeting the consumer challenge through genetically customized wine-yeast strains. Trends Biotechnol. 20:426–478.

Ribéreau-Gayon, P., D. Dubourdieu, B. Donèche, and A. Lonvaud. 2000. *Handbook of Enology, Volume 1: The Microbiology of Wine and Vinifications.* John Wiley and Sons, Ltd., West Essex, England.

Ribéreau-Gayon, P., Y. Glories, A. Maujean, and D. Dubourdieu. 2000. *Handbook of Enology, Volume 2: The Chemistry of Wine Stabilization and Treatments.* John Wiley and Sons, Ltd. West Essex, England.

Romano, P., C. Fiore, M. Paraggio, M. Caruso, and A. Capece. 2003. Function of yeast species and strains in wine flavour. Int. J. Food Microbiol. 86: 169–180.

11

Vinegar Fermentation

"Vinegar, son of wine."
Hebrew proverb

History

Although it is not possible to know precisely when human beings first began to produce and consume fermented foods, the origin for some products can reasonably be estimated. Vinegar, for example, was likely discovered shortly after (like about a week) the advent of the first successful wine fermentation. As excited as that first enologist must have been to have somehow turned grape juice into wine, one can imagine the disappointment that must have followed when the wine itself subsequently turned into a sour, unpalatable liquid, seemingly devoid of any redeeming virtues. Even though that sour wine, *vin aigre* in French, obviously couldn't be drunk with the same enjoyment or enthusiasm as the wine, it was not, it turned out, without value. Indeed, vinegar has a long history of use, and is now one of the most widely used ingredients in the food industry, with world-wide production of about 1 million L per year.

The actual history of vinegar consumption dates back several thousand years. Although other fermented foods, such as wine, beer, and cheese, evolved, in part, because of their enhanced preservation status, vinegar was likely used for its ability to preserve other non-fermented, perishable foods, such as meats and vegetables. Thus, vinegar can be considered as the first biologically-produced preservative. However, in addition to its use as a so-called pickling agent, it was also consumed directly as a flavoring agent, and, in a diluted form, as a beverage.

Vinegar consumption is first noted in the Bible (Numbers 6:1 and Ruth 2:14), and, according to the New Testament, was given to Jesus during the crucifixion (John 19:23). Diluted vinegar was a popular beverage throughout the Greek and Roman eras, where it gained favor as a therapeutic beverage. The production of vinegar was not confined to Europe; it was also produced throughout Asia. Although the substrates were different, the processes were remarkably similar to those that evolved in Europe. Finally, in addition to its use as a food or food ingredient, vinegar has also long been used as a topical disinfectant, as a cleaning agent, and as an industrial chemical due to its strong demineralization properties.

In the food industry, vinegar is used mainly as an acidulent, a flavoring agent, and a preservative, but it also has many other food processing applications. It is found in hundreds of different processed foods, including salad dressings, mayonnaise, mustard and ketchup, bread and bakery products, pickled foods, canned foods, and marinades and sauces. It is used in almost every culture and is part of nearly every cuisine. And although many of the vinegars produced around the world are made from ordinary substrates and often have rather nondescript sensory properties, others are produced from premium wines, carefully aged, and prized (and priced) based on their unique organoleptic attributes.

Returning to our early wine maker, there is a biological reason, of course, to account for the unfortunate fate of that ancient wine. Wine, we now know, turns into vinegar because naturally-occurring bacteria exist, namely species of *Acetobacter*, that oxidize the ethanol in the wine to form acetic acid. In fact, any ethanol-containing material can serve as a substrate for the vinegar fermentation, and, as will be discussed below, there are numerous types of vinegars that are produced from a variety of ethanolic substrates. While wine producers take special measures to prevent contamination by acetic acid-producing bacteria and to avoid the oxygen supply that is necessary for the ethanol-to-acetic-acid conversion, manufacturers of vinegar do just the opposite. In other words, the vinegar fermentation is conducted in the presence of *Acetobacter* and under conditions that favor growth and oxidative metabolism by this organism.

Definitions

There is no standard of identity for vinegar in the United States, but there are guidelines that define the starting material, the finished specifications, and the labeling declaration. First, as will be described in more detail later, vinegar must be made from one of various types of ethanol-containing solutions. The most common starting materials, whose identity must be indicated on the label, are grape and rice wine, fermented grain or malt mashes, and fermented apple cider. Distilled ethanol is also permitted as a substrate for vinegar manufacture. Importantly, vinegar must, by definition, result from the "acetous fermentation" of ethanol. In other words, acetic acid made via chemical synthesis

cannot be labeled as vinegar. Vinegar must contain at least 4% acetic acid or at least forty grains (where one grain = 0.1% acid). Usually, the ethanol concentration is less than 0.5%, and the pH is between 2.0 and 3.5. In countries where identity standards do exist, they generally are consistent with the U.S. definition.

As noted above, the raw material determines the name of the vinegar (e.g., red wine vinegar, apple cider vinegar, malt vinegar). However, the starting material also has a profound influence on the flavor and overall quality attributes of the vinegar. Although the predominant flavor of all vinegars is due to acetic acid, other flavors, specific to the ethanol source or the means of its manufacture, may also be present. In addition, some vinegars may also contain herbs (e.g., tarragon), added before or after the fermentation.

Vinegar Manufacturing Principles

The manufacture of vinegar consists of two distinct processes. The first step is an ethanolic fermentation performed mostly by yeasts. In the second step, an acetogenic fermentation is carried out by acetic acid bacteria. Whereas the ethanol fermentation is anaerobic, the latter is conducted under highly aerobic conditions. In fact, many of the technological advances in the vinegar fermentation have focused on ways to introduce more air or oxygen into the fermentation system.

Microorganisms

Aside from those bacteria that produce acetic acid as an overflow or side reaction from sugar metabolism (e.g., lactic acid bacteria), there are several genera of bacteria that produce acetic acid as the primary metabolic end product. It is important, however, to distinguish between those that produce acetic acid from one-carbon precursors and those that produce acetic acid via oxidation of ethanol. The former group are referred to as acetogens and include species of *Clostridium, Eubac-*

terium, Acetobacterium, Peptostreptococcus, and other Gram positive anaerobic rods and cocci. Acetogens rely on the acetogenesis pathway, in which energy is derived from other sources and CO_2 merely serves as the electron acceptor. The latter becomes reduced to acetic acid under strict anaerobic conditions. In contrast, in the ethanol oxidation pathway, acetic acid bacteria actually use the electrons from ethanol, with oxygen serving as the electron acceptor. Thus the same product is formed, but for two totally different biochemical reasons. The acetic acid bacteria are represented by four genera: *Acetobacter, Gluconobacter, Gluconoacetobacter,* and *Acidomonas* (a fifth genus, *Asaia*, has recently been described, but oxidizes ethanol weakly or not at all). These genera belong to the Phylum *Proteobacteria* and the family *Acetobacteraceae*. The phylogeny of these and related bacteria is illustrated in Figure 11-1. Acetic acid bacteria are Gram negative, obligate aerobes, motile or non-motile, and have an ellipsoidal to rod-like shape that appear singly, in pairs or in chains. They have a G+C base composition of 53% to 66%.

Acetic acid bacteria are widely distributed in plant materials rich in sugars, such as fruits and nectars. They are also common inhabitants of alcohol-containing solutions, including wine, beer, hard cider, and other ethanolic beverages. Not surprisingly, acetic acid bacteria often share habits with ethanol-producing yeasts. Although some acetic acid bacteria, notably *Acetobacter,* are more efficient at oxidizing ethanol, rather than glucose, to acetic acid, *Gluconobacter* oxidizes and grows especially well on glucose compared to ethanol.

Three genera, *Acetobacter, Gluconobacter,* and the recently named *Gluconoacetobacter,* contain most of the species used industrially for vinegar manufacture (see below). Although *Acetobacter aceti* has long been considered to be the primary organism involved in the vinegar fermentation, many other species have been identified in vinegar and in production facilities, including *Acetobacter pasteurianus,*

Gluconoacetobacter europaeus (formerly *Acetobacter europaeus*), *Gluconacetobacter xylinus* (formerly *Acetobacter xylinus*), and *Gluconobacter oxydans* (Table 11-1).

Many vinegar fermentations are conducted using mixed or wild cultures, accounting for the large number of strains isolated from different production facilities. For example, *Gluconobacter entanii* was the predominant organism isolated from a single high-acid vinegar production facility in Germany, and *A. pasteurianus* was identified as the main species involved in production of rice wine vinegars produced in Japan. In Germany, *A. europaeus* is now considered to be the most common organism isolated from vinegar factories that use submerged-type fermentation processes (see below). It seems clear, therefore, that no single species can be considered to be solely responsible for the vinegar fermentation.

Although all species of *Acetobacter* and *Gluconobacter* produce acetic acid from ethanol, they differ, metabolically, in several respects (Table 11-2). First, whereas *Acetobacter* can, under certain conditions, completely oxidize ethanol to CO_2, metabolism of ethanol by *Gluconobacter* stops at acetic acid (discussed below). In contrast, oxidative metabolism of glucose and other sugars is much greater in *Gluconobacter* compared to *Acetobacter,* which produces little acid from sugars. Finally, these organisms can be distinguished on the basis of the quinone co-factor systems used as electron acceptors (see below) during the ethanol oxidation reactions, because *Gluconobacter* possesses a G_{10} type quinone system and *Acetobacter* has a Q_9 system.

As noted above, several species of *Acetobacter* are used in industrial production of vinegar. In general, the criteria used to distinguish between species have been based on those biochemical and physiological properties most relevant to the acetic acid fermentation (Table 11-3). The phylogeny and taxonomy of acetic acid bacteria have undergone significant changes in recent years. Until 2000, eight species of *Acetobacter* were recognized,

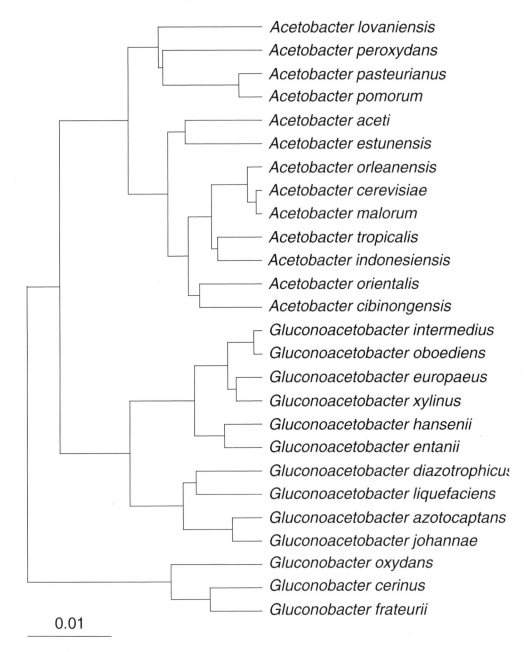

Figure 11–1. Phylogenetic relationship of *Acetobacter, Gluconobacter,* and *Gluconoacetobacter* sp. based on 16S rRNA sequences. The tree was generated using the UPGMA clustering method. For most of the nodes, bootstrap values were 90% to 100%.

Table 11.1. Species of *Acetobacter* and *Gluconoacetobacter* isolated from vinegar[1].

Acetobacter aceti	*Gluconoacetobacter xylinus*[2]
Acetobacter cerevisiae	*Gluconoacetobacter hansenii*[2]
Acetobacter cibinongensis	*Gluconoacetobacter europaeus*[2]
Acetobacter estunensis	*Gluconoacetobacter intermedius*[2]
Acetobacter indonesiensis	*Gluconoacetobacter liquefaciens*[2]
Acetobacter lovaniensis	
Acetobacter malorum	
Acetobacter orleanensis	
Acetobacter orientalis	
Acetobacter pasteurianus	
Acetobacter peroxydans	
Acetobacter pomorum	
Acetobacter tropicalis	

[1]From Cleenwerck et al. 2002; Garrity et al., 2004; and Lisdiyanti et al., 2000
[2]These species were previously classified as *Acetobacter*

Table 11.2. Physiological properties and characteristics of acetic acid bacteria.

Property	*Acetobacter*	*Gluconobacter*	*Gluconoacetobacter*
Temperature optimum (°C)	25–30	25–30	25–30
pH optimum	5.4–6.3	5.5–6.0	5.4–6.3
Acetic acid oxidation	+	–	+/–
Lactic acid oxidation	+	–	+/–
Acid from glucose	+	+	+
Ubiquinone type	Q-9	Q-10	Q-10

Adapted from Yamada et al., 2000 and Holt et al., 1994

based on the characteristics listed in Table 3, as well as on DNA homology and G+C content. Recently, however, it was proposed, on the basis of additional DNA-DNA and phylogenetic relatedness, that several *Acetobacter* species be moved to the genus *Gluconoacetobacter*, and that several other new species be added (Table 11-1).

Metabolism and Fermentation

The acetic acid pathway used by *Acetobacter*, *Gluconobacter* and other acetic acid bacteria is an example of what is referred to as an incomplete oxidation. Whereas in most oxidative pathways (e.g., the Krebs or citric acid cycle), organic substrates are ordinarily oxidized all the way to CO_2 and H_2O, in the vinegar fermentation, acetic acid bacteria usually oxidize the substrate, ethanol, only to acetic acid. However, as described below, exceptions exist where complete oxidation to CO_2 can occur.

The actual pathway consists of just two main steps (Figure 11-2). First, ethanol is oxidized to acetaldehyde and then the acetaldehyde is oxidized to acetic acid. An intermediate step, in which acetaldehyde hydrate is formed, may also occur. Oxygen is required as the terminal electron acceptor (see below). It should also be noted that the conversion of ethanol to acetic acid occurs on an equimolar basis, such that after the reactions are complete, the final concentration of acetic acid will be equal to that of the ethanol in the starting material (assuming negligible loss from evaporation). Only a minor amount of the carbon from ethanol is used for biomass or converted to other products.

Table 11.3. Biochemical and physiological properties of species of acetic acid bacteria.[1]

Property	Acetobacter aceti	Acetobacter pasteurianus	Gluconoacetobacter diazotrophicus	Gluconoacetobacter europaeus	Gluconoacetobacter hansenii	Gluconoacetobacter liquefaciens	Gluconoacetobacter xylinus	Acidomonas methanolicus
Growth on 3% ethanol in 4-8% acetic acid	–	–	–	+	–	–	–	–
Growth in absence of acetic acid	+	+	u[2]	–	+	+	+	+
Growth on acetate at pH 2.5 in absence of ethanol	–	–	–	+	–	–	–	–
Growth on 1.7 M glucose	–	–	+	–	–	–	–	–
Growth in presence of:								
lactic acid	+	d	w[3]	d[4]	d	+	u	–
ethanol	+	d	+	+	–	+	–	w
30% glucose	–	–	+	–	–	–	–	–
Biosynthesis of cellulose	–	–	–	d	–	–	+	–
DNA homology (%) with:								
G. europeus	<10	<25	u	70–100	<15	<15	30	<10
G.xylinus	u	<10	u	37	u	u	100	u
A.pasteurianus	u	100	u	17	u	u	<10	u
G + C content (%)	56–60	53–63	61–63	56–58	58–63	62–65	55–63	62

[1] Adapted from Sievers and Teuber, 1995 and Sokollek et al., 1998

[2] undetermined

[3] weak

[4] some strains positive

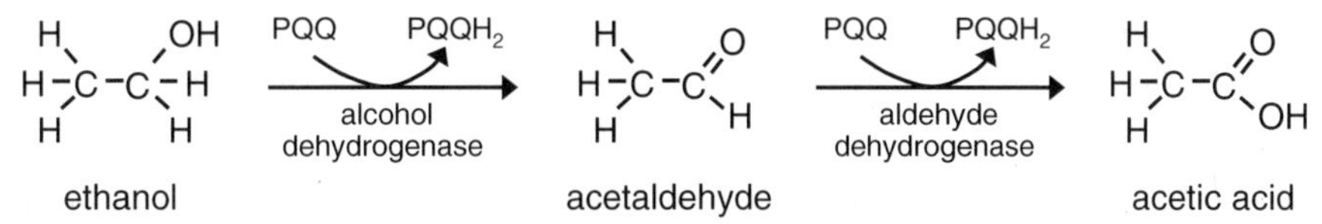

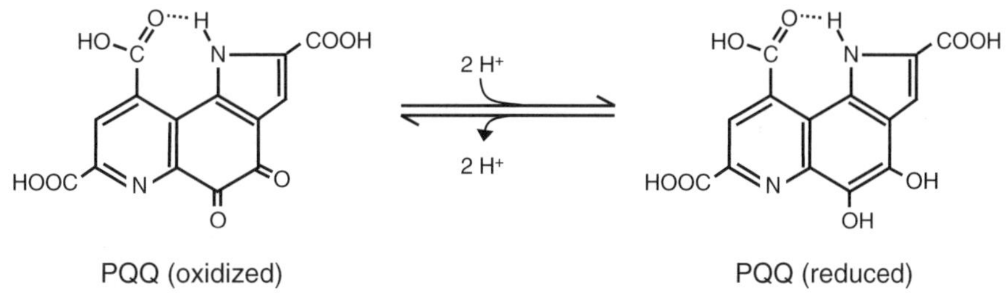

Figure 11–2. Formation of acetic acid from ethanol by *Acetobacter aceti*. The oxidation of acetic acid to ethanol involves two enzymatic reactions (upper panel). In the first reaction, ethanol is oxidized to acetaldehyde by alcohol dehydrogenase. Next, acetaldehyde dehydrogenase further oxidizes acetaldehyde to acetic acid. An intermediate hydration reaction may also occur, in which acetaldehyde hydrate is formed (not shown). Both alcohol dehydrogenase and acetaldehyde dehydrogenase enzymes require pyrroloquinoline quinone (PQQ) as a co-factor, which serves as an electron acceptor (lower panel). The reduced PQQ then supplies electrons to the electron transport chain, eventually leading to the formation of ATP via oxidative phosphorylation. Adapted from Fuchs, G., 1999.

The biochemical and physiological processes involved in the acetic acid fermentation are quite unlike those described for the lactic and ethanolic fermentations. The acetic acid fermentation is performed by obligate aerobes, and metabolism occurs not in the cytoplasm, but rather within the periplasmic space and cytoplasmic membrane. The acetic acid pathway yields energy, but not by substrate level phosphorylation. Instead, the oxidation reactions are coupled to the respiratory chain which then generates ATP via electron transport and oxidative phosphorylation reactions. Finally, both the substrate and product are toxic, which no doubt contributes to the relative lack of competitors during what is essentially an open fermentation. In fact, acetic acid bacteria are rather remarkable for their ability to tolerate low pH and high acetic acid concentrations (Box 11-1).

The two main enzymes involved in the acetic acid fermentation, alcohol dehydrogenase (AlDH) and acetaldehyde dehydrogenase (AcDH), have been characterized, and the genes coding for their expression have been cloned and sequenced (Figure 11-3). These enzymes are the "primary dehydrogenases," responsible for nearly all of the detectable oxidation activity. They are located within the cytoplasmic membrane and face into the periplasm (Figure 11-4). This means that the ethanol and acetaldehyde oxidation reactions occur in the periplasm and that ethanol can be used without having to be accumulated within the cytoplasm. Both AlDH and

Box 11–1. Acetic Acid Tolerance in *Acetobacter*

Microorganisms that produce peroxides, alcohols, acids, or other inhibitory agents must also be able to tolerate such products. In the case of acetic acid bacteria, tolerance to both ethanol and acetic acid is a necessary feature of their physiology.

During the vinegar fermentation, cells typically encounter ethanol at concentrations near 12% (more than 2.5 M). Later, as ethanol is oxidized to acetic acid, vigorous growth is maintained even when the concentration of acid reaches more than 7%. Some strains of *Acetobacter* can produce (and tolerate) as much as 20% acetic acid (3.3 M). To provide some perspective, *Escherichia coli* is inhibited by 1% or less acetic acid (Lasko et al., 2000). Thus, the means by which acetic acid bacteria deal with acetic acid stress is not only of fundamental interest, but it also has practical implications. Strains that are more acid tolerant, for example, would be expected to be more productive during commercial vinegar fermentations.

To fully appreciate the challenge faced by these cells, consider first the actual problem that exists when *Acetobacter* grows during the acetic acid fermentation. In a 6% (1 M) acetic acid solution at a pH of 4.5, most of the acetic acid will be in the undissociated or acid form (given that the pKa of acetate is 4.76). Since the cell membrane is permeable to the acid form, cells at that pH could accumulate, in theory, more than 0.6 M acetic acid within the cytoplasm (via simple diffusion). Accumulation of the acid will continue, in fact, until the inside and outside concentrations are equal. Of course, because the physiological pH within the cytoplasm is generally higher than the external pH (at least for most bacteria), the accumulated acid will immediately dissociate to form the anion and a free proton (in accordance to the Henderson-Hasselbach equation). This will further shift the [acid$_{in}$]/[acid$_{out}$] equilibrium such that even more acid will be accumulated.

At some point one of several outcomes are possible. First, unless the accumulated protons are expelled from the cytoplasm, their accumulation will eventually cause the intracellular pH to decrease to an inhibitory level. Second, if the cells relied on proton extrusion to maintain a more conducive intracellular pH, then substantial amounts of ATP (or its equivalent) would be required to drive the proton pumping apparatus, leading to a major drain on the energy resources available to the cell. Finally, if a high cytoplasmic pH is maintained, the acetate anion could, in theory, reach concentrations of 1 M or higher (under the conditions stated above). This would create severe toxicity and osmotic problems for the cell, which may be even worse than acidity problems (Russell, 1992).

Despite these apparent hurdles, researchers, primarily in Switzerland and Japan, have revealed that there may be several mechanisms that account for the ability of *Acetobacter* and other acetic acid bacteria to tolerate high acetic-acid and low-pH environments. These mechanisms involve both physiological and genetic responses (Table 1). As noted above, acetic acid accumulation depends on the pH gradient across the cell membrane. The higher the gradient (i.e., low pH outside and high pH inside), the greater will be the acetate that will be trapped inside. If, on the other hand, the cytoplasmic pH is allowed to drop (decreasing the gradient), then less acetate is accumulated.

Table 1. Summary of mechanisms contributing to acetic acid tolerance.

Mechanisms	Possible effect
Low pH gradient	Reduced anion accumulation
groESL operon induced	General stress response
Expression of Aar proteins	Stimulate acetate assimilation
Expression of Aap proteins	Stimulate transport systems
Expression of Asp proteins	Unknown
Aconitase expression	Stimulate acetate assimilation
Acetic acid efflux pump	Stimulate acetic acid efflux

Box 11–1. Acetic Acid Tolerance in *Acetobacter (Continued)*

In fact, this indeed seems to be the case for *Acetobacter aceti*, because this organism maintains only a very small pH gradient (Menzel and Gottschalk, 1985). Thus, the cell is spared the problem caused by anion (acetate) accumulation (of course, this implies that *Acetobacter* must still be physiologically able to tolerate a low intracellular pH). That acetate accumulation might contribute to cell inhibition is supported by the observation that resistant strains accumulate less acetate than wild-type strains that are only moderately resistant (Steiner and Sauer, 2003). The response to high acetate concentrations at a near neutral pH may, however, be quite different. These same researchers showed, for example, that *A. aceti* at pH 6.5 accumulated more than 3 M acetate, at an outside concentration of less than 0.4 M. However, these high-pH conditions clearly do not reflect those that this organism would likely experience.

As indicated above, intracellular acetic acid also could, in theory, be pumped out directly, provided the energy sources were available. Recently, the presence of an acetic acid efflux system was reported in *Acetobacter aceti* (Matsushita et al., 2005). The system was sensitive to protonophores and ionophores, agents that dissipate proton and ion gradients across the cell membrane and by a respiration inhibitor. Therefore, it appears that acetic acid efflux is energized by a proton motive force that is itself dependent on respiration activity.

Transcriptional analyses of *A. aceti* also have revealed the presence of genes whose expression is induced by acetic acid and that may be associated with acetic acid resistance. Examples include the *groESL* and the *aar* (*a*cetic *a*cid *r*esistance) operons (Fukaya et al., 1990; Fukaya et al., 1993; Okamoto-Kainuma et al., 2002). The GroES and GroEL proteins (encoded by *groESL*) are part of the chaperonin family of stress-induced proteins that protect cells by stabilizing proteins and preventing their mis-folding. In *A. aceti, groESL* was induced by heat as well as acetic acid and ethanol. Furthermore, a *groESL* over-expressing strain was shown to be even more resistant to these stresses, suggesting that these gene products contribute to overall acetic acid resistance (Okamoto-Kainuma et al., 2002).

The *aar* operon consists of three genes, *aarA*, *aarB*, and *aarC,* that encode for proteins with homology to enzymes involved in the citric acid pathway. Specifically, *aarA* encodes for the citrate-forming enzyme citrate synthase and *aarC* appears to encode for coenzyme A transferase, an enzyme also involved in acetate assimilation. Thus, assimilation of acetic acid (i.e., via its oxidation) would be one way to not only de-toxify the acid, but also enable the cell to use acetate as an energy source.

Recently, another assimilation or acetate oxidation pathway was suggested to be responsible for acetate resistance (Nakano et al., 2004). These researchers used two-dimensional (2-D) gel electrophoresis to show that expression of the enzyme aconitase by *A. aceti* was significantly increased in response to an acetic acid shock (Nakano et al., 2004). Further, a strain harboring multiple copies of the aconitase gene (and that over-expressed aconitase) produced more acetic acid and was more acetic acid resistant compared to the parent strain. Apparently, the increased expression of aconitase, an enzyme involved in both the citric acid cycle and glyoxylate pathway, confers resistance by stimulating consumption and detoxification of intracellular acetic acid. Although effective for the cells, these strategies (the *aar* and aconitase) are not ones that would likely be exploited for vinegar production, since the product is consumed and the overall yield would not be enhanced.

Finally, as shown by proteome analyses (using 2-D gels), a number of proteins are synthesized by *Acetobacter* and *Gluconobacter* in response to both short-term and long-term exposure to acetic acid (Lasko et al., 1997; Steiner and Sauer, 2001). These proteins are distinct from the general stress-response proteins induced by heat shock and are thus referred to as either acetate adaptation proteins (Aaps) for adapted cells or acetate-specific stress proteins (Asps) for unadapted cells. Although the function of most of these proteins has not yet been established, it appears (based on partial sequence analyses) that at least several of the Aaps are associated with

(Continued)

Box 11–1. Acetic Acid Tolerance in *Acetobacter (Continued)*

membranes and may be involved in transport processes. Indeed, acetate transport and membrane-associated respiratory proteins could both be envisioned to promote acetate resistance (Steiner and Sauer, 2001).

References

Fukaya, M., H. Takemura, H. Okumura, Y. Kawamura, S. Horinouchi, and T. Beppu. 1990. Cloning of genes responsible for acetic acid resistance in *Acetobacter aceti*. J. Bacteriol. 172:2096-2104.

Fukaya M., H. Takemura, K. Tayama, H. Okumura, Y. Kawamura, S. Horinouchi, and T. Beppu. 1993. The *aarC* gene responsible for acetic acid assimilation confers acetic acid resistance on *Acetobacter aceti*. J. Ferm. Bioeng. 76:270-275.

Lasko, D.R., N. Zamboni, and U. Sauer. 2001. Bacterial response to acetate challenge: a comparison of tolerance among species. Appl. Microbiol. Biotechnol. 54:243-247.

Lasko, D.R., C. Schwerdel, J.E. Bailey, and U. Sauer. 1997. Acetate-specific stress response in acetate-resistant bacteria: an analysis of protein patterns. Biotechnol. Prog. 13:519-523.

Matsushita, K., T. Inoue, O. Adachi, and H. Toyama. 2005. *Acetobacter aceti* possesses a proton motive force-dependent efflux system for acetic acid. J. Bacteriol. 187:4346-4352.

Menzel, U., and G. Gottschalk. 1985. The internal pH of *Acetobacterium wieringae* and *Acetobacter aceti* during growth and production of acetic acid. Arch. Microbiol. 143:47-51.

Nakano, S., M. Fukaya, and S. Horinouchi. 2004. Enhanced expression of aconitase raises acetic acid resistance in *Acetobacter aceti*. FEMS Microbiol. Lett. 235:315-322.

Okamoto-Kainuma, A., W. Yan, S. Kadono, K. Tayama, Y. Koizumi, and F. Yanagida. 2002. Cloning and characterization of *groESL* operon in *Acetobacter aceti*. J. Biosci. Bioeng. 94:140-147.

Russell, J.B. 1992. Another explanation for the toxicity of fermentation acids at low pH: anion accumulation versus uncoupling. J. Appl. Bacteriol. 73:363-370.

Steiner, P., and U. Sauer. 2001. Proteins induced during adaptation of *Acetobacter aceti* to high acetate concentrations. Appl. Environ. Microbiol. 67:5474-5481.

Steiner, P., and U. Sauer. 2003. Long-term continuous evolution of acetate resistant *Acetobacter aceti*. Biotechnol. Bioeng. 84:40-44.

AcDH also contain pyrroloquinoline quinone (PQQ) as a prosthetic group (Figure 11–2). The latter serves as the primary electron acceptor during the oxidation reactions, and is responsible for the transfer of electrons to the cytochromes of the respiratory chain.

Protein analyses of purified PQQ-dependent AlDH and AcDH from several *Acetobacter and Gluconobacter* species have shown that these enzymes consist either of two or three subunits. These subunits contain heme-binding moieties that mediate intramolecular transport of electrons, in addition to the PQQ prosthetic groups. Acetic acid bacteria also have NAD- (or NADP)-dependent dehydrogenases present within the cytoplasm. However, the specific activities of these enzymes are up to 300 times lower than the membrane-bound, PQQ-dependent dehydrogenases. In addition, the pH optima of the latter

enzymes in *Acetobacter* is between 4.0 and 5.0, which is much closer to the normal physiological pH than the NAD-dependent dehydrogenase enzymes, whose pH optima is above 7.0. Although the function of these enzymes has not yet been established, it has been suggested that they are involved in acetaldehyde and acetate assimilation. Finally, it should be noted that alcohols and aldehydes other than ethanol and acetaldehyde can serve as substrates for both of these dehydrogenases. Thus, primary alcohols such as propanol, and secondary alcohols and polyols, such as isopropanol and glycerol, all of which can be present in mashes used for vinegar production, can be oxidized and converted into acids, ketones, and other organic end products. Many of these compounds make important contributions to the aroma and flavor characteristics of vinegar (see below).

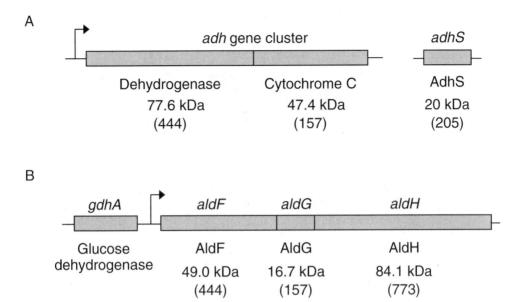

Figure 11–3. Organization of genes coding for ethanol oxidation in *Acetobacter*. The alcohol dehydrogenase complex (A) consists of three subunits. The dehydrogenase subunit and the cytochrome c subunit are encoded by two co-transcribed genes. The third gene, *adhS*, is not located near the *adh* gene cluster. The function of its gene product, AdhS, is not known, although it is required for enzyme activity (except for those strains that apparently contain only subunits I and II). The aldehyde dehydrogenase gene cluster (B) also contains three genes coding for three subunits of aldehyde dehydrogenase. Based on sequence analysis, the gene products all contain domains that bind co-factors involved in electron transport. AldF and AldH are heme-binding proteins, and AldG binds an iron-sulfur cluster. A gene encoding glucose dehydrogenase is upstream of the *aldFGH* promoter (arrow). The predicted molecular mass (and amino acid residues) are shown for each protein. From Takemura, et al., 1993; Kondo, et al., 1995; and Thurner, et al., 1997.

As noted earlier, the acetic acid pathway is usually considered an incomplete oxidation, since the substrate, ethanol, is only partially oxidized and the acetic acid that is formed is not oxidized further. However, while this is true for some acetic acid bacteria, such as *Gluconobacter* (a member of the so-called suboxydans group), most species of *Acetobacter* can oxidize acetic acid, provided conditions are suitable. The latter include bacteria of the oxydans group, represented by the common vinegar-producing species *A. aceti* and *A. pasteurianus*. The absence of ethanol in the medium is the main condition necessary for acetic acid oxidation. Ethanol apparently represses synthesis of citric acid cycle enzymes; when ethanol is absent, those enzymes are induced and complete oxidation of acetic acid to CO_2 and H_2O can

occur. Of course, in actual vinegar production, this so-called over-oxidation of acetic acid is particularly undesirable, because it causes the literal disappearance of the end product to the atmosphere. Thus, this is likely one reason why vinegar fermentations have historically been conducted in a semi-continuous mode (see below), in which a minimum amount of ethanol is always present.

Although vinegar-producing cultures are frequently used repeatedly without loss of viability or performance, genetic instability is not uncommon. Spontaneous mutations, at relatively high frequencies, have been reported for several *Acetobacter* sp., resulting in defects in several important functions. For example, mutants having lost the ability to oxidize ethanol, tolerate acetic acid, and form surface film (see

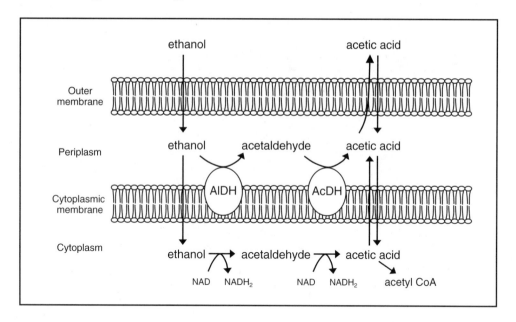

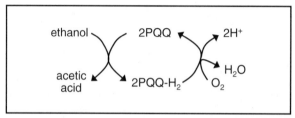

Figure 11–4. Membrane biology, oxidation of ethanol, and acetate assimilation by *Acetobacter aceti*. The oxidation reactions (upper panel), catalyzed by the PQQ-dependent enzymes alcohol dehydrogenase (AlDH) and acetaldehyde dehydrogenase (AcDH) occur within the periplasm. Ethanol can also be oxidized by cytoplasmic NAD-dependent dehydrogenases, but at very low rates. Acetic acid assimilation to acetyl Co-A and subsequent oxidation to CO_2 via the citric acid cycle occurs only when ethanol is absent. The oxidation reactions are accompanied by transfer of electrons from PQQ to cytochromes of the respiratory chain (lower panel), leading to formation of a proton motive force and synthesis of ATP via oxidative phosphorylation. Oxygen serves as the terminal electron acceptor. Adapted from Matsushita, K., H. Toyama, and O. Adachi. 1994. Respiratory chains and bioenergetics of acetic acid bacteria, p. 247–301. In A.H. Rose and D.W. Tempest (ed.), *Advances in Microbial Physiology, vol. 36*. Academic Press, Ltd., London.

below) have been obtained. Most industrial strains of *Acetobacter* contain plasmids, whose loss could conceivably account for phenotypic changes. However, except for the presence of antibiotic resistance genes, functions for other plasmid-borne genes have not been identified, and most plasmids are considered to be cryptic and not essential for acetic acid formation.

In contrast, the presence of insertion sequences (IS) in *Acetobacter* has been associated

with several loss of function mutations. Insertion sequences are small DNA sequences, usually less than 3 kb, that contain inverted repeats and other characteristic features. They can transpose or integrate, randomly or at specific sites within the genome, resulting in the disruption of gene encoding regions. For example, in *A. pasteurianus, A. aceti, G. xylinus*, and other *Acetobacter* sp., the high-copy number IS element, IS*1380*, was found to integrate within the gene coding for cytochrome c, which, along with alcohol dehydrogenase, is essential for ethanol oxidation. In addition, transposition of another insertion sequence, IS*1031*, in *A. xylinum,* led to the appearance of mutants that were unable to synthesize the cellulose-containing surface film necessary during surface type fermentations (see below). Thus, it appears that the genetic instability associated with acetic acid bacteria is likely due to inactivation of genes necessary for growth and acetic acid formation by transposable IS elements.

Vinegar Technology

The first step in the manufacture of vinegar, as noted earlier, is the production of an ethanolic substrate. For the most part, the relevant fermentations (i.e., wine and beer) are discussed separately (Chapters 9 and 10). However, it is worth noting that ethanol from musts, ciders, or malt mashes are usually prepared with the eventual end product (i.e., vinegar) in mind. Thus, the starting materials do not necessarily have to be of the highest quality. However, this does not mean that poor quality materials should be used, since many of the flavor and aroma properties of the finished vinegar are derived from the starting material. Of course, if distilled ethanol is used as the substrate, then the resulting vinegar will lack most of these flavor and aroma characteristics.

The second step of vinegar production, conversion of ethanol substrate to acetic acid, can be performed by one of several methods, all of which rely on oxidative fermentation of ethanol by acetic acid bacteria. These processes are generally referred to as the open vat method, the trickling generator process, and the submerged fermentation process. The latter two are now the more widely used, especially for larger vinegar manufacturers, since they can be performed in continuous mode and can reduce the fermentation time from several weeks to a matter of days. However, even the traditional open vat method can be run in a semi-continuous fashion and is still used in many parts of the world. Moreover, the open vat process is the method of choice for the premium types of vinegar, including various wine vinegars, such as Balsamic (Box 11-2).

Open vat process

The open vat process relies on surface growth of acetic acid bacteria in vats, barrels, jars, or

Box 11–2. The Special Appeal of Specialty Vinegars (or, "You Paid How Much for That Vinegar?")

By far, most of the vinegar produced in the United States is used by the food industry in the manufacture of salad dressings, pickles, catsup, mustard, and a variety of other processed foods. In general, the vinegar used for these products should be inexpensive and plain-tasting. Even the more flavorful wine, malt, and cider vinegars available at the local grocery store and used as table vinegar carry rather modest retail prices.

A visit to the gourmet section, however, will reveal that some specialty vinegar products are as expensive as fine wines. Among the most well-known of the specialty vinegars are the sherry and balsamic vinegars. They are made, respectively, in the Jerez region of Spain and the Modena region of Italy. Although versions are now made in the United States and elsewhere, these are protected names in Europe, where their production and authenticity are highly regulated.

Balsamic vinegar is striking in several respects. It has a very dark brown appearance, a strong complex aroma, and a sour, but slightly sweet flavor. Importantly, the fermentation of balsamic vinegar is quite different than traditional vinegars. The starting material is musts obtained from a variety of grapes, mainly Trebbiano, but also Lambrusco and Sauvignon. The musts are concentrated two-fold by raising the temperature to near boiling, followed by simmering until the mixture is syrup-like. Thus, step-wise process can take as long as two to three days and results in a

(Continued)

sweet solution with a sugar concentration of 20% to 24%. Next, the material is inoculated with a "mother culture" (from a previous batch) and transferred into premium quality wooden barrels (examples include oak, chestnut, mulberry, and cherry wood). These wild cultures contain various osmophilic yeasts, including strains of *Saccharomyces* and *Zygosaccharomyces*, and the acetic acid bacterium *Gluconobacter*. An alcoholic fermentation and the acetic acid fermentation essentially occur at very near the same time, since *Gluconobacter* can oxidize sugars directly to acetic acid.

Another unique feature of the balsamic fermentation process concerns the manner in which the vinegar is aged. Rather than aging the product in bulk in a single barrel, balsamic vinegar is aged in a series of steps in which the product is transferred, via decantation, from barrel to barrel (Figure 1). Each barrel is usually constructed from different woods, which contributes to the flavor and aroma of the finished product. During each transfer step, about half of the volume from the preceding barrel is moved to the next barrel.

The barrels (usually in sets of five or seven) are held in attics that are exposed both to very warm and very cool temperatures, depending on the season. Warm temperatures stimulate fermentation, whereas cool temperatures promote sediment formation (which improves clarity). The rich and aromatic flavors associated with balsamic vinegar are due, in part, to the grape constituents and the volatile and non-volatile components extracted from the wood, but also to the acids, esters and other metabolic end products produced by the microflora. To achieve Denominazione di Origine Controllate (DOC) status and Protected Designation of Origin (PDO) certification, the vinegar must be aged for at least twelve years, but some balsamic vinegars are aged for as long as twenty-five to fifty years. Less than 12,000 liters of authentic balsamic vinegar (Aceto Balsamico Tradizionale) are released for sale per year, thus it is not surprising that retail prices for 90 ml bottles can be more than $150. While more moderately priced versions exist (Aceto Balsamico di Modena), they usually are diluted with wine vinegars or aged less.

Sherry vinegars are produced via the "solera" process in a manner similar to that used for Sherry wines (Chapter 10). Specifically, a portion (less than one-third of the total volume) of the vinegar is transferred from cask (or "butt") to cask. Each successive portion is more aged than the preceding material, such that the final product has a uniform and consistent quality. Aging typically occurs for a minimum of six months, to as long as two years or more (to earn "Reserva" status). Flavors are, as for Balsamic vinegar, derived from fermentation, as well as the wood used

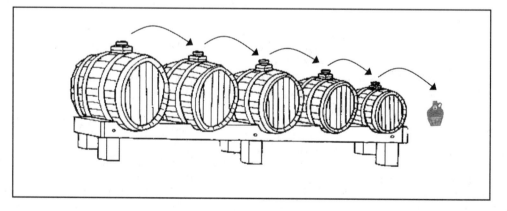

Figure 1.   Sequential barrel formation used during manufacture and aging of balsamic vinegar. Successive barrels are made from different woods, and as aging progresses, the size of the barrels decreases.

Box 11–2. The Special Appeal of Specialty Vinegars (or, "You Paid How Much for That Vinegar?") *(Continued)*

during aging. The latter contribute important phenolic compounds associated with aged sherry vinegar.

Finally, it should be noted that distinctive specialty wine vinegars are not just produced in Modena, Jerez, or Orleans, but also in the United States. A growing market for varietal vinegars, made from single variety wines, has emerged, especially in the wine-producing regions in California. These vinegars are fermented in barrels; some producers have adopted traditional Orleans-style fermentors and long aging periods.

References

Cocchi, M., P. Lambertini, D. Manzini, A. Marchetti, and A. Ulrici. 2002. Determination of carboxylic acids in vinegars and in Aceto Balsamico Tradizionale di Modena by HPLC and GC methods. J. Agric. Food Chem. 50:5255–5261.

García-Parilla, M.C., F.J. Heredia, and A.M. Troncoso. 1999. Sherry wine vinegars: phenolic composition changes during aging. Food Res. Int. 32:433–440.

trays. This was undoubtedly the first method used for vinegar production, since it can be easily performed and involves naturally-occurring or wild cultures. Although many variations exist, barrel-type fermentations are typical of the processes that evolved in Europe and that are still used today. The most well-known example is the Orleans process, which takes its name from the French city where this particular method was originally developed. Similar processes also evolved independently in the Far East, but the latter were based on rice wine and other types of alcohol-containing substrates, rather than on grape wine or cider. The principle of the process is simple and straightforward. The ethanolic substrate is placed in a suitable vessel, and the fermentation is initiated either by acetic acid bacteria that naturally contaminate the vessel or by a portion of vinegar from a recent batch. The material is exposed to the atmosphere, but is otherwise left undisturbed (hence, this process is sometimes referred to as the "let alone" method).

In the open vat process, acetic acid bacteria grow only at the surface. Growth is accompanied by production of a polysaccharide-containing film or pellicle that forms at the liquid-air interface. Although most strains of *Acetobacter* are capable of pellicle formation, this property can occasionally be lost if cells are grown under non-static (i.e., shaking) con-

ditions (see above). This is important during open vat processes, because an intact film is essential to the success of the fermentation. Any disturbance or disruption of the film may delay the fermentation. Thus, open vat processes may take several weeks before the fermentation is established, and several months before the fermentation is complete.

Traditional open vat or surface film fermentations ordinarily operate in batch mode. After the fermentation is complete, the vessel is emptied, then refilled with substrate. The acetic acid bacteria, left over or added, must then re-establish surface growth. It is possible, nonetheless, to perform this fermentation in a semi-continuous mode, provided steps are taken to maintain the surface film during substrate addition and product removal steps.

In the Orleans process, for example, the barrels are constructed such that a filling device (i.e., a pipe) extends from just outside the top of the barrel all the way to the inside bottom (Figure 11–5). Aeration is provided by holes (covered with cheesecloth) drilled into the sides of the vessel. A tap is positioned at the bottom end that is used to withdraw the product. The barrel is filled with wine to about 60% to 70% capacity and inoculated with a fresh vinegar culture ("mother of vinegar"). When the acetic acid fermentation is completed, it is then possible to remove a portion (from one-third to as much as

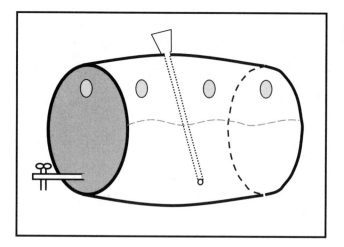

Figure 11–5. Schematic illustration of vinegar barrel used for the Orleans process. Adapted from Adams and Moss, 1995.

two-thirds) of the fermented vinegar from the tap end, and to replace it with fresh wine or cider, through the filling device, without disrupting the film. The process takes two to three weeks, or longer, for each completed cycle, depending on the temperature and the rate of volume exchange. Although other methods require significantly less time, the quality of vinegar produced by the Orleans and similar barrel methods is considered to be far superior to those produced by more rapid methods.

Trickling generator processes

In the vinegar fermentation, the rate at which ethanol is oxidized to acetic acid depends on the presence and availability of oxygen and the surface area represented by the air-liquid interface. In other words, the ability of acetic acid bacteria to perform the acetic acid fermentation is limited primarily by the diffusion or transport of oxygen from the atmosphere to the cell surface. It is possible to significantly accelerate the oxidation of ethanol by increasing the surface area to which oxygen is exposed. This observation, which was originally described nearly 300 years ago by the famous Dutch physician-chemist Hermann Boerhaave, forms the basis for the trickling generator processes, which are sometimes referred to as simply quick vinegar processes.

The earliest examples of trickling generators used on an industrial scale were the Schutzenbach and Ham processes, introduced in 1823 and 1824, respectively. In these systems, which are still widely used, ethanolic substrates are circulated or trickled through cylindrical fermentation vessels or vats containing inert packing materials, such as curled wood shaving, wood staves, or corn cobs (Figure 11-6). Many other materials can be used, provided they do not contribute flavor or are degraded. The main requirement is that they have a large surface area.

As the inoculated substrate or feedstock passes from the top to the bottom of the vessel, the total surface area of the liquid material increases as it moves around and between the particulate packing material. Growth of acetic acid bacteria will then occur at the air-liquid interface, such that the ethanol concentration decreases and the acetic acid concentration increases during the transit of substrate from top to bottom. Holes can be drilled into the side of the vessel to ensure that aeration is adequate. When the substrate reaches the bottom section (the collection chamber), it may be returned to the top until the effluent is sufficiently acidic to be called vinegar. Alternatively, a second tank can be used to increase the oxidation rate and product throughput. The inoculated mash is fermented first in Tank 1, then

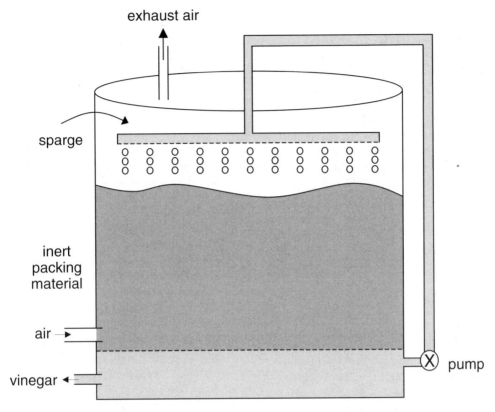

Figure 11–6. Trickling generator system.

passed, with aeration, into Tank 2, and back and forth until all of the ethanol is converted to vinegar. About three days are required to convert a 12% (v/v) ethanol solution to a vinegar containing 10% to 12% acetic acid.

Many modern trickling generator-type systems have been developed over the past seventy years based on the same principle as described above, but with more control and operational features. The Frings generator, introduced in the 1930s, and still widely used today in various modified formats, provides a means for incorporating air, in a counter-current direction, directly into the fermentation vessel. These systems can also accommodate a heat-exchange step, so that a constant temperature can be maintained. In addition, the feedstock is ordinarily dispensed via a sparging arm that breaks up the liquid into smaller droplets

that are more easily distributed within the interior matrix.

Submerged fermentation

Advances in biotechnology, and in particular, the fermentation industry, have led to the development of modern industrial fermentors capable of rapid, high throughput bioconversion processes. Although submerged fermentation systems are now widely used for many industrial fermentation processes, they were actually developed by the vinegar industry (and the Heinrich Frings Company, in particular) more than fifty years ago. The Acetator (the Frings fermentor), Cavitator, Bubble fermentor, and other similar units are constructed of stainless steel, can be easily cleaned and sanitized, and can operate in batch, semi-continuous, or continuous mode. Most of the vinegar produced worldwide is now made using submerged fermentation systems.

The most important feature of the submerged fermentation systems is their ability to provide rapid and efficient aeration. For example, the Frings Acetator is equipped with turbines that mix the liquid with air or oxygen and deliver the aerated mixture at very high rates inside the fermentor. The aeration system's ability to break up air bubbles and facilitate transfer of oxygen molecules from the gas phase to the liquid phase is essential, since the success of submerged fermentation systems relies on the transfer of oxygen from the medium to the bacteria. The viability of *Aceto-bacter* cells can be quickly compromised if aeration is even momentarily lost.

In addition to the aeration system, process controls for propeller speed (up to 1,750 rpm), nutrient feed, foam-handling, and temperature further provide for consistent, semi-continuous operation. Temperature control may be of particular concern in large-scale fermentations because ethanol oxidation is an exothermic reaction and heat is released. The medium temperature, if not controlled, can increase from an optimum of 28°C to 30°C to as high as 33°C. Even this modest increase in temperature can result in a serious decrease in fermentation productivity.

The size and design of submerged fermentation systems vary considerably. Production capacities of 25 liters per day of 10% acetic acid for very small-scale fermentors, up to 30,000 liters per day for large-scale reactors, are possible. In general, the starting material contains both ethanol and acetic acid (from a prior fermentation). After a single cycle of sixteen to twenty-four hours, the ethanol concentration will be reduced from about 5% to less than 0.5%, and the acetic acid will increase from around 7% to 12%. Up to half of the vinegar is then removed and replaced with fresh ethanol feedstock. Cell growth during a typical fermentation cycle is modest, with only a single doubling of the initial population.

Post-fermentation processing

Depending on the manner in which the fermentation occurs, the vinegar may be relatively clear, in the case of barrel-aged product, or very turbid, in the case of submerged fermentation-produced product. Thus, vinegar obtained by the Orleans method or after aging and subsequent transfer in barrels may require little, if any, filtration. In contrast, removing suspended cells from vinegar produced by submerged fermentation generally requires more elaborate filtration treatments, including the use of inert filtration aids. Vinegar produced by trickling fermentation processes may also require a filtration treatment.

Although it may seem counter-intuitive to pasteurize vinegar (after all, what can grow in a 1 M acetic acid solution?), heat treatments (up to 80°C for thirty to forty seconds) are common in the vinegar industry. Usually the targets are acetic acid bacteria, lactic acid bacteria, and wild yeasts and fungi. As noted earlier, some vinegars are aged, usually in wooden barrels, although this practice is usually reserved for high quality wine vinegars (Box 11–2).

Spoilage, bacteriophages, and other problems

Spoilage of vinegar by microorganisms is rare, although aceto-tolerant fungi, such as *Moniliella acetoabutens*, have occasionally been known to grow in raw vinegar. The more common problem, at least for open vat or trickling type processes, is the occasional contamination with mites and flies. The vinegar eel, in particular, was once quite common, especially in traditional open vat or trickling fermentation systems, but is now infrequently present. The eel is actually a small worm (i.e., a nematode classified as *Anguillula aceti*) whose main effect is aesthetic rather than influencing the fermentation or product quality. Vinegar eels are readily removed by filtration and easily killed by heat.

Given that vinegar fermentations are conducted in the presence of ethanol and acetic acid, one might expect that bacteriophages would not be a serious problem. However, the absence of a heating step for the substrate, the repeated use of the same culture, and the generally open manner in which vinegar fermentations are conducted, certainly provides the right opportunities for phage proliferation. Even submerged fermentations in enclosed stainless steel fermentors do not preclude phage infections, unless pasteurized substrates are used or other phage control measures are applied. Nonetheless, it appears that although phage problems have occasionally been reported, the occurrence is apparently low, and there are only a few published accounts of phage infections in vinegar production facilities (although the incidence of unpublished phage problems may be higher).

In general, phages are more commonly associated with trickling generator processes, because those fermentations are usually conducted by a diverse mixture of strains where phages would likely proliferate. In fact, trickling processes are considered to be the main reservoir for *Acetobacter* phages. However, submerged fermentation systems are likely more sensitive to phage problems, since they often rely on a single strain whose infection could conceivably result in a slow or failed fermentation.

Perhaps the most frequent microbial problem associated with vinegar spoilage is caused by oxidation of acetic acid by *Acetobacter*. As described earlier, these bacteria have the metabolic capacity to oxidize the acetic acid to CO_2 and water. However, the genes that encode for citric acid cycle enzymes necessary for acetic acid oxidation are not induced in the presence of ethanol. Moreover, acetic acid oxidation is apparently repressed when the acetic acid concentration is below 6% to 7%. Thus, under the ordinary conditions in which the acetic acid fermentation occurs (i.e., when the ethanol concentration ranges between 2% and 10% and the acetic acid concentration is above 6%), the oxidation pathway is not active. Only under prolonged fermentations, when the ethanol is completely dissipated and when the

acetic acid concentration is low, would overoxidation occur.

Vinegar Quality

At least two criteria must be considered to assess the quality of vinegar. In contrast to other fermented foods, where authenticity is rarely disputed, vinegar can be adulterated, either with less expensive vinegar or with other acidic agents. In rare instances, adulteration is caused by even more dubious means. For example, acetic acid produced via chemical synthesis can be diluted and then marketed (illegally) as fermentation-derived vinegar. Thus, the first quality criterion is based on establishing that the vinegar is actually vinegar (and not just diluted acetic acid) and that the type of vinegar indicated on the label accurately represents what is in the product. Satisfying this requirement, however, is no easy matter, and requires a statistical multivariate approach (reviewed nicely by Tesfaye et. al., 2002). Discriminating analyses that are used to distinguish samples of vinegar involve measuring selected chemical constituents, including minerals, alcohols, acids, phenolics, and other volatile compounds. Using such an approach, it is even possible to differentiate the fermentation process used to produce the vinegar (e.g., traditional versus quick acetification methods).

The other main characteristics on which vinegar quality is based are those that relate to flavor, aroma, and other organoleptic properties. Vinegar flavor is particularly influenced by the raw ethanolic material from which it was made. Thus, wine vinegars contain a mixture of phenolic compounds that are ordinarily present in grapes. And although acetic acid is by far the predominant flavor present in vinegar, other volatile flavor compounds are also present and contribute to the overall flavor profile of vinegar. The enzymes—alcohol dehydrogenase and aldehyde dehydrogenase—that oxidize ethanol and acetaldehyde, respectively, also oxidize other alcohols and aldehydes. Thus, if the must, mash, or other starting material contains these substrates, a diverse array of end products can be obtained. These products include formic, propionic, and butyric acids, and glycerol. Other volatile compounds commonly found in vinegar include diacetyl, acetoin, and ethyl acetate.

Of course, the ultimate measure of vinegar quality, as for other fermented food products, is based on sensory analysis. Performing these analyses for vinegar, however, is particularly challenging since acetic acid has a strong, overpowering flavor. Therefore, samples can be diluted, mixed with lettuce, or even served in small amounts in wine glasses. In fact, sensory analysis of vinegar is not unlike that used for wine analysis, in that both generally require trained tasters and both rely on similar descriptors (e.g., woody, fruity, etc.).

Bibliography

Adams, M.R. Vinegar. p. 1–44. In B.J.B. Wood (ed.) *Microbiology of Fermented Foods, Volume 1, 1998.* Blackie Academic and Professional (Chapman and Hall). London, United Kingdom.

Adams, M.R., and M.O. Moss. 1995. In Food Microbiology. Royal Society of Chemistry. Cambridge, U.K.

Beppu, T. 1993. Genetic organization of *Acetobacter* for acetic acid fermentation. Antonie Van Leeuwenhoek 64:121–135.

Cleenwerck, I., K. Vandemeulebroecke, D. Janssens, and J. Swings. 2002. Re-examination of the genus *Acetobacter*, with descriptions of *Acetobacter cerevisiae* sp. nov. and Acetobacter malorum sp. nov. Int. J. Syst. Evol. Microbiol. 52:1551–1558.

Coucheron, D. 1991. An *Acetobacter xylinum* insertion sequence element associated with inactivation of cellulose production. J. Bacteriol. 173: 7523–5731.

Fuchs, G. 1999. Oxidation of organic compounds, p. 187–233. In J. W. Lengeler, G. Drews, and A. S. Schlegel (ed.), *Biology of the Prokaryotes.* Blackwell Science, Stuttgart, Germany.

Garrity, G.M., J. Bell, and T.G. Lilburn. 2004. Taxonomic Outline of the Prokaryotes. *Bergey's Manual of Systematic Bacteriology, 2nd ed.* Release 5.0, Springer-Verlag. DOI: 10.1007/bergeysonline.

Holt, J.G., N.R. Krieg, P.H.A. Sneath, J.T. Staley, and S.T. Williams. 1994. Gram-negative aerobic/microaerophilic rods and cocci, p. 71–174. In *Bergey's Manual of Determinative Bacteriology, 9th Ed.,* Williams and Wilkins Co. Baltimore, Maryland.

Kondo, K., T. Beppu, and S. Horinouchi. 1995. Cloning, sequencing, and characterization of the gene encoding the smallest subunit of the three-component membrane-bound alcohol dehydrogenase from *Acetobacter pasteurianus.* J. Bacteriol. 177:5048–5055.

Krahulec, J., M. Kretová, and J. Grones. 2003. Characterization of plasmids purified from *Acetobacter pasteurianus 2374.* Biochem. Biophys. Res. Comm. 310:94–97.

Lisdiyanti, P., H. Kawasaki, T. Seki, Y. Yamada, T. Uchimura, and K. Komagata. 2000. Systematic study of the genus *Acetobacter* with descriptions of *Acetobacter indonesiensis* sp. nov., *Acetobacter tropicalis* sp. nov., *Acetobacter orleanensis* (Henneberg 1906) comb. nov., *Acetobacter lovaniensis* (Frateur 1950) comb. nov., and *Acetobacter estunensis* (Carr 1958) comb. nov. J. Gen. Appl. Microbiol. 46:147–165.

Matsushita, K., H. Toyama, and O. Adachi. 1994. Respiratory chains and bioenergetics of acetic acid bacteria, p. 247–301. In A.H. Rose and D.W. Tempest (ed.), *Advances in Microbial Physiology, vol. 36.* Academic Press, Ltd., London.

Sievers, M., and M. Teuber. 1995. The microbiology and taxonomy of *Acetobacter europaeus* in commercial vinegar production. J. Appl. Bacteriol. Symp. Suppl. 79:84S–95S.

Sokollek, S.J., C. Hertel, and W.P. Hammes. 1998. Description of *Acetobacter oboediens* sp. nov. and *Acetobacter pomorum* sp. nov., two new species isolated from industrial vinegar fermentations. Int. J. Syst. Bacteriol. 48:935–940.

Stamm, W.W., M. Kittelmann, H. Follmann, and H.G. Trüper. 1989. The occurrence of bacteriophages in spirit vinegar fermentation. Appl. Microbiol. Biotechnol. 30:41–46.

Takemura, H., S. Horinouchi, and T. Beppu. 1991. Novel insertion sequence IS*1380* from *Acetobacter pasteurianus* is involved in loss of ethanol-oxidizing ability. J. Bacteriol. 173:7070–7076.

Takemura, H., K. Kondo, S. Horinouchi, and T. Beppu. 1993. Induction by ethanol of alcohol dehydrogenase activity in *Acetobacter pasteurianus.* J. Bacteriol. 175:6857–6866.

Tesfaye, W., M.L. Morales, M.C. García-Parrilla, and A.M. Troncoso. 2002. Wine vinegar: technology, authenticity and quality evaluation. Trends Food Sci. Technol. 13:12–21.

Thurner, C., C. Vela, L. Thöny-Meyer, L. Meile, and M. Teuber. 1997. Biochemical and genetic characterization of the acetaldehyde dehydrogenase complex from *Acetobacter europaeus*. Arch. Microbiol. 168:81–91.

Yamada, Y., K. Katsura, H. Kawasaki, Y. Widyastuti, S. Saono, T. Seki, T. Uchimura, and K. Komagata. 2000. Asaia bogorensis gen. nov., sp. nov., an unusual acetic acid bacterium in the α-Proteobacteria. Int. J. Syst. Evol. Microbiol. 50:823–829.

12

Fermentation of Foods in the Orient

"Chiang (fermented soybean paste) is to food, what a general is to an army."
Yen Shiu-Ku, seventh century writer

History

Like the bread, wine, dairy, and other fermentations described in earlier chapters, Asian-type fermented foods also evolved thousands of years ago. In addition, these Asian fermented foods had many of the same general characteristics and properties as those that developed in Middle Eastern and Western cultures. That is, the products were comprised of ingredients native to their geography, they had enhanced functional properties, and they were well preserved. Likewise, Asian fermented foods were also subject to cultural, economical, and religious influences.

Despite these similarities, the starting materials, the specific types of products produced in the Far East, and the means by which fermentation occurs are dramatically different from products that evolved in the West. Also, whereas the fermented foods industry in Westernized countries, and especially in the United States, have become large, mechanized, and technologically well defined, production methods for fermented foods in the Orient vary greatly, with many small manufacturers still in operation. The trend toward large-scale production, however, has now become evident in China, Japan, Malaysia, Thailand, Korea, and many other Far East countries.

One of the main differences between the fermented foods that evolved in West and those that evolved in the Far East was that the latter populations depended far less on animal products. Several reasons likely accounted for this difference. First, religions practiced in the Orient often excluded meat products and instead promoted diets based on plant proteins. Buddhism, for example, which developed in the sixth century B.C.E., prohibited animal foods in the diet. Second, that the Far East was generally more densely populated meant that less space could be devoted to animal agriculture so greater emphasis was placed on growing plant foods for human consumption. In addition, economic pressures led to greater reliance on inexpensive food commodities, such as cereals, grains, and legumes. Finally, the ready availability of fish and seafood provided an abundant source of inexpensive, high-quality protein.

The other major differences between fermented foods of the West and East relate directly to the fermentation substrates, the manner in which the fermentations occur, and the types of microorganisms that are involved. The substrates used for the fermented foods of the West, for example, are usually simple sugars, whereas in many of the Asian-type fermentations, the substrates consist largely of carbohydrates in the form of starch and other polysaccharides. There are, for example, little if any free fermentable sugars present in rice and soybeans. Therefore, the manufacture of the Asian fermented products must start with a step that converts the starch to sugars that can be used by fermentative organisms. In the West, there is one product that shares this problem—beer.

As described in Chapter 9, the conversion of starch to free sugars is catalyzed by various amylolytic enzymes present in malt during a step called mashing. The free sugars, mainly maltose and glucose, are then readily fermented to ethanol and carbon dioxide by yeasts. In the production of Asian fermented products, a very similar mashing process occurs, with one significant difference. The enzymes necessary for saccharification (i.e., starch hydrolysis) are supplied not by malted cereals, but rather by fungal organisms, previously grown on cereal matter. Like malt, the production of this enzyme-laden fungal material is itself an important part of the overall manufacturing process.

Another important difference between fermented foods of the East and West, as noted above, concerns the very nature of the fermentation. For the most part, dairy, meat, vegetable, and ethanolic fermentations are conducted under predominantly anaerobic conditions. In contrast, aerobic conditions are necessary (at least, for certain critical parts of the process) to produce many of the Asian fermented products. Moreover, whereas pure starter cultures, containing defined strains, are now well accepted and widely used for most fermented food products in the West, defined cultures are infrequently used (except by large manufacturers) in the production of Asian fermented foods. Finally, although fungi are used in the manufacture of a few Western-type fermented foods, such as some cheeses and sausages, they are essential in most of the Asian fermented foods. In fact, the food products described in this chapter are often referred to simply as fungal-fermented foods.

It should be noted that the consumption of Asian fermented or fungal-derived foods is no longer confined to Asia or Asian populations. Chinese foods have long been part of the Western cuisine, but in the past two decades, Japanese, Thai, Vietnamese, Indonesian, Korean, and other cuisines from the Far East are now widely consumed throughout the world. It is remarkable, however, that few consumers realize that so many of these foods contain, as primary ingredients or constituents, fermented foods products. Perhaps even more surprising is that many of the Asian fermented foods have captured a large part of the U.S. ethnic foods market, such that many of these products, including tempeh, miso, kimchi, and fish sauce, are available in American grocery and specialty food stores. Due to the savory, meat-like flavor some of the products impart, they have also become popular as vegetable-based meat substitutes (as "faux" meats). There are now several manufacturing facilities in the United States that produce these products, and their popularity is predicted to increase in coming years.

Types of Asian Fermented Foods

There are hundreds of different types of fermented foods produced in China, Japan, the Philippines, and throughout Asia (Beuchat, 2001). However, there are two general types of fermented foods that are associated with or are indigenous to Eastern or Asian cuisines: those that are plant-based and those that are fish-based. The former are made using primarily soy and rice as substrates, but other grains and legumes are also used. For the most part, these soy-based fermentations have been industrialized and the products are now produced on a large scale. In addition, many of the microbiological and biochemical events that occur during the fermentation have been studied and defined. The fish-based products use various types of fish and seafood that are indigenous to the Orient. However, in contrast to the soy- or plant-based fermentations, the fish fermentations are generally not well defined, nor has there been, until recently, very much published research (at least in English-language journals) on the microorganisms and biochemical reactions involved in their manufacture.

Plant-based Fermentations

Soy and rice are the most frequent substrates for Asian fermented foods. Wheat flour is also often included as an ingredient in many of these products. With few exceptions, however, there is a common starting material—koji—that is essential for most Asian fermented foods. As will be described in more detail below, koji is simply a moldy mass of grain, and is derived from the Chinese word meaning "moldy grain." In some cases, the koji mold is added to a portion of the raw material which is later added to the remaining substrate (analogous, in a way, to a bulk culture or bread sponge). In other applications, the koji mold is added to the entire raw material. Koji is used not only for production of the many soy sauce-type products that will be described later, but also for sake and related rice wines. Because of the important role koji plays in so many of the products that will described, its manufacture and its microbiological and enzymatic properties are treated separately.

Koji and Tane Koji Manufacture

First, it is necessary to recognize there are many types of koji used in the Far East, and that each fermented food requires a specific type of koji. Thus, Japanese soy sauces generally use a koji that is different from the one used to make Chinese-style soy sauces, and both are different from the koji used for sake manufacture. Generally, koji can be referred to by its intended product (e.g., sake koji or shoyu koji) or by the substrate from which the koji is prepared (e.g., rice koji, barley koji, or soybean koji).

Despite the type of koji or how it is made, the purpose and function are always the same. Namely, the koji provides a source of enzymes necessary to convert solid, raw materials containing complex and non-fermentable substrates into soluble, simple, metabolizable products that can be easily fermented by suitable microorganisms. It is also important to realize that koji contains not only amylolytic enzymes, but an array of enzymes capable of hydrolyzing proteins and peptides, lipids, cellulose, pectin, and other complex substrates. The koji also serves as a source (often the main source) of substrates for the enzymes produced by the koji fungi.

Raw materials preparation

The manufacture of koji starts by treating the raw materials. When soybeans are used, they are first soaked for about twelve hours in several changes of water. For traditional koji-making, they are then cooked, usually with steam under pressure, for one hour—a process called puffing. For some products, soybean koji also contains wheat, which is prepared by roasting wheat kernels (or wheat flour) at a high temperature, followed by a crushing step. Rice koji for sake manufacture is prepared from polished rice (rice minus the bran) that is steeped and steamed, similar to soybean koji.

Despite the seemingly simple process involved in preparing the koji substrate, numerous innovations have been designed to enhance the digestibility of the raw soy beans and ultimately improve yield and product quality. For soy sauce, digestibility or yield is based, in large part, on the amount of total protein nitrogen that is converted to amino nitrogen during the mashing step. However, how the soy beans are initially heat processed influences the rate and extent that the soy proteins are hydrolyzed. This is because only denatured soy proteins are hydrolyzed by fungal proteinases, whereas native proteins are not. Thus, a cooking step is essential. Yields of about 82% can be achieved when a traditional cooking process (e.g., 118°C at 0.9 kg/cm^2 for 45 minutes) is used. In contrast, high-temperature, short-time cooking not only enhances protein extraction, but also gives amino acid yields greater than 90%. It is now possible, for example, to perform the cooking step in less than two minutes, using temperatures and pressures above 150°C and 5 kg/cm^2, respectively.

Microorganisms

Once the koji substrates are prepared, there are essentially two means by which they can be inoculated. One is simply to use a pure culture containing spores of *Aspergillus oryzae* and/or *Aspergillus sojae*. Alternatively, the mixture can be inoculated with 0.1% to 0.2% of a seed culture called tane koji. Tane koji is made by inoculating soaked, steamed, and cooled rice (usually brown rice) with spores of *A. oryzae* or *A. sojae*. The inoculated rice is incubated overnight, then transferred and distributed evenly into shallow trays and incubated at 30°C for five to seven days. Intermittent mixing provides aeration necessary for fungi growth. Eventually, the moldy rice mixture is dried and packaged. This tane koji material is then used to inoculate other koji mixtures.

Ordinarily, multiple strains are used to make tane koji, depending on the final product that is manufactured. In general, koji mold strains should produce proteolytic, peptidolytic, and amylolytic enzymes with high activity, and they should sporulate and grow well on their substrate. Some species of *Aspergillus* are known to produce aflatoxin and other mycotoxins. However, none of the koji strains that have been isolated and studied produce these toxins under production conditions. Finally, koji strains should be genetically stable and produce products having consistent flavor and color properties.

After the koji substrate (rice, soybeans, wheat, barley) has been inoculated with either a pure spore culture or the tane koji, it is mixed and incubated in large rectangular trays or boxes (5 × 12 meters) with a depth of about 30 to 40 cm. Very large producers may use vats that can accommodate even greater quantities. Perforations in the trays enhance air and moisture circulation. Temperature control is particularly important because some proteolytic enzymes tend to be produced at lower temperatures (25°C to 30°C), and amylases are produced at higher temperatures (>30°C to 35°C). Modern manufacturers now rely on temperature programs to maximize enzymatic activities. In fact, the incorporation of continuous cookers, automated mixers, and other modern devices has made it possible to automate the entire koji-making operation so that as much as 3,000 kg of koji can be produced per hour. After two to four days at 30°C, the mass should be completely covered throughout with mold growth.

Manufacture of Soy Sauce and Related Products

Soy sauce is one of the most widely consumed products in Asia. In Japan, per capita consumption is more than 10 liters per person per year, or more than 30 g per day. Dozens of different types of soy sauces are manufactured in Asia. In fact, even within the same country, there may be several distinct products, each having their own particular qualities and each made according to specific manufacturing procedures (Table 12–1).

Moreover, quality standards further distinguish one product type from another. For example, in Japan (where the name for soy sauce is shoyu), three shoyu production methods and five types of shoyu products are recog-

Table 12.1. Types of soy sauces.

Product	Country	Description
Shoyu	Japan	Five types based on soybean and wheat content
Chiang-yiu	China	Similar to shoyu tamari (mostly soybeans)
Kecap	Indonesia	Two types, manis (sweetened) and asin (salty)
Kicap	Malaysia	Tamari type
Kanjang	Korea	Tamari type
See-iew	Thailand	Two types, sweetened and salty
Toyu mansi	Philippines	Lemon-like flavored

Table 12.2. Compositions of different types of shoyu[1].

Product	Total N[2] (g/100 ml)	Reducing sugars[3] (g/100 ml)	Ethanol (v/v)	pH	Color
Koikuchi	1.6–2.0	2.8–5.5	2.1–2.5	4.7–4.9	Dark brown
Usukuchi	1.2	4.0–5.0	2.1–2.6	4.8	Light brown
Tamari	1.8–2.6	2.4–5.3	0.1–2.4	4.8–5.0	Dark red-brown
Shiro	0.5	14–20	0–0.1	4.5–4.6	Yellow/tan
Saishikomi	2.0–2.4	7.5–9.0	0–2.2	4.6–4.8	Dark brown

[1] Adapted from Yokotsuka and Sasaki, 1998 and Fukushima, 1989
[2] Total nitrogen, comprised of amino acids, small peptides, and ammonium salts
[3] Consisting of about 85% glucose, 10% galactose + mannose, and 5% pentoses

nized by the Japan Agricultural Standard (Table 12–2). In addition, there are three quality grades for each Japanese shoyu—Special, Upper, and Standard—that are assigned based on manufacturing methods, composition, flavor, and color. The most widely consumed variety of shoyu in Japan is koikuchi, which contains 50% wheat. In contrast, tamari, another Japanese shoyu, contains little or no wheat. Similar versions of these soy sauce products, containing either part wheat or no wheat, exist in China, Taiwan, and other Asian countries.

Koji

The manufacture of traditional soy sauce or shoyu starts with preparation of the raw materials and addition of the koji (Figure 12–1). The soy fraction consists of either whole beans, soy meals, or soy flakes. It is now common to use de-fatted flakes rather than whole beans or flakes to improve yield and reduce fermentation times. When beans are used, they are washed, sorted, and soaked overnight, then cooked under pressure and cooled rapidly. Flakes are simply soaked and cooked (as for whole beans). At the same time, whole wheat kernels, if included, are roasted and crushed. The wheat adds flavor and color, reduces the moisture so that growth of undesirable bacteria is minimized, enhances mold growth, and contributes glutamic acid-rich proteins. When wheat is used, the soy-wheat ratio depends on the manufacturer's preference and can range from 50:50 to 67:33.

The soy bean-wheat mixture is then inoculated with either tane koji or a pure culture of suitable fungal strains of *A. oryzae* or *A. sojae* to start the koji fermentation. As described above, the inoculated material is incubated in large trays, boxes, or vats that are perforated to allow air and moisture to circulate throughout the material and to enhance fungal growth. The temperature is maintained at about 30°C, which means that the incubation rooms or vats must have cooling capacity, since fungal growth generates heat. Once the koji is mature (i.e., covered completely with mold), it is fully developed and ready to be used for the fermentation.

Mashing

It is during the next step, mashing, where the koji enzymes begin to hydrolyze proteins, polysaccharides, and other substrates, and where microorganisms begin to use the products of these reactions. First, however, a high salt brine containing 20% to 25% sodium chloride is added to the solid material. The volume added may vary, depending on the manufacturer's specifications, but a ratio of about 1:1.2 to 1:1.5 (solid to brine) is normal. The high salt concentration in the mash (ranging from 16% to 19%) restricts growth of microorganisms to only those that are especially halo- or osmotolerant. The mash material, referred to as "moromi" or the "moromi mash," is allowed to ferment in large (300,000 L) tanks for up to a year. Whereas wooden tanks were commonly used, these have largely been replaced first by concrete and then resin-coated

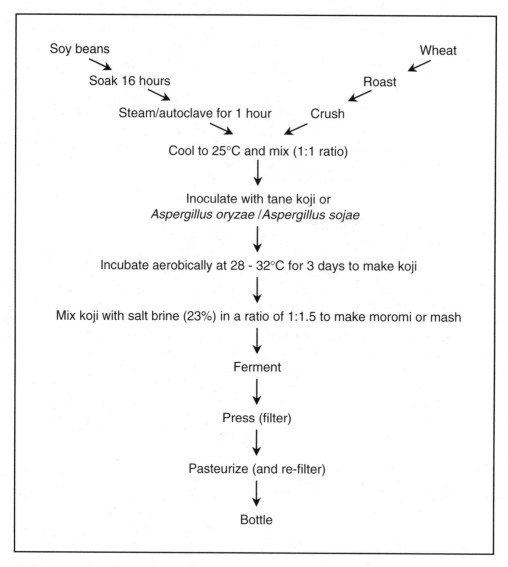

Figure 12–1. Shoyu flow chart.

iron tanks. Moromi held at lower temperatures (15°C) are thought to produce a better quality product, but take longer. In contrast, higher temperatures (as high as 35°C) can be used to complete the fermentation in as little as three to four months. Agitation and forced aeration generally enhances enzymatic activity and microbial growth during mashing.

Moromi enzymology

The koji molds, *A. oryzae* and *A. sojae*, produce a wide assortment of enzymes. They are especially prolific at secreting proteinases, peptidases, cellulases, and amylases as they grow in the koji. During the mashing step, these enzymes degrade soy (and wheat) proteins, starches, and other macromolecules to end products that are either necessary as nutrients or that can be fermented by the mash microflora. As noted above, the actual amounts and specific types of enzymes produced by koji *Aspergillus* can vary, depending on the incubation temperature, strain, and substrate used during koji manufacture. In general, the lower the temperature, the greater will be the enzyme activity (within a range of 20°C to 35°C).

Protein hydrolysis occurs via one of several different fungal proteinases and peptidases, each having specific pH optima and substrate preferences. In terms of enzyme activity, alkaline proteinase has more than ten times the activity of other proteinases. This enzyme is active between pH 6.0 and 11.0, although its pH optima is about pH 10.0. A semi-alkaline proteinase (pH optima 8.3), as well as at least two neutral- proteases and three acid-proteases (pH optima around 7.0 and 3.0, respectively), all contribute to hydrolysis of soy proteins. The peptides that are generated are further hydrolyzed by peptidases that also have individual pH optima and are specific for particular amino or carboxy amino acid residues. Enzymes that release glutamic acid or that convert glutamine to glutamic acid are especially important, because this amino acid is responsible for much of the flavor-potentiating characteristics of the soy sauces. Salt-tolerance is another important property of koji proteinases, since they will still retain activity in the presence of high salt concentrations. A nearly complete hydrolysis of soy proteins is desirable, since the subsequent lactic acid and yeast fermentation, color and flavor development, and overall organoleptic quality of the

finished soy sauce depend on amino acid formation during the mashing step.

Most of the amylolytic activity present during mashing is due to fungal α- and β-amylases that effectively convert starch to simple sugars. Although some sugars are used to support mold growth, most remain in the mash where they will later serve as substrates during the fermentation stage of the process. This simple sugar fraction is comprised of several reducing sugars, including glucose and maltose, as well as xylose, galactose, and arabinose that are released from other polysaccharides present in the soy or wheat mixture (discussed below). The reducing sugars play an especially important role in color and flavor development, since they react with free amines during nonenzymatic browning reactions.

The other main group of enzymes important during mashing are the cellulases, pectinases, hemicellulases, and other tissue-degrading enzymes. They enhance extraction of substrates from the soy bean and wheat tissues, thereby increasing yield and nutrient availability. Formation of pentoses and other sugars also provide additional substrates for microorganisms and for browning reactions.

Fermentation

Immediately after the addition of the salt brine to the koji, the mash will contain not only the *Aspergillus* strains, but also bacteria and yeasts that were present as part of the natural koji microflora (Table 12–3). The fungi are salt-sensitive and quickly lyse and die (Figure 12–2). Other organisms initially present include mostly *Micrococcus* and *Bacillus* species, with

Table 12.3. Microorganisms important in soy sauce.

Fungi	Bacteria	Yeast
Aspergillus oryzae	*Tetragenococcus halophilus*	*Zygosaccharomyces rouxii*
Aspergillus sojae	*Lactobacillus delbrueckii*	*Zygosaccharomyces soya*
Mucor sp.	*Leuconostoc mesenteroides*	*Candida versitalis*
Rhizopus sp.		*Torulopsis* sp.
		Hansenula sp.

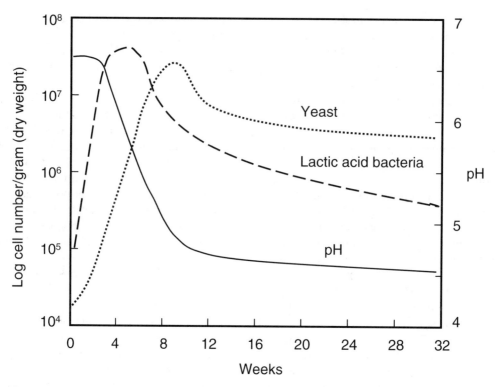

Figure 12–2. Idealized model for successive growth of lactic acid bacteria and yeast during the soy sauce fermentation.

starting levels as high as 10^8 cells per gram. In contrast, lactic acid bacteria and various yeasts are present at much lower initial cell concentrations (10^1 to 10^4 per gram).

Although the initial pH of the mash (6.5 to 7.0) is perfectly fine for all of these organisms, the high salt concentration poses a substantial barrier. A mash containing 18% salt will have an osmolality of more than 6 Osm and a water activity (a_w) of about 0.88, which is lower than most of the moromi organisms can tolerate. Thus, the salt-sensitive organisms, including the *Micrococcus*, wild yeasts, and most of the *Bacillus*, will not survive for long and will be undetectable in the mash within a month or two. In contrast, the restrictive conditions will select for those organisms capable of tolerating a high salt, high osmotic pressure, low a_w environment. Among the lactic acid bacteria that can tolerate these conditions and predom-

inate the early stages of the mash fermentation are strains of *Lactobacillus delbrueckii* and *Tetragenococcus halophilus* (formerly *Pediococcus halophilus*). Although they may be present initially at low levels ($<10^3$/ml) in the mash, they can reach cell densities of 10^7 cells/ml or higher after six to eight weeks. It is now common to add selected strains of these organisms to the moromi, rather than rely on their natural occurrence.

Salt-tolerant yeasts, *Zygosaccharomyces rouxii* and *Candida versitalis,* are also typically added to the mash. These yeasts are even more salt- and acid-tolerant than the lactic acid bacteria. When the mash pH falls below 5.0, growth of *L. delbrueckii* and *T. halophilus* (in the presence of more than 18% salt) will be inhibited, and yeasts will dominate the rest of the fermentation. Eventually, when the pH reaches 4.0, even *Z. rouxii* will stop growing.

The manner in which the lactic acid bacteria and yeasts grow during the soy sauce fermentation, therefore, resembles that of a succession-type fermentation.

A wide variety of fermentation end products accumulates in the mash and is responsible for the complex flavor of the finished soy sauce (Table 12–4). Growth of the homofermentative *T. halophilus* on glucose results primarily in lactic acid formation. Fermentation of other sugars, particularly pentoses, however, may yield heterofermentative products including ethanol, acetate, and CO_2, as well as other organic acids and alcohols. Acids, amines, ammonia, and other products may be produced from amino acid metabolism. The subsequent sugar fermentation by the yeasts also generates ethanol and CO_2 via the ethanolic pathway, as well as many other higher alcohols, esters, furanones, and other flavor volatiles.

Pressing and refining

After the fermentation is complete and the moromi is adequately aged, the soluble soy sauce must be separated from the solid residue. This is done using hydraulic filter presses that force the moromi through multi-layered sheets of cloth or nylon filters. Pressing can last for several days, with pressures increasing incrementally, up to 100 kg per cm^2. About 90 liters or more of liquid material can be obtained from 100 liters of moromi. The solid filter cake byproduct ordinarily is used as animal feed. Greater filtration efficiency and overall shoyu yield can be achieved by enhancing protein hydrolysis and pectinase activities during mashing.

Refining is an important step, since the liquid material that is collected after pressing still contains components that need to be removed. In particular, an oil layer that appears in the upper phase, and a sediment material that collects in the bottom phase, are both undesirable and are separated. The middle phase, containing the shoyu, can be re-filtered and standardized to a desired composition (based on salt and nitrogen concentrations and color and flavor).

Pasteurization and packaging

The refined, raw soy sauce is usually heat pasteurized, either in bulk or, more commonly, in a plate-type heat exchange system (at 70°C to 80°C). Flash pasteurization at 115°C or higher for three to five seconds may also be used.

Table 12.4. Concentration of important flavor compounds in shoyu[1]

Compound	Concentration (ppm)	Compound	Concentration (ppm)
Alcohols		**Aldehydes**	
Ethanol	31,501	Isovaleraldehyde	233
Methanol	62	Isobutyraldehyde	15
2-phenolethanol	4	acetaldehyde	5
n-propanol	4	Furanones	
Isobutyl alcohol	12	HMF[2]	256
Isoamyl alcohol	10	HEMF[2]	232
Furfuryl alcohol	12		
Acids		**Polyols**	
Lactic acid	14,347	Glycerol	10,209
Acetic acid	2,107	Acetoin	10
Esters		2,3-butanediol	238
Ethyl lactate	24	**Phenols**	
Ethyl acetate	15	4-EG[2]	3
Methyl acetate	14	4-EP[2]	0.3

[1]Adapted from Nunomora and Sasaki, 1992
[2]See text for chemical name

Aside from inactivating enzymes and microorganisms and thereby increasing shelf-life and stability, pasteurization contributes other important properties (Table 12-5). In particular, heating promotes color and flavor development by accelerating the non-enzymatic browning reaction. In addition, heat induces precipitation of insoluble proteins and other components that can then be removed by filtration to give greater product clarity. Finally, the heat step concentrates acids, phenols, and other compounds that either contribute desirable flavors or that have antimicrobial activity. Heating, however, may cause loss of volatile components.

Most soy sauces are packaged in glass bottles varying in size from 0.5 liters to 2 liters. Plastic bottles and 20 liter pails (for institutional use) are also used. Various chemical preservatives, including benzoic acid, benzoate salts, and ethanol, are often added.

Product characteristics

The final chemical composition of soy sauces depends on the specific type being produced (Table 12-2). In general, the greater the proportion of soy beans, relative to wheat, the greater will be the total nitrogen concentration. Conversely, products made using higher levels of wheat will contain less nitrogenous material, but more reducing sugars. Thus, shoyu tamari, which is produced mainly from soy beans, contains nearly four times more total nitrogen, but six times fewer reducing sug-

Table 12.5. Beneficial effects of pasteurization on soy sauce.

- Enhances color formation via the Maillard non-enzymatic browning reaction
- Inactivates enzymes
- Kills microorganisms
- Accelerates precipitation of protein complexes
- Concentrates anti-fungal agents (phenols and organic acids)
- Concentrates flavor compounds (phenolic compounds, aldehydes, pyrazines, ketones, and other organic compounds)

ars, than shiro, a type of shoyu made from wheat with very few soy beans. The amount of nitrogenous material and reducing sugars, as reflected by the wheat-to-soy bean ratio, has a major influence on color. Soy sauce products containing greater levels of wheat are generally more light-colored and products containing more soy beans are dark-colored. Shiro, therefore, has a more tan appearance whereas tamari has a dark red-brown color.

The final composition of soy sauce depends on the specific type and the manufacturer's specifications. In general, soy sauce contains (on a weight/volume basis) about 1.5% total nitrogen, 1% sugars, 1% lactic acid, 2% to 2.5% ethanol, and 14% to 18% salt. The pH is usually between 4.5 and 4.8. Due to the association of high salt diets with hypertension and other human health problems, there has been a considerable effort to reduce the salt concentration in soy sauce products. Among the strategies that have been considered are the use of salt substitutes, salt removal systems, and manufacturing modifications (Box 12–1).

The Flavor of Soy Sauce

Considering that most soy sauces contain up to 18% NaCl, the most immediate and obvious flavor one detects is saltiness. However, the flavor of soy sauce is far more complex than simply saltiness. In fact, the shoyu products listed in Table 12–2 all contain between 16% and 19% salt, yet their flavor profiles, and the concentrations of volatile flavor compounds, can vary considerably. Nearly 200 volatile flavor components have been identified in shoyu, using GC or GC/MS analysis. Several of these, in particular, various furanones and phenolic compounds, are considered to be the most important, based on their relatively high concentrations and their characteristic soy sauce-like flavor (Table 12–4).

Furanones include 4-hydroxy-2-(or 5) ethyl-5 (or 2) methyl-3-(2H) furanone (HEMF) and 4-hydroxy-5-methyl-3-(2H) furanone (HMF), which confer sweet or roasted flavor notes, respectively. HEMF has been considered to be the main "character-impact" flavor of most

Box 12–1. Taking Salt out of Soy Sauce and Other Soy-fermented Foods

Salt is an essential ingredient in many of the Asian-type fermented foods. For example, shoyu and other soy sauces contain between 14% and 18% salt. Likewise, miso contains as much as 13% to 14% salt. Even when used in moderation, these high-salt foods may still account for a relatively high intake of salt. A single 1 tablespoon serving of soy sauce (about 17 g) contains about 950 mg of sodium or 2.4 grams of salt, and a cup of miso soup contains nearly 2 g of salt.

Not surprisingly, those countries that regularly consume these products have high salt consumption rates. In Japan, for example, per capita salt consumption has averaged (from 1975 to 1997) around 13 g per person per day, 30% more than the 10 g per day that is recommended as a maximum. In contrast, salt consumption in the United States is somewhat less, about 10 g per person per day, but still nearly twice that recommended by health authorities.

The problem with high-salt diets is that there is a strong association between sodium or salt consumption and various disease states, including hypertension, stroke, and gastric cancer. This has led public health agencies to recommend that salt intake be reduced, especially for at-risk individuals. Since soy sauce and related products account for as much as 20% of total salt intake in Japan and other Asian countries, decreasing their salt content could contribute to a significant overall reduction in salt consumption.

There are a number of possible ways to reduce the sodium content in soy sauce types of products. The most common means is to manufacture the product in the usual way (i.e., adding the same amount of salt to the mash as would be done for a conventional fermentation), followed by removal of a portion of the salt via one of several separation technologies. The latter include ion exchange chromatography using de-salting resins, ion-exchange membrane processing, and electrolysis. Potassium chloride can then be added to satisfy many of the functions played by sodium chloride (e.g., preservation and flavor).

A second way to make low-salt soy products is to replace some of the sodium chloride with potassium chloride prior to the fermentation. The potassium salt will still influence the microflora, as does sodium chloride, by selecting for salt-tolerant lactic acid bacteria and yeast. The downside to both of these approaches, however, is that potassium imparts an undesirable bitter or metallic flavor that consumers generally dislike. Another approach is simply to reduce the salt content and add another antimicrobial agent, such as ethanol, to control the microflora (Chiou et al., 1999).

A quite different route to reduce the salt content has been to identify other compounds that contribute "saltiness" (without affecting the overall soy sauce flavor) and then to use those substances as salt mimics. Researchers in Japan recently discovered that several amino acid derivatives and peptides, such as ornithyltaurine, glycine ethyl ester hydrochloride, and lysine hydrochloride, were effective substitutes for potassium chloride (Kuramitsu, 2004).

In addition to affecting flavor, reducing the salt content of soy sauce and other fermented soy products may also have an effect on the microbiological stability of these products. However, pathogen challenge studies on low-salt miso revealed that even when salt levels were reduced to less than 3%, *Staphylococcus aureus*, *Clostridium botulinum*, *Salmonella typhimurium*, and *Yersinia enterocolitica* were unable to grow or survive, and toxin production was absent (Tanaka et al., 1985). Still, as an additional barrier, some manufacturers add ethanol to the product to control spoilage or pathogenic organisms.

References

Chiou, R.Y.-Y., S. Ferng, and L.R. Beuchat. 1999. Fermentation of low-salt miso as affected by supplementation with ethanol. Int. J. Food Microbiol. 48:11–20.

Kuramitsu, R., 2004. Quality assessment of a low-salt soy sauce made of a salty peptide or its related compounds. Adv. Exp. Med. Biol. 542:227–238.

Tanaka, N., S.K. Kovats, J.A. Guggisberg, L.M. Meske, and M.P. Doyle. 1985. Evaluation of the bacteriological safety of low-salt miso. J. Food Prot. 48:435–437.

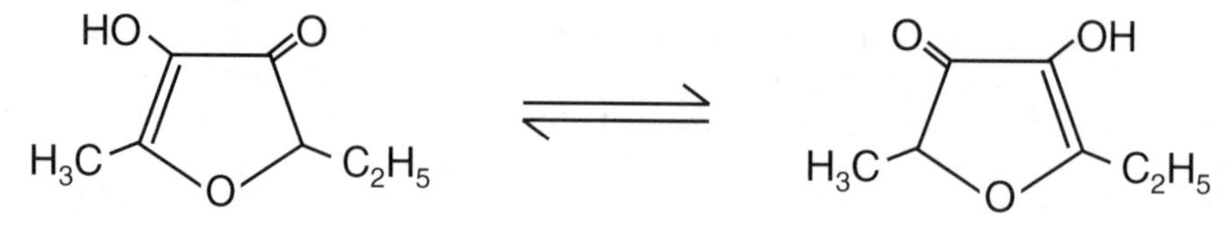

Figure 12–3. Chemical structure of HEMF. Adapted from Nunomura and Sasaki, 1992.

Japanese-type soy sauce (Figure 12–3). However, although it has been suggested that furanones are derived from fermentation, their presence and concentration in shoyu depends mostly on the composition of the raw material. For example, HEMF occurs in koikuchi, but not in tamari, indicating that wheat is necessary for its synthesis. Among the phenolic compounds that contribute to soy sauce or shoyu flavor are 4-ethylguaiacol (4-EG) and 4-ethylphenol (4-EP). They are also produced during fermentation by yeasts, but are derived from precursors formed by koji molds grown on wheat. They confer typical soy sauce-like flavors.

In addition to the furanone and phenols, a large number of acids, aldehydes, alcohols, and esters are produced during the shoyu fermentation. Although the concentrations of some of these compounds, i.e., ethanol and lactic and acetic acids, and 2,3-butanediol, can reach very high levels, most are in the order of parts per million. They are generally produced by lactic acid bacteria and yeasts. Other important flavors are generated, along with color, via the Maillard reaction. Although the formation of Maillard reaction products is accelerated by heating (i.e., pasteurization conditions), many end products are also formed during the mashing and moromi aging steps.

Another group of flavor constituents exists that is very important in soy sauce products, but whose flavor is not so easily described. Included in this group are several nitrogenous compounds, especially the amino acid glutamic acid, and the 5′ nucleotides inosine 5′-monophosphate (IMP) and guanosine 5′-monophosphate (GMP). The flavor or taste of these compounds (and their sodium salts) fall outside what are ordinarily considered to be the four basic flavors—sweet, sour, bitter, and salty. Rather, sodium glutamate, sodium inosinate, and sodium guanylate impart a totally unique flavor known as umami, a Japanese term meaning "deliciousness." Umami has been described as conferring a meaty, savory, brothy flavor to foods. The glutamate salt monosodium glutamate is also considered to be a flavor enhancer. And although umami is usually associated with Asian cuisines, it is present in a wide array of foods, including cured ham, Parmesan cheese, and shitake mushrooms.

Non-fermented soy sauce

Given the many steps involved in preparing the raw materials, making the koji, and performing the fermentation, it is not surprising that an alternative, non-biological process for making soy sauce exists. This process is based on an acid hydrolysis of soy beans, and although the finished product is less expensive to make and takes much less time, it lacks the flavor and complexity of fermented or brewed soy sauce.

Spoilage and defects

Although the low pH and high salt concentration inhibits most spoilage organisms, benzoate is sometimes added as a preservative, mainly against fungi. Ethanol may also be added for preservation. As noted above, excessive browning during mashing or aging may also be considered a defect, especially when the soy sauce is destined for use as a flavor in-gredient in light-colored products (Box 12-2). In addition, some undesirable flavors, such as isobutyric acid and isovaleric acid, may form during prolonged storage.

Miso

Miso is another popular fermented soy product in the Orient. Although miso originated in China and Korea more than a thousand years

Box 12–2. Reducing the Dark Color in Soy Sauce

The dark brown color of shoyu and other types of soy sauce is ordinarily an expected property. A brown color is certainly a major part of the overall appearance of many of these products. However, for many applications—in particular, when soy sauce is used as an ingredient in processed foods—a lighter, amber-like color is more desirable. Although it is certainly possible to produce such products by adding more wheat to the formulation (e.g., shiro has a lighter color than tamari), there is much interest in developing manufacturing processes that produce traditional soy sauce products with a lighter, less brown appearance.

Strategies that have been devised to "lighten" the color of soy sauce are based mainly on controlling the Maillard reaction responsible for brown pigment formation, which occurs primarily during the mashing and pasteurization steps. The two reactants in the Maillard, non-enzymatic browning reaction are reducing sugars and primary amino acids. Due to the extensive hydrolysis of soy protein and wheat starches and polysaccharides, both are present at very high concentrations in the moromi mash. For example, the amino acid and reducing sugar concentrations may be as high as 1.5% and 7%, respectively (although average levels are usually somewhat less). Among the total free sugars present in the mash are various pentoses, including arabinose (30 mM) and xylose (20 mM). Both of these sugars are very reactive in the Maillard reaction. In fact, the order in which these sugars "brown" is xylose > arabinose > hexoses. Thus, reducing the concentration of xylose in the mash could have a significant impact on brown pigment formation.

One way to decrease xylose levels would be to block or inhibit the xylan hydrolysis reactions that release free xylose from the xylan polysaccharides that are present in the wheat and soybean cell walls. Xylan hydrolysis during mashing occurs by the action of one or more of several xylanases and β-xylosidases produced by the koji mold *Aspergillus oryzae*. Xylanases hydrolyze xylan to xylobiose and xylan-containing tri- and tetra-saccharides, which are then hydrolyzed by β-xylosidases to give free xylose.

Researchers in Japan identified and cloned the *xynF1* gene encoding for the major xylanase in an industrial strain of *A. oryzae* (Kitamoto et al., 1998). Then, by re-introducing (in *trans*) multiple copies of the *xynF1* promoter region back into the parent strain, transcription of *xynF1* was reduced simply by a titration effect. As expected, expression of the xylanase was also decreased. The same group also identified the *xylA* gene that encodes for the major β-xylosidase produced by this organism (Kitamoto et al., 1999). An antisense mRNA strategy was subsequently used to reduce expression of the enzyme by 80%. In theory, then, if this modified *A. oryzae* strain were used for koji-making, xylose formation and browning reactions would be significantly reduced.

A second, altogether different, approach was adopted by another group of Japanese researchers (Abe and Uchida, 1989). They observed that, despite the metabolic capacity of *Tetragenococcus halophilus* (formerly *Pediococcus halophilus*) to metabolize free xylose, nearly

(Continued)

Box 12–2.   Reducing the Dark Color in Soy Sauce *(Continued)*

all of the xylose present in the moromi mash remained unfermented. As noted above, the mash contains plenty of glucose, which is the preferred energy and carbon source for this organism. During the soy sauce fermentation, however, use of xylose and other sugars is catabolite repressed by glucose. That is, the *xyl* genes coding for transport (*xylE*), isomerization (*xylA*), and phosphorylation of xylose (*xylB*) are present (Figure 1), but are not induced as long as glucose is available. These researchers reasoned, however, that *T. halophilus* would ferment xylose, even in the presence of glucose, if catabolite repression could be lifted or de-repressed. Spontaneous

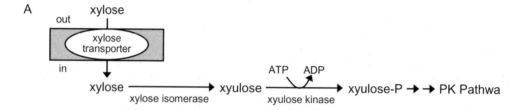

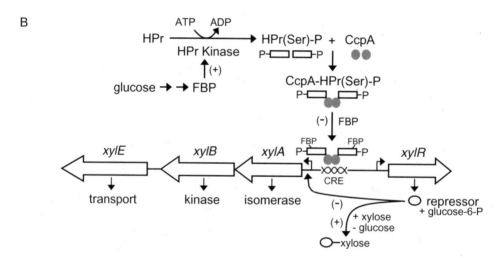

Figure 1.   Xylose metabolism (A) and catabolite repression (B) of the xylose operon in *Tetragenococcus halophilus*. The xylose metabolic pathway consists of a transporter, an isomerase, and a kinase, encoded by *xylE*, *xylA* and *xylB*, respectively. The product of this pathway, xyulose phosphate, then feeds directly into the pentose phosphoketolase (PK) pathway. Expression of the *xyl* operon is mediated by positive (+) and negative (−) regulation. During active growth on glucose, the glycolytic metabolite fructose-1,6-bisphosphate (FBP) accumulates inside the cell. FBP then activates the enzyme HPr kinase, leading to the phosphorylation of the phosphotransferase protein HPr at a specific serine residue. HPr(Ser)-P then forms a complex with another protein called Catabolite-control protein A (CcpA). This complex, along with FDP, binds to a DNA sequence called catabolite repression element (CRE) located upstream of the promoter region ◀┐ of the xylose operon, preventing its transcription (indicated by the negative sign). In the absence of glucose (and its metabolites), CcpA repression is lifted. In addition, a xylose repressor that is expressed during growth on glucose binds to a putative operator region positioned near the *xyl* promoter, further repressing transcription. However, when xylose is present and glucose is absent, xylose can bind directly to the repressor, causing its release from the operator and allowing the operon to be transcribed (+). Adapted from Abe and Higuchi, 1998 and Takeda et al., 1998.

Box 12–2. Reducing the Dark Color in Soy Sauce *(Continued)*

catabolite repression mutants were subsequently isolated that fermented xylose concurrently with glucose. Biochemical analyses revealed that the mutant strain was defective in glucose transport. Specifically, the mannose phosphotransferase system (PTS), which is primarily responsible for glucose transport in this strain, was found to be non-functional. Thus, if a product or component of the mannose PTS is necessary for catabolite repression (as is the case for other Gram positive bacteria), then a non-functional PTS would result in the xylose-fermenting phenotype these researchers observed.

Indeed, detailed analysis of the xylose operon revealed that catabolite repression of the xylose pathway was mediated, in part, by the phosphorylated form of the PTS protein HPr. Thus, repression required an intact PTS, which would explain why a PTS mutant would be de-repressed for xylose use. Furthermore, these researchers also identified an *xylR* gene, encoding a xylose repressor, whose expression (and effectiveness as a repressor) was activated by glucose-6-phosphate (Figure 1).

References

Abe, K., and K. Uchida. 1989. Correlation between depression of catabolite control of xylose metabolism and a defect in the phosphoenolpyruvate:mannose phosphotransferase system in *Pediococcus halophilus*. J. Bacteriol. 171:1793–1800.

Kitamoto, N., S. Yoshino, M. Ito, T. Kimura, K. Ohmiya, and N. Tsukagoshi. 1998. Repression of the expression of genes encoding xylanolytic enzymes in *Aspergillus oryzae* by introduction of multiple copies of the *xynF1* promoter. Appl. Microbiol. Biotechnol. 50:558–563.

Kitamoto, N., S. Yoshino, K. Ohmiya, and N. Tsukagoshi. 1999. Sequence analysis, overexpression, and antisense inhibition of a β-xylosidase gene, *xylA*, from *Aspergillus oryzae* KBN616. Appl. Environ. Microbiol. 65:20–24.

Takeda, Y., K. Takase, I. Yamoto, and K. Abe. 1998. Sequencing and characterization of the *xyl* operon of a Gram-positive bacterium, *Tetragenococcus halophila*. Appl. Environ. Microbiol. 64:2513–2519.

ago, Japan is the leading producer and consumer, with per capita consumption at about 5 Kg per person per year or 14 g/day. Miso has a flavor similar to soy sauce, except instead of being a liquid, it is paste-like, with a texture like that of a thick peanut butter. It can be used like soy sauce as a seasoning or flavoring agent, but is more commonly used, at least in Japan, to make soups and broths. Similar products are produced in Korea (doenjang), China (jang), Indonesia (taoco), and the Philippines (taosi).

In Japan, there are three general types of miso Each is based mainly on composition, specifically the raw ingredients used to make the miso koji (Table 12-6). Rice miso is made using a rice koji, barley miso is made using a

Table 12.6. Types and properties of different types of miso[1].

Product	Moisture (%)	Protein (%)	Reducing sugar (%)	Fat (%)	pH	Color
Rice miso						
Sweet	43–49	10–11	15	3–5	5.4	White or red
Salty	45	10–13	12	6	5.3	Yellow or red
Barley miso						
Sweet	44–47	10	17	4–5	5.2	Yellow
Salty	46–48	13	11	5–6	5.1	Red
Soy bean miso	45–46	17–20	4	11	5.0	Red or brown

[1]Adapted from Ebine, 1986 and Baens-Arcega et al., 1996

barley koji, and soybean miso is made using a soybean-only koji. There are, however, variations of these three miso types, depending on the salt content, color, and flavor. Because color development occurs mostly as a function of how much rice (or barley) is added and how long the product is aged, lighter colored misos typically contain more rice (or barley) and are aged for only a few months. They also have a more mild flavor, due to shorter fermentation times. For example, white or shiro miso is made using a rice koji, contains soy beans and salt, and is usually light in appearance. This type of miso has a mild, sweet flavor. Red miso or "akamiso" is made from a barley koji and has a stronger, more robust flavor. Finally, hatcho miso contains only soybeans, is aged for as long as three years, and has a very dark brown color and a strong, complex flavor.

Manufacture of miso

Miso and related products are manufactured much like soy sauce, except for one major difference. In miso manufacture, dry salt, rather than a brine, is added directly to the koji-soy bean mixture. Therefore, the resulting product has approximately twice the total solids of soy sauce (50% to 60% versus 24% to 28%). The process starts with the manufacture of a koji (Figure 12–4). As noted above, the koji substrate can be rice, barley, or soy beans. The rice (usually polished rice is used) or barley is soaked in water overnight at 15°C, then steamed in a batch or continuous cooker. After cooling, a tane koji (or a spore culture), containing strains of *A. oryzae* and *A. sojae* with defined properties, is used as the inoculum (at 0.1%). The koji is then incubated at 30°C to 40°C for forty to forty-eight hours in fermentation chambers equipped with aerating devices.

The soy beans (usually yellow soy beans are used) are similarly prepared, first by sorting and soaking and then by cooking under pressure (0.5 kg/cm^2 to 1.0 kg/cm^2) for about fifteen to forty-five minutes. At this point in the process, the cooked soy beans are ground or extruded, then mixed with the koji and salt in automated mixing machines. Alternatively, the whole cooked soybeans, koji, and salt can be mixed, and then the entire mixture is mashed. This mashing step is performed using an extrusion-like device that grinds the mixture to a chunky homogenous paste. The mixture is transferred to large tanks or fermentors (stainless steel or epoxy resin-lined steel) with capacities of 1,000 kg to more than 100,000 kg. Depending on the type of miso being made, the salt concentration may range from 6% to 13%.

Fermentation

The miso fermentation occurs in a manner similar to that of soy sauce, in that the hydrolysis of complex macromolecules to form simple nutrients and the subsequent fermentation and metabolism of those nutrients occurs essentially at the same time. As it does with soy sauce, the koji serves both as the source of proteolytic and amylolytic enzymes, as well as a substrate source for those enzymes. About 50% of the total protein and 75% of the polysaccharides are completely hydrolyzed to amino acids and monosaccharides, respectively. Because miso is made from whole soy beans, the lipid portion (about 3% to 10%) is also present in the miso mash. Fungal lipases (from the koji) hydrolyze triglycerides to di- and monoglycerides, glycerol, and free fatty acids. The latter may accumulate to high levels (as much as 1% to 3% in the miso). Collectively, the amino acids, sugars, and fatty acids provide a rich source of nutrients for fermentative organisms.

In traditional miso manufacture, a miso seed culture, obtained from previous raw miso product, is used to initiate the fermentation. Although many small manufacturers continue this practice, miso starter cultures are now often used, especially by large modern manufacturers. These cultures ordinarily contain yeast strains of *Z. rouxii* and *C. versatilis,* along with *T. halophilus, P. acidilactici, L. delbrueckii,* and other lactic acid bacteria. Fermentation of sugars present in miso results in the use of glucose and other free sugars and the formation of several organic acids, including lactic, acetic, and

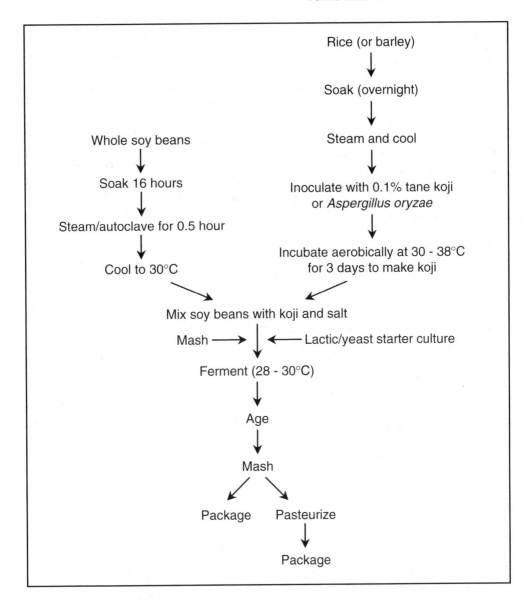

Figure 12–4. Miso flow chart.

succinic acids. The longer the miso ferments, the lower the sugar concentration. However, the total acid concentrations in the finished miso generally are less than 1% and the pH is usually not less than 5.0. About 0.1% to 1% ethanol and 1% to 2% glycerol may be formed during the yeast fermentation.

Miso fermentations are conducted in tanks at 25°C to 30°C. Temperature control and mechanical stirring are common features in many factories. Fermentation can be as short as one month or less to more than two years. Transferring the miso from one tank to another is often done to promote a homogenous fermentation. When the fermentation and aging period is complete, the miso is re-mashed and packaged into small plastic tubs or bags (about 400 g to 500 g capacity). The miso can then either be left in its raw state or subjected to a short-time heat treatment. While pasteurization does enhance shelf-life by preventing package swelling (due to gas formation), this is strictly an optional practice. To many consumers, raw miso is much preferred, based on the putative health properties conferred by the live microorganisms. In this respect, miso is viewed much like yogurt, in that any heat step that inactivates microorganisms is tantamount to blasphemy.

Spoilage and defects

Despite the high salt concentration and relatively low pH, growth of spoilage organisms in miso can occur, resulting in gas, off-odors, over-acidification, and surface slime. Yeasts and bacteria responsible for these defects include *Hansenula* and *Z. rouxii*, *Pediococcus acidilactici*, and *Bacillus* sp. Spoilage is more likely to occur when salt levels are reduced (<12%) or when the koji molds are inhibited. Pasteurization, either before or after packaging, inactivates these organisms. However, as indicated above, many consumers prefer raw, nonheat-treated miso.

Natto

Natto is another soybean-fermented product consumed mainly in Japan, but similar products are also produced in China, Thailand, and the Philippines. Per capita consumption in Japan is about 1.2 Kg per person per year or 3 g per day. Natto is used as a condiment or flavoring agent, usually for rice and vegetables or as an ingredient in sushi. Nutritionally, natto is comparable to other fermented soybean products. It contains 16% to 18% protein (45% on a dry basis), with good digestibility and biological value, compared to cooked soybeans. It is considered to be an excellent source of isoflavones and readily adsorbed Vitamin K. The latter is thought to contribute to bone-strength and reduced rates of fractures among natto-consuming populations.

Manufacture of natto

The manufacturing procedure for natto begins like that for miso; however, the organisms involved in the fermentation are different and the final product bears little resemblance to miso. Natto is made from whole, somewhat small-sized soybeans that are cleaned, soaked for twelve to twenty hours at ambient temperature, and steamed at 121°C for twenty to forty minutes. The thoroughly cooked and cooled beans are then inoculated with about 10^6 to 10^7 spores per Kg of *Bacillus subtilis* var. *natto* (formerly *Bacillus natto*), and the material is well mixed. The incubated beans are then divided into 100 g portions and placed into packages. According to traditional practices, bundled rice straw was used as the container (some manufactures still use straw); however, polyethylene bags are now common. In either case, the beans are moved into aerobic incubators at 40°C at 85% relative humidity for sixteen to twenty hours. Growth of *B. subtilis* var. *natto* occurs primarily at the surface and is accompanied by a change in color (from yellow to white) and synthesis of a highly viscous polysaccharide material. The latter can completely cover the entire bean surface, accounting for nearly 1% of the total dry weight of natto. After incubation, the natto is held at 2°C to 4°C to minimize further growth.

The final product is quite different from miso and shoyu. Its main flavor attribute is sweetness, due to the presence of the polysaccharide material, which is composed of a fructose-containing polymer. The polysaccharide also confers a definite sticky and stringy texture to the product, which, although characteristic of natto, nonetheless causes some consumers to consider other menu options.

Tempeh

Tempeh is a mold-fermented soy bean product that originated many centuries ago in Indonesia, where it remains a major food staple and an inexpensive source of dietary protein. Unlike other fermented soy products, tempeh production has spread to only a few other countries, including Malaysia, the Netherlands (Indonesia was once under Dutch rule), Canada, and the United States. However, Indonesia is by far the main producer and consumer of tempeh. Current per capita consumption in Indonesia is about 15 grams per person per day. Although tempeh production has been industrialized, a substantial amount of the more than 500 million kg of tempeh produced in Indonesia per year is still made in the home or in small "cottage-sized" production facilities.

It is remarkable, given the fact that tempeh was not discovered by American consumers until the last twenty to thirty years, that it has become quite popular (relatively speaking) in the United States. Undoubtedly, this sudden popularity is due, in part, to interest in vegetarian cuisine and non-meat alternative food products ("faux" meats). However, the popularity of tempeh in the United States may also be due to its nutritional properties. In particular, tempeh is a rich source of protein (19%) and, in addition, is one of only a few plant-based foods that contains vitamin B_{12} (discussed below). Since this vitamin is often lacking in vegetarian diets, tempeh serves as a modest source of this essential nutrient (a 100 g serving provide 6% of the 2.4 μg of B_{12} recommended for a healthy adult, according to USDA estimates).

Another reason, perhaps, why tempeh has attracted interest in the United States relates to its versatile applications and organoleptic properties. In its raw state, tempeh has a bland, mushroom-like flavor. However, cooking transforms this plain-tasting material into a pleasant, nutty, flavorful product. If one can get past the fact that tempeh consists entirely of moldy beans, its flavor, especially when it is sauteed or fried, resembles that of cooked or roasted meat (sort of). After all, tempeh and muscle protein (i.e., meat) both derive much of their flavor from the Maillard reaction products that result when amino acids and reducing sugars are heated at high temperature. The lipid component of soy beans may also serve as a precursor for meaty flavor and aroma development. Finally, the development of a tem-

peh "industry" in the United States and consumer interest in tempeh as a food have likely been driven by academic researchers, particularly Keith Steinkraus at Cornell University and C. W. Hesseltine and H.L. Wang at the USDA, who played key roles in understanding many of the microbiological and technical issues related to tempeh manufacture.

Manufacture of tempeh

The industrial manufacture of tempeh is quite simple, although numerous variations exist, depending largely on the scale of production, geographical and climatic considerations, and manufacturer preferences. The only raw material is soy beans, the fermentation time is short, and there is no aging or ripening period involved. In fact, the entire start-to-finish process is less than forty-eight hours. Tempeh can be considered as solid-state fermentation in that it consists of soy beans that are essentially held together by the mold mycelium that grows throughout and in between the individual beans. It should be noted that although other legumes, cereal grains, vegetables, and even seafood can be incorporated into tempeh (and are commercially available), soy beans are by far the most common starting material.

Substrate preparation

The process starts by sorting the soy beans to remove damaged or moldy beans (Figure 12-5). The beans are usually given a quick heat treatment (about 70°C to 100°C for ten to thirty minutes), and the hulls are removed either manually or mechanically. The former method, practiced by very small traditional manufacturers, involves rubbing the beans with hands (or even feet!) and then separating the hulls by floatation. Manual de-hulling, however, has been largely replaced by the use of various types of mills, which can be used on dry or wet beans. Some manufacturers omit the first boiling step, although this is considered by some tempeh experts to be an essential part of the process, facilitating hydration and hull removal.

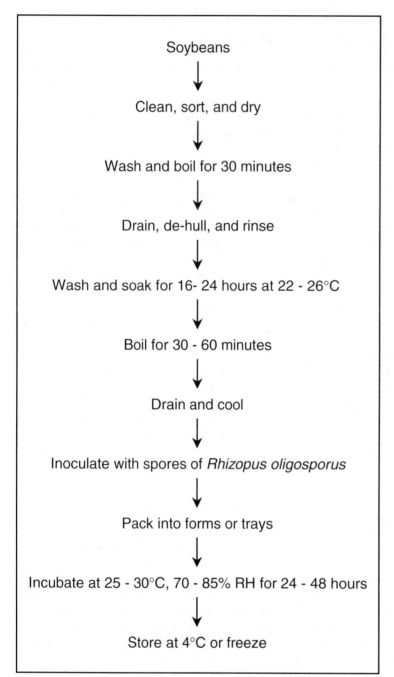

Soybeans

↓

Clean, sort, and dry

↓

Wash and boil for 30 minutes

↓

Drain, de-hull, and rinse

↓

Wash and soak for 16- 24 hours at 22 - 26°C

↓

Boil for 30 - 60 minutes

↓

Drain and cool

↓

Inoculate with spores of *Rhizopus oligosporus*

↓

Pack into forms or trays

↓

Incubate at 25 - 30°C, 70 - 85% RH for 24 - 48 hours

↓

Store at 4°C or freeze

Figure 12–5. Tempeh flow chart. Adapted from Shurtleff and Aoyagi, 1979.

The de-hulled beans are then soaked in water, a seemingly simple step, but one which has very important implications for tempeh quality. It is during this steeping period (sixteen to twenty-four hours at warm ambient temperature), that endogenous lactic acid bacteria grow and produce acids that lower the pH. The low pH generally restricts growth of undesirable spoilage bacteria, as well as potential pathogens. It should be noted, however, that some of these undesirable bacteria can survive during the soaking step even if acid conditions are established. To promote acidification, the steep water can be acidified directly with lactic or acetic acid. Finally, potential mold inhibitors may also be extracted from the soy beans during the steeping period.

After the soaking step, the remaining hulls are removed from the beans and the dehulled beans or cotyledons are heated a second time, either by steaming or boiling (for anywhere from thirty minutes to two hours). This heat treatment enhances extraction of soluble nutrients and inactivates microorganisms that might otherwise interfere with the subsequent fermentation. This step is also necessary to denature trypsin inhibitor, a native soy protein that acts as an anti-nutritional factor.

Inoculation

After cooking, the beans are drained and dried. Next, they are inoculated with either a portion from a previous batch of fully developed tempeh, a wild, mixed strain culture called "usar," or, as is more common in modern manufacture, with a spore culture of *Rhizopus microsporus* var. *oligosporus* (hereafter referred to as *R. oligosporus*). Typical culture inocula levels range from 10^7 to 10^8 spores (about 1 g) per kg of beans. Pure *R. oligosporus* cultures can be added directly to the beans, or the spores can be inoculated onto steamed rice, incubated until well-grown, and then used to inoculate the soy beans.

Fermentation

According to traditional Indonesian manufacturing practices, the inoculated beans are then shaped into cakes and wrapped in banana leaves. The use of banana leaves as the packing chamber is not merely for natural aesthetics; they provide a moist, microaerophilic environment that supports rapid growth of *R. oligosporus*. However, for large-scale manufacture of tempeh (and as conducted in the United States), the inoculated beans are distributed on trays 1 cm to 3 cm deep and ranging in length

and width from 1 meter to several meters. After one to two days of incubation in a warm room (ranging from 25°C to 37°C), the beans are covered with white mycelium, and the fermentation is considered complete. The mycelia also will have grown in between the individual beans such that solid soy bean cakes will have formed. Detailed analyses have revealed that the fungal hyphae actually penetrate nearly 1 mm into the bean, or about 25% of the diameter of the cotyledon. It is important that the fermentation ends, however, before the mold begins to sporulate, since the appearance of the dark-colored black or grey sporangia is generally unattractive to consumers (see below).

Tempeh Microbiology

The surface of raw soy beans contains an assortment of Gram positive and Gram negative bacteria, including *Lactobacillus casei* and other lactic acid bacteria; enterococci, staphylococci; streptococci; bacilli; *Enterobacter*, *Klebsiella*, and other coliforms. Yeasts, such as *Pichia*, *Saccharomyces*, and *Candida*, may also be present. During the soaking step, sucrose, stachyose, and raffinose diffuse out of the beans and into the water. Their subsequent hydrolysis by invertases and glucosidases releases glucose and fructose, which can then be used to support growth of the resident microflora. Consequently, at the end of the soaking period (twenty to twenty-four hours at 20°C), the total microflora may reach levels of 10^9 cfu per ml or higher. Most of the organisms isolated after soaking are lactobacilli, enterococci, and streptococci. However, the specific species that predominate appears to depend on the temperature and the pH of the soak water (i.e., in those applications where lactic or acetic acid is added to the water). Importantly, the pH values of the soak water, whether acidified or not, generally will decrease to 4.5 to 5.0 by virtue of lactic and mixed acid fermentations.

Although the manufacture of tempeh clearly depends on the growth of *R. oligosporus* (discussed below), the fermentation that occurs during the soaking of the soy beans is also essential. This is because the formation of organic acids and the decrease in pH are necessary to control pathogens, including *Salmonella typhimurium, Yersinia enterocolitica, Staphylococcus aureus,* and *Clostridium botulinum,* that might otherwise grow in non-acidified soy beans. Low pH also inhibits *Bacillus, Enterobacter,* and other microorganisms capable of causing spoilage effects. It is important to note that even if the beans are heated prior to soaking, the lactic fermentation will still occur, although the rate and extent of the fermentation may be affected.

Tempeh cultures

As noted above, the primary fermentation is mediated by growth of *R. oligosporus,* which can be added to the soy beans in one of several different forms. First, it can be added as a pure spore culture. Recommended strains include NRRL 2710 and DSM 1964, both isolated from Indonesian tempeh and both available from public culture collections. Like the commercial strains used for other fungal-fermented products, tempeh starter cultures should be selected based on specific phenotypic traits (Table 12–7). Alternatively, a backslop material consisting of a dried tempeh culture can be used. Finally, a third form, used in traditional tempeh manufacture, is called *usar,* and is made by inoculating wild *Rhizopus* spores onto the surface of leaves obtained from the indigenous Indonesian *Hibiscus* plant. After two to three days of incubation, the leaves contain a dense spore crop that can be dried and used to inoculate the soy beans. Species other than *R. oligosporus* may be present when wild cultures are used. Other species isolated from

tempeh include *Rhizopus oryzae, Rhizopus stolonifer,* and *Rhizopus microsporus* var. *chinensis.*

Regardless of source or strain, tempeh cultures have a limited shelf life, as little as three to four months. This is because the *Rhizopus* spores enter into a dormancy phase during storage that reduces viability and germability (the ability of dormant spores to become activated and produce biomass). Moreover, even if the soy beans are inoculated with a pure spore culture of *R. oligosporus,* the tempeh is unlikely to be maintained in a pure state for too long, since the substrate itself will likely be contaminated with an array of different organisms. In fact, as discussed below, other microorganisms may play important nutritional and organoleptic roles during the tempeh fermentation.

Tempeh biochemistry

In addition to producing the mycelia mass that literally holds the soy beans together, *R. oligosporus* is also responsible for causing major biochemical changes in the composition of the soy bean substrate (Table 12–8). In particular, lipids and proteins serve as substrates for fungi-excreted lipases and proteinases, respectively. During the incubation period, about a third of the lipid and a fourth of the protein fractions are degraded. Lipid hydrolysis results mainly in mono- and diglycerides, free fatty acids, and only a small amount of free glycerol.

Table 12.7. Properties of *Rhizopus microsporus* var. *oligosporus.*

Unable to metabolize major soy carbohydrates (sucrose, stachyose, or raffinose)
Aerobic
Rapid growth and mycelia production at 30–42°C
Proteolytic and lipolytic
Uses free fatty acids, derived from lipids, as an energy source

Table 12.8. Composition of soybeans and tempeh[1].

Constituent	Soybeans[2] (g/100 g)	Tempeh[3] (g/100 g)
Moisture	7–9	60–65
Protein	30–40	18–20
Soluble Nitrogen	<1	2–4
Carbohydrate	28–32	10–14
Fiber	4–6	1–2
Fat	18–22	4–12
pH	6–7	6–7

[1]Adapted from Hachmeisterand Fung, 1993 and Winarno and Reddy, 1986
[2]Whole raw soybeans (prior to soaking)
[3]Fresh (wet) weight basis

Most of the free fatty acids are subsequently oxidized by *R. oligosporus*, resulting in a 10% decrease in the total dry matter in the finished tempeh. In fact, the relative inability *R. oligosporus* to metabolize the available soy carbohydrates (mainly stachyose, raffinose, and sucrose) means that fatty acids serve as the primary energy and carbon source.

In contrast, only about 10% of the released amino acids and peptides are oxidized by *R. oligosporus*. Of the remaining portion, about 25% are assimilated into biomass, and the rest is left in the tempeh. The soluble nitrogen concentration, therefore, increases four-fold, from about 0.5% to 2%. Despite the limited metabolism of proteins and amino acids, ammonia accumulates during the tempeh fermentation, such that the pH rises from 5.0 to above 7.0. This increase in pH during the fungal fermentation stage underscores the importance of acidification during bean soaking, since the latter step is responsible for inactivating potential pathogens that may have been present in the starting material. Once the low pH barrier no longer exists, neutrophilic pathogens could theoretically grow and cause problems.

Polysaccharide-hydrolyzing enzymes are also produced by *R. oligosporus*. These enzymes attack pectin, cellulose, and other fiber constituents, releasing various pentoses (xylose, arabinose) and hexoses (glucose, galactose). However, the activity of these enzymes during the tempeh fermentation is modest, and only minor amounts of free sugars are found in the final product.

Tempeh nutrition and safety

Among the most important changes that occur during the tempeh fermentation are those that affect the nutritional quality of tempeh. As noted above, the concentration of the major macronutrients (i.e., protein, fat, and carbohydrates) decreases as the soybeans are converted to tempeh, due to enzymatic hydrolysis. These changes may account, in part, for an improvement in nutritional quality. For example, it has been suggested that protein hydrolysis

makes tempeh more digestible, compared to soybeans, although the protein efficiency ratio (used as a measure of protein quality) of tempeh is no higher than an equivalent amount of cooked soy beans. There is also a decrease in the amount of soy oligosaccharides (mainly stachyose and raffinose) during the conversion of soybeans into tempeh. These sugars, which are quite undesirable due to their ability to cause flatulence, are removed from soybeans not by fermentation, but rather by diffusion during the soaking and cooking steps.

Probably the most important nutritional improvement that occurs during tempeh manufacture is the increase in vitamin content. Of particular interest is vitamin B_{12}, whose concentration in cooked tempeh generally ranges from 0.1 µg to 0.2 µg per 100 g. Since soybeans contain negligible levels of this vitamin, it is evident that its presence in tempeh must occur as a result of biosynthesis by microorganisms in the tempeh. What is surprising, however, is that vitamin B_{12} is made not by *R. oligosporus* nor by lactic acid bacteria, but rather by bacteria that are essentially chance contaminants in the tempeh-making process (Box 12–3). For example, *Klebsiella pneumoniae*, a Gram negative member of the family *Enterobacteriaceae*, is a known B_{12} producer, and is also present in Indonesian tempeh. Other vitamins, including riboflavin (B_2), niacin, pyroxidin (B_6), biotin, pantothenic acid, and folic acid, also increase in concentration by up to five-fold during tempeh production. The concentration of thiamine (B_1), however, decreases, and there is no effect on the level of fat-soluble vitamins.

In addition to the increase in vitamin content, the processing and conversion of soybeans into tempeh also results in a decrease in several anti-nutritional factors ordinarily present in soybeans. These compounds are anti-nutritional because they either interfere with digestion (trypsin inhibitor), reduce protein quality (tannins), reduce mineral adsorption (phytic acid), cause blood to form clumps (hemagglutenins), or cause metabolic disturbances (goitrogens). Some of the water-soluble, low-molecular-

Box 12–3. Vitamin B_{12} in Tempeh

Tempeh is one of only a few non-meat or non-animal-containing foods that contain vitamin B_{12}. However, despite considerable research interest, the actual means by which synthesis of B_{12} occurs has yet to be resolved. In part, this is because measuring the actual amount of this vitamin and its biological activity in tempeh have not been easy tasks. Reported concentrations range from 0.002 μg/100 g to 8 μg/100 g, a 4,000-fold difference (Nout and Rombouts, 1990), depending on both the particular assay method (radio-immunoassay versus bioassay) and the sample extraction method. In general, tempeh is considered to contain 0.1 μg to 0.2 μg of B_{12}/100 g. According to the USDA Nutrient Database, the Vitamin B_{12} content is 0.14 μg per 100 grams of cooked tempeh.

The main questions regarding the B_{12} content of tempeh, however, concern the organisms and pathways involved in B_{12} biosynthesis. The tempeh mold, *Rhizopus oligosporus*, does not produce any B_{12}. In fact, the vitamin B_{12} (or cyanocobalamin) biosynthetic pathway is absent in eukaryotic organisms (Keuth and Bisping, 1993; Rodionov, 2003). In contrast, production of vitamin B_{12} by several bacteria, including aerobes (e.g., *Pseudomonas*), anaerobes (e.g., *Propionibacterium* and *Clostridium*), and facultative anaerobes (*Citrobacter, Lactobacillus*) has been reported (Martens et al., 2002; Taranto et al., 2003). Furthermore, a recent genomic analysis has revealed that the biosynthetic machinery for making B_{12} is widely distributed in bacteria (Rodionov, 2003).

In tempeh, the main organisms responsible for vitamin B_{12} biosynthesis are *Klebsiella pneumoniae* and *Citrobacter freundii* (Keuth and Bisping, 1994). Although these organisms are present in the soy beans and soaking water, growth and vitamin production do not occur until later, after the overnight soaking step and during the primary fungal fermentation. Interestingly, there are reports that B_{12} production by *C. freundii* may be up to three times higher when these organisms are grown in the presence of *R. oligosporus* (Keuth and Bisping, 1993; Wiesel et al., 1997).

The reader may be wondering about the presence of *K. pneumoniae* in tempeh and its suggested use as a starter culture. This organism is an opportunistic pathogen and is capable of producing several toxins, including three enterotoxins. However, not all strains are toxigenic; indeed it does not appear that strains isolated from tempeh contain enterotoxin genes (Keuth and Bisping, 1994).

References

Keuth, S., and B. Bisping. 1994. Vitamin B_{12} production by *Citrobacter freundii* or *Klebsiella pneumoniae* during tempeh fermentation and proof of enterotoxin absence by PCR. Appl. Environ. Microbiol. 60:1495–1499.

Keuth, S., and B. Bisping. 1993. Formation of vitamins by pure cultures of tempe moulds and bacteria during the tempe solid substrate fermentation. J. Appl. Bacteriol. 75:427–434.

Martens, J.-H., H. Barg, M.J. Warren, and D. Jahn. 2002. Microbial production of vitamin B_{12}. Appl. Microbiol. Biotechnol. 58:275–285.

Nout, M.J.R., and F.M. Rombouts. 1990. Recent developments in tempe research. J. Appl. Bacteriol. 69:609–633.

Rodionov, D.A., A.G. Vitreschak, A.A. Mironov, and M.S. Gelfand. 2003. Comparative genomics of the vitamin B_{12} metabolism regulation in prokaryotes. J. Biol. Chem. 278:41148–41159.

Wiesel, I., H.J. Rehm, and B. Bisping. 1997. Improvement of tempe fermentations by application of mixed cultures consisting of *Rhizopus* sp. and bacterial strains. Appl. Microbiol. Biotechnol. 47:218–225.

Taranto, M.P., J.L. Vera, J. Hugenholtz, G.F. De Valdez, and F. Sesma. 2003. *Lactobacillus reuteri* CRL1098 produces cobalamin. J. Bacteriol. 158:5643–5647.

weight compounds, such as phytic acid and polyphenolic tannins, are removed during soaking and washing. In contrast, proteinaceous compounds, including trypsin inhibitor and other protease inhibitors, and hemagglutenins are inactivated by heating steps. It is also possible that some of these compounds are degraded by enzymes produced during fermentation. For example, phytic acid can be hydrolyzed by phytases produced by *R. oligosporus*.

Finally, there has long been concern about the microbiological safety and the possible presence of mycotoxins in tempeh. As is the case with other mold-fermented foods (e.g., soy sauce and miso), the wild or pure culture strains used for tempeh production do not produce mycotoxins.

Spoilage and defects

In Indonesia, where tempeh is consumed on a near-daily basis, spoilage is not much of an issue, provided the product is eaten within a day or two of manufacture. However, the shelf-life of tempeh held at room temperature is very short, owing to the continued growth of the mold and bacteria. Once *R. oligosporus* begins to sporulate and produce colored sporangia, the product's shelf-life is essentially finished. Even when stored at refrigeration temperatures, mold growth is slowed but not stopped. Therefore, some form of preservation is necessary. In the United States, tempeh is most often vacuum packaged in oxygen impermeable plastic to restrict growth of aerobic fungi and bacteria. Another effective way to preserve tempeh is freezing, which halts fungal growth. Finally, tempeh can be dehydrated or cooked prior to packaging or made into various processed products, such as vegetarian meat-like foods.

Manufacture of Sake and Rice Wines

As noted in Chapter 10, most wines produced throughout the world rely on grapes as the starting raw material. Grapes used for wine making ordinarily contain an ample amount of glucose and fructose, which are readily fermented by the endogenous yeasts or added pure yeast cultures. In contrast, when starchy substrates, such as rice, are used as the raw material, the complex polysaccharides (mainly amylose and amylopectin) must first be hydrolyzed to produce fermentable sugars. When rice wines were first developed, this hydrolysis step was done by chewing and masticating the rice. As will be discussed below, saccharification is now performed by a rice koji, not unlike the koji used in the soy sauce and miso fermentations.

The most well-known rice wine is *sake* (rhymes with hockey), a Japanese rice wine that likely originated in China, probably several millennia ago. Many other rice-derived wines and alcoholic beverages exist throughout the Far East. Examples include shaosing (Chinese rice wine), awamori (a Japanese product distilled from rice), makgeolli (Korean rice wine), and ruou (Vietnamese rice wine). Historically, many of these wines have long been associated with Shintoism and Buddhism religious rites. For many years, until relatively recently, rice wines were the most popular alcoholic products consumed in Japan, China, and other Asian countries. Current (year 2000) per capita consumption of sake in Japan is 8.2 L, nearly 50% less than was consumed in 1970. Consumption of rice wine in China is now (2003) only 1 L per person per year (of course, that is still a lot of sake, considering there are nearly 1.3 billion people living in China). The decrease in sake popularity is due mainly to increased beer consumption (with per capita consumption of about 20 L in both Japan and China). In contrast, in the United States, sake is on the upswing, with several modern sake manufacturing facilities, although the total U.S. market is still very small.

Manufacture of sake

Sake is different from other wine fermentations in at least two main respects. First, as noted above, fermentable sugars are absent in rice, the sake substrate. Thus, it is necessary to provide exogenous enzymes, in the form of a koji, that can hydrolyze starch to simple sugars that the yeasts can ferment. This part of the sake manufacturing process, therefore, shares similarity with the beer-brewing process, in which malt is used to convert the starch (in barley) to simple sugars. The other major difference that distinguishes sake production from wine is that the saccharification step just described and the actual ethanolic fermentation step occur simultaneously or in parallel. In other words, nearly as soon as sugars are made available by action of koji enzymes, they are quickly fermented by sake yeasts. The implications of these parallel processes will be discussed below.

The sake-making process starts by preparing a rice koji (Figure 12-6). There are actually several different types of sake produced in Japan, based in part on whether alcohol is added, but mainly on how the rice used to make the koji is milled (or polished). In general, the whole or brown rice is milled to remove 25% to 50% of the surface material (containing the germ and bran), which is necessary since the fat and protein components are undesirable. Next, the rice is rinsed and soaked for several hours to achieve about 30% moisture. The moist rice is then briefly steamed (<1 hour) and cooled to 30°C to 35°C. About three-fourths of this rice is removed and cooled further to 5°C to 10°C for later use (see below). The remaining fourth is used for making koji.

To make the koji, the rice is inoculated with about 0.01% of a tane koji (see above) or an *A. oryzae* spore culture. The material is incubated

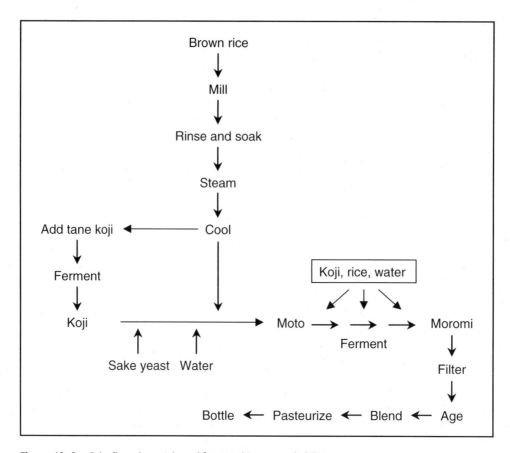

Figure 12–6.   Sake flow chart. Adapted from Yoshiwawa and Ishikawa, 2004.

for forty to forty-eight hours at 30°C to 35°C under high humidity. The koji is mixed at intervals to re-distribute the growing fungi, maintain aerobiosis, and prevent excessive heat build-up. When finished, the rice should be covered with fungal mycelia and should contain high amylolytic and proteolytic activities.

Next, the koji is moved to a tank, and the steamed rice (from above) and water are added. This material, called moto, is essentially a pre-culture whose purpose is to increase the population of endogenous yeast and lactic acid bacteria (see below) and to initiate a fermentation. According to traditional sake brewing practices, this moto seed culture incubates for about two weeks at 15°C, and then defined sake yeast strains are added. It is now more common, however, to add sake yeast at the outset, rather than rely on wild yeast for the initial fermentation. As the moto incubates, enzymatic hydrolysis of rice starch and proteins occur, releasing sugars and nutrients that support growth of the yeast. The moto is eventually transferred to large tanks (capable of accommodating more than 10,000 kg of rice) containing more koji, steamed rice, and water for combined mashing and fermentation.

The fermenting mixture, now called moromi, is then diluted several times with an equal amount of koji-rice-water, at two-to-three day intervals, such that the now thick , moromi

mash increases in volume three- to four-fold. This process ensures that the yeast population remains high throughout the fermentation and provides, on a step-wise basis, adequate amounts of substrates to support an extended, semi-continuous fermentation. This unique fermentation process also serves to maintain lactic acid levels and a moderately high solids content, such that contaminating organisms are inhibited. Thus, even though the sugar concentration never reaches much more than 6% to 8%, the ethanol concentration in the finished product may be 18% or higher. To achieve such a high yield of ethanol by a simple batch fermentation (i.e., where all of the fermentable sugar was initially present) would require a mash containing as much as 30% glucose. This is far more than could be osmotically tolerated by the yeast and would result in a stuck fermentation.

The *S. cerevisiae* strains used for sake manufacture are different from typical wine and beer strains, in that they generally have higher osmotic, acid, and ethanol tolerance. They also produce copious amounts of foam, which occupies about a third of the volume in the fermentation tank, reducing the efficiency of the fermentation. Strains that are defective in making foam (i.e., foamless mutants) could have industrial advantages (Box 12–4). In addition to yeast, other organisms are also involved in the

Box 12–4. Improving Sake Yeasts

Although the origin of sake manufacture, like most other fermentations, dates back thousands of years, ways to improve the fermentation are now actively being explored. In particular, there has been interest in modifying sake yeasts to enhance fermentation efficiency, flavor, and the overall quality of the finished product. One specific approach, described below, involves isolating a particular class of foam-defective mutants that have been used successfully for more than thirty years in the Japanese sake industry. The molecular basis for this phenotype of these strains, however, has only recently been studied.

Growth of sake strains of *Saccharomyces cerevisiae* results in formation of a thick froth or foam layer. The foam occupies so much volume in the fermentation tank (as much 50%) that less space is available for the moromi. High throughput processes are required in modern, large scale sake manufacturing facilities. High-foaming fermentations decrease production capacity. Fermentation efficiency could be improved if foam formation could be reduced using non-foaming yeast strains.

(Continued)

Box 12–4. Improving Sake Yeasts *(Continued)*

This rationale led researchers in Japan to establish screening protocols so that non-foaming mutants could be isolated (Ouchi and Akiyama, 1971). The screening strategies were based on the observation that froth or foam is formed either by strains that agglutinate to one another via a capsular material or by strains that have affinity to the surface of gas bubbles. One method involved separating agglutinating wild-type cells from spontaneous non-agglutinating mutants during a series of successive fermentations, thereby concentrating the non-agglutinating, non-foaming cells. Alternatively, a similar approach was used to isolate *S. cerevisiae* cells that did not have bubble affinity. In this case, cells were grown in broth in the presence of sparged air (to form bubbles). Wild-type cells, having affinity for bubbles, stuck to the bubbles and accumulated in the froth layer. The non-foaming mutants could then be retrieved from beneath the froth layer and re-inoculated into fresh broth, with the entire cycle repeated several times. The mutants were then isolated by plating, characterized to ensure other desirable properties had not been lost, and then used successfully in commercial production (and are still in use).

A more recent investigation on the molecular differences between foaming and non-foaming sake strains has led to the identification of a gene that confers the foaming phenotype (Shimoi et al., 2002). The *AWA1* gene ("awa" is Japanese for foam) encodes for a cell wall protein that is covalently attached to glucans located within the cell wall. Insertion mutants that did not express Awa1 did not form foam in a sake mash. Interestingly, this gene is not present in the *S. cerevisiae* genome database (built from laboratory strains), indicating it is likely specific to sake strains. Using a method that measures cell surface hydrophobicity (i.e., adherence ability), it was also shown that wild-type cells were significantly more "hydrophobic" than non-foaming cells. Thus, the latter cells are less likely to attach to other cells or to the surface of gas bubbles that form during the sake fermentation.

A second protein involved in the foaming phenotype was also recently identified by proteome analysis using two-dimensional electrophoresis. The protein, YIL169c, was qualitatively and quantitatively different between the foaming parental strain and a non-foaming mutant. Protein sequence comparison of YIL169c with the *S. cerevisiae* genome data base revealed that this protein was similar (71.5% identity) to YOL155c, a cell wall protein with glucosidase activity. Thus, although the function of YIL169c has not been determined, it appears that this protein likely protrudes from the cell wall, where it can interact with bubbles or other hydrophobic surfaces.

References

Ouchi, K., and H. Akiyama. 1971. Non-foaming mutants of sake yeasts: selection by cell agglutination method and by froth flotation method. Agri. Biol. Chem. 35:1024–1032.

Ouchi, K., and Y. Nunokawa. 1973. Non-foaming mutants of sake yeasts: their physio-chemical characteristics. J. Ferment. Technol. 51:85–95.

Shimoi, H., K. Sakamoto, M. Okuda, R. Atthi, K. Iwashita, and K. Ito. 2002. The *AWA1* gene is required for the foam-forming phenotype and cell surface hydrophobicity of sake yeast. Appl. Environ. Microbiol. 68:2018–2025.

Yamamoto, Y., K. Hirooka, Y. Nishiya, and N. Tsutsui. 2003. The protein which brings about foam-forming by the sake yeast. Seibutsu-Kogaku Kaishi 81:461–467 (Abstract only).

sake fermentation. In particular, endogenous lactic acid bacteria, including *Lactobacillus sake* and other *Lactobacillus* and *Leuconostoc* sp., grow and produce acid early in the fermentation. This early acidification evidently helps to control other adventitious organisms, although the lactics are eventually inhibited by the high ethanol, high acid, and high osmotic pressure that accumulates in the moromi. Lactic acid can be added directly to promote acidification, a now-common practice in modern sake facilities.

After about three weeks, the fermentation is ended, and the moromi is separated by settling and filtration, yielding a sake cake and a very clear, light yellow liquid, which is then called sake. The sake can be aged (usually only a few months), adjusted to a desirable ethanol content, pasteurized by heat (65°C) or ultrafiltration, and bottled. Like grape-derived wine, the finished sake can be sweet or dry. However, this classification is not entirely based on the amount of residual sugar that is present, but rather on a combination of the sugar, acid, and alcohol content. Depending on the type of sake and consumer preference, sake can be served warmed (35°C to 40°C) or slightly chilled. In general, higher quality sake is usually served at the lower temperature to retain more of the aroma and flavor volatiles.

Fermented Fish-type Foods

Although fish and shrimp sauces and pastes are mostly unknown to Western consumers, they are staple items in much of Southeast Asia, including Thailand, Vietnam, Malaysia, and the Philippines (from where such products were thought to have evolved). Although fresh fish has long been widely available throughout this region, refrigeration has not. Thus, fish would likely spoil before it could be consumed. In contrast, fish-type sauces and pastes not only have a long shelf-life, they also serve as an inexpensive source of high quality protein and other nutrients. In addition, these products significantly enhance the flavor of rice, noodles, and other bland-tasting foods. Per capita consumption ranges from 10 ml to 15 ml per day in Thailand to about 1 ml per day in the Philippines.

The manufacture of fish sauces in Southeast Asia is a major industry; over 40 million liters of fish sauce are produced annually in Thailand alone. The popularity of cuisines from those countries in the United States (in food service as well as in processed foods) has undoubtedly led to consumption of fish sauces by U.S. consumers, even if they aren't aware of what actually contributes to the unique flavors of those foods. In regions where these products are pro-

duced and consumed, they are considered indispensable flavoring agents, much like salt in the United States or shoyu in Japan.

There are many types of fermented fish products, ranging from light- or dark-colored pourable sauces to very thick pastes (Table 12-9). In some cases, the same manufacturing process is used to make both a liquid and a paste. For example, in the Philippines, a salted and fermented fish mixture, when allowed to settle, yields a supernatant liquid called patis and a sediment that, when dried, gives a paste called bagoong. Just as with soy sauce and other fermented foods from Asia and the Far East, there are countless versions of fish sauces and pastes, many of which are made in the home or on a very small cottage scale using traditional techniques. Thus, only rather generic descriptions will be given (see below).

Although fish sauces are considered to be fermented foods, this is true only in a rather broad sense. Microorganisms do, indeed, grow and produce end products during the manufacture of these products (see below). However, most of the reactions that are responsible for flavor and aroma development occur as a result of endogenous fish enzymes, released during autolysis of fish tissue, rather than from microbial activities.

Manufacture of fish sauces and pastes

The general procedure for the production of fish sauces is not complicated. The starting material can be small (<15 cm in diameter) fish,

Table 12.9. Types of fish sauces and pastes.

Product	Country	Description
Nam-pla	Thailand	Fish sauce
Budu	Malaysia	Fish sauce
Patis	Philippines	Fish or shrimp sauce
Ishiru	Japan	Fish sauce
Nouc-mam	Vietnam	Fish sauce
Nam-pa	Laos	Fish sauce
Bagoong	Philippines	Fish or shrimp paste
Mam	Vietnam	Fish paste
Trassi	Indonesia	Shrimp or fish paste
Belachan	Malaysia	Shrimp paste

such as sardines (that otherwise have minimal commercial value), small shrimp, squid, or oysters (Figure 12–7). Fish is usually used whole and uneviscerated, although de-headed, eviscerated, ground, or cut-up pieces can also be used. The only other ingredient necessary to make these products is salt. The fish-to-salt ratio varies, depending on the product, but usually ranges from 3:1 to 5:1. Fish sauces do not undergo a lactic fermentation, per se, and are preserved mainly by salt and low water activity, rather than by pH. Thus, high salt concentrations are necessary.

After the salt is added to the fish (on wooden or concrete floors), the mixture is moved into tanks (often built into the ground) and covered. The material is held for about six months (or longer) at ambient temperature. At various times, the mixture may be uncovered, stirred, and exposed to air and sunlight, all of which are thought to improve flavor and color and accelerate enzyme activity. During this incubation period, the solid fish material is transformed, or more precisely, liquefied by the action of endogenous fish enzymes. These enzymes, primarily trypsin-like acid-proteases and various endo- and exo-peptidases, are ordinarily present within the intact cells of various fish tissues. However, in the non-living animal, the cells soon autolyze and those enzymes are released, result-

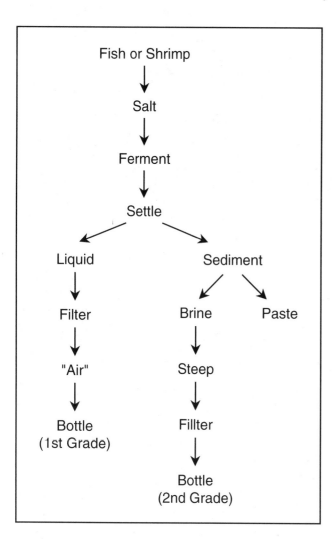

Figure 12–7. Fish sauce flow chart.

ing in extensive hydrolysis of muscle tissue. In fresh fish, autolysis and proteolysis result in tissue softening and spoilage; in fish sauce production, the result is liquefaction.

In addition to the physical transformation from solid to liquid, proteolytic digestion of the fish substrates results in formation of free amino acids and peptides. In intact tissue, for example, the soluble nitrogen concentration is essentially nil, but in nam-pla and nuoc-mam (Thai and Vietnamese fish sauce, respectively) there is more than 2% soluble nitrogen (mostly amino nitrogen). Glutamic acid, which, like in soy sauce products, is responsible for flavor enhancement, is among the amino acids that accumulate in fish sauce. Likewise, $5'$-nucleotides may also be formed, providing a source of umami or meaty-like flavors (as described above).

Further hydrolysis of peptides and amino acids by enzymes that are either endogenous or microbial in origin (see below) eventually results in a large number of volatile aroma and flavor products. Among those that are most prominent and that confer "fish sauce flavor" are ammonia, triethylamine, and various alcohols, aldehydes, ketones, and lactones. Lipolysis also occurs during fish sauce manufacture, resulting in formation of volatile fatty acids, including acetic, butanoic, and propanoic acids. These compounds are particularly characteristic of fish sauce flavor, which is sometimes described as "cheesy."

After several months of enzymolysis and fermentation, the liquid is separated from the sedimented material by decanting or filtering the liquid directly through the fish solids. This "first run" product has the highest quality. Additional brine can be added to the solid material, the mixture aged for several more weeks (or simply boiled), and then a second, lower quality, liquid is obtained. The remaining solids can then be recovered and used as a paste. Some fish sauces and pastes are aged in the open (and exposed to the sun) for several more weeks to allow partial dissipation of the strong fish aroma. The sauce or paste is then bottled. The final composition (weight basis) of fish sauces is usually about 60% moisture, 30% salt,

10% protein (including amino nitrogen), with a final pH of about 6.5. Pastes contain about 30% moisture, 20% salt, 30% protein, and 20% ash.

Fish sauce microbiology

While it is evident that microorganisms are present during the production of fermented fish sauces, it is not clear to what extent these organisms contribute to the finished product. The microbial population in raw, unsalted fish and shellfish is high in number and rich in diversity. Considering the fact that whole uneviscerated fish (guts and all) are usually used to make fish sauces, the initial load of organisms is significant.

In addition, the manufacturing environment is not aseptic, and even the salt (usually obtained by solar drying of sea water) may contribute microorganisms. However, the high salt concentration established early on in the fermentation provides strong selective pressure for halotolerant organisms. Not surprisingly, therefore, there is a shift in the initial bacterial population, from a wide variety of aerobic and anaerobic organisms to a more narrow microflora consisting mainly of salt-tolerant species of *Bacillus*, *Halobacillus*, *Staphylococcus*, and *Micrococcus*. Lactic acid bacteria are also widely present in fish sauce (Table 12–10), including some strains that are capable of growing in media containing 12% (2.1 M) salt.

Table 12.10. Lactic acid bacteria isolated from fish sauces[1].

Organism
Lactobacillus plantarum
Lactobacillus farciminis
Lactobacillus acidipiscis
Lactobacillus pentosus
Weissella thailandensis
Leuconsotoc citreum
Lactococcus lactic subsp. *lactis*
Tetragenococcus muriaticus

[1]From Tanasupawat et al., 1998, Tanasupawat et al., 2000, Paludan-Muller et al., 1999, and Kimura et al., 2001

Establishing the function of the microflora during the fish sauce fermentation is not easy. It has been suggested that there is a succession that occurs during the fermentation, leading to the eventual dominance of the more salt- and acid-tolerant organisms. While this may be true for some products, in which considerable microbial growth is evident, for other products, microbial growth hardly occurs. In either case, the population eventually decreases, such that at the end of the fermentation (six to twelve months), there are usually less than 10^3 cells per ml. Many of the organisms isolated from fermented fish products have proteolytic activity and likely contribute, at least in part, to the overall proteolysis of fish protein. Lactic acid bacteria and other anaerobes also can metabolize amino acids, producing volatile fatty acids, amines, ammonia, and other volatile end-products.

Safety of Fungal Fermented Foods

There are two main reasons why the safety of Asian, fungal-fermented foods has been questioned. First, the *Aspergillus* sp. used in the production of soy sauce, miso, sake, and related products are taxonomically similar to the mycotoxigenic aspergilli that produce aflatoxins, ochratoxin, and other toxins. Despite these similarities, however, surveys in which these foods have been analyzed for the presence of mycotoxins indicate that mycotoxins are not present. Recent studies have shown that very subtle genetic differences exist between industrial strains and mycotoxin-producing strains that would account for these findings (Box 12-5).

A second reason for the concern is due to the suggested epidemiological relationship between consumption of indigenous foods from the Far East and the development of certain

Box 12–5.   Mycotoxins in Fungal Fermented Products: Cause for Concern?

In nature, growth of fungi is often accompanied by production of mycotoxins, produced as secondary metabolites. Some of these mycotoxins—and aflatoxins, in particular—are extremely carcinogenic and mutagenic. In fact, aflatoxins are considered to be among the most toxic of all naturally occurring compounds found on the planet.

There are as many as sixteen different types of aflatoxins, of which B1, B2, G1, and G2 are most commonly found in human and animal foods. They are produced mainly by *Aspergillus flavus* and *Aspergillus parasiticus*, members of the *Aspergillus* section *Flavi* (or the *A. flavus* group). This taxonomic group also includes *Aspergillus sojae* and *Aspergillus oryzae*, species that are used in the manufacture of koji, soy sauce, miso, sake, and other Oriental fermented products.

Given the morphological and genetic similarities between these four different species, questions have been raised regarding the potential ability of *A. sojae* and *A. oryzae* to produce mycotoxins in foods in which these fungi are used. Indeed, several recent studies have revealed that some of the genes (or homologs) that encode for proteins involved in the aflatoxin biosynthesis pathway may be present in these food production strains (see below). Despite the apparent metabolic potential of *A. sojae* and *A. oryzae* to produce aflatoxins, however, there are no reports of aflatoxins being produced by these fungi or for the presence of aflatoxins in fermented soy products (Watson et al., 1999). These observations have led investigators in Japan, the United Kingdom, and the United States to explore, in detail, why *A. sojae* and *A. oryzae* do not synthesize aflatoxins.

The biochemical pathway for aflatoxin biosynthesis consists of at least twenty-three enzymatic reactions and more than fifteen intermediates (Yu et al, 2004A and 2004B). In 2004, the complete gene cluster encoding for aflatoxin biosynthesis in *A. flavus* and *A. parasiticus* was identified and sequenced (Yu et al., 2004A; Figures 1 and 2). There are twenty-five genes within this 70 Kb cluster (four additional sugar utilization genes are located immediately downstream).

Because enzymatic activities associated with the aflatoxin pathway are generally not detectable in *A. sojae*, it would appear that the relevant genes must also be absent. However, according to one

Box 12–5.   Mycotoxins in Fungal Fermented Products: Cause for Concern? *(Continued)*

report (Matsushima et al., 2001B), three aflatoxin genes (*aflR*, *nor1*, and *pksA*) were present in a commercial soy sauce strain (strain 477), based on Southern hybridization analysis. Another study showed that *omtA*, *nor1*, and *ver1* were also present in *A. sojae* and *A. oryzae* and that these genes had high sequence similarity to those from aflatoxin-producing strains of *A. parasiticus* (Watson et al., 1999). However, none of these genes were transcribed, even when cells were grown in alfatoxin-conducive media. Two other genes, *avfA* and *omtB*, were also reported to be present in *A. sojae* and were 99% identical to those found in *A. parasiticus* (Yu et al., 2000). Furthermore, complementation assays showed that the *A. sojae avfA* gene was fully functional. Thus, if aflatoxin genes are present in *A. sojae* and *A. oryzae*, and they appear to be functional, why don't these fungi produce aflatoxin in foods?

The answer to this question can now be revealed. The aflatoxin pathway in *A. parasiticus* is regulated mainly by the *aflR* gene product AflR. An adjacent gene, *aflS*, appears to modulate *aflR* expression, but its exact role is not known. AflR is a 47 kDa protein that acts as a positive regulator by binding to the promoter region of the *afl* gene cluster, activating transcription of the structural genes. Mutant strains of *A. flavus*, defective in *aflR*, do not express alfatoxin genes, indicating that *AflR* is required for aflatoxin biosynthesis. Although a homolog of the *aflR* gene is present in *A. sojae* and *A. oryzae*, *afl* genes are still not transcribed in these strains. The non-aflatoxin-producing phenotype is not due to the absence of a promoter-binding region, since that region has been shown to be functional (Takahashi et al., 2002). Rather, it is AflR that is non-functional, not only in the native *A. sojae* and *A. oryzae* backgrounds, but also in an *A. parasiticus aflR* deletion mutant transformed with *aflR* from *A. sojae* (Takahashi et al., 2002). Also, whereas introduction of an additional copy of the *A. parasiticus aflR* gene into *A. parasiticus* increased aflatoxin production, no such effect was observed when the *aflR* from *A. sojae* was introduced into *A. parasiticus* (Matsushima et al., 2001A).

Sequence analysis has shown that there are two mutations located within the *A. sojae aflA* gene (Watson et al., 1999). One mutation consists of a duplication of histidine (H) and alanine (A) residues at amino acid positions 111 to 114, resulting in a HAHA motif (no joke). The second mutation consists of a C– T transition, and subsequent conversion of an arginine codon at position 385 to a premature stop codon. This results in a truncated protein that is short sixty-two amino acid residues at the carboxy end, which is where the transcription-activating domain is located. The HAHA mutation, it turns out, however, does not appear to affect AflR activity or synthesis of aflatoxin, because cells harboring a hybrid gene containing the intact carboxy-terminal region—but with the HAHA motif—have a normal aflatoxin phenotype (Matsushima et al., 2001A). Thus, the defect in *aflR* must be due to the missing sixty-two residue region. The truncated AflR is then unable to activate transcription of *alf* genes, resulting in the lack of aflatoxin biosynthesis.

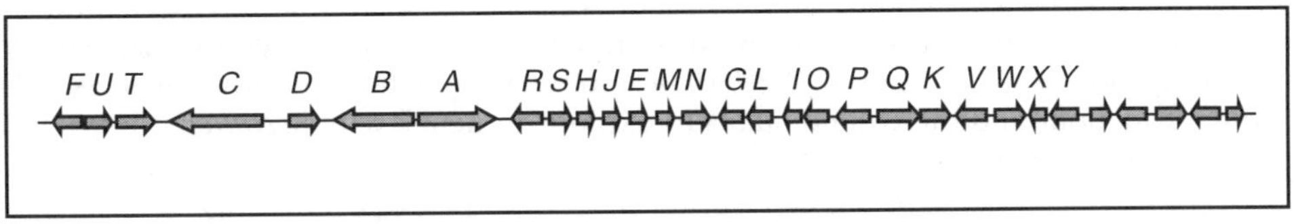

Figure 1.   Linear arrangement of the *afl* gene cluster from *Aspergillus parasiticus*. Individual gene names are abbreviated by a single letter (where *aflR = R*), and are named according to the scheme proposed by Yu et al., 2004B. Genes described in the text include *pksA* (*aflC*), *aflJ* (*aflS*), *norA* (*aflE*), *avfA* (*aflI*), and *omtB* (*aflO*). Four genes at the 3′ end of the *afl* cluster encode for a sugar utilization pathway, followed by a fifth gene whose function is unknown. Adapted from Yu et al., 2004A and 2004B.

(Continued)

Box 12–5. Mycotoxins in Fungal Fermented Products: Cause for Concern? *(Continued)*

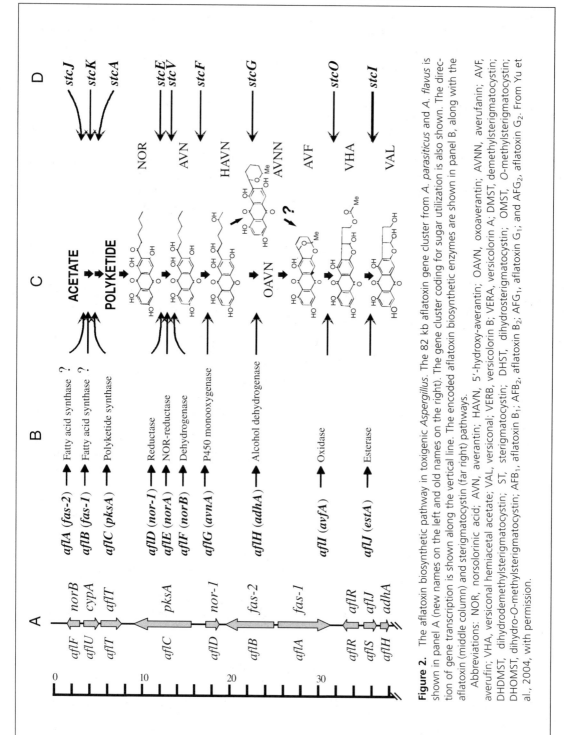

Figure 2. The aflatoxin biosynthetic pathway in toxigenic *Aspergillus*. The 82 kb aflatoxin gene cluster from *A. parasiticus* and *A. flavus* is shown in panel A (new names on the left and old names on the right). The gene cluster coding for sugar utilization is also shown. The direction of gene transcription is shown along the vertical line. The encoded aflatoxin biosynthetic enzymes are shown in panel B, along with the aflatoxin (middle column) and sterigmatocystin (far right) pathways.

Abbreviations: NOR, norsolorinic acid; AVN, averantin; HAVN, 5'-hydroxy-averantin; OAVN, oxoaverantin; AVNN, averufanin; AVF, averufin; VHA, versiconal hemiacetal acetate; VAL, versiconal; VERB, versicolorin B; VERA, versicolorin A; DMST, demethylsterigmatocystin; DHDMST, dihydrodemethylsterigmatocystin; ST, sterigmatocystin; DHST, dihydrosterigmatocystin; OMST, O-methylsterigmatocystin; DHOMST, dihydro-O-methylsterigmatocystin; AFB₁, aflatoxin B₁; AFB₂, aflatoxin B₂; AFG₁, aflatoxin G₁; and AFG₂, aflatoxin G₂. From Yu et al., 2004, with permission.

Box 12–5. Mycotoxins in Fungal Fermented Products: Cause for Concern? *(Continued)*

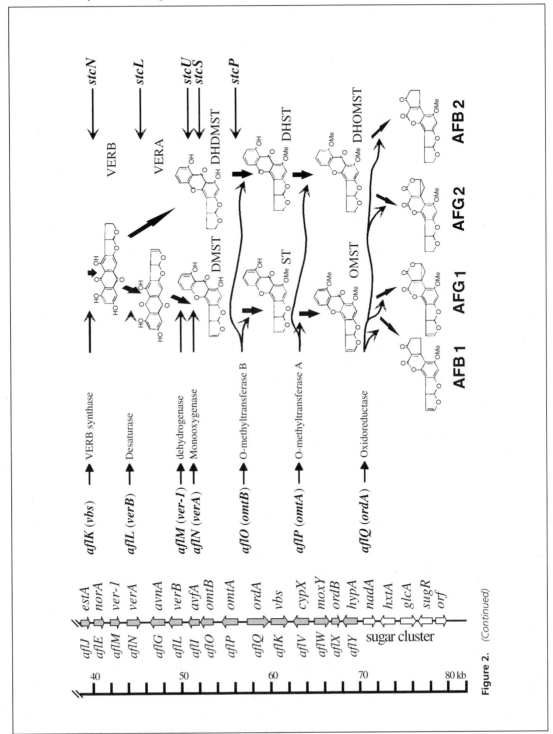

Figure 2. *(Continued)*

(Continued)

According to another recent investigation, mutations in a second regulatory gene, *aflJ* (see above), may also be responsible, in part, for non-expression of *afl* genes in *A. sojae*. In *A. parasiticus, aflJ* gene product, AflJ, upregulates synthesis of aflatoxin. Mutations in *aflJ* decreased transcription of both *afl* structural genes and aflatoxin synthesis, although *alfR* expression was not affected (Chang, 2003). Thus, it appears that AflJ is a co-activator that interacts with AflR to activate gene transcription. However, analogous to the *aflR* situation, the *aflJ* gene in *A. sojae* is defective, owing to two amino acid substitutions that reduce this interaction and reduce transcription of *afl* genes (Chang, 2004).

References

Chang, P.-K. 2004. Lack of interaction between AFLR and AFLJ contributes to nonaflatoxigenicity of *Aspergillus sojae*. J. Biotechnol. 107:245–253.

Chang, P.-K. 2003. The *Aspergillus parasiticus* protein AFLJ interacts with the aflatoxin pathway-specific regulator AFLR. Mol. Gen. Genomics 268:711–719.

Matsushima, K., P.-K. Chang, J. Yu, K. Abe, D. Bhatnagar, and T.E. Cleveland. 2001A. Pre-termination in *aflR* of *Aspergillus sojae* inhibits aflatoxin biosynthesis. Appl. Microbiol. Biotechnol. 55:585–589.

Matsushima, K., K. Yashiro, Y. Hanya, K. Abe, K. Yabe, and T. Hamasaki. 2001B. Absence of aflatoxin biosynthesis in koji mold (*Aspergillus sojae*). Appl. Microbiol. Biotechnol. 55:771-776.

Takahashi, T., P.-K. Chang, K. Matsushima, J. Yu, K. Abe, D. Bhatnagar, T.E. Cleveland, and Y. Koyama. 2002. Nonfunctionality of *Aspergillus sojae aflR* in a strain of *Aspergillus parasiticus* with a disrupted *aflR* gene. Appl. Environ. Microbiol. 68:3737–3743.

Watson, A.J., L.J. Fuller, D.J. Jeenes, and D.A. Archer. 1999. Homologs of aflatoxin biosynthesis genes and sequences of *aflR* in *Aspergillus oryzae* and *Aspergillus sojae*. Appl. Environ. Microbiol. 65:307–310.

Yu, J., D. Bhatnagar, T.E. Cleveland. 2004A. Completed sequence of aflatoxin pathway gene cluster in *Aspergillus parasiticus*. FEBS Lett. 564:126–130.

Yu, J., P.-K. Chang, K.C. Ehrlich, J.W. Cary, D. Bhatnagar, T.E. Cleveland, G.A. Payne, J.E. Linz, C.P. Woloshuk, and J.W. Bennett. 2004B. Clustered pathway genes in aflatoxin biosynthesis. Appl. Environ. Microbiol. 70:1253–1262.

Yu, J., C.P. Woloshuk, D. Bhatnagar, and T.E. Cleveland. 2000. Cloning and characterization of *avfA* and *omtB* genes involved in aflatoxin biosynthesis in three *Aspergillus* species. Gene 248:157–167.

cancers, including gastric cancer and esophageal cancer. Indeed, although some cancer rates are higher in China, Japan, and other Far East countries, the evidence suggests that dietary habits other than consumption of fungal-fermented foods may be more likely to be responsible. For example, the Asian diet is high in salt, which may predispose individuals to these cancers. In specific provinces in China, where cancer rates are especially high, consumption of vegetables is very low.

Finally, it is also noteworthy that positive nutritional factors have been associated with fungal fermented foods. kIn particular, consumption of soy beans (and fermented soy products) have been promoted for contributing many health benefits, including reduced rates of coronary heart disease and mortality due to heart disease. Soy beans contain several flavones, isoflavones, flavanols, and other flavonoids that may have anticarcinogenic activity and whose consumption may lower the risk of certain cancers, including breast cancer and prostate cancer. Although not all of these substances are present in fermented soy products, several specific compounds have been shown to have anticancer activity. For example, the flavor furanones HEMF and HMF (discussed previously) that are found in soy sauce inhibit carcinogenesis in laboratory animals.

Bibliography

Baens-Arcega, L., A.G., Ardisher, C.G. Bellows, and 31 other authors. 1996. Indigenous amino acid/peptide sauces and pastes with meatlike flavor. p. 509-654. In K.H. Steinkraus (ed.), *Handbook of Indigenous Fermented Foods, 2nd ed.* Marcel Dekker, Inc. New York, New York.

Beddows, C.G. 1998. Fermented fish and fish products, p. 416-440. In B.J.B. Woods, (ed.) *Microbiology of Fermented Foods, Volume 1*. Blackie Academic and Professional. London.

Beuchat, L.R. 2001. Traditional fermented foods, p. 701-719. In M.P. Doyle, L.R. Beuchat, and T.J. Montville (eds.). *Food Microbiology: Fundamentals and Frontiers, 2nd Ed.* ASM Press, Washington, D.C.

Ebine, H. 2004. Industrialization of Japanese miso fermentation, p. 99-147. In K.H. Steinkraus (ed.), *Industrialization of indigenous fermented foods, 2nd ed.* Marcel Dekker, Inc., New York, New York.

Ebine, H. 1986. Miso, p. 47-68. In N.R. Reddy, M.P. Pierson, and D.K. Salunkhe (eds.), Legume-based fermented foods. CRC Press. Boca Raton, Florida.

Fukushima, D. 2004. Industrialization of fermented soy sauce production centering around Japanese shoyu, p. 1-98. In K.H. Steinkraus (ed.), *Industrialization of indigenous fermented foods, 2nd ed.* Marcel Dekker, Inc., New York, New York.

Hachmeister, K.A., and D.Y.C. Fung. 1993. Tempeh: a mold-modified indigenous fermented food made from soybeans and/or cereal grains. Crit. Rev. Microbiol. 19:137-188.

Keuth, S., and B. Bisping. 1993. Formation of vitamins by pure cultures of tempe moulds and bacteria during the tempe solid substrate fermentation. J. Appl. Bacteriol. 75:427-434.

Kimura, B., Y. Konagaya, and T. Fujii. 2001. Histamine formation by *Tetragenococcus muriaticus*, a halophilic lactic acid bacterium isolated from fish sauce. Int. J. Food Microbiol. 70:71-77.

Kiuchi, S., and S. Watanabe. 2004. Industrialization of Japanese natto. p. 193-246. In K.H. Steinkraus (ed.), *Industrialization of Indigenous Fermented Foods, 2nd ed.* Marcel Dekker. New York, New York.

Mulyowidarso, R.K., G.H. Fleet, and K.A. Buckle. 1989. The microbial ecology of soybean soaking for tempe production. Int. J. Food Microbiol. 8:35-46.

Nout, M.J.R., and F.M. Rombouts. 1990. Recent developments in tempe research. J. Appl. Bacteriol. 69:609-633.

Nout, M.J.R., and J.K. Kiers. 1995. Tempe fermentation, innovation and functionality: update into the third millennium. J. Appl. Microbiol. 98:789-805.

Nunomura, N., and M. Sasaki. 1992. Japanese soy sauce with emphasis on off-flavors, p. 287-312. In G. Charalambous (ed.). *Off-flavors in foods and beverages.* Elsevier Science Publishers. Amsterdam, Netherlands.

Ohta, T. 1986. Natto, p. 85-93. In N.R. Reddy, M.P. Pierson, and K. Salunkhe (eds.), *Legume-based fermented foods.* CRC Press. Boca Raton, Florida.

Paludan-Muller, C., H.H. Huss, and L. Gram. 1999. Characterization of lactic acid bacteria isolated from a Thai low-salt fermented fish product and the role of garlic as substitute for fermentation. Int. J. Food Microbiol. 46:219-229.

Ruiz-Terán, F., and J.D. Owens. 1996. Chemical and enzymic changes during the fermentation of bacteria-free soya bean tempe. J. Sci. Food Agri. 71:523-530.

Schipper, M.A.A., and J.A. Stalpers. 1984. A revision of the genus *Rhizopus*, II. The *Rhizopus microsporus*-group. Stud. Mycol. 25:20-34.

Shurtleff, W., and A. Aoyagi. 1979. *Tempeh Production. The Book of Tempeh: Volume II.* New Age Foods, Lafayette, California.

Sparringa, R.A., and J.D. Owens. 1999. Protein utilization during soybean tempe fermentation. J. Agric. Food Chem. 47:4375-4378.

Tanasupawat, S., S. Okada, and K. Komagata. 1998. Lactic acid bacteria found in fermented fish in Thailand. J. Gen. Appl. Microbiol. 44:193-200.

Tanasupawat, S., O. Shida, S. Okada, and K. Komagata. 2000. *Lactobacillus acidipiscis* sp. nov. and *Weisella thailandensis* sp. nov. isolated from fermented fish in Thailand. Int. J. Syst. Evol. Microbiol. 50:1479-1485.

Thanh, N.V., and M.J.R. Nout. 2004. Dormancy, activation and viability of *Rhizopus oligosporus* sporangiospores. Int. J. Food Microbiol. 92:171-179.

Winarno, F.G., and N.R. Reddy. 1986. Tempe. p. 95-117. In N.R. Reddy, M.P. Pierson, and K. Salunkhe (eds.), *Legume-based Fermented Foods.* CRC Press. Boca Raton, Florida.

Yokotsuka, T. and M. Sasaki. 1998. Fermented protein foods in the Orient: shoyu and miso in Japan, p. 351-415. In B.J.B. Woods (ed.), *Microbiology of Fermented Foods, Volume 1.* Blackie Academic and Professional. London.

Yoshizawa, K., and T. Ishikawa. 2004. Industrialization of Sake manufacture p. 149—192. *Industrialization of indigenous fermented foods, 2nd ed.* Marcel Dekker, Inc., New York, New York.

Index